全国科学技术名词审定委员会

海峡两岸化学工程名词

（第二版）

海峡两岸化学工程名词工作委员会

国家自然科学基金资助项目

科学出版社

北京

内 容 简 介

本书是由海峡两岸化学工程界专家会审的海峡两岸化学工程名词对照本第二版,是在 2007 年出版的《海峡两岸化学工程名词》的基础上加以增补修订而成。共收词约 12 600 余条。供海峡两岸化学工程界和相关领域的人士使用。

图书在版编目(CIP)数据

海峡两岸化学工程名词/海峡两岸化学工程名词工作委员会编. –2 版.
—北京:科学出版社,2017.9
ISBN 978–7–03–054663–0

Ⅰ.①海… Ⅱ.①海… Ⅲ.①化学工程–名词术语 Ⅳ.①TQ02–61

中国版本图书馆 CIP 数据核字(2017)第 236660 号

责任编辑:才 磊 顾英利/责任校对:陈玉凤
责任印制:张 伟/封面设计:槐寿明

科 学 出 版 社 出版
北京东黄城根北街 16 号
邮政编码:100717
http://www.sciencep.com
北京建宏印刷有限公司 印刷
科学出版社发行 各地新华书店经销

*

2007 年 6 月第 一 版 开本:787×1 092 1/16
2017 年 9 月第 二 版 印张:44 5/8
2018 年 6 月第二次印刷 字数:1 038 000
定价:298.00 元
(如有印装质量问题,我社负责调换)

海峡两岸化学工程名词工作委员会委员名单

第一届委员(2004—2013)

大陆召集人:蒋楚生

大 陆 委 员:苏健民　萧成基　罗北辰

大 陆 秘 书:才　磊

臺灣召集人:石延平

臺 灣 委 員(以姓名筆畫爲序):

萬其超　紀榮昌　曾憲政　蔡信行　蔣孝澈　劉清田　賴君義

第二届委员(2014—)

大陆召集人:洪定一　杨元一

大 陆 委 员(以姓名笔画为序):

王　燕　朱慎林　乔金梁　胡云光　钱鸿元　郭　锴　康小洪

臺灣召集人:萬其超

臺 灣 委 員(以姓名筆畫爲序):

王釿鎊　汪上曉　陳慧英　楊美惠　劉懷勝　簡政展

序

　　科学技术名词作为科技交流和知识传播的载体,在科技发展和社会进步中起着重要作用。规范和统一科技名词,对于一个国家的科技发展和文化传承是一项重要的基础性工作和长期性任务,是实现科技现代化的一项支撑性系统工程。没有这样一个系统的规范化的基础条件,不仅现代科技的协调发展将遇到困难,而且,在科技广泛渗入人们生活各个方面、各个环节的今天,还将会给教育、传播、交流等方面带来困难。

　　科技名词浩如烟海,门类繁多,规范和统一科技名词是一项十分繁复和困难的工作,而海峡两岸的科技名词要想取得一致更需两岸同仁作出坚韧不拔的努力。由于历史的原因,海峡两岸分隔逾50年。这期间正是现代科技大发展时期,两岸对于科技新名词各自按照自己的理解和方式定名,因此,科技名词,尤其是新兴学科的名词,海峡两岸存在着比较严重的不一致。同文同种,却一国两词,一物多名。这里称"软件",那里叫"软体";这里称"导弹",那里叫"飞弹";这里写"空间",那里写"太空";如果这些还可以沟通的话,这里称"等离子体",那里称"电浆";这里称"信息",那里称"资讯",相互间就不知所云而难以交流了。"一国两词"较之"一国两字"造成的后果更为严峻。"一国两字"无非是两岸有用简体字的,有用繁体字的,但读音是一样的,看不懂,还可以听懂。而"一国两词"、"一物多名"就使对方既看不明白,也听不懂了。台湾清华大学的一位教授前几年曾给时任中国科学院院长周光召院士写过一封信,信中说:"1993年底两岸电子显微学专家在台北举办两岸电子显微学研讨会,会上两岸专家是以台湾国语、大陆普通话和英语三种语言进行的。"这说明两岸在汉语科技名词上存在着差异和障碍,不得不借助英语来判断对方所说的概念。这种状况已经影响两岸科技、经贸、文教方面的交流和发展。

　　海峡两岸各界对两岸名词不一致所造成的语言障碍有着深刻的认识和感受。具有历史意义的"汪辜会谈"把探讨海峡两岸科技名词的统一列入了共同协议之中,此举顺应两岸民意,尤其反映了科技界的愿望。两岸科技名词要取得统一,首先是需要了解对方。而了解对方的一种好的方式就是编订名词对照本,在编订过程中以及编订后,经过多次的研讨,逐步取得一致。

　　全国科学技术名词审定委员会(简称全国科技名词委)根据自己的宗旨和任务,始终把海峡两岸科技名词的对照统一工作作为责无旁贷的历史性任务。近些年一直本着积极推进,增进了解;择优选用,统一为上;求同存异,逐步一致的精神来开展这项工作。先后接待和安排了许多台湾同仁来访,也组织了多批专家赴台参加有关学科的名词对照研讨会。工作中,按照先急后缓、先易后难的精神来安排。对于那些与"三通"

有关的学科,以及名词混乱现象严重的学科和条件成熟、容易开展的学科先行开展名词对照。

在两岸科技名词对照统一工作中,全国科技名词委采取了"老词老办法,新词新办法",即对于两岸已各自公布、约定俗成的科技名词以对照为主,逐步取得统一,编订两岸名词对照本即属此例。而对于新产生的名词,则争取及早在协商的基础上共同定名,避免以后再行对照。例如101～109号元素,从9个元素的定名到9个汉字的创造,都是在两岸专家的及时沟通、协商的基础上达成共识和一致,两岸同时分别公布的。这是两岸科技名词统一工作的一个很好的范例。

海峡两岸科技名词对照统一是一项长期的工作,只要我们坚持不懈地开展下去,两岸的科技名词必将能够逐步取得一致。这项工作对两岸的科技、经贸、文教的交流与发展,对中华民族的团结和兴旺,对祖国的和平统一与繁荣富强有着不可替代的价值和意义。这里,我代表全国科技名词委,向所有参与这项工作的专家们致以崇高的敬意和衷心的感谢!

值此两岸科技名词对照本问世之际,写了以上这些,权当作序。

2002 年 3 月 6 日

第二版前言

众所周知,海峡两岸化学工程名词存在不一致,主要由三方面的原因引起。首先是两岸汉字的书写不同,大陆通用简化汉字,而台湾一直采用繁体汉字,如矾的繁体为礬,同样窑和窰、酰与醯、碱和鹼、霉和黴等也需要对照才容易识别;其次是有的同样事物大陆和台湾用词不同,如硅与矽、钚与铈、液泛与氾流、衍射与繞射、卷积与摺积、纳米与奈米、等离子与电浆、丙烯酸树脂与壓克力树脂等;三是英文人名翻译不一致引起,如傅里叶与傅利葉、伯努利与白努利等。

两岸化学工程名词的不一致势必引起交流、沟通方面出现一系列问题,两岸化工学界都迫切希望出版两岸化学工程名词对照本。全国科学技术名词审定委员会为此于 2005 年及时启动了《海峡两岸化学工程名词》的编撰工作,并得到了台湾李国鼎科技发展基金会的大力支持。第一版《海峡两岸化学工程名词》的编撰,采用全国科学技术名词审定委员会公布的《化学工程名词》(1995)和台湾化工学会提供的《化工名词》作为对照蓝本,由大陆专家先行完成《海峡两岸化学工程名词》的初稿,共收词约 13000 条,后通过电子邮件由台湾专家进行了补充对照工作,提交了修订稿,最终在两岸专家共同努力下,全国科学技术名词审定委员会组织完成了《海峡两岸化学工程名词》第一版的审定稿,于 2007 年 6 月由科学出版社正式出版。

《海峡两岸化学工程名词》第一版的出版,对于缓解因两岸化学工程名词不一致而引起的交流、沟通方面的一系列问题,起到了重要的作用,惠及两岸化学工程领域的产业合作、学术交流、知识传播以及相关文献的编撰和检索。但是,第一版《海峡两岸化学工程名词》出版以来,化工学科又有了新的发展,尤其是化工产业技术又有了新进展,因此有必要对第一版进行修订。第二版《海峡两岸化学工程名词》的编撰工作,是 2014 年 4 月 21 日在北京召开的海峡两岸化学工程名词审定座谈会上正式启动的,全国科学技术名词审定委员会、中国化工学会和台湾教育研究院再次合作,组织起两岸专家共同完成。

第二版的编撰先由台湾教育研究院组织台湾方面专家在《海峡两岸化学工程名词》第一版的基础上,先行提交了《海峡两岸化学工程名词》第二版的初稿,共收词约 13798 条,后通过电子邮件由大陆 7 名专家进行了补充对照工作,并通过 3 次审定会,完成了修订稿,最终在两岸专家共同努力下,全国科学技术名词审定委员会于 2016 年 7 月组织完成了《海峡两岸化学工程名词》第二版的审定稿。《海峡两岸化学工程名词》第二版与第一版之间有着明显的传承关系,其中部分名词做了一些修改,例如,第一版中台湾词汇"固有动力学"在第二版中改为"本徵动力学";第一版中台湾词汇"不銹鋼"在第二版中改为"不鏽鋼";第一版中台湾词汇"矽橡膠"在第二版中改为"矽氧橡膠"等等。此外,第二版还增加了一些新名词,如新增加了与乙烯原料紧密相关的词条"页岩气",相应的台湾名词是"页岩氣";又如新增了"比强度",相应的台湾名词是"比强强度、强度重量比",

等等。《海峡两岸化学工程名词》第二版将于2017年10月由科学出版社正式出版，届时也正好是第一版出版10周年。在《海峡两岸化学工程名词》第二版的编撰工作中，虽然两岸化学工程专家反复推敲、认真讨论，但不妥之处仍属难免，尤其是对照本中大陆使用的不少名词尚未审定公布，希望得到两岸读者的指正。

海峡两岸化学工程名词工作委员会

2017 年 8 月

第 一 版 前 言

随着海峡两岸学术交流的不断加强,两岸化学工程名词不一致的问题,给两岸的学术交流带来的障碍逐渐显露出来。两岸化学工程界的同行们认为共同开展海峡两岸化学工程名词的交流和研讨,并出版对照本是非常必要的。在两岸化学工程名词对照研讨过程中两岸专家可以增进了解,取长补短,逐步达成共识,从而促进名词的统一。这项工作将惠及两岸化学工程领域的学术交流、知识传播以及相关文献的编纂和检索。鉴于此,全国科学技术名词审定委员会、中国化工学会和台湾李国鼎科技发展基金会、台湾化工学会的有关负责人协商,决定在两岸分别邀请有关专家组成"海峡两岸化学工程名词工作委员会"承担此项重任。该委员会决定以台湾化工学会提供的《化工名词》和全国科学技术名词审定委员会(原全国自然科学名词审定委员会)公布的《化学工程名词》(1995)作为对照蓝本,并由大陆专家先行完成《海峡两岸化学工程名词》补充对照工作,后通过电子邮件请台湾专家补充修订,完成对照初稿。

2005 年 11 月 18 日在北京召开了"海峡两岸化学工程名词研讨会"。专家们进一步明确了收词原则,对初稿中存疑的问题和不一致的定名进行了认真、细致地讨论。此外,还对学术上存在争议的名词或以其他学科为主定名的名词,也进行了讨论,在尊重习惯、择优选用、求同存异、逐步一致的原则指导下,专家的认识趋于接近。会后两岸专家又分别对相关部分进行了审核,完成了定稿。

通过对化工名词的对照研讨,两岸专家认识到,名词对照统一工作是一项长期而细致的工作,应该长期地进行下去。这项工作对海峡两岸的学术交流和知识传播都会直到积极的促进作用和支撑作用。今后两岸化学工程专家应加强交流与沟通,共同确定本学科领域的新名词,以减少新名词定名不一致的现象。我们的工作仅仅是一个开始,在化工名词的对照研讨及本书的编辑过程中,虽经两岸化学工程专家反复推敲、认真讨论,但不妥之处尚属难免,尤其是对照本中大陆使用的许多名词尚未审定公布,希望得到两岸读者的指正。

海峡两岸化学工程名词工作委员会

2007 年 1 月

编 排 说 明

一、本书是海峡两岸化学工程名词对照本。

二、本书分正篇和副篇两部分。正篇分大陆名、台湾名、英文名三项,按大陆名的汉语拼音顺序编排;副篇分英文名、大陆名、台湾名三项,按英文名的字母顺序编排。大陆名使用简体字,台湾名使用繁体字。

　　正篇

三、[ ]中的字使用时可以省略。

四、大陆名正名和异名分别排序,在异名处用(=)注明大陆名的正名和英文名同义词。

　　副篇

五、对应的英文名为多个时(包括缩写词)用",”分隔。

六、英文名所对应相同概念的汉文名用",”分隔,不同概念的用① ② ③分别注明。

七、英文名的同义词用(=)注明。

八、英文缩写词排在全称后的( )内。

目　录

正 篇

A

大 陆 名	台 湾 名	英 文 名
阿伦尼乌斯方程	阿瑞尼斯方程式	Arrhenius equation
阿马加定律	愛瑪哥定律	Amagat law
安全玻璃	安全玻璃	safety glass
安全阀	安全閥	safety valve
安全放空	安全排氣孔	safety vent
安全工作压强	安全操作壓力	safe working pressure
安全工作应力	安全工作應力	safe working stress
安全检查清单	安全檢查報表	safety check list
安全门	安全門	safety door
安全系数	安全因數	safety factor
安全泄压阀	安全洩壓閥	safety-relief valve
安全泄压面积	安全釋放面積	safety-relief area
安全泄液阀	液體釋放閥	liquid relief valve
安全性试验	安全性試驗法	safety testing
安全裕度	安全裕量	safety margin
安全炸药	安全炸藥	safety explosive
安全装置	安全裝置	safety device
安托万方程	安東尼方程式	Antoine equation
安息香	安息香	benzoin
安装	安裝	installation
安装成本	安裝成本	installation cost
氨饱和器	氨飽和器	ammonia saturator
氨苄青霉素	胺苄青黴素	ampicillin
氨萃取过程	氨萃取法,氨萃取程序	ammonia extraction process
氨反应器	氨反應器	ammonia reactor
氨合成	氨合成	ammonia synthesis
氨化肥料	氨化肥料	ammoniated fertilizer
氨化[作用]	氨化[作用]	ammoniation

大　陆　名	台　湾　名	英　文　名
氨基树脂	胺基樹脂	amino resin
氨基酸	胺基酸	amino acid
氨碱法	氨鹼法	ammonia-soda process
氨解[作用]	氨解	ammonolysis
氨水	氨水,氨液	aqueous ammonia, ammonia liquor, ammonia water, aqua ammonia
氨洗涤器	洗氨器	ammonia washer
氨盐水	氨化鹽水	ammoniated brine
鞍点共沸物	鞍點共沸液	saddle azeotrope
铵矾	銨礬	ammonium alum
铵皂	銨皂	ammonium soap
胺化[作用]	胺化[作用]	amination
昂萨格倒易关系	翁沙格倒易關係	Onsager reciprocal relation
凹槽	凹槽,V形槽	notch
凹槽式滤板	凹槽式濾板,帶框濾板	recessed plate
凹痕	縮痕	sink mark
凹坑	坑,槽,池	pit
螯合剂	螯合劑,鉗合劑	chelating agent
奥长石	鈉鈣長石	oligoclase
奥纶	奧綸	orlon
奥斯陆蒸发结晶器	奧斯陸蒸發結晶器	Oslo evaporative crystallizer

B

大　陆　名	台　湾　名	英　文　名
八田数	八田數	Hatta number
巴龙霉素	巴龍黴素	paromomycin
巴氏灭菌器	低溫殺菌器	pasteurizer
巴氏消毒法	低溫殺菌法,巴氏殺菌法	pasteurization
巴西棕榈蜡	棕櫚蠟,卡拿巴蠟	carnauba wax
拔顶[蒸馏]	直餾	topping
钯催化剂	鈀觸媒	palladium catalyst
靶管	靶管	target tube
靶效率	靶效率	target efficiency
白炽灯光	白熱燈光	incandescent light
白垩	白堊	chalk
白放线菌素	放線菌素	actinomycetin

大　陆　名	台　湾　名	英　文　名
白金	白色[飾]金	white gold
白磷	白磷,黄磷	white phosphorus
白榴石	白榴石	leucite
白水泥	白水泥	white cement
白炭	白碳,白煙	white charcoal
白铁矿	白鐵礦	marcasite
白液	白液,燒鹼液	white liquor
白云母	白雲母	muscovite
白云石	白雲石	dolomite
白云石石灰	白雲石石灰	dolomitic lime
白云石水泥	白雲石水泥	dolomite cement
白云石陶器	白雲石陶器	dolomite earthenware
白云石质石灰石	白雲石石灰石	dolomitic limestone
百分绝对湿度	百分絕對濕度	percentage absolute humidity
百分湿度	百分濕度	percentage humidity
百叶板填充塔	填板塔	slat packed tower
百叶窗挡板	百葉窗型擋板	louver type baffle
摆	[鐘]擺	pendulum
摆动	鏈段運動	segmental motion
摆动连续结晶槽	擺動連續結晶槽,沃柏 結晶槽	Wulff-Bock crystallizer
摆式黏度计	擺式黏度計	pendulum viscometer
摆式张力试验机	擺式張力試驗機	pendulum-type tension testing machine
摆线鼓风机	擺旋鼓風機	cycloidal blower
[班伯里]密[闭式混] 炼机	密閉式混煉機	Banbury mixer
斑点	斑點	spot
斑点反应	斑點反應	spot reaction
斑点分析	斑點分析	spot analysis
斑点试验	斑點試驗	spot test
斑染	斑染	speck dyeing
斑贴试验	斑貼試驗	patch test
斑铜矿	斑銅礦	bornite
板波纹填料	板波紋填料	Mellapak packing
板翅换热器	板翅熱交換器	plate-fin heat exchanger
板框过滤机	板框過濾機	frame and plate filter
板框式过滤器	板框式過濾器	plate and frame filter
板框式压滤机	板框壓濾機	plate-and-frame filter press, plate-frame

大　陆　名	台　湾　名	英　文　名
		filter press
板框组件	板框組件	plate-and-frame module
板片搅拌器	葉片攪拌器	blade agitator
板式换热器	板式熱交換器	plate heat exchanger
板式输送机	板式運送機,板運機	slat conveyor
板式塔	板[式]塔	plate column, plate tower, tray column
板式通气器	板式通氣器	plate aerator
板式蒸发器	板式蒸發器	plate-type evaporator
板数	板數	plate number
板效率	板效率	tray efficiency
半补强炉黑	半強化爐黑	semi-reinforcing furnace black
半导体	半導體	semiconductor
半导体催化剂	半導體觸媒	semiconductor catalyst
半导体聚合物	半導電聚合物	semiconductive polymer
半导体涂料	半導電塗料	semiconductive coating
半对数坐标纸	半對數[坐標]紙	semi-logarithmic paper
半分批法	半分批法	semi-batch process
半分批反应器	半批次反應器	semi-batch reactor
半分批式操作	半批式操作	semi-batch operation
半分批选择性	半批式選擇性	semi-batch selectivity
半封闭叶轮	半封閉葉輪	semi-enclosed impeller
半干性油	半乾性油	semidrying oil
半光涂料	半光塗層	semigloss coating
半合成[机]油	半合成[機]油	semisynthetic oil
半合成纤维	半合成纖維	semisynthetic fiber
半糊状漆	去漆膏	semi-paste paint
半互穿透聚合物网络	半互穿聚合物網絡	semi-interpenetrating polymer network
半化学纸浆	半化學紙漿	semi-chemical pulp
半化学制浆法	半化學製漿法	semi-chemical pulping process
半极性键	半極性鍵	semipolar bond
半结晶聚合物	半結晶聚合物	semi-crystalline polymer
半晶体	半晶體	semi-crystal
半连续过程	半連續程序	semi-continuous process
半连续聚合	半連續聚合	semi-continuous polymerization
半连续窑	半連續窯	semi-continuous kiln
半流动式反应器	半流動式反應器	semiflow reactor
半硫化	半硬化,半交聯	semi-cure, semi-curing
半耐久性黏合剂	半耐久性黏合劑	semi-durable adhesive

大　陆　名	台　湾　名	英　文　名
半挠性链	半可撓鏈	semi-flexible chain
半区间法	半區間法	interval halving
半衰期	半衰期,半生期	half life
半衰期法	半衰期法,半生期法	half life method
半体	半體	half body
半透明的	半透明的,半蔽光的	semi-opaque
半透明度	半透明性	translucence
半透膜	半透膜	semipermeable membrane
半微量分析	半微量分析	semi-microanalysis
半无光	半消光織物	semi-dull
半无烟煤	半無煙煤	semi-anthracite
半纤维素	半纖維素	hemicellulose
半烟煤	半煙煤	semi-bituminous coal
半硬泡沫	半硬發泡體	semi-rigid foam
半煮法	半煮法	semi-boiling process
半自动系统	半自動系統	semi-automatic system
半自动压机	半自動壓機	semi-automatic press
拌和机	拌和機	mingler
棒磨机	桿磨	rod mill
磅达	磅達	poundal
包藏	包藏	occlusion
包藏树脂	包藏樹脂	occluded resin
包合物	晶籠化合物	clathrate
包合作用	晶籠［作用］	clathration
包埋	包埋	entrapment
包装	包装	packaging
包装纸	包裝紙	packing paper
胞内酶	［細］胞内酵素	intracellular enzyme
胞腔模型	細胞模型	cell model
胞外酶	胞外酶	extracellular enzyme
饱和	飽和	saturation
饱和度	飽和度	degree of saturation
饱和极限	飽和極限	saturation limit
饱和聚酯	飽和聚酯	saturated polyester
饱和空气	飽和空氣	saturated air
饱和器	飽和器	saturator
饱和曲线	飽和曲線	saturation curve
饱和溶液	飽和溶液	saturated solution

大　陆　名	台　湾　名	英　文　名
饱和湿度	飽和濕度	saturated humidity
饱和水蒸气	飽和蒸汽	saturated steam
饱和态	飽和態	saturated state
饱和烃聚合物	飽和碳氫聚合物	saturated hydrocarbon polymer
饱和温度	飽和溫度	saturation temperature
饱和压力	飽和壓力	saturation pressure
饱和液体	飽和液	saturated liquid
饱和蒸汽	飽和蒸氣	saturated vapor
饱和蒸汽压	飽和蒸氣壓	saturated vapor pressure
保持期间	貯留期間	holding period
保持时间	貯留時間	holding time
保留(=截留)		
保留时间	滯留時間	retention time
保留体积	滯留體積	retention volume
保留周期	滯留期間	retention period
保湿剂	保濕劑	humectant
保温	絕緣	insulation
保温材料(=绝热材料)		
保温管	隔熱管	insulated pipe, lagged pipe
保温效率	絕緣效率,隔熱效率	insulating efficiency
保温砖	隔熱磚	insulating brick
保险	保險	insurance
保证质量	保證品質	guaranteed quality
报告	報告	report
报警器	警報器	alarm
鲍尔环	鮑爾環	Pall ring
暴晒蒸发	太陽能蒸發	solar evaporation
爆聚[合]	爆聚合	explosive polymerization
爆裂	爆裂	burst
爆裂强度	爆裂強度	bursting strength
爆裂压力	爆裂壓力	bursting pressure
爆燃	爆燃	deflagration
爆燃器	爆燃器	deflagrator
爆炸	爆炸	explosion,blast
爆炸化合物	爆炸化合物	blasting compound
爆炸极限	爆炸界限	explosion limit
爆炸时间	爆炸時間	detonation time
爆炸危险	爆炸傷害	explosion hazard

大 陆 名	台 湾 名	英 文 名
爆炸压[力]	爆炸壓[力]	explosion pressure
爆炸指数	爆炸指數	explosion index
爆震试验	爆震試驗	knock characteristic test
北极鲸蜡油	北極鯨蠟油,抹香鯨油	arctic sperm oil
备择假设	備擇假設,對立假設	alternative hypothesis
背压	背壓	back pressure
背压阀	背壓閥	back pressure-valve
钡矾	鋇礬	barium alum
倍示压力计	倍示壓力計	multiplying manometer
倍增时间	倍增時間	doubling time
焙烧炉	焙燒爐	roaster
焙用碱	焙鹼	baking soda
本构方程	物性方程式	constitutive equation
本森溶解度系数	本森溶解係數	Benson solubility coefficient
本体聚合	本體聚合[反應]	mass polymerization
本体模量	總體模數	bulk modulus
本体温度	總體溫度	bulk temperature
本体性质	總體性質	bulk property
本征安全	固有安全	intrinsic safety
本征动力学	本徵動力學	intrinsic kinetics
苯胺橙	苯胺橙	aniline orange
苯胺点	苯胺點	aniline point
苯胺黑	苯胺黑	aniline black, nigrosine
苯胺红	苯胺紅	aniline red
苯胺蓝	苯胺藍	aniline blue
苯胺绿	苯胺綠	aniline green
苯胺染料	苯胺染料	aniline dye
苯胺紫	苯胺紫	aniline violet
苯并呋喃	苯并呋喃,薰草質,香豆質	coumarone, benzofuran
苯并呋喃-茚树脂	苯并呋喃-茚樹脂	coumarone-indene resin
苯乙烯	苯乙烯	styrene
苯乙烯-丙烯腈[共聚物]	苯乙烯–丙烯腈[共聚物]	styrene acrylonitrile copolymer,SAN
苯乙烯-丙烯酸酯共聚涂料	苯乙烯–丙烯酸酯共聚物塗料	styrene acrylate copolymer coating
苯乙烯-二乙烯基苯共聚物	苯乙烯二乙烯苯共聚物	styrene divinylbenzene copolymer

大 陆 名	台 湾 名	英 文 名
苯乙烯基化油	苯乙烯化油	styrenated oil
苯乙烯-甲基丙烯酸甲酯树脂	苯乙烯-甲基丙烯酸甲酯樹脂	styrene methyl methacrylate resin
苯乙烯树脂	苯乙烯樹脂	styrene resin
苯乙烯-顺丁烯二酸酐共聚物	苯乙烯-順丁烯二[酸]酐共聚物	styrene maleic anhydride copolymer
苯乙烯橡胶	苯乙烯橡膠	styrene rubber
泵	泵	pump
泵功率	泵馬力	pump horsepower
泵轮	泵葉輪	pump impeller
泵送功	泵功	pumping work
泵送损耗	泵抽損失	pumping loss
泵特性	泵特性	pump characteristics
泵效率	泵效率	pump efficiency
泵压降	泵壓[力]降	pump pressure drop
泵轴	泵軸	pump shaft
泵转子	泵轉子	pump rotor
比表面	比表面積	specific surface
比表面积	比表面積	specific surface area
比冲量	比衝量	specific impulse
比电阻	比阻力,電阻率	specific resistance
比反应速率	比反應速率	specific reaction rate
比反应速率常数	比反應速率常數	specific reaction rate constant
比刚度	比剛度	stiffness-weight ratio
比刚性	比剛性	specific rigidity
比焓	比焓	specific enthalpy
比较器	比較器	comparator
比绝缘电阻	比絕緣電阻	specific insulation resistance
比例-重调-比率作用	比例-重設-速率作用	proportional plus reset plus rate action
比例-重调作用	比例-重設作用	proportional plus reset action
比例带调整	比例帶調整	proportional band adjustment
比例度	比例帶	proportional band
比例-积分控制	比例-積分控制	proportional-integral control
比例-积分-微分作用	比例-積分-微分作用	proportional-integral-derivative action
比例-积分作用	比例-積分作用	proportional-integral action
比例控制	比例控制	proportional control
比例控制偏离	比例控制偏離	bias in proportional
比例控制器	比例控制器	proportional controller

大　陆　名	台　湾　名	英　文　名
比例灵敏度	比例靈敏度	proportional sensitivity
比例-微分控制	比例-微分控制	proportional-derivative control
比例-微分作用	比例-微分作用	proportional-derivative action
比例元件	比例元件	proportional element
比例增益	比例增益	proportional gain
比例作用	比例作用	proportional action
比滤饼阻力	比濾餅阻力	specific cake resistance
比黏度	比黏度	specific viscosity
比浓对数黏度	比濃對數黏度	inherent viscosity
比强度	比強度,強度重量比	specific strength, strength-to-weight ratio
比热容	比熱容	specific heat capacity
比色计	比色計	chromometer, colorimeter
比色密度计	比色密度計	color densitometer
比色器	比色器	color comparator
比熵	比熵	specific entropy
比生长速率	比生長速率	specific growth rate
比湿	比濕度	specific humidity
比速率	比速率	specific rate
比体积	比容	specific volume
比推力	比推力	specific thrust
比消光系数	比消光係數	specific extinction coefficient
比旋光度	比旋光度,旋光率	specific rotatory power
比重	比重	specific gravity, specific weight
比重瓶	比重瓶	pycnometer, specific gravity bottle
比转速	比轉速	specific speed
比浊法	濁度計	turbidimeter
比阻尼容量	比制震能	specific damping capacity
吡唑啉酮染料	吡唑啉染料	pyrazolone dye
毕奥数	畢奧數	Biot number
闭工叶轮	封閉式葉輪	enclosed impeller
闭环	閉環	closed loop
闭环传递函数	閉環轉移函數	closed loop transfer function
闭环控制	閉環控制,閉路控制	closed loop control
闭环频率应答	閉環頻率應答	closed loop frequency response
闭环稳定性	閉環穩定性	closed loop stability
闭环系统	閉環系統,閉路系統	closed loop system
闭路	閉路	closed circuit
闭路压碎	閉路壓碎,循環壓碎	closed circuit crushing

大　陆　名	台　湾　名	英　文　名
闭路研磨	閉路研磨,循環研磨	closed circuit grinding
闭式边界	閉邊界	closed boundary
闭式容器	密閉容器	closed vessel
闭锁式料斗	閉鎖式料斗	lock hopper
蓖麻子油	蓖麻油	castor oil
壁厚	壁厚	wall thickness
壁剪应力	表面摩擦	skin friction
壁摩擦角	壁摩擦角	angle of wall friction
壁湍流	壁紊流	wall turbulence
壁效应	壁效應	wall effect
边际利润	邊際利潤	marginal profit
边界层	邊界層	boundary layer
边界层方程	邊界層方程[式]	boundary layer equation
边界层浓度	邊界層濃度	boundary layer concentration
边界条件	邊界條件	boundary condition
边界相	邊限相位	margin phase
边界增益	邊限增益	margin gain
边缘效应	邊緣效應	edge effect
编程	規畫	programming
编织	梭織	weaving
编织机	梭織機	weaving machine
鞭毛	鞭毛	flagella
变动成本	變動成本	variable cost
变动系数	變異係數	coefficient of variation
变分微积分	變分學	variational calculus
变分学	變分學,變分法	calculus of variations
变换	變換[式]	transform
变换反应	[水煤氣]轉化反應	shift reaction
变换炉	[水煤氣]轉化器	shift converter
变阶反应	變階反應	shifting-order reaction
变量	變數	variable
变送器	傳送器,[信號]發射器	transmitter
变送器增益	傳送器增益	transmitter gain
变速泵	變速泵	variable-speed pump
变体	變異體	variant
变温吸附	變溫吸附	temperature swing adsorption
变形	變形	deformation
变形低共熔物	變形共熔物	deformation eutectic

大　陆　名	台　湾　名	英　文　名
变形点	變形點	deformation point
变形范围	變形範圍	deformation range
变形功	變形功	deformation work, work of deformation
变形速率	變形速率	deformation rate
变形坐标	變形坐標	deformation coordinates
变性	變性	denaturation
变性剂	變性劑	denaturant
变性酒精	變性酒精	denatured alcohol, methylated spirit
变压器	變壓器	transformer
变压吸附	變壓吸附	pressure swing adsorption, PSA
变异性	變異性	variability
辨识	識別	
标称值	標稱值	nominal value
标度律	標度律	scaling law
标度因子	比例因數	scale factor
标度指示器	標度指示器	scale indicator
标量	純量	scalar
标准	標準	standard
标准操作	標準操作	standard operation
标准操作步骤	標準操作步驟	standard operation procedure
标准差	標準[偏]差	standard deviation
标准稠度	標準稠度	standard consistency
标准大气压	標準大氣壓力	standard atmospheric pressure
标准电池	標準電池	standard cell
标准反应焓变化	標準反應焓變	standard enthalpy change of reaction
标准反应热	標準反應熱	standard heat of reaction
标准反应条件	標準反應條件	standard reaction condition
标准符号	標準符號	standard symbol
标准化	標準化	standardization
标准化合物	標準化合物	standard compound
标准化流程图	標準化流程圖	standardization flow-sheet
标准浇口	標準澆口	standard gate
标准孔口	標準孔口	standard orifice
标准量规	標準規	standard gage
标准螺纹弯头	標準螺旋彎頭	standard screwed elbow
标准命名法	標準命名法	standard nomenclature
标准喷嘴	標準噴嘴	standard nozzle
标准燃烧热	標準燃燒熱	standard heat of combustion

大 陆 名	台 湾 名	英 文 名
标准溶液	標準溶液	standard solution
标准筛	標準篩	standard sieve
标准生成焓	標準生成焓	standard enthalpy of formation
标准生成焓变化	標準生成焓變化	standard enthalpy change of formation
标准生成热	標準生成熱	standard heat of formation
标准试验	標準試驗	standard test
标准试验筛	標準試驗篩	standard testing sieve
标准态逸度	標準狀態逸壓	standard state fugacity
标准温度	標準溫度	standard temperature
标准温度和压力	標準溫壓	standard temperature and pressure
标准温度计	標準溫度計	standard thermometer
标准误差	標準誤差	standard error
标准纤维素	標準纖維素	standard cellulose
标准线性固体	標準線性固體	standard linear solid
标准状态	標準狀態	standard state
表	表	gauge
表观比热	視比熱	apparent specific heat
表观比重	視比重	apparent specific gravity
表观纯度	視純度	apparent purity
表观活度	視活性	apparent activity
表观活化能	視活化能	apparent activation energy
表观级数	視階	apparent order
表观剪切率	視剪[切]率	apparent shear rate
表观剪切应力	視剪[切]應力	apparent shear stress
表观孔隙率	視孔隙度	apparent porosity
表观扩散系数	視擴散係數	apparent diffusivity
表观密度	視密度	apparent density
表观黏度	視黏度	apparent viscosity
表观溶解度	視溶解度	apparent solubility
表观速度	表觀速度	superficial velocity
表观体积	視體積	apparent volume
表观应力	視應力	apparent stress
表观质量	視質量	apparent mass
表观重力	視重力	apparent gravity
表观重量	視重量	apparent weight
表观组成	視組成	apparent composition
表面层	表層	surface layer
表面处理	表面處理	surface treatment

大　陆　名	台　湾　名	英　文　名
表面粗糙度	表面粗糙度	surface roughness
表面粗化处理	表面粗化處理	surface roughening treatment
表面催化	表面催化[作用]	surface catalysis
表面电导	表面電導	surface conductance
表面电荷	表面電荷	surface charge
表面电位	表面電位	surface potential
表面电阻	表面電阻,表面阻力	surface resistance
表面电阻率	表面電阻率	surface resistivity
表面发酵	表面發酵	surface fermentation
表面发射率	表面發射率	surface emissivity
表面反应阻力	表面反應阻力	surface reaction resistance
表面覆盖度	表面覆蓋率	surface coverage
表面改性	表面改質	surface modification
表面改性剂	表面改質劑	surface modifier
表面改性纤维	表面改質纖維	surface modified fiber
表面更新理论	表面更新理論	surface renewal theory
表面更新因数	表面更新因數	surface renewal factor
表面工程	表面工程	surface engineering
表面横向结晶	表面橫列結晶	surface transcrystallinity
表面化学	表面化學	surface chemistry
表面活性	表面活性	surface activity
表面活性剂	表面活性劑,界面活性劑	surface-active agent, surfactant
表面活性剂泛点	界面活性劑氾流[法]	surfactant flooding
表面积	表面積	surface area
表面接枝	表面接枝	surface grafting
表面结构	表面結構	surface structure
表面卷取机	表面捲取機	surface winder
表面科学	表面科學	surface science
表面孔隙度	表面孔隙度	surface porosity
表面扩散	表面擴散	surface diffusion
表面力	表面力	surface force
表面裂纹	表面裂紋	surface crack
表面流变学	表面流變學	surface rheology
表面流动	表面流動	surface flow
表面轮廓	表面輪廓	surface profile
表面络合物	表面錯合物	surface complex
表面密度	表面密度	surface density

大　陆　名	台　湾　名	英　文　名
表面磨蚀	表面磨耗	surface abrasion
表面能	表面能	surface energy
表面黏度	表面黏度	surface viscosity
表面黏度计	表面黏度計	surface viscometer
表面黏接	表面黏接	surface bonding
表面浓度	表面濃度	surface concentration
表面培养	表面培養	surface cultivation
表面缺陷	表面缺陷	surface imperfection
表面施胶	表面上膠	surface sizing
表面式冷凝器	表面冷凝器	surface condenser
表面水分	表面水分	surface moisture
表面速率	表面速率	surface rate
表面酸性部位	表面酸性部位	surface acid site
表面涂层	表面塗裝,表面塗層	surface coating
表面位错	表面移位	surface dislocation
表面温度	表面溫度	skin temperature, surface temperature
表面吸附	表面吸附	surface adsorption
表面吸收器	表面吸收器	surface absorber
表面现象	表面現象	surface phenomenon
表面形态	表面形態	surface morphology
表面压力	表面壓力	surface pressure
表面曳引	表面阻力	surface drag
表面硬化	表面硬化	case hardening, surface hardening
表面张力	表面張力	surface tension
表面张力计	表面張力計	surface tensiometer
表面制备	表面製備	surface preparation
表面状况	表面狀態	surface condition
表面自由能	表面自由能	surface free energy
表皮效应	表皮效應,集膚效應	skin effect
表皮型包装	密著包裝	skin packaging
表压	錶壓	gauge pressure
宾厄姆流体	賓漢流體	Bingham fluid
冰醋酸	冰醋酸	glacial acetic acid
冰点	冰點	ice point
冰晶石	冰晶石	cryolite
冰糖	冰糖	rock sugar
冰浴	冰浴	ice bath
丙二酸	丙二酸	malonic acid

大　陆　名	台　湾　名	英　文　名
丙酸纤维素	丙酸纖維素	cellulose propionate
丙烯低聚物	丙烯寡聚物	propylene oligomer
丙烯腈	丙烯腈	acrylonitrile
丙烯腈-丁二烯-苯乙烯树脂	丙烯腈–丁二烯–苯乙烯樹脂	acrylonitrile-butadiene-styrene resin，ABS resin
丙烯酸酯	丙烯酸酯	acrylic ester
丙烯酸[酯]树脂	丙烯酸樹脂,壓克力樹脂	acrylic resin
丙烯酸酯橡胶	丙烯酸酯橡膠	acrylic ester rubber
并发反应	併發反應	concurrent reaction
并联补偿	並聯補償	parallel compensation
并联电路	並聯電路	parallel circuits
并联运行	並聯運轉	parallel running
并列复式涡轮机	並列複式渦輪機	parallel compound turbine
并流	同向流	cocurrent flow
并流操作	同向流操作	cocurrent operation
并流过程	同向流程序	cocurrent process
并流热交换	並流熱交換	paraflow heat exchange
并流,同向流	並流	paraflow
并行操作	並聯操作	parallel operation
并行过程	並聯程序	parallel processes
病毒	病毒	virus
病原体	病原體	pathogen
波长	波長	wavelength
波传播	波傳播	wave propagation
波动的	梳麻	rippling
波动力学	波動力學	wave mechanics
波数	波數	wave number
波纹板	浪板	corrugated sheet
波纹管	波形管	corrugated tube
波纹管式流量计	伸縮式流量計	bellows type flowmeter
波纹管式压力计	伸縮式壓力計	bellows manometer
波纹塔板	波紋塔板	ripple tray
波状流	波狀流	wavy flow
波状运动	波動	wave motion
玻耳兹曼分布	波茲曼分佈	Boltzmann distribution
玻化瓷器	玻化瓷器	vitreous china
玻化相	玻化相	vitreous phase

大 陆 名	台 湾 名	英 文 名
玻璃	玻璃	glass
玻璃衬里	玻璃襯裏	glass lining
玻璃吹制	玻璃吹製	glass blowing
玻璃电极	玻璃電極	glass electrode
玻璃固化	玻[璃]化	vitrification
玻璃固化期	玻[璃]化期	vitrification period
玻璃固化砖	玻[璃]化磚	vitrified brick
玻璃管	玻璃管	glass pipe, glass tube
玻璃管及配件	玻璃管子及配件	glass pipe and fitting
玻璃过滤器	玻璃[過]濾器,玻璃濾光片	glass filter
玻璃化温度	玻璃轉移溫度	glass transition temperature
玻璃拉制	玻璃拉製	glass drawing
玻璃料	玻璃料	frit
玻璃棉	玻璃綿,玻璃絨	glass wool
玻璃强化塑料	玻璃強化塑膠	glass reinforced plastics
玻璃区	玻璃區	glassy zone
玻璃态	玻璃態	glassy state
玻璃搪瓷	玻璃搪瓷	glass enamel
玻璃陶瓷	玻璃陶瓷	glass ceramics
玻璃温度计	玻璃溫度計	glass stem thermometer
玻璃纤维	玻璃纖維	glass fiber
玻璃纤维强化塑料	玻璃纖維強化塑膠	glass fiber reinforced plastics
玻璃液位计	玻璃液位計,玻璃量規	glass level gage
玻色-爱因斯坦分布	玻[色]-愛[因斯坦]分佈	Bose-Einstein distribution
剥离	剝離	peeling
剥离试验	剝離試驗	stripping test
剥落	剝落物,剝落	spall, spallation, spalling
剥落试验	散裂試驗	spalling test
剥落试验板	散裂試驗板台	spalling test panel
剥蚀	麻點,凹痕	pitting
剥撕	剝撕	split-tear
伯努利方程	白努利方程[式]	Bernoulli equation
泊	泊	poise
铂重整	鉑媒重組	platforming
铂重整产品	鉑媒重組油	platformate
铂重整过程	鉑媒重組程序	platforming process

大　陆　名	台　湾　名	英　文　名
铂电阻球	鉑電阻球	platinum resistance bulb
铂电阻温度计	鉑電阻溫度計	platinum resistance thermometer
铂坩埚	鉑坩堝,白金坩堝	platinum crucible
铂海绵	鉑海綿	platinum sponge
铂黑催化剂	鉑黑觸媒	platinum black catalyst
铂接触过程	鉑接觸法	platinum contact process
铂-镍催化剂	鉑鎳觸媒	platinum nickel catalyst
铂网	鉑網	platinum gauze
铂氧化铝催化剂	鉑鋁氧觸媒	platinum-alumina catalyst
博登斯坦数	博登施泰數	Bodenstein number
[博拉德]双层网环	[博拉德]雙層網環	Borad ring
薄层	薄層	lamina
薄层色谱法	薄層層析法	thin layer chromatography
薄膜闪蒸器	薄膜閃蒸器	flash film evaporator
薄膜蒸发器	薄膜蒸發器	thin-film evaporator
薄片	片	slice
簸动	簸選	jigging
簸动筛	簸選篩	jigging screen
薄荷醇	薄荷腦	menthol
薄荷油	薄荷油	peppermint oil
补偿	補償	compensation
补偿器	補償器	compensator
补偿温度计系统	補償溫度計系統	compensated thermometer system
补偿效应	補償效應	compensation effect
补强剂	強化劑	strengthening agent
捕获剂	捕獲劑	scavenger
捕集效率	收集效率	collection efficiency
捕液器	液阱	liquid trap
不摆应答	不擺應答	dead beat response
不饱和	不飽和	unsaturation
不饱和聚酯	不飽和聚酯	unsaturated polyester
不饱和烃	不飽和烴	unsaturated hydrocarbon
不饱和脂肪酸	不飽和脂肪酸	unsaturated fatty acid
不变时间系统	非時變系統	time invarying system
不变式	不變式	invariant
不对称流动	不對稱流動	asymmetrical flow
不合格产品	不合[規]格產品	off-spec product
不互溶流动	不互溶流動	immiscible flow

大　陆　名	台　湾　名	英　文　名
不互溶溶剂	不互溶溶劑	immiscible solvent
不互溶系统	不互溶系統	immiscible system
不互溶相	不互溶相	immiscible phase
不互溶性	不互溶性	immiscibility
不互溶液体	不互溶液體	immiscible liquid
不挥发性	不揮發性	nonvolatility
不挥发油	不揮發油	fixed oil
不均匀性	不匀性	heterogeneity
不可逆催化	不可逆催化[作用]	irreversible catalysis
不可逆电池	不可逆電池	irreversible cell
不可逆反应	不可逆反應	irreversible reaction
不可逆功	不可逆功	irreversible work
不可逆过程	不可逆程序,不可逆過程	irreversible process
不可逆[过程]热力学	不可逆熱力學	irreversible thermodynamics
不可逆化学反应	不可逆化學反應	irreversible chemical reaction
不可逆性	不可逆性	irreversibility
不可压缩流动	不可壓縮流動	incompressible flow
不可压缩流体	不可壓縮流體	incompressible fluid
不可压缩滤饼	不可壓縮濾餅	incompressible filter cake
不可压缩性	不可壓縮性	incompressibility
不冷凝气体	不冷凝氣體	incondensable gas
不连续沉降	個別沈降	discrete settling
不连续过程	不連續程序	discontinuous process
不连续加工	不連續加工[法]	discontinuous processing
不良分布	不良分佈	maldistribution
不凝气体	不冷凝氣體	noncondensable gas
不确定型决策	不確定性決策制定	decision making under uncertainty
不溶解化	不溶解化	insolubilization
不渗透性	不透性	impermeability
不透明度	不透明度	opacity
不透明釉	不透明釉	opaque glaze
不透水的	防水	waterproof
不透性隔板	不透性隔板	impermeable partition
不稳定操作点	不穩定操作點	unstable operating point
不稳定常数	不穩定常數,不安定常數	instability constant
不稳定区	不穩定區	labile region, labile zone

大　陆　名	台　湾　名	英　文　名
不稳定系统	不穩定系統	unstable system
不稳定响应	不穩定應答	unstable response
不稳定性	不穩定性,不安定性	instability
不稳定状态	不安定穩態	unstable steady state
不锈钢	不鏽鋼	stainless steel
不锈钢纤维	不鏽鋼纖維	stainless-steel fiber
不皂化物	不皂化物	unsaponified matter
布朗扩散	布朗擴散	Brownian diffusion
布滤机	布濾機	cloth filter
布罗伊登法	布羅伊登法	Broyden method
布置	布置	layout
布置检查表	布置檢查表	layout check list
布置图	布置圖	layout diagram
步进电机	步進馬達	step motor
钚反应器	鈽反應器	plutonium reactor
部分反应	部分反應	partial reaction
部分分隔流动	部分分隔流[動]	partially segregated flow
部分分解	部分分解	partial decomposition
部分分离	部分分離	partial separation
部分分式	部分分式	partial fraction
部分分式展开式	部分分式展開[法]	partial-fraction expansion
部分互溶	部分互溶性	partial miscibility
部分互溶系统	部分互溶系統	partially miscible system
部分回流	部分回流	partial reflux
部分甲基化	部分甲基化	partial methylation
部分解耦	部分解偶	partial decoupling
部分冷凝	部分冷凝	partial condensation
部分裂化	部分裂解	partial cracking
部分燃烧	部分燃燒	partial combustion
部分寿命	部分壽命	fractional life
部分衰减法	部分衰減法	fractional-life method
部分氧化	部分氧化	partial oxidation
部分氧化反应	部分氧化反應	partial oxidation reaction
部分游离	部分游離	partial ionization
部分再循环过程	部分循環程序	partial recycle process
部分真空	部分真空	partial vacuum
部分转化	部分轉化	partial conversion

C

大　陆　名	台　湾　名	英　文　名
擦涂法	擦抹取樣過程	swabbing process
擦洗剂	擦潔劑	abrasive cleaner
材料成本	材料成本	material cost
材料成本指数	材料成本指數	material cost index
材料工程	材料工程	materials engineering
材料合成	材料合成	material synthesis
材料加工	材料加工	material processing
材料科学	材料科學	materials science
材料设计	材料設計	material design
材料制造	材料製造	material manufacturing
财产税	財產稅	property tax
财务会计成本	資金成本	financing cost
采样	採樣	sampling
采样数据	採樣數據	sampled data
采样数据控制	採樣數據控制	sampled data control
采样数据控制系统	採樣數據控制系統	sampled data control system
采样数据系统	採樣數據系統	sampled data system
采样数据信号	採樣數據信號	sampled data signal
采样系统	採樣系統	sampling system
采油	油回收	oil recovery
参比电极	參考電極	reference electrode
参比接点	參考接點	reference junction
参比条件	參考條件	reference condition
参考标准	參考標準	reference standard
参考输入	參考輸入	reference input
参考温度	參考溫度	reference temperature
参考值	參考值	reference value
参考坐标	參考坐標	frame of reference
参数	參數	parameter
参数泵	參數泵	parametric pump
参数法分离	參數法分離	parametric pumping
参数集总	集總	lumping
参数灵敏度	參數靈敏度	parameter sensitivity

大　陆　名	台　湾　名	英　文　名
参数选择	參數選擇	parameter selection
残差	殘餘誤差	residual error
残差分析	殘差分析	residual analysis
残油	殘留產物	residual product
残余贡献	剩餘貢獻	residual contribution
残余焓	剩餘焓	residual enthalpy
残余馏分	殘留分率	residual fraction
残余气体	殘留氣體	residual gas
残余熵	剩餘熵	residual entropy
残余体积	剩餘體積	residual volume
残余项	殘留項	residual term
残余性质	剩餘性質	residual property
残余应变	殘留應變	residual strain
残值	殘餘價值	salvage value
蚕胶丝	蠶膠	silkworm gut
蚕丝	蠶絲	silk
操纵变量	操縱變數	manipulating variable
操作变量	操作變數	operating variable, operational variable
操作标准	操作標準	operation standard
操作参量	操作參數	operating parameter
操作成本	操作成本	operating cost
操作窗口	操作窗	operating window
操作点	操作點	operating point
操作放大器	操作放大器	operation amplifier
操作费用	操作成本	operation cost
操作极限	操作極限	operating limit
操作流程图	操作流程圖	operation flowsheet
操作模式	操作方式	operation mode
操作频率	操作頻率	operating frequency
操作曲线	操作曲線	operating curve
操作时间	操作時間,操作期	operating time, operation length
操作手册	操作手冊	operating manual
操作数据	操作數據	operating data
操作弹性	操作彈性	operating flexibility
操作条件	操作條件	operation condition
操作图	操作圖	operating diagram
操作温度	操作溫度	operating temperature
操作稳定性	操作穩定性	operating stability

大　陆　名	台　湾　名	英　文　名
操作线	操作線	operating line
操作效率	操作效率	operation efficiency
[操作]性能控制	性能控制	performance control
[操作]性能图	性能圖	performance chart
操作压力	操作壓力	operating pressure
操作原理	操作原理	operating principle
操作再生循环	操作再生循環	operation regeneration cycle
操作周期	操作周期	operation cycle
槽	槽	channel
槽车	液罐車,油罐車	tankcar
槽法炭黑	槽黑	channel black
槽炉	槽爐	tank furnace
槽式反应器	槽式反應器	tank reactor
槽式结晶器	槽式結晶器	tank crystallizer
槽式流量计	槽式流量計	channel type flowmeter
草纸浆	草[類]紙漿	straw pulp
侧回流	側回流	side reflux
侧基	側基	side group
侧链基	側鏈[自由]基	side chain radical
侧链运动	側鏈運動	side chain motion
侧[链]支化	側支,側鏈	side branch
侧流	側流,支流	side stream
侧馏分	側流餾出物	sidecut distillate
侧线[馏分]汽提塔	側流汽提塔	side stream stripper
侧线汽提塔	側流汽提塔	side stripper
侧线塔	側流塔	sidestream column
侧向应力	側向應力	lateral stress
测尘器	測塵計	konimeter
测定	測定	determination
测量	量測,測定	measurement
测量元件	量測元件	measuring element
测量滞后	量測滯延	measurement lag
测量装置	量測裝置	measuring device
测黏流动	測黏[度]流動	viscometric flow
测声计	測聲計	acousimeter
测湿法	測濕法	hygrometry
测试	試驗法	testing
测试点	測試點	test point

大 陆 名	台 湾 名	英 文 名
测试过程	測試步驟	test procedure
测试仪	試驗器	tester
测微器	測微計	micrometer
测温法	測溫法	thermometry
测温组件	測溫元件	temperature measuring element
测压孔	測壓分接頭	pressure tap
测压液	測壓液	manometric fluid
测振仪	測振儀	vibrometer
层间应力	層間應力	interlaminar stress
层离黏土	剝層黏土	delaminated clay
层流	層流,流線流	laminar flow, streamline flow
层流边界层	層流邊界層	laminar boundary layer
层流底层	層流次層	laminar sublayer
层流反应器	層流反應器	laminar flow reactor
层流区	層流區[域]	laminar region
层流式喷流吸收器	層流式噴射吸收器	laminar jet absorber
层压	積層	lamination
层压板壁	積層板壁	laminated wall
层压材料	積層板	laminate
层压塑料	分層塑膠	stratified plastic
层压压制机	積層壓製機	lamination press
层状的	層列狀	smectic
层状态	[液晶]層列狀態	smectic state
层状相	[液晶]層列狀相	smectic phase
插塞	塞[子]	plug
插线	插線	patching
差动阀	[微]差壓閥	differential valve
差分方程	差分方程[式]	difference equation
差降	差降	drop
差热分析[法]	微差熱分析[法]	differential thermal analysis
差示热机械分析法	微差熱機械分析[法]	differential thermomechanical analysis, DTMA
差示热重分析法	微差熱重分析[法]	differential thermogravimetric analysis
差示 U 形管	微差 U[形]管	differential U-tube
差示压力计	微差壓力計	differential manometer
差式风压计	差式風壓計	draft gage
差速离心分离	差速離心分離	differential centrifugation
差压池	微差壓計	differential pressure cell

大　陆　名	台　湾　名	英　文　名
差压传感器	微差壓轉換器	differential pressure transducer
差压极限	微差壓極限	differential pressure limit
差异压头	微分高差	differential head
柴油	柴油[燃料]	diesel fuel
柴油爆震	柴油震爆	diesel knock
柴油机	柴油引擎,柴油機	diesel engine
掺合	掺合	blending
掺合物	掺合物	blend
掺和槽	掺配槽	blending tank
掺和过程	掺配程序	blending process
掺和机	掺合機	blender
掺和剂	掺合劑	blending agent
掺和式混合器	掺合式混合器	blending mixer
掺和系统	掺配系統	blending system
掺气流	掺氣流,通氣流,曝氣流	aerated flow
掺入	掺雜	dope
掺杂	掺雜	doping
掺杂剂	掺雜劑	dopant
缠结	纏結	entanglement
产量	產量,生產量	production, output, throughput
产量分率	產量分率	fractional yield
产品成本	產品成本	product cost
产品储存	產品儲存	product storage
产品船运清单	產品運送清單	product shipment check list
产品分布变异	產品分佈變異	product distribution distortion
产品分离	產物分離	product separation
产品分离器	產物分離器	product separator
产品工程	產品工程	product engineering
产品规格	產品規格	product specification
产品规划	產品規畫	product planning
产品回收	產物回收	product recovery
产品开发	產品開發	product development
产品批量工厂	多產品批式工廠	multiproduct batch plant
产品设计	產品設計	product design
产品物流	產物流	product stream
产品质量	產品品質	product quality
产酸菌	產酸[細]菌	acid-forming bacteria
产物分布	產物分佈	product distribution

大　陆　名	台　湾　名	英　文　名
产物抑制	產物抑制[作用]	product inhibition
长管蒸发器	長管蒸發器	long tube evaporator
长宽比	長寬比	aspect ratio
长石	長石	feldspar
长石釉	長石釉	feldspathic glaze
长条记录纸	長條記錄紙	strip chart
长弯头	長彎頭	long sweep elbow
长网成型机	長綱[製紙]機	fourdrinier machine
长效剂	長效劑,持久劑	persistent agent
肠蛋白酶	腸蛋白酶	erepsin
常数	常數	constant
常微分方程	常微分方程[式]	ordinary differential equation
常压操作	常壓操作	atmospheric operation
常压分馏塔	常壓分餾塔,常壓分餾器	atmospheric fractionator
常压干燥器	常壓乾燥器	atmospheric drier
常压平衡蒸馏	常壓平衡蒸餾	air equilibrium distillation
常压闪蒸塔	常壓驟餾塔	atmospheric flash tower
常压芽浆糖化	常壓製醪	atmospheric mashing
常压蒸馏	常壓蒸餾	air distillation
常压蒸汽	常壓蒸汽	atmospheric steam
厂址	廠址	plant location
厂址选择	廠址選擇	plant site selection
敞开系统	開放系統	open system
敞式反应器	開敞反應器	open reactor
敞式沙滤器	開敞砂濾器,開敞過濾器	open sand filter, open filter
敞式序列反应	非連鎖反應	open sequence reaction
钞票纸	鈔票紙	bank-note paper
超薄膜	超薄膜	ultrathin film
超大规模集成电路	超大型積體電路	very large scale integrated circuit, VLSI
超导薄膜	超導薄膜	superconducting thin film
超导发电机	超導發電機	superconducting generator
超导合金	超導合金	superconducting alloy
超导聚合物	超導聚合物	superconductive polymer
超导态	超導態	superconducting state
超导碳氮化物	超導碳氮化物	superconducting carbonitride
超导特性	超導特性	superconducting characteristic

大　陆　名	台　湾　名	英　文　名
超导体	超導體	superconductor
超导性	超導性	superconductivity
超导转变温度	超導轉變溫度	superconducting transition temperature
超额函数	過剩函數	excess function
超额焓	超額焓,過剩焓	excess enthalpy
超额化学势	過剩化學勢	excess chemical potential
超额吉布斯自由能	過剩吉布斯自由能	excess Gibbs free energy
超额浓度法	過量濃度法	excess concentration technique
超额熵	超額熵,過剩熵	excess entropy
超额体积	超額體積,過剩體積	excess volume
超额性质	過剩性質	excess property
超分子	超分子	supermolecule
超分子结构	超分子結構	supermolecular structure
超分子有序	超分子次序	supermolecular order
超分子织态结构	超分子紋理	supermolecular texture
超分子转变	超分子轉變	supermolecular transition
超高聚物	超聚合物	superpolymer
超级计算机	超級計算機	supercomputer
超加速剂	超催速劑	ultra accelerator
超晶格	超晶格	superlattice
超净化	超淨化	ultrapurification
超拉伸	超延伸	super drawing
超离子导电聚合物	超離子導電聚合物	super ion-conductive polymer
超临界萃取	超臨界萃取	supercritical extraction
超临界流	超臨界流動	supercritical flow
超临界气体	超臨界氣體	supercritical gas
超流体	超流體	superfluid
超滤	超過濾,超微過濾	ultrafiltration, hyperfiltration
超滤机	超過濾器	ultrafilter
超前补偿	超前補償	lead compensation
超声波沉淀	超音波沈澱	ultrasonic precipitation
超声波分解	超音波分解	sonolysis
超声波计	超音波計	ultrasonic meter
超声波流动	超音波流	ultrasonic flow
超声波黏度计	超音波黏度計	ultrasonic viscometer
超声波学	超音波學	ultrasonics
超声波振动器	超音波振動器	ultrasonic vibrator
超声附聚	超音波團聚	ultrasonic agglomeration

大　陆　名	台　湾　名	英　文　名
超声净化	超音波清潔	ultrasonic cleaning
超声速喷嘴	超音速噴嘴	supersonic nozzle
超松弛法	過鬆弛法	overrelaxation method
超速离心	超離心［作用］	ultracentrifugation
超［速］离心机	超離心機	ultracentrifuge
超速离心机	高速離心機	supercentrifuge
超弹性	超彈性	super-elasticity
超弹性材料	超彈性物質	hyperelastic materials
超调	超越量,超調量	overshoot
超微量化学	超微量化學	ultra microchemistry
超吸附	超吸附［法］	hypersorption
超吸附装置	超吸附裝置	hypersorber
超细研磨机	超細研磨機	ultragrinder
超显微法	超顯微法	ultra microscopy
超显微镜	超顯微鏡	ultra microscope
超限保护	超限保護	over range protection
超氧化物	超氧化物	superoxide
巢式控制环路	巢式控制環路	nested control loop
潮解	潮解	deliquescence
潮霉素	潮黴素	hygromycin
车用汽油	車用汽油	petrol
沉底酵母	沈底酵母	bottom yeast
沉淀	沈澱	precipitation
沉淀槽	沈降槽	sedimentation chamber
沉淀池	沈降池	sedimentation basin
沉淀滴定［法］	沈澱滴定［法］	precipitation titration
沉淀点	設定點	set point
沉淀反应	沈澱反應	precipitation reaction
沉淀聚合	沈澱聚合［反應］	precipitation polymerization
沉淀瓶	沈降瓶	sedimentation bottle
沉淀器	沈澱器	precipitator
沉淀物	沈澱物	precipitate
沉积池	沈積槽	sedimentation tank
沉积电势	沈降電位	sedimentation potential
沉积高岭土	沈積高嶺土	sedimentary kaolin
沉积黏土	沈積黏土	sedimentary clay
沉积速率	沈積速率	sedimentation rate
沉积物	沈積物	sediment

大　陆　名	台　湾　名	英　文　名
沉积岩	沈積岩	sedimentary rock
沉降	沈降	sedimentation, settling
沉降槽	沈降槽	settling tank
沉降常数	沈降常數	sedimentation constant
沉降池	沈降池	settling basin
沉降法	沈降法	sedimentation method
沉降过滤	沈降過濾	settling filtration
沉降盘	沈降盤	sedimentation pan
沉降平衡	沈降平衡	sedimentation equilibrium
沉降平衡法	沈降平衡法	sedimentation equilibrium method
沉降平均分子量	沈降平均分子量	sedimentation average molecular weight
沉降器	沈降器,沈降槽	settler
沉降区	沈降區	settling zone
沉降时间	沈降時間,安定時間	settling time
沉降室	沈降室	settling chamber
沉降速度	沈降速度	sedimentation velocity, settling velocity
沉降速度法	沈降速度法	sedimentation velocity method
沉降系数	沈降係數	sedimentation coefficient
沉降柱	沈降柱	settling column
沉降柱分析	沈降柱分析	settling column analysis
衬玻璃管	玻璃襯管	glass-lined pipe
衬里	襯裏	lining
衬里管	襯管	lined pipe
成本	成本	cost
成本工程师	成本工程師	cost engineer
成本估计	成本估計[值]	cost estimate, cost estimation
成本核算	成本會計	cost accounting
成本曲线	成本曲線	cost curve
成本效益	成本效益	cost effectiveness
成本效益分析	成本效益分析	cost benefit analysis
成本效益管理	成本效益管理	cost-effective management
成本因子	成本因數	cost factor
成本指数	成本指數	cost index
成核	成核[作用]	nucleation
成核结晶	成核結晶	nucleation crystallization
成气反应	成氣反應	gas-forming reaction
成球	成球,球狀化,製粒	balling, prilling
成套设备	套裝設備	packaged equipment

大　陆　名	台　湾　名	英　文　名
成套设计文件	設計套件	design package
承插接头	插承接頭	bell and spigot joint
城市垃圾	都市廢棄物	municipal waste
城市污水	都市污水	municipal wastewater
乘法器	乘法器	multiplier
程长	路徑長度	path length
程序变温	程序變溫	temperature programming
程序变温器	程序變溫器	temperature programmer
程序控制	程式化控制	programmed control
程序控制器	程式化控制器	programmed controller
程序框图	程序方塊圖	block flow diagram
澄清	澄清[作用]	clarification, defecation
澄清过程	澄清程序	clarification process
澄清过滤	澄清過濾	clarification filtration
澄清过滤器	澄清過濾器	clarifying filter
澄清能力	澄清能力	clarification capacity, clarifying capacity
澄清器	澄清器	clarifier, defecator
澄清区	澄清區	clarification zone
澄清汁	澄清汁	defecated juice
弛豫法(＝松弛法)		
弛豫模量	鬆弛模數	relaxation modulus
弛豫时间	鬆弛時間	relaxation time
池沸腾	池沸騰	pool boiling
池沸腾曲线	池沸騰曲線	pool boiling curve
持气率	持氣量	gas holdup
尺寸比	尺寸比	size ratio
尺寸稳定性	尺寸穩定性	dimensional stability
尺寸因子	尺寸因子	size factor
齿轮	齒輪	gear
齿轮泵	齒輪泵	gear pump
齿轮式混合器	齒輪式混合器	gear mixer
齿轮组	齒輪組	gear train
齿隙	背隙	backlash
赤磷	赤磷,紅磷	red phosphorus
赤铁矿	赤鐵礦	hematite
赤铜矿	赤銅礦	cuprite
翅管式热交换器	鰭片管熱交換器	finned-tube heat exchanger
翅片	①鰭片 ②散熱片	fin

大　陆　名	台　湾　名	英　文　名
翅片表面	鰭片表面	finned surface
翅片管	鰭片管	finned pipe, finned tube
充电器	充電器	charger
充分发展流	完全發展流動	fully developed flow
充气	通氣	aeration
充气槽	通氣槽	aeration tank
充气管	通氣管	gasfilled tube
充气搅拌	空氣攪拌,氣動攪拌	air agitation
充气搅拌器	空氣攪拌器,氣動攪拌器	air agitator
充气率	通氣速率	aeration rate
充气轮	曝氣輪	aeration wheel
充气器	通氣機	aerator
充气室	充氣室	plenum chamber
充气数	通氣數	aeration number
充气装置	通氣裝置	aeration device
充填率	裝填分率	packing fraction
充氧器	充氧器	oxygenator, oxygenizer
冲程	衝程	stroke
冲击	衝擊	impact, impingement
冲击波	震波	shock wave
冲击波反应器	震波反應器	shock wave reactor
冲击波聚合	震波聚合[作用]	shock wave polymerization
冲击腐蚀	衝擊腐蝕	impingement corrosion
冲击机	衝擊機	impactor
冲击凝固	衝擊凝聚[作用]	shock coagulation
冲击破碎机	衝擊粉碎機	impact crusher
冲击强度	衝擊強度	impact strength
冲击[式]涤气器	衝擊洗氣器	impingement scrubber
冲击[式]离心分离器	衝擊離心分離器	impingement centrifugal separator
冲击试验	衝擊試驗	impact test
冲击数	衝擊數	impact number
冲击弹性	衝擊彈性	shock elasticity
冲击压[力]	衝擊壓[力]	impact pressure
冲击压头	衝擊高差	impact head
冲击载荷	衝擊負載	shock load
冲角	攻角,衝角	angle of attack
冲量	①脈衝 ②衝量	impulse

大　陆　名	台　湾　名	英　文　名
冲流电压	冲流電壓,冲流電位	streaming potential
冲洗	冲盡	washout
重氮化	重氮化[反應]	diazotization
重氮化合物	重氮化合物	diazo compound
重氮染料	重氮染料	diazo dye
重叠操作	重疊操作	overlapping operation
重沸器(＝再沸器)		
重复操作	重複操作	repetitive operation
重复性聚合物	重複性聚合物	repetitive polymer
重合压延机	重合壓延機	doubling calender
重整	重組	reforming
重整催化剂	重組觸媒	reforming catalyst
重整反应	重組反應	reforming reaction
重整汽油	重組汽油	reformed gasoline
重整器	重組器	reformer
重组	重組	recombination
重组 DNA	重組 DNA	recombinant DNA
抽水泵	吸取泵	suction pump
抽丝	紡絲	spinning
抽提塔	萃取器	extractor
稠度	①稠度②厚度	consistency, thickness
稠度试验	稠度試驗[法]	consistency test
臭氧	臭氧	ozone
臭氧龟裂	臭氧龜裂	ozone cracking
臭氧化	臭氧化[作用]	ozonization
臭氧计	臭氧計	ozonometer
出发角	發射角	angle of departure
出口	出口	outlet
出口阀	出口閥	outlet valve
出口气体	出口氣體	exit gas
出口温度	出口溫度	outlet temperature
出气	除氣	outgassing
初步概算	初估值	preliminary estimate
初步设计	初步設計,前期設計	preliminary design
初沸点	初沸點	initial boiling point
初级沉降槽	初級沈降槽	primary settler
初级澄清器	初級澄清器	primary clarifier
初级净化	初級純化	primary purification

大　陆　名	台　湾　名	英　文　名
初[级蒸]馏	初[級蒸]餾	primary distillation
初馏塔	初餾塔	primary tower
初滤机	初濾機	preliminary filter
初始松弛系统	初始鬆弛系統	initial relaxed system
初始条件	初始條件	initial condition
初[始状]态	初[始狀]態	initial state
初速[度]	初速[度]	initial velocity
初速率	初速率	initial rate
初速率法	初速率法	method of initial rate
初值定理	初值定理	initial value theorem
除草剂	除草劑	herbicide, weed killer
除虫菊酯	除蟲菊酯	pyrethrin
除臭	除臭	deodorization
除臭剂	除臭劑	deodorant
除臭器	除臭器	deodorizer
除焦	除焦,去焦	decoking
除焦机	除焦機	coke knocker
除焦时间	除焦時間,去焦時間	decoking time
除沫器	除霧器	demister
除湿器	除濕器	dehumidifier
除锈剂	除鏽劑	rust remover
储槽	儲槽	storage tank
储存量	儲存容量	storage capacity
储存期限	儲存期限,儲存壽命	shelf life
储存设备	儲存裝置	storage equipment
储存设施	儲存設施	storage facility
储存寿命	保存期限	storage life
储存损失	儲存損失	storage loss
储存温度	儲存溫度	storage temperature
储存稳定性	儲存穩定性	storage stability
储存装置流程图	儲存裝置流程圖	storage equipment flowsheet
储料堆	儲料[堆]	stock pile
储能函数	貯能函數	stored energy function
储能模量	儲存模數	storage modulus
储能柔量	儲存柔量	storage compliance
储热器	貯熱器	heat reservoir
储酸罐	酸液儲槽	acid tank
储油罐	油箱,油槽	oil tank

大　陆　名	台　湾　名	英　文　名
处理工厂	處理工場	treatment plant
处理设备	處理設備	treatment device
触变胶	觸變減黏膠,搖變減黏膠	thixotrope
触变性	觸變減黏性,搖變減黏性	thixotropy
触变性流体	觸變減黏流體,搖變減黏流體	thixotropic fluid
触变性质	觸變減黏性質,搖變減黏性質	thixotropic property
穿孔	穿孔,打洞	perforation
穿流塔板	雙流塔板	dual-flow tray
穿流栅板	穿流栅板	turbo-grid tray
穿透点	貫流點	breakthrough point
穿透概率	穿透機率	penetration probability
穿透理论	穿透理論	penetration theory
穿透曲线	貫流曲線	breakthrough curve
穿透容量	貫流容量	breakthrough capacity
穿透时间	貫流時間	breakthrough time
穿透性	穿透性	penetrability
传播	傳播	propagation
传播函数	傳播函數	propagator
传导	傳導	conduction
传导性	傳導性	conductibility
传递	傳送,轉移	transfer
传递成型	轉注成型	transfer molding
传递单元	傳送單元	transfer unit
传递函数	轉換函數	transfer function
传递扩散系数	輸送擴散係數	transport diffusivity
传递系数	傳送係數	transfer coefficient
传递现象	輸送現象	transport phenomenon
传递滞后	傳送滯延	transfer lag
传动滞后	傳送滯延	transmission lag
传感器	感測器,[信號]轉換器	sensor, transducer
传热 j 因子	j_H 因數,熱傳 j 因數	j_H-factor
传热单元高度	熱傳單元高度	height of a [heat] transfer unit
传热单元数	熱傳單元數	number of heat transfer unit
传热分系数	個別熱傳係數	individual heat transfer coefficient

大　陆　名	台　湾　名	英　文　名
传热介质	熱傳介質	heat transfer medium
传热面积	熱傳面積	heat transfer area
传热(=热量传递)		
传热设备	熱傳裝置	heat transfer equipment
传热水力半径	水力半徑	hydraulic radius for heat transfer
传热速率	熱傳速率	heat transfer rate
传热污垢因子	熱傳積垢因數	scaling factor for heat transfer
传热系数	熱傳係數	heat transfer coefficient
传热阻力	熱傳阻力	heat transfer resistance
传输线	傳輸線	transmission line
传送	運送	conveying
传送带	輸送帶	conveyor belt
传送机	運送機	conveyer
传氧速率	氧傳[送]速率	oxygen transfer rate
传氧系数	氧傳[送]係數	oxygen transfer coefficient
传制控制	質傳控制	mass transfer control
传质 j 因子	j_D-因數	j_D-factor
传质操作	質傳操作	mass transfer operation
传质单元高度	傳送單元高度	height of a [mass] transfer unit, HTU
传质单元数	傳送單元數	number of transfer unit, NTU
传质分系数	個別質傳係數	individual mass transfer coefficient
传质区	質傳區	mass transfer zone, MTZ
传质速率	質傳速率	mass transfer rate
传质系数	質傳係數	mass transfer coefficient
传质,质量传递	質傳,質量傳送	mass transfer
传质装置	質傳裝置	mass transfer equipment
传质阻力	質傳阻力	mass transfer resistance
船底漆	船底漆	ship bottom paint
喘振	①衝擊 ②突波	surging
串-并联反应	串並聯反應	series-parallel reactions
串行过程	串聯程序	serial process
串级	串級	cascade
串级补偿	串級補償	cascade compensation
串级反应器	串級反應器	cascade reactor
串级分馏板	串型分餾板	cascade-type plate
串级控制	串級控制	cascade control
串级冷冻	串級冷凍	cascade refrigeration
串级冷却器	串級冷卻器	cascade cooler

大　陆　名	台　湾　名	英　文　名
串级浓缩板	串級濃縮板	cascade concentration plate
串级浓缩器	串級濃縮器	cascade concentrator
串级塔	串級塔	cascade tower
串级系统	串級系統	cascade system
串级循环	串級循環	cascade cycle
串级蒸发器	串級蒸發器	cascade evaporator
串级组合	串級組合	cascade combination
串晶	串晶	shish-kebab
串晶结构	串晶結構	shish-kebab crystalline structure
串联泵组	串聯泵組	pump series
串联补偿	串聯補償	series compensation
串联槽模型	串聯槽模式	tanks-in-series model
串联反应	串聯反應	series reaction
串联反应器	串聯反應器	series reactor
串联流动	串流	series flow
床层塌落技术	床層塌落技術	bed-collapsing technique
床层支架	床支架	bed support
床栅支架	床柵支架	bed grid support
吹管	吹管	blow pipe
吹模	吹模	blow mold
吹扫扩散	掃掠擴散	sweep diffusion
吹扫气体	吹掃氣,排放氣	purge gas
吹塑成型	吹氣成型法	blow molding
吹制玻璃	吹製玻璃	blown glass
吹制油	吹製油,吹煉油	blown oil
垂熔玻璃滤器	燒結玻璃過濾器	sintered glass filter
垂直筛板	垂直篩板	vertical sieve tray
锤[式]磨[碎]机	鎚磨機	hammer mill
锤[式破]碎机	鎚碎機	hammer crusher
纯度	純度	purity
纯氧活性污泥法	純氧活性污泥法	unox process
纯组分	純成分	pure component
醇解	醇解	alcoholysis
醇酶	醇酶,酒精酶	alcoholase
醇酸树脂	醇酸樹脂	alkyd resin
醇质清漆	酒精清漆	spirit varnish
瓷	瓷	porcelain
瓷坩埚	瓷坩堝	porcelain crucible

大　陆　名	台　湾　名	英　文　名
瓷绝缘子	瓷絕緣體	porcelain insulator
瓷漆	瓷漆	enamel paint
瓷漆干燥室	瓷漆乾燥劑	enamel drier
瓷器	瓷器	china, chinaware
瓷土	瓷土	china clay
瓷釉	瓷釉	porcelain glaze
瓷砖	瓷磚	ceramic tile
磁场	磁場	magnetic field
磁带	磁帶	magnetic tape
磁分离	磁力分離	magnetic separation
磁化率	磁化率	magnetic susceptibility
磁化学	磁化學	magnetochemistry
磁黄铁矿	磁黃鐵礦	pyrrhotite
磁搅拌器	磁攪拌器	magnetic stirrer
磁矩	磁矩	magnetic moment
磁力分离器	磁力分離器,磁選機	magnetic separator
磁力过滤	磁力過濾	magnetic filtration
磁力流态化	磁力流體化	magneto fluidization
磁流量计	磁力流量計	magnetic flowmeter
磁流体动力学	磁性流體力學	magnetohydrodynamics
磁盘	磁碟	magnetic disk
磁式数据储存	磁式資料儲存	magnetic data storage
磁铁矿	磁鐵礦	magnetite
磁稳流化床	磁穩定流[體]化床	magnetically stabilized fluidized bed
磁性不纯物	磁性不純物	magnetic impurity
次层	次層	sublayer
次级产品	次要產物	secondary product
次级弛豫	次級鬆弛	secondary relaxation
次级弛豫温度	次級鬆弛溫度	secondary relaxation temperature
次级转变	次級轉變	secondary transition
次氯酸盐漂白	次氯酸鹽漂白	hypochlorite bleaching
次生结构	二級結構	secondary structure
次生细胞壁	次生細胞壁	secondary cell wall
次烟煤	次煙煤	subbituminous coal
次要因次	次要因次	secondary dimension
次优控制	次優控制	suboptimal control
次优控制器	次優控制器	suboptimal controller
次优性	次優性	suboptimality

大　陆　名	台　湾　名	英　文　名
次正应力系数	次正應力係數	secondary normal stress coefficient
刺激物	刺激物	irritant
粗糙标度	粗糙標度	roughness scale
粗糙度	粗糙度	roughness
粗糙因子	粗糙因數	roughness factor
粗滴乳状液	粗滴乳液	macroemulsion
粗级压碎机	粗級壓碎機	coarse crusher
粗颗粒	粗粒子	coarse particle
粗滤	粗濾	straining
粗滤床	粗濾床	roughing bed, roughing filter
粗滤过程	粗濾程序	straining process
粗略分离	粗略分離	sloppy separation
粗筛网	粗篩	coarse screen
粗筛选	初級篩選,初上漿,尺寸初估	preliminary sizing
粗陶器	陶器	earthenware
粗悬浮液	粗懸浮液	coarse suspension
促进化催化剂	促進化觸媒	promoted catalyst
促进剂的早熟性	催速劑早熟性	precocity of accelerator
醋酸菌	醋酸菌	acetic acid bacteria
醋酸嫘萦(＝乙酸人造丝)		
醋酸纤维素(＝乙酸纤维素)		
醋酸酯膜(＝乙酸酯膜)		
催化	催化[作用]	catalysis
催化表面	催化表面	catalytic surface
催化部位	催化部位	catalytic site
催化重整	催化重組,觸媒重組	catalytic reforming
催化重整过程	催化重組法,觸媒重組法	catalytic reforming process
催化促进剂	催化助劑	catalytic promoter
催化动力学	催化動力學	catalytic kinetics
催化反应	催化反應	catalytic reaction
催化反应器	催化反應器	catalytic reactor
催化分解	催化分解	catalytic decomposition
催化过程	催化程序	catalytic process
催化化学吸附	催化化學吸附	catalytic chemisorption

大　陆　名	台　湾　名	英　文　名
催化活性	催化活性	catalytic activity
催化剂	①催化劑　②觸媒	①catalytic agent ②catalyst
催化剂补充剂	觸媒增效劑	catalyst extender
催化剂床	觸媒床	catalyst bed
催化剂活化	觸媒活化	catalyst activation
催化剂活性	觸媒活性	catalyst activity
催化剂几何形状	觸媒[幾何]形狀	catalyst geometry
催化剂接收器	觸媒接受器	catalyst receiver
催化剂结焦	觸媒結焦	catalyst coking
催化剂浸渍	觸媒含浸	catalyst impregnation
催化剂颗粒	觸媒粒	catalyst pellet
催化剂孔隙率	觸媒孔隙率,觸媒孔隙度	catalyst porosity
催化剂粒子	觸媒粒子	catalyst particle
催化剂磨损	觸媒磨耗	attrition of catalyst, catalyst attrition
催化剂黏结剂	觸媒黏結劑	catalyst binder
催化剂配方	觸媒配方	catalyst formulation
催化剂烧尽釜	觸媒燒盡釜	catalyst burn-off still
催化剂失活	觸媒失活	catalyst deactivation
催化剂寿命	觸媒壽命	catalyst life
催化剂衰退	觸媒衰退	catalyst decay
催化剂网	觸媒網	catalyst gauze
催化剂网反应器	觸媒網反應器	catalyst gauze reactor
催化剂稳定剂	觸媒安定劑	catalyst stabilizer
催化剂污染	觸媒積垢	catalyst fouling
催化剂污染指数	觸媒污染指數	catalyst contamination index
催化剂稀释剂	觸媒稀釋劑	catalyst diluent
催化剂细孔结构	觸媒細孔結構	catalyst pore structure
催化剂循环	觸媒循環	catalyst circulation
催化剂抑制剂	觸媒抑制劑	catalyst inhibitor
催化剂有效性	觸媒有效性	catalyst effectiveness
催化剂载体	觸媒載體	catalyst carrier, catalyst support
催化剂再生	觸媒再生,觸媒再活化	catalyst regeneration, catalyst reactivation
催化剂中毒	觸媒中毒	catalyst poisoning
催化金属	催化性金屬	catalytic metal
催化聚合	催化聚合	catalytic polymerization
催化裂化	催化裂解	catalytic cracking
催化裂化过程	催化裂解程序	catalytic cracking process

大　陆　名	台　湾　名	英　文　名
催化面积	催化面積	catalytic area
催化系统	催化系統	catalytic system
催化循环	催化循環	catalytic cycle
催化氧化	催化氧化[反應]	catalytic oxidation
催化载体	催化載體	catalytic carrier
催化专一性	催化特異性	catalytic specificity
催化转化	催化轉化	catalytic conversion
催化转化器	催化轉化器	catalytic converter
催化作用	催化作用	catalytic action
催化作用机理	催化機構	mechanism of catalysis
催泪[毒]气	催淚[毒]氣	tear gas
脆点	脆化點	brittle point
脆性	脆性,易碎性	brittleness, friability
脆性物料	脆性材料	brittle materials
淬火	淬火,驟冷	quenching
淬火水	淬火水,驟冷水	quenching water
淬火油	淬火油,驟冷油	quenching oil
萃取	萃取	extraction
萃取超滤	萃取性超過濾	extractive ultrafiltration
萃取过程	萃取程序	extraction process
萃取结晶	萃取結晶	extractive crystallization
萃取结晶器	萃取結晶器	extractive crystallizer
萃取率	萃取速率	extraction rate
萃取器	萃取器	extractor
萃取器组	萃取器組	extraction battery
萃取塔	萃取塔	extraction tower, extraction column
萃取物	萃取物	extract
萃取因子	萃取因子	extraction factor
萃取蒸馏	萃取蒸餾	extractive distillation
萃余物	萃餘物	raffinate
存储单元	儲存單元	memory cell
存储器	儲存	storage
存量	存量	inventory
错流	[交]錯流	crosscurrent, crossflow
错流操作	[交]錯流操作	crosscurrent operation, crossflow operation
错流萃取	[交]錯流萃取	crosscurrent extraction
错流吸附器	[交]錯流吸附器	crossflow adsorber

D

大 陆 名	台 湾 名	英 文 名
搭接接头	搭接接頭	lap joint
打浆机	打漿機	hollander, pulp beater
打泡	攪打	whipping
打印墨	印墨	stamp pad ink
大豆蛋白纤维	大豆蛋白纖維	soybean fiber
大豆蛋白质	大豆蛋白[質]	soybean protein
大豆油	大豆油	soybean oil
大分子	巨分子,大分子	macromolecule
大孔	大孔	macropore
大气	大氣	atmosphere
大气扩散	大氣擴散	atmospheric diffusion
大气冷凝器	氣壓冷凝器	barometric condenser
大气排放管	氣壓排洩管	barometric discharge pipe
大气腿	氣壓眞空柱,氣壓管	barometric leg, barometric pipe
大气压	大氣壓(壓力單位),大氣壓力,常壓	atmosphere,barometric pressure,atmospheric pressure
大气质量	空氣品質	air quality
大气质量指数	空氣品質指數	air quality index
大小比值	大小比值,量比值	magnitude ratio
大小标度	大小量度	magnitude scaling
大小头阻力	漸縮管阻力	reducer resistance
大中取大判据	大中取大準則	maximax criterion
大宗化学品	大宗化學品	commodity chemical
大宗聚合物	大宗聚合物	commodity polymer
代谢	代謝[作用]	metabolism
代谢控制	代謝控制	metabolic control
代谢速率	代謝速率	metabolic rate
代谢物	代謝物	metabolite
代用天然气	合成天然氣	substitute natural gas
带出	帶出	carryover
带静电[作用]	靜電起電	static electrification
带宽	帶寬	bandwidth
带式干燥器	帶式乾燥機	belt dryer

大 陆 名	台 湾 名	英 文 名
带式过滤机	帶[式過]濾機	belt filter
带状矩阵	帶式矩陣	banded matrix
袋滤器	袋[狀]濾器	bag filter
袋式成型	袋式成型	bag molding
袋式过滤分离器	袋式過濾分離器	bag filter separator
袋式集尘器	袋式集塵器	bag type dust collector
袋式研磨机	袋式研磨機	pocket grinder
单部位机理	單部位機構	single-site mechanism
单程操作	單程操作	once-through operation
单程过程	單程程序	once-through process
单程加热器	單程加熱器	single-pass heater
单程收率	單程產率	yield per path
单程转化率	單程轉化率	single pass conversion
单冲程预塑机	單衝程預塑機	single-stroke preforming press
单纯浸渍法	直浸漬過程	straight dipping process
单纯形法	單體[演算]法	simplex method
单纯形算法	單體演算法	simplex algorithm
单道分析仪	單道分析儀	single channel analyzer
单底物反应	單受質反應	single substrate reaction
单点测定法	單點測定[法]	single-point determination
单动泵	單動泵	single acting pump
单反应	單一反應	single reaction
单分布组分	單分佈成分	single distributed component
单分隔点	單分隔點	single-split point
单分散	單分散	monodisperse
单分散性	單分散性	monodispersity
单分子反应	單分子反應	monomolecular reaction, unimolecular reaction
单缸往复泵	單缸往復泵	simplex reciprocating pump
单股聚合物	單股聚合物	single-strand polymer
单辊压碎机	單輥壓碎機	single roll crusher
单环路控制	單環路控制	single-loop control
单级操作	單階操作	single stage operation
单级萃取	單階萃取	single stage extraction
单级导向阀	單階導引閥	single stage pilot valve
单级过程	單階程序	single-stage process
单级离心泵	單級離心泵	single stage centrifugal pump
单级离心机	單階離心機	single-stage centrifuge

大　陆　名	台　湾　名	英　文　名
单级蒸馏	單階蒸餾	single stage distillation
单极点	單極點	simple pole
单晶	單晶	single crystal
单晶纤维	單晶纖維	single-crystal fiber
单克隆抗体	單源抗體,單株抗體	monoclonal antibody
单利	單利	simple interest
单零点	單零點	simple zero
单螺杆混合机	單螺桿混合器	single-screw mixer
单螺杆挤出机	單螺桿擠出機	single-screw extruder
单宁(=鞣质)		
单宁酸(=鞣酸)		
单偶氮染料	單偶氮染料	monoazo dye
单腔模具	單穴模具	single cavity mold
单桥聚合物	單橋高分子	single-bridged polymer
单球菌	單球菌屬	monococcus
单色发射能力	單色發射能力	monochromatic emissive power
单纱	單紗	single yarn
单纱强力试验机	單紗強度試驗機	single thread tester
单输入单输出系统	單輸入單輸出系統	single-input-single-output system
单丝	單絲,絲狀纖維,單絲纖維	filament, monofilament
单丝加捻	單撚	single twist
单糖	單醣	monosaccharide
单体	單體	monomer
单位成本	單位成本	unit cost
单位成本估计	單位成本估計	unit cost estimation
单位回馈	單位回饋	unit feedback
单细胞蛋白	單細胞蛋白	single cell protein
单细胞生长	單細胞生長	unicellular growth
单向阀(=止逆阀)		
单向反应	單向反應	unidirectional reaction
单向狭缝	單向狹縫	unilateral slit
单相反应器	單相反應器	single-phase reactor
单相系统	單相系	single-phase system
单效蒸发器	單效蒸發器	single effect evaporator
单斜钠钙石	單斜鈉鈣石	gaylussite
单型腔模	單穴模	single impression mold
单一稳态	單一穩態	unique steady state

大　陆　名	台　湾　名	英　文　名
单一液体近似法	單一液體近似法	single liquid approximation
单元操作	單元操作	unit operation
单元过程	單元程序	unit process
单元计算	單元計算	unit computation
单元折旧率	單元折舊率	single unit depreciation
单轴[系]汽轮机	串接雙軸汽輪機	tandem compound turbine
单轴应力	單軸應力	uniaxial stress
单组分	單成分	single component
单组分系统	單成分系	one-component system
胆固醇	膽固醇,膽甾醇	cholesterol
胆甾状态	膽甾態,膽固醇態	cholesteric state
蛋白黑素	梅納汀	melanoidin
蛋白酶	蛋白酶	protease，proteinase
蛋白酶解	蛋白質水解	proteolysis
蛋白[水解]酶	蛋白[水解]酶	proteolytic enzyme
蛋白质	蛋白質	protein
蛋白质分级	蛋白質份化	protein fractionation
蛋白质工程	蛋白質工程	protein engineering
氮肥	氮肥	nitrogenous fertilizer
氮化硅	氮化矽	silicon nitride
氮芥子气	氮芥子氣	nitrogen mustard gas
氮气吹扫	氮氣沖洗	nitrogen purge
氮循环	氮循環	nitrogen cycle
氮氧需要量	氮氧需要量	nitrogen oxygen demand
当量	當量	equivalent
当量长度	等效長度	equivalent length
当量电导	當量電導度	equivalent conductance
当量粒径	等效粒徑	equivalent particle diameter
当量浓度	當量濃度	normal concentration
当量溶液	當量溶液	normal solution
当量收缩管长	縮管相當管長	contraction equivalent pipe length
当量直径	等效直徑	equivalent diameter
挡板	擋板,壓片	flapper,baffle,sheeting
挡板痕	擋板痕	baffle mark
挡板混合器	擋板式混合器	baffle mixer
挡板混合塔	擋板混合塔	baffle plate column mixer
挡板喷嘴机理	擋板噴嘴機構	flapper-nozzle mechanism
挡板喷嘴系统	擋板噴嘴系統	baffle nozzle system

大　陆　名	台　湾　名	英　文　名
挡板室	擋板室	baffle chamber
挡板塔	擋板塔	baffle plate tower, baffle column
挡板塔混合器	擋板塔混合器	baffle column mixer
挡板塔盘	擋板盤	baffle pan
刀式去皮机	刀式去皮機	knife barker
导出单位	導出單位	derived unit
导电高分子	導電聚合物	electroconductive polymer
导电计	電導計	electrolytic conductance meter
导电聚合物	導電聚合物	conductive polymer
导管	導管	duct
导流筒	導流管	draft tube
导流筒挡板	導流管擋板	draft tube baffle
导流筒挡板结晶器	導流筒擋板結晶器	draft-tube-baffled crystallizer
导流筒混合器	導流管混合器	draft tube mixer
导热系数	熱傳導係數	thermal conductivity coefficient
导数	導數	derivative
导体催化剂	導體觸媒	conductor catalyst
导线	導線,引線	lead wire
导线管	導管	conduit
导线误差	引線誤差	lead wire error
导向阀	導引閥	pilot valve
倒风窑	倒焰窯	downdraft kiln
捣矿杆组	搗礦杆組	stamp battery
捣磨机	搗碎機	stamp mill
到达角	到達角	angle of arrival
倒数标绘	倒數圖	reciprocal plot
倒数法则	倒數法則	reciprocal rule
倒数关系	倒數關係	reciprocal relation
灯黑	燈黑,燈煙	lamp black
灯笼填函盖	燈龍填函蓋	lantern gland
等边三角图	等邊三角圖	equilateral triangular diagram
等沉降粒子	等沈降粒子	equal-settling particle
等当点	當量點	equivalent point
等电沉淀	等電沈澱	isoelectric precipitation
等电点	等電點	isoelectric point
等电点分离	等電分離	isoelectric separation
等电聚焦	等電聚焦	isoelectric focusing
等动力学点	等動力點	isokinetic point

大　陆　名	台　湾　名	英　文　名
等动力学温度	等動力溫度	isokinetic temperature
等高线图	等高線圖	contour plot
等规聚丙烯	同排聚丙烯	isotactic polypropylene
等规聚合物(＝全同立构聚合物)		
等焓过程	等焓過程,等焓程序	isenthalpic process
等焓膨胀	等焓膨脹	isenthalpic expansion
等焓压缩	等焓壓縮	isenthalpic compression
等离点	等離子點,等電點	isoionic point
等离子体	電漿	plasma
等离子[体]刻蚀	電漿蝕刻[法]	plasma etching
等离子[体]增强刻蚀	電漿加強蝕刻[法]	plasma-enhanced etching
等[理论]板高度,理论板当量高度	理論板相當高度	height equivalent of a theoretical plate
等摩尔逆向扩散	等莫耳逆向擴散	equimolar counter diffusion
等浓度曲线	等濃度曲線	isoconcentration curve
等容过程	等容過程,等容程序	isochoric process, isometric process
等容热容	定容熱容	heat capacity at constant volume
等容线	等容線,等量線	isometrics, isochore
等熵过程	等熵過程,等熵程序	isentropic process
等熵流	等熵流動	isentropic flow
等熵膨胀	等熵膨脹	isentropic expansion
等熵压缩	等熵壓縮	isentropic compression
等渗溶液	等滲壓溶液	isotonic solution
等渗压浓度	等滲壓濃度	isotonic concentration
等式约束	等式約束	equality constraint
等势线	等勢線,等位線	equipotential line
等速电泳	等速電泳	isotachophoresis
等温操作	恆溫操作	isothermal operation
等温多重性	等溫多重性	isothermal multiplicity
等温反应	恆溫反應	isothermal reaction
等温反应器	恆溫反應器	isothermal reactor
等温过程	等溫過程,等溫程序	isothermal process
等温环境	等溫環境	isothermal environment
等温扩散	等溫擴散	isothermal diffusion
等温流动	等溫流動	isothermal flow
等温膨胀	等溫膨脹	isothermal expansion
等温热重量分析法	等溫熱重量法	isothermal thermogravimetry

大　陆　名	台　湾　名	英　文　名
等温温度相当值	相當溫度	equivalent isothermal temperature
等温吸附	等溫吸附	isothermal adsorption
等温线	等溫線	isotherm, isothermal line
等温线迁移	等溫線遷移	isotherm migration
等温压缩	等溫壓縮	isothermal compression
等温压缩性	等溫壓縮性	isothermal compressibility
等温有效因子	等溫有效度因數	isothermal effectiveness
等温蒸馏	等溫蒸餾	isothermal distillation
等相角曲线	等相角曲線	constant phase angle curve
等相位轨迹	等相位軌跡	locus of constant phase
等效	等效	equivalent
等效电导	當量電導度	equivalent conductance
等效径	等效直徑	equivalent diameter
等效自由沉降直径	等效自由沈降直徑	equivalent free-falling diameter
等斜线	等斜率線	isocline
等压法	等壓法	isopiestic method
等压过程	等壓過程, 等壓程序	isobaric process
等压过热	等壓過熱	isobaric superheating
等压热容	等壓熱容[量]	heat capacity at constant pressure
等压线	等壓線	isobar
等值管长度	等效管長	equivalent pipe length
低聚反应	低聚合[反應]	oligomerization
低聚物	寡聚物	oligomer
低硫天然气	脫臭天然氣	sweet gas
低密度聚乙烯	低密度聚乙稀	low density polyethylene
低热值煤气	低熱值燃氣	low-BTU gas
低渗溶液	低滲壓溶液	hypotonic solution
低通滤波器	低通濾波器	low pass filter
低温工业	低溫工業	cryogenic industry
[低温]焊料	軟焊料, 焊錫	solder
低温碳化	低溫碳化[作用]	low temperature carbonization
低选择器开关	低選擇器開關	low selector switch
低于额定值	低於额定值, 不足量	undershoot
滴	滴	drop
滴滴涕	DDT 殺蟲劑	dichloro-diphenyl-trichloroethane, DDT
滴定	滴定[法]	titration
滴定[分析]法	滴定[分析]法	titrimetric method
滴定计	滴定計	titrimeter

大　陆　名	台　湾　名	英　文　名
滴干法	滴乾	drip dry
滴流	滴流	trickling
滴流床	滴流床	trickle bed
滴流床反应器	滴流床反應器	trickle-bed reactor
滴滤	滴濾	trickling filtration
滴滤池	滴濾池	trickling filter
滴相	滴相	drop phase
滴注器	滴口	drip
滴状冷凝	滴式冷凝	dropwise condensation
底部产物	底部產物	bottom product
底部卸料	底部卸料	bottom discharge
底部液位控制	塔底液位控制	bottom-level control
底部蒸汽	底部蒸汽	bottom steam
底阀	底部閥,底閥,腳閥	bottom valve, foot valve
底流	底流	underflow
底流轨迹	底流軌跡	underflow locus
底流速度	底流速度	underflow velocity
底流速率	底流速率	underflow rate
底漆	①底漆 ②雷管 ③起動 注給器	primer, undercoat
底物	受質,基質	substrate
底物抑制	受質抑制[作用]	substrate inhibition
底相,下相	底相	bottom phase
底座	底座	pedestal
地蜡	地蠟	earth wax
地下排水系统	排水系統	underdrainage system
地下水	地下水	groundwater
地下水处理	地下水處理	groundwater treatment
地下水污染	地下水污染	groundwater pollution
递归	遞回	recursion
递归函数	遞回函數	recursive function
递减均衡法	遞減均衡法	declining balance method
递阶控制	階層控制	hierarchical control
递推公式	遞回公式	recursion formula
递推关系式	遞回關係[式]	recursive relation
第二位力系数	位力係數	second virial coefficient
缔合溶液模型	締合溶液模型	associated solution model
蒂勒模数	蒂勒模數	Thiele modulus

大　陆　名	台　湾　名	英　文　名
典范矩阵	正則矩陣	canonical matrix
点焊	點銲	spot welding
点焊机	點焊機	spot welder
点火	點火	ignition
点火时间	點火時間	ignition time
点胶合	點膠合	spot gluing
点蚀	麻點腐蝕	pitting corrosion
点效率	點效率	point efficiency
点选择性	點選擇性	point selectivity
点源	點源	point source
点阵	晶格	lattice
碘值	碘值	iodine value
电槽室	電槽室	cell house
电场	電場	electric field
电沉积	電沈積[法]	electrodeposition
电沉积物	電沈積物	electrodeposit
电除尘器(=静电沉降 　器)		
电瓷	絕緣瓷	electric porcelain
电磁磁选机	電磁分離器	electromagnetic separator
电磁阀	電磁閥	solenoid valve
电磁分离[法]	電磁分離	electromagnetic separation
电磁辐射	電磁輻射	electromagnetic radiation
电磁流量计	電磁流量計	electromagnetic flowmeter
电磁谱	電磁[波]譜	electromagnetic spectrum
电导	電導	conductance
电导池	電導[測定]槽	conductivity cell
电导滴定	電導滴定[法]	conductometric titration
电导计	電導計,熱導計	conductometer
电导率	電導係數	electric conductivity
电导系数	電導係數	electric conductivity coefficient
电动传送	電[力]傳送	electric transmission
电动传送器	電[力]傳送器	electric transmitter
电动机	電動機	motor
电动机常数	馬達常數	motor constant
电动机驱动器	馬達致動器	motor actuator
电动控制	電[動]控制	electric control
电动-气动致动器	電動氣動[式]致動器	electropneumatic actuator

大　陆　名	台　湾　名	英　文　名
电动势	電動勢	electromotive force，EMF
电动天平	電動天平	electrobalance
电动现象	電動力現象	electrokinetic phenomenon
电动信号	電[動]信號	electric signal
电动液压引动器	電動液壓致動器	electrohydraulic actuator
电动致动器	電致動器	electric actuator
电镀	電鍍	electroplating
电镀工程	電鍍工程	electroplating engineering
电镀工业	電鍍工業	plating industry
电分散	電分散	electrodispersion
电浮选法	電浮選[法]	electroflotation
电弧炉	電弧爐	arc furnace
电化当量	電化當量	electrochemical equivalent
电化学	電化學	electrochemistry
电化学反应	電化[學]反應	electrochemical reaction
电化学反应工程	電化[學]反應工程	electrochemical reaction engineering
电化学反应器	電化學反應器	electrochemical reactor
电化学工业	電化[學]工業	electrochemical industry
电化学势	電化[學]電勢,電化[學]電位	electrochemical potential
电极	電極	electrode
pH 电极	pH 電極	pH electrode
电极板除尘器	電板集塵器,電板沈積器	electric plate precipitator
电解	電解	electrolysis
电解槽	電解槽	electrolysis bath，electrolyzer
电解池	電解槽	electrolytic cell
电解分离	電解分離	electrolytic separation
电解分析	電[解]分析	electroanalysis
电解氟化	電氟化[反應]	electrofluorination
电解工业	電解工業	electrolytic industry
电解过程	電解程序	electrolytic process
电解加氢二聚合	電解加氫二聚合[反應]	electrohydrodimerization
电解氧化	電解氧化	electrolytic oxidation
电解质	電解質	electrolyte
电控制器	電控制器	electric controller
电类比	電類比	electric analogy

大 陆 名	台 湾 名	英 文 名
电离	游離	ionization
电离常数	游離常數	ionization constant
电离电位	游離電位	ionization potential
电离度	游離度	degree of ionization
电离能	游離能	ionization energy
电流滴定[法]	電流滴定	amperometric titration
电流作用	迦凡尼效應	galvanic action
电氯化反应	電氯化[反應]	electrochlorination
电毛细管现象	電毛細現象	electrocapillarity
电能	電能	electric energy
电黏效应	電黏性效應	electroviscous effect
电偶腐蚀	電流腐蝕	galvanic corrosion
电气管道除尘器	電管集塵器,電管沈積器	electric pipe precipitator
电气设施	電力裝置	electric installation
电迁徙	電遷移[法]	electromigration
电桥	電橋	bridge
电亲和性	電親和力	electroaffinity
电清洗	電解清洗	electrocleaning
电热反应器	電[阻抗]熱反應器	electric-impedance heated reactor
电热过程	電熱法	electrothermal process
电容	電容	capacitance
电容脉动式转速计	電容器脈衝[式]轉速計	capacitor impulse type tachometer
电容器	電容器	capacitor, condenser
电容器纸	電容器紙	condenser paper
电容湿度计	電容[式]濕度計	capacitance moisture meter
电容探针	電容探針	capacitance probe
电容压力转换器	電容[式]壓力轉換器	capacitance pressure transducer
电容液面计	電容[式]液面計	capacitance liquid level gage
电容元件	電容元件	capacitance element
电融合	電融合	electrofusion
电色谱	電層析法	electrochromatography
电渗	電滲[透]	electrosmosis
电渗析	電透析[作用]	electrodialysis
电渗析过程	電透析程序	electrodialysis process
电渗现象	電滲透	electro-osmosis
电石(=碳化钙)		

大　陆　名	台　湾　名	英　文　名
电势	電位	potential
电势梯度	[電]位梯度	potential gradient
电枢	電樞	armature
电涂	電塗裝	electrocoating
电涂层	電鑄	electrocasting
电涂装	電鑄	electrocasting
电位滴定[法]	電位滴定[法]	potentiometric titration
电位法	電位法	potentiometric method
电位分析[法]	電位分析	potentiometric analysis
[电]位降	位降	potential drop
电位器	電位計	potentiometer
电稳分散作用	電穩分散[作用],電穩 　　分散液	electrocratic dispersion
电压	電壓	voltage
电冶金	電冶金學	electrometallurgy
电泳	電泳	electrophoresis
电致变色显示	電致變色顯示器	electrochromic display
电致发光	電致發光	electroluminescence
电铸	電鑄	electroforming
电铸莫来石	電鑄富鋁紅柱石	electrocast mullite
电子材料	電子材料	electronic materials
电子放大器	電子放大器	electronic amplifier
电子计算机	電子計算機	electronic computer
电子控制	電子控制	electronic control
电子流量计	電子流量計	electric type flowmeter, electronic flowme- 　　ter
电子模拟计算机	類比電子計算機	electronic analog computer
电子模拟器	電子模擬器	electronic simulator
电子偶产生	成對生成	pair creation
电子偶生成	成對產生	pair production
电子配分函数	電子分配函數	electronic partition function
电子信号	電子訊號	electronic signal
电阻	電阻	resistance
电阻高温计	電阻高溫計	electric resistance pyrometer
电阻炉	電阻爐	resistance furnace
电阻温度计	電阻溫度計	resistance thermometer
垫料	襯墊料	padding
垫片材料	墊片材料	gasket materials

大　陆　名	台　湾　名	英　文　名
垫圈沟	墊圈溝	gasket groove
淀粉	澱粉	starch
淀粉碘化物试验	澱粉碘試驗	starch iodide test
淀粉碘化物试纸	澱粉碘試紙	starch iodide paper
淀粉碘试纸	澱粉碘試紙	starch iodine paper
淀粉基聚合物	澱粉基聚合物	starch based polymer
淀粉酶	澱粉酶	amylase
淀粉酶制剂	[澱粉]糖化酶	diastase
淀粉黏合剂	澱粉接著劑	starch adhesive
淀粉试纸	澱粉試紙	starch paper
淀粉水解	澱粉水解	amylolysis
靛白	靛白	indigo white
靛蓝	靛藍	indigo blue
靛染	靛染	indigo dyeing
靛染缸	靛染缸	indigo dye vat
靛洋红	食用藍色二號,靛胭脂	indigo carmine
靛棕	靛棕	indigo brown
吊篮式反应器	籃式反應器	basket reactor
迭代法	疊代法	iterative method
迭式反应器	疊式反應器	stack-up reactor
叠合重整	疊合重組[作用]	polyforming
叠合汽油	疊合汽油	polymer gasoline
叠加	疊加,疊合	superposition
叠加原理	疊加原理	superposition principle
蝶[形]阀	蝶型閥	butterfly valve
丁苯无规共聚物	苯乙烯丁二烯隨機共聚物	styrene butadiene random copolymer
丁基纤维素	丁基纖維素	butyl cellulose
丁基橡胶	丁基橡膠	butyl rubber
丁腈橡胶	腈橡膠	nitrile rubber
丁钠橡胶	丁二烯鈉橡膠	sodium-butadiene rubber
顶板	頂板	top plate
顶冷凝器	頂冷凝器	head condenser
顶相	頂[層]相	top phase
定比定律	定比定律	law of definite proportions
定长玻璃短纤维	玻璃短纖維	staple glass fiber
定常态(=稳态)		
定常运动	穩定運動	steady motion

大 陆 名	台 湾 名	英 文 名
定氮仪	氮量計	azotometer
定理	定理	theorem
π 定理	π 定理,pi 定理	π-theorem, pi-theorem
定量分析	定量分析	quantitative analysis
定律	定律	law
定时开关	定時開關	time switch
定态	定態	stationary state
定位槽	定高差槽	constant head tank
定香剂	香料固定劑	perfume fixing agent, perfume fixative
定向	定向	orientation
定向分布	位向分佈	orientation distribution
定向平面	定向平面	orientation flat
定向吸附	位向吸附	oriented adsorption
定性分析	定性分析	qualitative analysis
定域粒子系集	定域粒子系集	assembly of localize particles
定子	定子	stator
锭剂	錠劑	pastille
锭子油	錠子油	spindle oil
冬化	冬化,低溫脫脂	winterizing
动电电位	動電勢,動電位	electrokinetic potential
动电势序	電動勢序	electromotive force series
动电学	電動力學	electrokinetics
动力发动机	發電機	power generator
动力分析	動力分析	kinetic analysis
动力模式	動力模式	kinetic model
动力黏度	[動力]黏度	dynamic viscosity
动力性质	動力性質	kinetic property
动力学	動力學	dynamics, kinetics
动力学参数	動力參數	kinetic parameter
动力学方程	動態方程式	dynamic equation
动力学控制	動力學控制	kinetic control
动力学理论	動力論	kinetic theory
动力学式	動力學式	kinetic expression
动力学相似性	動態相似性	dynamic similarity
动力装置	動力廠	power plant
动量	動量	momentum
动量传递	動量傳送	momentum transfer
动量传递系数	動量傳送係數	momentum transfer coefficient

大　陆　名	台　湾　名	英　文　名
动量方程	動量方程式	momentum equation
动量分离器	動量分離器	momentum separator
动量矩守恒原理	角動量守恆原理	principle of conservation of moment of momentum
动量扩散系数	動量擴散係數	momentum diffusivity
动量平衡	動量平衡	momentum balance
动量守恒	動量守恆	momentum conservation
动量守恒原理	動量守恆原理	principle of conservation of momentum
动量输送	動量輸送	momentum transport
动量通量	動量通量	momentum flux
动量原理	動量原理	momentum principle
动能	動能	kinetic energy
动水压力	流體壓力	hydrodynamic pressure
动态补偿	動態補償	dynamic compensation
动态反射光谱法	動態反射光譜法,動態反射光譜學	dynamic reflectance spectroscopy
动态分析	動態分析	dynamic analysis
动态刚性	動態剛性	dynamic rigidity
动态规划	動態規劃	dynamic programming
动态过程	動態程序	dynamic process
动态模量	動態模數	dynamic modulus
动态模拟	動態模擬	dynamic simulation
动态模型	動態模型	dynamic model
动态平衡	動態平衡	dynamic equilibrium
动态去偶器	動態解偶器	dynamic decoupler
动态热机械法	動態熱機械法	dynamic thermal mechanical method
动态热重量分析法	動態熱重量法	dynamic thermogravimetry
动态试验	動態試驗	dynamic test
动态误差	動態誤差	dynamic error
动态误差系数	動態誤差係數	dynamic error coefficient
动态系统	動態系統	dynamic system
动态响应	動態應答	dynamic response
动态行为	動態行為	dynamic behavior
动态元素	動態元素	dynamic element
动态滞后	動態滯延	dynamic lag
动态最优化	動態最適化	dynamic optimization
动物胶	動物膠	animal glue
动物厩肥	動物廢肥	animal manure

大　陆　名	台　湾　名	英　文　名
动物皮	動物皮	animal hide
动物油	動物油	animal oil
动压[力]	動壓[力]	dynamic pressure
动压头	動高差	kinetic head
冻凝点	凍凝點	congealing point, congelation point
冻凝温度	凍凝溫度	congealing temperature
冻凝作用	凍凝[作用]	congelation
胨	腖	peptone
胨化	腖化[作用]	peptonization
斗式提升机	斗式升降機	bucket elevator
斗式运输机	斗式運送機	bucket conveyor
毒气	毒氣	toxic gas
毒素	毒素	toxin
毒物	毒物,毒劑	poison
毒性	毒性	toxicity
毒性化学品	毒性化學品	toxic chemical
毒烟	毒煙	toxic smoke
独居石	獨居石	monazite
独立反应	獨立反應	independent reaction
堵塞	阻塞	plugging
镀铂碳电极	鍍鉑碳[電]極	platinized carbon electrode
镀铬	鍍鉻	chrome plating
镀镍	鍍鎳	nickel plating
镀锡	鍍錫	tinning, tin plating
镀锡板	馬口鐵	tin plate
镀锌	鍍鋅	galvanization, zinc plating
镀锌保护	鍍鋅保護	galvanic protection
端点效应	端[點]效應	end effect
短杆菌酪肽	短桿菌酪肽	tyrocidine
短杆菌素	短桿菌素	tyrothricin
短杆菌肽	短桿菌素	gramicidin
短链支化	短鏈支化	short-chain branching
短期熟成	短期熟成	short ripening
短纤纱	機紡紗,短纖維加撚紗	spun yarn
短纤维	短纖維	short fiber
短纤维复合材料	短纖維複合材料	short fiber composite
短纤维增强	短纖強化	short fiber reinforcement
短纤维增强塑料	短纖強化塑膠	short fiber reinforced plastic

大　陆　名	台　湾　名	英　文　名
短效加速剂	短效催速劑	fugitive accelerator
短支链	短支鏈	short-chain branch
断键	斷裂	scission
断裂表面能	斷裂表面能	surface energy of fracture
断裂力学	破壞力學	fracture mechanics
断裂强度	破裂強度	fracture strength
断裂统计学	斷裂統計學	statistics of rupture
断裂应力	斷裂應力,破裂應力	rupture stress, fracture stress
煅富铁黄土	煅富鐵黃土	burnt sienna
煅烧	煅燒	calcination
煅烧焦	煅燒焦	calcined coke
煅烧炉	煅燒爐	calcinator, calciner
煅烧黏土	煅燒黏土	calcined clay
煅烧区	煅燒區	calcining zone
煅石膏	煅石膏	calcined plaster
煅赭石	煅赭石	burnt ocher
堆垛	堆垛	stacking
堆垛效应	堆垛效應	stacking effect
堆肥	堆肥	compost
堆[积]浸[取]	堆積浸取	heap and dump leaching
堆密度	[總]體密度	bulk density
堆芯活性区	[核]反應器爐心	reactor core
对苯乙烯磺酸	對苯乙烯磺酸	p-styrene sulfonic acid
对比饱和蒸汽压	對比飽和蒸汽壓	reduced saturated vapor pressure
对比变量	約化變數	reduced variable
对比分子量	約化分子量	reduced molecular weight
对比密度	對比密度	reduced density
对比体积	對比體積	reduced volume
对比温度	對比溫度	reduced temperature
对比性质	約化性質	reduced property
对比压力	對比壓力	reduced pressure
对比状态方程	約化狀態方程式	reduced equation of state
对比状态原理	對應狀態原理	principle of corresponding state
对比浊度	對比濁度	reduced turbidity
对比坐标	對比坐標	reduced coordinates
对称膜	對稱透膜	symmetric membrane
对称性	對稱	symmetry
对称中心	對稱中心	symmetry center

大　陆　名	台　湾　名	英　文　名
对称轴	對稱軸	axis of symmetry
对分搜索	二分法搜尋	dichotomous search
对接[头]	對接[頭]	butt joint
对流	對流	convection
对流传热	對流熱傳	convection heat transfer, convective heat transfer
对流传热系数	熱對流係數	convective heat transfer coefficient
对流传质	對流質傳	convection mass transfer
对流导数	對流導數	convective derivative
对流段	對流段	convection section
对流加速度	對流加速	convective acceleration
对流通量	對流通量	convective flux
对流坐标	對流坐標	convected coordinates
对硫磷	巴拉松	parathion
对偶变量	變數配對	pairing variables
对偶[性]	對偶[性],二元	duality
对数平均	對數平均[數]	logarithmic mean
对数平均半径	對數平均半徑	logarithmic mean radius
对数平均面积	對數平均面積	logarithmic mean area
对数平均摩尔分率	對數平均莫耳分率	logarithmic mean mole fraction
对数平均温差	對數平均溫差	logarithmic mean temperature difference
对数生长期	對數[生長]期	logarithmic phase
对数图	對數圖	logarithmic plot
对数正态分布函数	對數常態分佈函數	log normal distribution function
对应态	對應[狀]態	corresponding state
对应态律	對應狀態定律	law of corresponding state
对峙反应	逆向反應	opposing reaction
钝化	鈍化	passivation
钝化铁	鈍化鐵	passive iron
钝态	鈍態	passive state
多变比热	多變比熱	polytropic specific heat
多变常数	多變常數	polytropic constant
多变过程,多方过程	多變程序	polytropic process
多变量控制系统	多變數控制系統	multivariable control system
多变量系统	多變數系統	multivariable system
多变量最优化	多變數最適化	multivariable optimization
多变膨胀	多變膨脹	polytropic expansion
多变性	多變性	polytropism

大　陆　名	台　湾　名	英　文　名
多变指数	多變指數	polytropic index
多层次优化法	多層次優化法	multilevel method of optimization
多层吸附理论	多層吸附理論	multilayer adsorption theory
多程换热器	多程熱交換器	multipass exchanger
多程加热器	多程加熱器	multipass heater
多程冷凝器	多程冷凝器	multipass condenser
多重产物塔	多重產物塔	multiple product column
多重反应	多重反應	multiple reactions
多重反应器系统	多反應器系統	multiple reactor system
多重进料流	多重進料流	multiple feed stream
多重输出控制	多重輸出控制	multiple output control
多重输出系统	多[重]輸出系統	multiple-output system
多重输入系统	多[重]輸入系統	multiple-input system
多重态	多重態,多重[譜]線	multiplet
多重稳态	多重穩態	multiple steady state, multiple stability
多重稳态性	多重穩態	steady-state multiplicity
多重再沸器	多重再沸器	multiple reboiler
多床反应器	多床反應器	multibed reactor, multiple bed reactor
多次萃取	多階萃取	multiple extraction
多道分析仪	多通道分析儀	multichannel analyzer
多点记录器	多點記錄器	multipoint recorder
多段过程	多階程序	multistage process
多段体系	多階系統	multistage system
多方过程(=多变过程)		
多分布组分	多分佈成分	multiple distributed component
多分隔点	多分隔點	multiple split point
多功能催化剂	多功能觸媒	multifunctional catalyst
多功能酶	多功能酶	multifunctional enzyme
多管反应器	多管反應器	multitube reactor
多管系统	多管系統	multiple pipe system
多管旋风分离器	多[管]旋風分離器	multicyclone
多辊压延机	多輥壓延機	supercalender
多回路控制系统	多環路控制系統	multiloop control system
多回路系统	多環路系統	multiloop system, multiple loop system
多级扩散	多階擴散	multistage diffusion
多级离心泵	多階離心泵	multistage centrifugal pump
多级离心机	多階離心機	multiple stage centrifuge
多级流化床	多階流[體]化床	multistage fluidized bed

大　陆　名	台　湾　名	英　文　名
多级模型	分階模式	stagewise model
多级喷射泵	多階噴射器	multistage ejector
多级喷射器	多階射出器	stage ejector
多级批式萃取	多階批式萃取	multiple batch extraction
多级压缩	多階壓縮	multiple stage compression, multistage compression
多级压缩机	多階壓縮機	multiple stage compressor, multistage compressor
多级研磨	多階研磨	multistage grinding
多级蒸馏	多階蒸餾	multistage distillation
多降液管塔板	多降液管塔板	multidowncomer tray, MD tray
多金属催化剂	多金屬觸媒	polymetallic catalyst
多进出口系统	多進出口系統	multiport system
多孔板	多孔板	perforated plate
多孔板筛	多孔[板]篩	perforated screen
多孔壁管	多孔壁管	porous-wall tube
多孔材料	多孔[性]材料	porous materials
多孔催化剂	多孔觸媒	porous catalyst
多孔固体	多孔固體	porous solid
多孔管	多孔管	perforated pipe
多孔假底	多孔假底	perforated false bottom
多孔介质	多孔介質	porous medium
多孔粒	多孔粒	porous pellet
多孔粒脉冲反应器	多孔粒脈衝反應器	porous pellet pulse reactor
多孔膜	多孔膜	porous membrane
多孔塑料	海綿塑膠	sponge plastic
多孔塔板	穿孔塔板	perforated tray
多孔填料	多孔填料	porous packing
多孔砖	多孔磚	perforated brick
多流体理论	多流體理論	poly-fluid theory
多硫聚合物	多硫聚合物	polysulfide polymer
多目标规划	多目標規劃	multi-objective programming
多黏菌素	多黏菌素	polymyxin
多嵌段共聚物	鏈段共聚物	segmental copolymer, segment copolymer, segmented copolymer
多区模型	多區模型	multi-region model
多室槽式反应器	分倉槽式反應器	compartmented tank reactor
多室反应器	複床爐	multiple hearth furnace

大 陆 名	台 湾 名	英 文 名
多室离心机	多室離心機	multichamber centrifuge
多室涡轮机	多室渦輪機	multicasing turbine
多输入多输出系统	多輸入多輸出系統	multiple-input-multiple-output-system
多态[现象]	多形性	polymorphism
多膛炉	複床爐	multiple hearth furnace
多糖	多醣	polysaccharide
多糖酶	多醣分解酶	polysaccharidase
多烷基化作用	多烷基化[作用]	polyalkylation
多相反应器	多相反應器	multiphase reactor
多相流	多相流	multiphase flow
多相平衡	多相平衡,異相平衡	heterogeneous equilibrium
多相系统	多相系統	multiphase system
多相性	不匀性	heterogeneity
多效蒸发	多效蒸發	multiple evaporation
多效蒸发器	多效蒸發器	multiple evaporator
多叶混合器	多葉混合器	multiple blade mixer
多叶片风机	多葉[片]風扇	multiblade fan
多叶片混合器	多葉[片]混合器	multiblade mixer
多元醇	多元醇	polyalcohol
多元分离序列	多成分分離順序	multicomponent separation sequence
多元混合物,多组分混合物	多成分混合物	multicomponent mixture
多元系[统],多组分系统	多成分系統	multicomponent system
多轴应力	多軸應力	multiaxial stress
多锥管分离器	多錐管分離器	multicone separator
多组分	多成分,多元	multicomponent
多组分共沸物	多成分共沸液	multicomponent azeotrope
多组分混合物(=多元混合物)		
多组分扩散	多成分擴散	multicomponent diffusion
多组分吸收	多成分吸收	multicomponent absorption
多组分系统(=多元系[统])		
惰化剂	惰化劑	inactivator
惰性气体	惰性氣體	inert gas
惰性物	惰性物	inert

E

大　陆　名	台　湾　名	英　文　名
扼流	扼流	choke flow
恶臭污染物	氣味污染物	odor pollutant
颚式破碎机	颚碎機	jaw crusher
颚碎机	颚碎機	jaw breaker
蒽醌还原染料	蒽醌缸染料	anthraquinone vat dye
蒽醌染料	蒽醌染料	anthraquinone dye
蒽油	蒽油	anthracene oil
二苯乙烯	二苯乙烯	stilbene
二次采油	二次石油回收	secondary oil recovery
二次成核	次成核	secondary nucleation
二次成核作用	二次成核	secondary nucleation
二次反应	次要反應	secondary reaction
二次夹带	二次夾帶	re-entrainment
二次胶合	二次膠合	secondary gluing
二次矩	二次[力]矩,二次動差	second order moment, second moment
二次色散	次生分散	secondary dispersion
二次滞后	二次滯延	quadratic lag
二次撞击	次要衝擊	secondary impact
二道底漆	光面漆	surfacer
二级胺	二級胺	secondary amine
二级标准	二級標準	secondary standard
二级沉降槽	二級沈降槽	secondary settler
二级成核作用	二級成核作用	second order nucleation
二级澄清器	二級澄清器	secondary clarifier
二级处理	二級處理	secondary treatment
二级滴滤池	二級滴濾池	secondary trickling filter
二级电离电位	二級游離電位	second order ionization potential
二级加料器	二級加料器	secondary feeder
二级结晶	二級結晶	second order crystallization
二级聚合反应	二級聚合反應	secondary polymerization reaction
二级滞后	二級落後	second order lag
二级转变	二級[相]轉變	second order transition
二级转变温度	二級[相]轉變溫度	second order transition temperature

大　陆　名	台　湾　名	英　文　名
二极管	二極體	diode
二极函数发生器	二極函數產生器	diode function generator
二甲苯	二甲苯	xylene
二阶流体	二階流體	second-order fluid
二阶系统	二階系統	second-order system
二阶响应	二階應答	second-order response
二聚	二聚合	dimerization
二肽酶	二肽酶	dipeptidase
二维模型	二維模式	two-dimensional model
二向色谱法	二維層析[法]	two-dimensional chromatography
二行程引擎	二行程引擎	two-cycle engine
二氧化锆	氧化鋯	zirconia
二氧化硅	二氧化矽,矽石	silica
二氧化钛	氧化鈦	titania
二氧化碳测定计	二氧化碳測定計	calcimeter
二元催化剂	二元觸媒	binary catalyst
二元共沸液	二元共沸物	binary azeotrope
二元混和物	雙成分混合物	binary mixture
二元系[统]	雙成分系統	binary system
二元蒸馏	二元蒸餾	binary distillation

F

大　陆　名	台　湾　名	英　文　名
发电厂循环	發電廠循環	power plant cycle
发电站	發電廠	power plant
发光	發光	luminescence
发光菌	發光菌	photogenic bacteria
发光染料	發光染料	luminescent dye
发光涂料	發光漆	luminous paint
发汗	發汗法	sweating
发汗工艺	發汗法	sweating process
发酵	發酵[作用]	fermentation
发酵槽	發酵槽,發酵缸	fermentation tank, fermentation vat
发酵管	發酵管	fermentation tube
发酵管试验	發酵管試驗	fermentation tube test
发酵罐	發酵槽	fermenter, fermentor
发酵过程	發酵程序	fermentation process

大　陆　名	台　湾　名	英　文　名
发酵机理	發酵機構	fermentation mechanism
发酵技术	發酵技術	fermentation technology
发酵窖	發酵窖	fermentation cellar
发酵模式	發酵方式	fermentation pattern
发酵培养基	發酵基, 發酵介質	fermentation medium
发酵器械	發酵裝置	fermentation apparatus
发酵室	發酵室	fermentation chamber
发酵途径	發酵途徑	fermentation pathway
发酵效率	發酵效率	fermentation efficiency
发明	發明	invention
发泡	發泡	foaming
发泡剂	發泡劑, 充氣劑	blowing agent, foam agent, inflating agent
发散冷却	蒸散冷卻	transpiration cooling
发色团	發色團	chromophore
发射光谱	發射光譜	emission spectrum
发射光谱仪	發射分光計	emission spectrometer
发射率	發射率	emissivity
发射能力	發射能力	emissive power
发射体	發射體	emitter
发生炉煤气	發生爐氣	producer gas
发烟弹	煙幕彈	smoke shell
发烟罐	發煙罐	smoke pot
发烟硫酸	發煙硫酸	oleum
发烟试验	發煙試驗	smoking test
罚函数	處罰函數	penalty function
阀当量管长	閥相當管長	valve equivalent pipe length
阀范围性	閥範圍性	valve rangeability
阀杆	閥桿	valve stem
阀鉴定编码	閥識別碼	valve identification code
阀控制	閥控制	valve control
阀门定位器	閥定位器	valve positioner
阀驱动器	制動閥	valve actuator
阀容量	閥容量	valve capacity
阀塞	閥塞	valve plug
阀损失	閥損失	valve loss
阀特性	閥特性	valve characteristics
阀体	閥[主]體	valve body
阀响应	閥回應	valve response

大　陆　名	台　湾　名	英　文　名
阀序列	閥定序	valve sequencing
阀增益	閥增益	valve gain
阀滞后	閥遲滯	valve hysteresis
阀阻力	閥阻力	valve resistance
法定责任	法定責任	legal liability
法兰	法蘭,凸緣	flange
法兰接头	凸緣接頭	flange coupling
法兰连接	凸緣連接	flange connection
法兰联管节	凸緣接頭	flange joint
法兰旋塞	凸緣分接頭	flange tap
法向应力	法向應力	normal stress
法向应力差	法向應力差	normal stress difference
法坐标	法坐標,正規坐標	normal coordinates
珐琅釉	琺瑯釉	enamel glaze
发夹管	髮夾管	hair-pin tube
发状弹簧	髮狀彈簧	hair spring
凡士林	凡士林	vaseline
矾鞣	明礬鞣製	alum tanning
矾土	礬土	alumina
钒钾铀矿	釩酸鉀鈾礦	carnotite
钒接触法	釩接觸法	vanadium contact process
繁殖率	繁殖速率	reproduction rate
繁殖周期	繁殖週期	reproductive cycle
反常黏度	反常黏度	anomalous viscosity
反萃取	逆萃取	reverse extraction
反电动势	反電動勢	counter electromotive force
反杠杆法则	反槓桿法則	inverse lever arm rule
反回作用	返回動作	return action
反胶团萃取	逆微胞萃取	reverse micelle extraction
反竞争性抑制	反競爭抑制［作用］	anti-competitive inhibition, uncompetitive inhibition
反馈	回饋	feedback
反馈补偿	回饋補償	feedback compensation
反馈回路	回饋環路	feedback loop
反馈机理	回饋機構	feedback mechanism
反馈控制	回饋控制	feedback control
反馈控制器	回饋控制器	feedback controller
反馈控制系统	回饋控制系統	feedback control system

大 陆 名	台 湾 名	英 文 名
反馈伸缩囊	回饋伸縮囊	feedback bellow
反馈系统	回饋系統	feedback system
反馈信号	回饋信號	feedback signal
反馈元件	回饋元件	feedback element
反扩散	逆擴散	reverse diffusion
反流	逆向流動	reverse flow
反流态化	未流體化	defluidization
反射分光光度计	反射分光光度針	reflection spectrophotometer, reflectance spectrophotometer
反射光谱法	反射光譜法	reflectance spectroscopy
反射计	反射計	reflectometer
反射聚光器	反射聚光器	reflecting condenser
反射率	反射率	reflectivity
反渗透	逆滲透	reverse osmosis
反式加成	反式加成[反應]	trans addition
反微分控制	反微分控制	inverse derivative control
反洗	反洗,逆洗	backflushing
反洗速率	逆洗速率	backwash rate
反向泵	反向泵	reverse-running pump
反向传播算法	反向傳播演算法	back-propagation algorithm
反向电流	反向電流,逆流	reverse current
反向扩散	逆擴散	back diffusion
反向压力计	反向壓力計	inverted manometer
反相乳液	反相乳液	inverse emulsion
反絮凝(=解絮凝)		
反应	反應	reaction
反应步骤	反應步驟	reaction step
反应萃取	反應萃取	reactive extraction
反应答	反應答	inverse response
反应动力学	反應動力學	reaction kinetics
反应对	反應對	reaction pair
反应分子数	分子數,分子性	molecularity
反应釜	反應釜	reaction kettle
反应概率	反應機率	reaction probability
反应工程	反應工程	reaction engineering
反应过程	反應程序	reaction process, reactive process
反应过程变量	反應進展變數	reaction progress variable
反应焓	反應焓	enthalpy of reaction

大　陆　名	台　湾　名	英　文　名
反应机理	反應機構	reaction mechanism
反应级数	反應級數	reaction order, order of reaction
反应溅射蚀刻法	反應濺射蝕刻[法]	reactive sputtering etching
反应进度	反應程度	extent of reaction
反应扩散网络	反應擴散網[路]	reaction diffusion network
反应离子蚀刻	反應離子蝕刻[法]	reactive ion etching
反应灵敏度	反應靈敏度	reaction sensitivity
反应器	反應器	reactor
反应器安全	反應器安全性	reactor safety
反应器参数	反應器參數	reactor parameter
反应器操作	反應器操作	reactor operation
反应器操作特性	反應器性能	reactor performance
反应器动力学	反應器動力學	reactor kinetics
反应器动态学	反應器動力學	reactor dynamics
反应器分析	反應器分析	reactor analysis
反应器工程	反應器工程	reactor engineering
反应器构型	反應器構型	reactor configuration
反应器规模	反應器容量	reactor capacity
反应器控制	反應器控制	reactor control, control-reactor
反应器理论	反應器理論	reactor theory
反应器模型	反應器模式	reactor model
反应器设计	反應器設計	reactor design
反应器网络	反應器網路	reactor network
反应器稳定性	反應器穩定性	reactor stability
反应器系统	反應器系統	reactor system
反应器优化	反應器最適化	reactor optimization
反应器中毒	反應器中毒	reactor poisoning
反应器装配	反應器設置	reactor setup
反应器组	反應器組	reactor train
反应亲和力	反應親和力	reaction affinity
反应区	反應區	reaction zone
反应曲线	反應曲線	reaction curve
反应曲线法	反應曲線法	reaction curve method
反应热	反應熱	heat of reaction
反应时间	反應時間	reaction time
反应式分配器	作用式分配器	reaction-type distributor
反应室	反應室	reaction chamber
反应速度	反應速度	reaction velocity

大　陆　名	台　湾　名	英　文　名
反应速率	反應速率	reaction rate
反应速率常数	反應速率常數	reaction rate constant
反应速率理论	反應速率理論	reaction rate theory
反应塔	反應塔	reaction tower
反应图	反應圖	reaction diagram
反应途径	反應途徑	reaction path
反应网络	反應網路	reaction network
反应网络分析	反應網路分析	analysis of reaction network, reaction network analysis
反应温度	反應溫度	reaction temperature
反应物	反應物	reactant
反应物比例	反應物比例	reactant ratio
反应系统	反應系統	reaction system
反应相	反應相	reacting phase
反应性	反應性	reactivity
反应循环	反應循環	reaction cycle
反应蒸馏	反應[性]蒸餾	reactive distillation
反应中心	反應中心	reactive center
反应周期	反應週期	reaction cycle time
反应专一性	反應特異性,反應特定性	reaction specificity
反应坐标	反應坐標	reaction coordinate
反转温度	反轉溫度	inversion temperature
返回率	報酬率	return rate
返混反应器	逆混反應器	backmix reactor
返位时间	重設時間,積分時間	reset time
泛点	溢流點	flooding point
泛点速度	溢流速度	flooding velocity
泛速度分布	泛速度分佈	universal velocity distribution
范德瓦耳斯常数	凡得瓦常數	van der Waals constant
范德瓦耳斯方程	凡得瓦方程式	van der Waals equation
范德瓦耳斯力	凡得瓦力	van der Waals force
范德瓦耳斯吸附	凡得瓦吸附[作用]	van der Waals adsorption
范德瓦耳斯状态方程	凡得瓦狀態方程式	van der Waals equation of state
范拉尔方程	范拉爾方程	van Laar equation
范托夫定律	凡特何夫定律	van't Hoff law
范托夫方程式	凡特何夫方程式	van't Hoff equation
范托夫因子	凡特何夫因子	van't Hoff factor

大 陆 名	台 湾 名	英 文 名
范围	範圍	scope
方波	方波	square wave
方程	方程式	equation
BET 方程	BET 方程[式]	Brunauer-Emmett-Teller equation,BET equation
BWR 方程	本[尼迪克特]韋[伯]魯[賓]方程	Benedict-Webb-Rubin equation,BWR equation
NRTL 方程	NRTL 方程	NRTL equation
RK 方程	RK 方程	Redlich-Kwong equation,RK equation
MESH 方程组	MESH 方程組	MESH equations
方法	方法	method
方解石	方解石	calcite
方块三对角矩阵	分塊三對角方陣	block tridiagonal matrix
方镁石	方鎂石	periclase
方硼石	方硼石	boracite
方石英	白矽石	cristobalite
芳构化	芳化[作用]	aromatization
芳[香]烃	芳[香]烴,芳香族[化合物]	aromatics
芳香油	香料油	perfume oil
防爆	防爆	explosion proof
防潮纸	油纸	saturated paper
防冲板	擋板	baffle plate
防喘振控制	抗激變控制	anti-surge control
防冻	防凍劑	antifreeze
防冻剂	防凍劑	antifreezing agent
防冻器	防凍器	antifreezer
防腐剂	防腐劑	preservative
防腐蚀	防[腐]蝕	anticorrosion, corrosion prevention
防护催化剂	防護觸媒	guard catalyst
防护罩	排氣罩	hood
防火	防火	fire protection
防火漆	防火漆	fire-retarding paint
防溅挡板	防濺板	splash plate
防溅盘	防濺盤	splash disc
防焦剂	防焦劑	scorch retarder
防结块剂	防結塊劑	anti-blocking agent
防沫剂,消泡剂	防沫劑,消泡劑	antifoam agent

大　陆　名	台　湾　名	英　文　名
防软[化]剂	防軟[化]劑	antisoftener
防蚀剂	防蝕劑	anticorrosive agent
防蚀漆	防蝕漆	anticorrosive paint
防水处理	防水加工	water proofing
防水剂	撥水劑	repellent
防水润滑脂	防水潤滑脂	waterproof grease
防水水泥	防水水泥	waterproof cement
防缩模	防縮模	shrinkage jigs
防缩器	防縮裝置	shrink fixture
防缩整理	防縮加工	shrink resistant finish
防污染剂	防污劑	antistaining agent
防锈脂	防鏽脂	antirust grease
防盐雾性	抗鹽霧性	salt-fog resistance
防震涂料	防震塗料	shock proof lacquer
纺前染色丝	色紡紗	spun-dyed yarn
纺丝	紡絲	spinning
纺丝仓	紡絲槽	spinning cell
纺丝罐	紡絲罐	spinning can, spinning pot
纺丝机	紡絲機	spinning machine
纺丝溶液	紡絲液	spinning solution
[纺丝]丝饼	紡絲餅	spinning cake
纺丝酸	紡絲酸	spinning acid
纺丝头	紡絲嘴	spinning nozzle
纺丝头拉伸	紡絲頭拉伸	spinning draft
纺丝甬道	紡絲通道	spinning channel
纺丝油	紡絲油	spinning oil
纺丝浴	紡絲浴	spinning bath
纺丝浴拉伸	紡絲浴拉伸	spinning bath stretch
纺织品(=纺织物)		
纺织物,纺织品	織物,紡織品	textile
纺织物整理	紡織物整理	textile finishing
纺织纤维	紡織纖維	textile fiber
放大	放大,等比增大	amplification, scale up
放大法则	放大法則	scale-up rule
放大器	放大器	amplifier
放大问题	放大問題	scale-up problem
放气阀	洩氣閥	release valve
放热反应	放熱反應	exothermic reaction, heat liberating reac-

大　陆　名	台　湾　名	英　文　名
		tion
放热过程	放熱程序	exothermic process
放热性	放熱度	exothermicity
放射化学	放射化學	radiochemistry
放射热分析	射氣熱分析[法]	emanation thermal analysis
放射性	放射性	radioactivity
放射性尘	放射[性]塵	radioactive dust
放射性沉降物	放射性落塵	radioactive fallout
放射性毒物	放射性毒	radioactive poison
放射性废物	放射性廢棄物	radioactive waste
放射性核	放射性核	radioactive nucleus
放射性核素	放射性核種	radioactive nuclide
放射性裂变	放射性分裂	radioactive fission
放射性示踪物	放射性示蹤劑	radioactive tracer
放射性衰变	放射性衰變	radioactive decay
放射性同位素	放射性同位素	radioactive isotope
放射性蜕变	放射性蛻變	radioactive disintegration
放射性污染	放射性污染	radioactive contamination
放射性污染物	放射性污染物	radioactive pollutant
放射性元素	放射元素	radioactive element
放射性指示剂	放射性指示劑	radioactive indicator
放射源	放射[性]源	radioactive source
放线菌素	放線菌黴素	actinomycin
放线酮	放線菌酮	cycloheximide
放泄阀	排洩閥	drain valve
飞温,温度失控	溫度失控	temperature runaway
非重叠操作	非重疊操作	nonoverlapping operation
非重复聚合物	非重複性聚合物	nonrepetitive polymer
非催化反应	非催化反應	noncatalytic reaction
非等温反应器	非等溫反應器	nonisothermal reactor
非等温吸收	非等溫吸收	non-isothermal absorption
非等温性	非等溫性	nonisothermality
非定常流	非穩態流動	unsteady flow
非定态	非穩[定]態	unsteady state
非定态操作	非穩態操作	unsteady state operation
非定态传热	非穩態熱傳	unsteady state heat transfer
非定态过程	非穩態程序	unsteady state process
非独立粒子系集	交互作用粒子系集	assembly of interacting particles

大　陆　名	台　湾　名	英　文　名
非对称膜	非對稱膜	asymmetric membrane
非多孔颗粒吸收器	無孔顆粒吸收器	nonporous pellet absorber
非多孔膜	無孔透膜	nonporous membrane
非干性油	非乾性油	nondrying oil
非隔离反应器	非隔離反應器	nonsegregated reactor
非隔离返混	非隔離逆混	nonsegregated backmixing
非隔离混合	非隔離混合	nonsegregated mixing
非惯性坐标系统	非慣性坐標系統	noninertial coordinate system
非活化吸附	非活化化學吸附	nonactivated chemisorption
非基本反应	非基本反應	nonelementary reaction
非结构模型	非結構化模式	unstructured model
非结合水分	非結合水分	unbound water
非金属	非金屬	nonmetal
非晶性的	非晶性的	noncrystalline
非竞争性抑制	非競爭性抑制[作用]	noncompetitive inhibition
非绝热反应	非絕熱反應	nonadiabatic reaction
非绝热反应器	非絕熱反應器	nonadiabatic reactor
非均相催化	非均相催化,異相催化	heterogeneous catalysis
非均相反应	非均相反應	heterogeneous reaction
非均相反应器	非均相反應器	heterogeneous reactor
非均相共沸混合物	異相共沸[混合]液	heteroazeotrope, heterogeneous azeotrope
非均相聚合	非均相聚合	heterogeneous polymerization
非均相系统	非均勻系統	heterogeneous system
非均相状态	非均勻狀態	heterogeneous state
非均一流化床	非均勻流[體]化床	heterogeneously fluidized bed
非均匀流	非均勻流動	nonuniform flow
非均匀性	非均勻性,不勻性	nonhomogeneity, inhomogeneity
非离子洗涤剂	非離子清潔劑	nonionic detergent
非离子型表面活性剂	非離子界面活性劑	nonionic surface active agent, nonionic surfactant
非理想表面	非理想表面	nonideal surface
非理想反应器	非理想反應器	nonideal reactor
非理想流动	非理想流動	nonideal flow
非理想气体	非理想氣體	nonideal gas
非理想溶液	非理想溶液	nonideal solution
非连锁反应	非連鎖反應	nonchain reaction
非炼焦煤	非煉焦煤	noncoking coal
非黏流动	非黏性流動	inviscid flow

大　陆　名	台　湾　名	英　文　名
非牛顿流体	非牛頓流體	non-Newtonian fluid
非平衡级模型	非平衡階模型	non-equilibrium stage model
非平衡热力学	非平衡[態]熱力學	non-equilibrium thermodynamics
非平衡系统	非平衡系統	non-equilibrium system
非破坏性试验	非破壞性試驗	nondestructive test
非亲和吸附	非親和吸附	non-affinity adsorption
非侵入式传感器	非侵入性感測器	noninvasive sensor
非侵入性仪器	非侵入性儀器	noninvasive instrument
非润湿表面	不潤濕表面	nonwetting surface
非时变系统	非時變系統	time invariant system
非随机两液方程	NRTL 方程	non-random-two-liquid equation，NRTL equation
非稳态	不穩定狀態	unstable state
非稳态过程	非穩態程序	nonsteady state process
非洗式板框	非洗式板框	nonwashing plate
非线性	非線性	nonlinearity
非线性动力学	非線性動力學	nonlinear kinetics
非线性规划	非線性規劃	nonlinear programming
非线性回归分析	非線性回歸分析[法]	nonlinear regression analysis
非线性控制阀	非線性控制閥	nonlinear control valve
非线性控制器	非線性控制器	nonlinear controller
非线性控制系统	非線性控制系統	nonlinear control system
非线性黏弹性	非線性黏彈性	nonlinear viscoelasticity
非线性稳定性	非線性穩定性	nonlinear stability
非线性系统	非線性系統	nonlinear system
非圆管当量直径	相當管徑	equivalent non-circular duct diameter
非再生能源	非再生性能源	nonrenewable energy sources
非正规溶液	非正規溶液	nonregular solution
非正式报告	非正式報告	informal report
非质子溶剂	非質子性溶劑	aprotic solvent
非自发反应	非自發反應	nonspontaneous reaction
非自发过程	非自發程序	nonspontaneous process
非最小相位滞后	非最小相位滯延	nonminimum phase lag
菲克定律	費克定律	Fick law
肥料	肥料	fertilizer
斐波那契搜索法	費布納西搜尋法	Fibonacci search method
榧子油	榧子油	kaya oil
翡翠	祖母綠	emerald

大　陆　名	台　湾　名	英　文　名
翡翠绿	翡翠綠	emerald green
废碱	廢鹼	spent caustic
废碱液	廢鹼液	spent lye
废气	廢氣	waste gas
废弃物	廢棄物	waste
废弃物陆上处理	［廢棄物］陸上處理［法］	land disposal
废燃料	廢燃料	waste fuel
废热	廢熱	waste heat
废热干燥器	廢熱乾燥器	waste heat drier
废热锅炉	廢熱鍋爐	waste heat boiler
废热回收	廢熱回收	waste heat recovery
废水	廢水	waste water
废水处理	廢水處理	waste water treatment
废酸	廢酸	waste acid
废物处理	廢棄物處理	waste treatment, waste disposal
废物管理	廢棄物管理	waste management
废物坑	廢棄物坑	refuse pit
废橡胶	廢橡膠	scrap rubber
废液	廢液	spent liquor, waste effluent, waste liquid, waste liquor
废渣埋填	［廢棄物］掩埋場	landfill
沸点	沸點	boiling point
沸点测定器	沸點測定器,沸點裝置	boiling point apparatus
沸点降低	沸點下降	boiling point depression, boiling point lowering
沸点曲线	沸點曲線	boiling point curve
沸点升高	沸點上升	boiling point elevation, boiling point rise
沸点图	沸點圖	boiling point diagram
沸点指数	沸點指數	boiling point index
沸点-组成图	沸點-組成圖	boiling point-composition diagram
沸石	沸石	zeolite
沸石催化劑	沸石觸媒	zeolite catalyst
沸腾	沸騰	boiling, ebullition
沸腾床	鼓泡床	bubbling bed, ebullated bed
沸腾床反应器	沸騰床反應器	ebullated bed reactor
沸腾淀粉	糊化澱粉	boiling starch
沸腾范围	沸點範圍	boiling range

大　陆　名	台　湾　名	英　文　名
费米-狄拉克分布	費米-狄拉克分佈	Fermi-Dirac distribution
分瓣模	對合鑄模	split mold
分贝	分貝	decibel, dB
分辨率	解析度	resolution
分辨能力	解析能力	resolving power
分布板	分佈板	distribution plate
分布参数模型	分佈參數模式	distributed-parameter model
分布参数系统	分散參數系統	distributed-parameter system
分布成分	分佈成分	distributed component
分布函数	分佈函數,分配函數	distribution function
分布器臂	分配器臂	distributor arm
分布器,分配器	分佈器,分配器	distributor
分布曲线	分佈曲線,分配曲線	distribution curve
分布式系统	分散系統	distributed system
分布性质	分佈性質	distributed property
分布滞后	分佈滯延	distributed lag
分步结晶	分段結晶	fractional crystallization
分步膨胀	分步膨脹	fractional expansion
分层	分層[作用],剝層[作用]	stratification, delamination
分层流	分層流	stratified flow
分叉	分歧	bifurcation
分叉分析	分歧分析	bifurcation analysis
分程调节	分程控制	split-range control
分段操作	分階操作	stage operation
分段床反应器	分段床反應器	segmental-bed reactor
分段分析	分段分析,分餾分析	fractional analysis
分段梯形聚合物	立梯型聚合物	stepladder polymer
分隔	分隔	partitioning
分光光度计	分光光度計,光譜儀	spectrophotometer
分光镜	分光鏡	spectroscope
分光浊度滴定	分光濁度滴定	spectroturbidimetric titration
分级	分類	classification
分级沉淀	分級沈澱	fractional precipitation
分级模型	分階模式	stage model
分级器	分類器	classifier
分级渗析	分級透析	fractional dialysis
分级吸附	部分吸附	fractional adsorption

大　陆　名	台　湾　名	英　文　名
分级吸收	部分吸收	fractional absorption
分级效率	分級效率	fractional efficiency
分解	分解	decomposition
分解程度	分解程度	extent of decomposition
分解代谢	分解代謝[作用]	catabolism
分解电势	分解勢,分解電位	decomposition potential
分解电压	分解電壓	decomposition voltage
分解剂	分解劑	decomposer
分解器	分解器	decomposer
分解热	分解熱	heat of decomposition
分解效率	分解效率	decomposition efficiency
分解压力	分解壓[力]	decomposition pressure
分解蒸馏	分解蒸餾	destructive distillation
分界表面,界面相	分界面	dividing surface
分界线	分模線	parting line
分块对角矩阵	分塊對角方陣	block diagonal matrix
分馈吸附	分饋吸附	split feed adsorption
分离	分離	separation
分离池	分離池	separation cell
分离点	分離點	separation point
分离度	解析度	resolution
分离高度	[流體化床]分離高度	transport disengaging height
分离鼓	液氣分離器	knockout drum
分离过程	分離程序	separation process
分离技术	分離技術	separation technology
分离技术规格表	分離規格表	separation specification table
分离流	分離流動	separated flow
分离膜	分離膜	separative membrane
分离因子	分離因數	separation factor
分流	分叉流	split flow
分流器	分流器,分離器	splitter
分馏	分餾	fractionation, fractional distillation
分馏萃取	分步萃取	fractional extraction
分馏器	分餾器,分餾塔	fractionator
分馏塔	分餾塔,分餾柱	fractionating tower, fractional column
分馏柱	分餾柱,分餾塔,分餾管	fractionating column
分馏装置	分餾裝置	fractionation assembly
分率	分率	fraction

大　陆　名	台　湾　名	英　文　名
分凝管	分凝管	fractional condensing tube
分凝器	分凝器,部分冷凝器	fractional condenser, partial condenser
分凝作用	分級冷凝	fractional condensation
分配比	分配率	distribution ratio
分配常数	分配常數	partition constant
分配定律	分配[定]律	distribution law, distributive law
分配器(＝分布器)		
分配色谱法	分配層析[法]	partition chromatography
分配系数	分配係數	distribution coefficient, partition coefficient
分批补料式培养	饋料批式培養	fed-batch culture
分批培养	批式培養[菌]	batch culture
分期偿还	分期償還	amortization
分熔点	分熔點	incongruent melting point
分散	分散[作用]	dispersion
分散本领	分散能力	dispersive power
分散度	分散度	degree of dispersion
分散剂	分散劑	dispersing agent, dispersant, dispersion reagent
分散剂	分散劑	disperser
分散介质	分散介質,分散媒	dispersed medium
分散控制	分散控制	decentralized control
分散流,弥散流	分散流	dispersed flow
分散磨	分散磨機	dispersion mill
分散器	分散器	disperser
分散塞流	分散塞流	dispersed slug flow
分散系数,弥散系数	分散係數	dispersion coefficient
分散系统	分散系統	dispersed system
分散相	分散相	dispersed phase, disperse phase
分散性	分散性	dispersibility
分散性染料	分散染料	dispersed dye
分时	分時系統	time sharing
分数	分數	fraction
分数级反应	分數級反應,分數階次反應	fractional-order reaction
分析模式	分析模式	analysis mode
分析器	分析器	analyzer
分析天平	分析天平	analytical balance

大　陆　名	台　湾　名	英　文　名
分析蒸馏	分析蒸餾	analytical distillation
分销成本	分銷成本	distribution cost
分形	碎形	fractal
分压力	分壓	partial pressure
分支定界法	分支定界法	branch and bound method
分子	分子	molecule
分子参数	分子參數	molecular parameter
分子重排	分子重排	molecular rearrangement
分子动态法	分子動力[學]法	molecular dynamic method, MD method
分子对称性	分子對稱性	molecular symmetry
分子分散	分子分散	molecular dispersion
分子构型	分子組態	molecular configuration
分子间力	分子間力	intermolecular force
分子结构	分子結構	molecular structure
分子扩散	分子擴散	molecular diffusion
分子扩散系数	分子擴散係數	molecular diffusivity
分子量	分子量	molecular weight
分子量分布	分子量分佈	molecular weight distribution, MWD
分子模拟	分子模擬	molecular simulation
分子内力	分子內力	intramolecular force
分子配分函数	分子配分函數	molecular partition function
分子热力学	分子熱力學	molecular thermodynamics
分子筛	分子篩	molecular sieve
分子筛催化剂	分子篩觸媒	molecular-sieve catalyst
分子筛沸石	分子篩沸石	molecular-sieve zeolite
分子生物学	分子生物學	molecular biology
分子输送	分子輸送	molecular transport
分子束沉积法	分子束沈積[法]	molecular beam deposition
分子束外延	分子束磊晶	molecular beam epitaxy
分子速度	分子速度	molecular velocity
分子物种	分子物種	molecular species
分子相互作用	分子相互作用	molecular interaction
分子蒸馏	分子蒸餾	molecular distillation
分子转换	分子變換	molecular transformation
分子自集	分子自組裝	molecular self-assembly
芬斯克方程	范氏方程式	Fenske equation
芬斯克填料	芬斯基填料	Fenske packing
酚红	酚紅	phenol red

大　陆　名	台　湾　名	英　文　名
酚醛树脂	酚[甲]醛樹脂,電木	bakelite, phenol formaldehyde resin, phenolic resin
酚醛塑料	酚醛樹脂,含酚物	phenolics
酚树脂黏合剂	酚樹脂黏合劑	phenol resin adhesive
酚酞	酚酞	phenolphthalein
焚化	焚化	incineration
焚化炉	焚化爐	incinerator
粉尘	粉塵	dust
粉尘分离器	粉塵分離器	dust separator
粉尘粒子	粉塵粒子	dust particle
粉尘粒子大小	粉塵粒子大小	dust particle size
粉尘旋风分离器	粉塵旋風分離器	dust cyclone separator
粉尘装置	粉塵裝置	dust equipment
粉煤	粉煤	pulverized coal
粉磨机	粉碎機	pulverizing mill, pulverizer
粉末冶金	粉末冶金[學]	powder metallurgy
粉碎	粉碎	comminution, pulverization, size reduction
粉碎比	粉碎比	size reduction ratio
粉碎机	粉碎機	comminuting machine, comminutor
粉[体]	粉[體]	powder
粉体技术	粉[粒]體技術	powder technology
粉体密度	粉體密度	powder density
粉状燃料	粉狀燃料	pulverized fuel
丰度	豐度	abundance
丰度曲线	豐度曲線	abundance curve
风洞	風洞	wind tunnel
风干	風乾	air drying, loft drying
风干失重	風乾損失	air-drying loss
风干收缩	風乾收縮	air shrinkage
风化	風化[作用]	efflorescence, weathering
风帽分布板	風帽分佈器	tuyere distributor
风煤气	風煤氣	air gas
风蚀试验	風蝕試驗	slacking test
风速计	風速計	anemometer
风险	風險	risk
风险分析	風險分析	risk analysis venture analysis
风险函数	風險函數	risk function
风险利润	風險利潤	venture profit

大　陆　名	台　湾　名	英　文　名
风险收益率	風險收益率	risk earning rate
风险型决策	風險性決策制定	decision making under risk
风险因子	風險因素	risk factor
风箱	風箱	bellows
风向仪	風向計	anemoscope
封闭导管	封閉導管	closed conduct
封闭式叶轮	閉式葉輪	closed impeller
封闭系统	封閉系統	closed system
封堵剂	封口機	sealer
封口	封口	sealing
封口机	封口機	sealing machine
峰度	峰態	kurtosis
峰负荷	尖峰負載	peak load
峰共振	尖峰共振	peak resonance
峰流量	尖峰流[量]	peak flow
峰时间	尖峰時間	peak time
峰增益	尖峰增益	peak gain
峰增益比	尖峰增益比	peak gain ratio
峰值	尖峰	peak
锋面分析	鋒面分析	frontal analysis
蜂蜡	蜂蠟	bees wax
蜂窝状催化剂	蜂巢狀觸媒	honeycomb catalyst
冯卡门边界层理论	馮卡門邊界層理論	von Karman boundary layer theory
冯卡门积分法	馮卡門積分法	von Karman integral method
冯卡门类比	馮卡門類比	von Karman analogy
冯卡门数	馮卡門數	von Karman number
冯卡门涡街	馮卡門渦列	von Karman vortex street
冯韦曼方程式	馮韋曼方程式	von Weimarn equation
缝编织物	縫編織物	stitch bonded fabric
缝焊	縫焊	seam welding
缝焊机	縫焊機	seam welder
缝隙腐蚀	裂隙腐蝕	crevice corrosion
呋喃树脂	呋喃樹脂	furan resin
呋喃塑料	呋喃塑膠	furan plastic
弗洛里-哈金斯理论	弗[洛里]-哈[金斯]理論	Flory-Huggins theory
伏特计	伏特計,電壓計	voltmeter
氟磷灰石	氟磷灰石	fluorapatite

大　陆　名	台　湾　名	英　文　名
氟氯烷	氟氯烷	freon
氟碳化物	氟碳化物	fluorocarbon
氟碳树脂	氟碳樹脂	fluorocarbon resin
浮标,浮子	浮標,浮子,浮筒	float
浮标指示计	浮標指示計	float indicator
浮雕漆	浮雕漆	relief paint
浮顶罐	浮頂槽	floating roof tank
浮动控制	浮動控制	floating control
浮动速率	浮[動]速率	floating rate, floating speed
浮动压力操作	浮動壓力操作	floating pressure operation
浮动压力控制	浮動壓力控制	floating pressure control
浮动作用	浮動作用	floating action
浮阀	浮閥	float valve
浮阀塔板	浮閥板	float-valve tray
浮法玻璃	浮法玻璃	float glass
浮盖消化槽	浮[動]蓋消化槽	floating-cover digester
浮力	浮力	buoyant force, buoyancy
浮力效应	浮力效應	buoyant effect
浮[力中]心	浮力中心	center of buoyancy
浮石	浮石	pumice
浮式钟型压力计	浮式鐘形壓力計	floating type bell gage
浮水皂	浮水皂	floating soap
浮头换热器	浮[動]頭熱交換器	floating head heat exchanger
浮选	浮選[法]	flotation
浮选槽	浮選槽	flotation tank
浮选过程	浮選法	flotation process
浮渣	浮渣	scum
浮渣收集装置	浮渣收集設備	scum-collecting device
浮钟压力计	鐘式壓力計	bell manometer
浮子(=浮标)		
符号	符號	symbol
符号规定	符號規定	sign convention
福尔马林	福馬林,甲醛水[溶液]	formalin
辐流式汽轮机	徑向流渦輪	radial flow turbine
辐射	輻射	radiation
辐射安全	輻射安全	radiation safety
辐射测量仪	輻射計	radiation meter
辐射常数	輻射常數	radiation constant

大　陆　名	台　湾　名	英　文　名
辐射穿透	輻射穿透	radiation penetration
辐射传热	輻射熱傳	radiation heat transfer
辐射定律	輻射定律	radiation law
辐射段	輻射段	radiation section
辐射防护	輻射防護	radiation protection
辐射高温计	輻射高溫計	radiation pyrometer
辐射隔屏	輻射屏	radiation screen
辐射沟槽滤板	輻射溝濾板	radial grooved filter plate
辐射化学	輻射化學	radiation chemistry
辐射剂量	輻射劑量	radiation dose
辐射剂量计	輻射劑量計	radiation dosimeter
辐射帽	輻射帽	radiation bonnet
辐射密度	輻射密度	radiation density
辐射能	輻射能	radiant energy, radiation energy
辐射能发射	輻射能發射	radiant energy emission
辐射能级	輻射級位	radiation level
辐射屏蔽	輻射屏蔽	radiation shield
辐射器	輻射器	radiator
辐射强度	輻射強度	radiation intensity
辐射热	輻射熱	radiation heat
辐射热计	輻射熱測定計	bolometer
辐射束	輻射束	radiant beam
辐射损耗	輻射損失	radiation loss
辐射损伤	輻射損害	radiation damage
辐射危害	輻射危害	radiation hazard
辐射系数	輻射發射係數	radiant emissivity
辐射消毒	輻射滅菌	radiation sterilization
辐射压[强]	輻射壓力	radiation pressure
釜	鍋	kettle
辅料(=填充剂)		
辅酶	輔酶	coenzyme
辅因子	輔因子	cofactor
辅助电极	輔助電極	auxiliary electrode
辅助吸收器	輔助吸收器	auxiliary absorber
辅助增塑剂	助塑化劑	secondary plasticizer
辅助周期	不生產期	nonproductive period
腐浆防治	黏泥控制	slime control
腐蚀	腐蝕	corrosion

大　陆　名	台　湾　名	英　文　名
腐蚀剂	腐蝕質	corrodent
腐蚀疲劳	腐蝕疲勞	corrosion fatigue
腐蚀试验	腐蝕試驗	corrosion test
腐蚀速率	腐蝕速率	corrosion rate
腐蚀裕量	允蝕度,腐蝕容許量	corrosion allowance
腐蚀作用	腐蝕作用	corrosive action
腐殖质	腐植質	humus
负催化剂	負觸媒	negative catalyst
负催化作用	負催化[作用]	negative catalysis
负电阻	負電阻	negative resistance
负反馈	負回饋	negative feedback
负共沸混合物	負共沸液	negative azeotrope
负荷变动	負載變化,負荷變化	load change
负荷不足	負荷不足	underloading
负荷搅动	負荷擾動	load disturbance
负荷效应	負荷效應	loading effect
负偏差	負偏差	negative deviation
负吸附	負吸附	negative adsorption
负压	負壓	subatmospheric pressure
负载	負載,負荷	load
负载点	負載點	loading point
负载因数	負載因子	duty factor
负增长反应	去傳遞反應	depropagation reaction
附铂浮石	附鉑浮石	platinized pumice
附铂硅胶催化剂	附鉑矽膠觸媒	platinized silica gel catalyst
附铂石棉	附鉑石綿	platinized asbestos
附铂石棉催化剂	附鉑石綿觸媒	platinized asbestos catalyst
附加值	附加價值	added value
附着空洞	附著空洞	clinging cavity
附着热	附著熱	heat of adhesion
复变量	複變數	complex variable
复分解	複分解	metathesis
复分解反应	複分解反應,複分解	metathetical reaction, double decomposition reaction
复合壁	複合壁	composite wall
复合材料	複合材料	composite materials
复合传递系数	複合傳送係數	composite transfer coefficient
复合管道系统	複雜管路系統	complex pipe system

大　陆　名	台　湾　名	英　文　名
复合过程	複雜程序	complex process
复合膜	複合薄膜	composite membrane
复合平均方程式	複合平均方程式	composite-averaged equation
复合球磨机	複合球磨機	compound ball mill
复合形法	複合形法	complex method
复合样品	複合樣品	composite sample
复合液体	複合液體	composite liquid
复活	再活化	reactivation
复利率	複利[率]	compound interest
复利因子	複利因數	compound interest factor
复平面	複數平面	complex plane
复式簿记	複式簿記	double entry bookkeeping
复式精馏塔	複式精餾塔	compound rectifying column
复式蒸馏器	複式蒸餾器	compound distillating apparatus
复数共轭根	複數共軛根	complex conjugate root
复数黏度	複數黏度	complex viscosity
复位	重設	reset
复位角	重設角	reset corner
复位势	複勢	complex potential
复位速率	重設速率	reset rate
复位响应	重設應答	reset response
复位终结	重設繞緊	reset windup
复位作用	重設作用	reset action
复写纸	複寫紙	carbon paper
复杂反应	複雜反應,錯合反應	complex reaction
复杂反应网络	複雜反應網路	complex reaction network
复制	繁殖	reproduction
副产物	副產物	by-product
副反应	副反應	side reaction
副环路	副環路	secondary loop
副回路	次環路	minor loop
副作用	副效應	side effect
傅里叶数	傅立葉數	Fourier number
富氮海鸟粪	富氮[海]鳥糞	nitrogen guano
富过磷酸钙	重過磷酸鹽	double superphosphate
富集	增濃,加強	enrichment
富集培养	增濃培養	enrichment culture
富集因子	增濃因數	enrichment factor

大　陆　名	台　湾　名	英　文　名
富相	富相	rich phase
覆盖率	覆蓋率	fraction of coverage

G

大　陆　名	台　湾　名	英　文　名
改性醇酸树脂	改質醇酸樹脂	modified alkyd resin
改性淀粉	改質澱粉	modified starch
改性聚丙烯腈纤维	改質[聚]丙烯腈纖維	modacrylic fiber
钙玻璃	鈣玻璃	lime glass
钙长石	鈣長石	lime feldspar
钙芒硝	鈣芒硝	glauberite
钙镁橄榄石	鈣橄欖石	monticellite
钙皂	鈣皂	calcium soap
概率	機率	probability
概率标绘图	機率圖	probability plot
概率分布	機率分佈	probability distribution
概率分布函数	機率分佈函數	probability distribution function
概率密度	機率密度	probability density
概率密度函数	機率密度函數	probability density function
概率曲线	機率曲線	probability curve
干草过滤器	乾草過濾器	hay filter
干电池	乾電池	dry cell
干纺	乾紡[絲]	dry spinning
干馏	乾餾	dry distillation
干馏釜(＝甑)		
干滤器	乾式過濾器	drying filter
干凝胶	乾凝膠	xerogel
干球温度	乾球溫度	dry-bulb temperature
干球温度计	乾球溫度計	dry-bulb thermometer
干扰素	干擾素	interferon
干热灭菌	乾熱滅菌	dry heat sterilization
干涉图样	干涉圖型	interference pattern
干湿比	濕度係數	psychrometric ratio
干湿表	[乾濕球]濕度計	psychrometer
干湿老化试验	乾濕老化試驗	admiralty test
干湿球湿度	乾濕球濕度計	wet and dry bulb hygrometer
干湿球温度表	濕球溫度	wet bulb temperature

大　陆　名	台　湾　名	英　文　名
干湿球温度计	乾濕球溫度計	dry-and-wet-bulb thermometer
干式燃烧	乾式燃燒	dry combustion
干式氧化	乾式氧化法	dry oxidation
干性油	乾性油	drying oil
干燥	乾燥	drying
干燥橱	乾燥櫥	drying cabinet
干燥房	乾燥房	drying house
干燥过程	乾法	dry process
干燥烘道	乾燥烘道	drying tunnel
干燥烘箱	乾燥烘箱	drying oven
干燥剂	乾燥劑	desiccant , drying agent , dryer
干燥架	乾燥架	drying hack
干燥炉	乾燥爐	drying stove
干燥面	乾燥面	dry surface
干燥盘	乾燥盤	drying tray
干燥棚	乾燥棚	drying shed
干燥器	乾燥器,乾燥機	desiccators , dryer
干燥器皿	乾燥器	desiccation apparatus
干燥区	乾燥區	drying zone
干燥室	乾燥室	drying room
干燥收缩	乾燥收縮	drying shrinkage
干燥速率	乾燥速率	drying rate
干燥速率曲线	乾燥速率曲線	drying-rate curve
干燥塔	乾燥塔	drying tower
干燥箱	乾燥箱	drier bin
干燥窑	烘窯	drying kiln
干燥转鼓	乾燥桶	drying drum
干蒸汽	乾蒸汽	dry steam
甘汞电池	甘汞電池	calomel cell
甘汞电极	甘汞電極	calomel electrode
甘酞树脂	甘[油]酞[酸]樹脂	glyptal resin
甘油	甘油	glycerine
甘油单酯	單甘油酯	monoglyceride
甘油二酯	雙甘油酯	diglyceride
甘油松香酯	松香硬酯,酯膠	ester gum
甘油酯	甘油酯	glyceride
甘蔗	甘蔗	sugarcane
甘蔗蜡	蔗蠟	sugarcane wax

大　陆　名	台　湾　名	英　文　名
矸石	廢棄物	refuse
杆菌	桿菌	bacillus
杆菌肽	枯草桿菌肽	bacitracin
感光减敏剂	感光減敏劑	photographic desensitizer
感光密度计	感光密度計	densitometer
感光乳剂	感光乳液	photographic emulsion
感光纸	感光紙	sensitized paper
感胶液晶	溶致液晶	lyotropic liquid crystal
感应炉	感應爐	induction furnace
橄榄仁油	橄欖仁油	olive-kernel oil
橄榄石	橄欖石	olivine
橄榄油	橄欖油	olive oil
干细胞	幹細胞	stem cell
刚度	剛性	rigidity
刚度试验	剛性試驗	stiffness test
刚铝石,刚玉	剛鋁石,剛玉	alundum
刚铝石过滤器,刚玉过滤器	剛鋁石過濾器,剛玉過濾器	alundum filter
刚性方程	剛性方程	stiff equation
刚性模量	剛性模量	stiffness modulus
刚性模数	剛性模數	rigidity modulus
刚玉(=刚铝石)		
刚玉过滤器(=刚铝石过滤器)		
缸内晶种	罐內晶種,罐內種晶	pan seed, pan seeding
缸式发酵槽	缸式發酵槽	jar fermenter
钢	鋼	steel
钢包	澆斗	ladle
钢管	鋼管	steel pipe
钢化玻璃	強化玻璃,韌化玻璃	tempered glass, toughened glass
钢纤维	鋼纖維	steel fiber
杠杆安全阀	槓桿安全閥	lever safety valve
杠杆法则	槓桿法則	lever rule
杠杆原理	槓桿原理	lever principle
高苯乙烯橡胶	高苯乙烯橡膠	high styrene rubber
高差效应	高差效應	head effect
高分子半导体	半導體聚合物	semiconducting polymer
高分子共混物	高分子摻合物	polyblend

大　陆　名	台　湾　名	英　文　名
高硅铁	高矽鐵	high silica iron
高硅氧玻璃	耐熱玻璃	shrunk glass，vycor glass
高级处理	高級處理	advanced treatment
高级废水处理	高級廢水處理	advanced wastewater treatment
高级糖蜜	高級糖蜜	high test molasses
高径比	長寬比	aspect ratio
高径比	細長比	slenderness ratio
高聚物	聚合物，高分子	high polymer
高抗爆性燃料	高抗震爆燃料	high antiknock fuel
高抗冲聚苯乙烯	耐衝擊聚苯乙烯	high impact polystyrene
高抗拉强度	高抗拉強度	high tensile strength
高宽比	高寬比	aspect number
高岭石	高嶺石	kaolinite
高岭土	高嶺土	kaolin
高炉	高爐	blast furnace
高炉焦	高爐焦	blast furnace coke
高炉煤气	高爐氣，鼓風爐氣	blast furnace gas
高炉渣	高爐渣，鼓風爐渣	blast furnace slag
高铝玻璃	高鋁玻璃	high alumina glass
高铝耐火材料	高鋁耐火物	high alumina refractory
高铝黏土	高鋁黏土	high alumina clay
高铝水泥	高鋁水泥	high alumina cement
高铝砖	高鋁磚	high alumina brick
高锰酸盐[滴定]值	過錳酸[鹽]值	permanganate value
高锰酸盐值	過錳酸[鹽]值	permanganate number
高密度聚乙烯	高密度聚乙烯	high density polyethylene
高能化分子	高能化分子	energized molecule
高强硅盐水泥	高強度[卜特蘭]水泥	high strength portland cement
高热值	高熱值	high heating value
高热值气体	高熱量氣體	high-BTU gas
高渗溶液	高滲壓溶液	hypertonic solution
高斯消元法	高斯消去法	Gaussian elimination
高速率消化槽	高速率消化槽	high-rate digester
高速碳钢	高速碳鋼	high-speed carbon steel
高速蒸发器	高速蒸發器	high-speed evaporator
高铁含量[波特兰]水泥	高鐵[卜特蘭]水泥	high-iron Portland cement
高通滤波器	高通濾波器	high-pass filter

大　陆　名	台　湾　名	英　文　名
高透光玻璃	高透光玻璃	high transmission glass
高温法	高溫法	hot process
高温分解	高溫分解,熱解	pyrogenic decomposition
高温计	高溫計	pyrometer
高温热裂解工艺	高熱裂解法	thermofor pyrolytic cracking process
高温橡胶	熱製橡膠	hot rubber
高效型洗涤剂	強效清潔劑	heavy duty detergent
高效液相色谱法	高效能液相層析儀	high performance liquid chromatography
高选择器开关	高選擇器開關	high selector switch
高压釜	高壓釜	autoclave
高压釜膨胀试验	高壓膨脹試驗	autoclave expansion test
高原	[曲線]平台區	plateau
高真空操作	高真空操作	high vacuum operation
高真空蒸馏	高真空蒸餾	high vacuum distillation
锆石	鋯英石	zircon
锆石瓷	鋯瓷	zircon porcelain
锆英石耐火材料	鋯英石耐火物	zircon refractory
割缝	狹縫	slit
格拉斯霍夫数	葛瑞斯何夫數	Grashof number
格雷茨数	格雷茲數	Graetz number
格筛	柵篩	grizzly
隔板	隔板	dummy plate
隔离法	隔離法,單離法	method of isolation
隔离流动	隔離流動	segregated flow
隔离流反应器	隔離流反應器	segregated reactor
隔离系统,孤立系统	隔離系統	isolated system
隔膜	隔膜	septum
隔膜泵	隔膜泵	diaphragm pump
隔膜操作阀	隔膜操作閥	diaphragm operated valve
隔膜电解槽	隔膜[電解]槽	diaphragm cell
隔膜阀	隔膜閥	diaphragm valve
隔膜盒	隔膜盒	diaphragm box
隔膜盒液位计	隔膜盒液位計	diaphragm box level gauge
隔膜控制阀	隔膜控制閥	diaphragm control valve
隔膜片	隔膜,膜片	diaphragm
隔膜筛	隔膜篩	diaphragm screen
隔膜式压力转换器	隔膜壓力轉換器	diaphragm type pressure transducer
隔膜式应变计	隔膜應變計	diaphragm type strain gage

大　陆　名	台　湾　名	英　文　名
隔膜室	隔膜室	diaphragm chamber
隔膜压缩机	隔膜壓縮機	diaphragm compressor
隔热	隔熱	thermal insulation
隔热材料	絕緣材料,隔熱材料	insulating materials
隔热层	保溫層	lagging
隔热衬里	絕緣襯裏,隔熱襯裏	insulating lining
隔热耐火材料	隔熱耐火物	insulating refractory
隔热耐火砖	隔熱耐火磚	insulating firebrick
隔热绕道	隔熱澆道	insulated runner
隔热柱	隔熱柱	insulated column
隔音材料	隔音材料	sound-insulating materials
隔音黏合剂	隔音黏合劑	sound-insulating adhesive
隔音涂料	隔音塗料	sound-proof coating
镉电池	鎘電池	cadmium cell
镉黄	鎘黃	cadmium yellow
镉皂	鎘皂	cadmium soap
各向同性	各向同性,等向性	isotropy
各向同性固体	各向同性固體,等向性固體	isotropic solid
各向同性流动	各向同性流動,等向性流動	isotropic flow
各向同性湍流	各向同性紊流	isotropic turbulence
各向异性	各向異性,異向性	anisotropy
各向异性膜	各向異性膜,異向性膜	anisotropic membrane
各向异性湍流	非等向性紊流	nonisotropic turbulence
各向异性吸收	各向異性吸收,異向性吸收	anisotropic absorption
铬橙	鉻橙	chrome orange
铬矾	鉻礬	chrome alum
铬钢	鉻鋼	chrome steel
铬革	鉻革,鞣皮	chrome leather
铬红	鉻紅	chrome red
铬黄	鉻黃	chrome yellow
铬绿	鉻綠	chrome green
铬媒染色	鉻媒染法	chrome mordant dyeing
铬镁砖	鉻鎂磚	chrome-magnesite brick
铬染料	鉻媒染料	chrome dye
铬鞣	鉻鞣法	chrome tannage

大　陆　名	台　湾　名	英　文　名
铬鞣制	鉻鞣[製]	chrome tanning
铬颜料	鉻顏料	chrome pigment
铬质耐火材料	鉻質耐火物	chrome refractory
铬砖	鉻磚	chrome brick
给硫剂	給硫劑	sulfur donor agent
给硫体	硫施體	sulfur donor
给体	[給]予體	donor
根轨迹	根軌跡	root locus
根轨迹法	根軌跡法	root locus method
根轨迹方法	根軌跡法	root locus technique
根轨迹角	根軌跡角	root locus angle
根轨迹图	根軌跡圖	root locus plot
更迭成本	更迭成本	changeover cost
更迭时间	更迭時間	changeover time
更换费用	置換成本	replacement cost
更换价值	置換價值	replacement value
更换评估	置換評估	replacement evaluation
工厂布置	工廠布置	plant layout
工厂成本估算	工廠成本估計值	plant cost estimate
工厂资产	工廠資產	plant asset
工程	工程	engineering
工程材料	工程材料	engineering materials
工程分析	工程分析	engineering analysis
工程流程图	機械流程圖	mechanical flow diagram, mechanical flowsheet
工程设计	工程設計	engineering design
工程设计成本	工程設計成本	engineering design cost
工程实验	工程實驗	engineering experiment
工程塑料	工程塑膠	engineering plastic
工时	工時	man hour
工时学	工時學	time and motion study
工业	工業,產業	industry
工业反应器	工業反應器	industrial reactor
工业废料	工業廢棄物	industrial waste
工业废水	工業廢水	industrial wastewater
工业废水处理	工業廢水處理	industrial wastewater treatment
工业过程	工業程序	industrial process
工业化学	工業化學	industrial chemistry

大　陆　名	台　湾　名	英　文　名
工业化学计量学	工業化學計量學	industrial stoichiometry
工业酒精	工業酒精	industrial alcohol
工业煤气	工業用氣體	industrial gas
工业色谱	工業級層析法	process-scale chromatography
工业水处理	工業用水處理	industrial water treatment
工业仪器	工業儀器	industrial instrument
工业用水	工業用水	industrial water
工业用洗涤剂	工業用清潔劑	industrial-use detergent
工艺安全	程序安全[性]	process safety
工艺布置图	程序配置[圖]	process layout
工艺参数	程序參數	process parameter
工艺单元	程序單元	process unit
工艺负荷	程序負載	process load
工艺负荷变化	程序負載變化	process load change
工艺工程师	製程工程師,方法工程師	process engineer
工艺规划	程序規畫	process planning
工艺过程设计	程序設計	process design
工艺计算	程序計算	process calculation
工艺技术	程序技術	process technology
工艺流程图	程序圖,程序流程圖,方法流程圖	process chart, process flow diagram, process flowsheet
工艺设备	製程設備	process equipment
工艺设计工程师	程序設計工程師	process design engineer
工艺设计师	程序設計師	process designer
工艺条件	程序條件	process condition
工艺物流	程序流	process stream
工艺系统	程序系統	process system
工艺性能	程序性能	process performance
工艺用真空[系统]	程序真空[系統]	process vacuum
工艺蒸气	製程蒸汽	process steam
工资	工資	wage
工资单	薪資單	payroll
工作单	表單	worksheet
工作流体	工作流體	working fluid
工作应力	工作應力	working stress
弓形降液管	弓形下導排管	segmental downtake calandria
弓形孔口	弓形孔口	segmental orifice

大　陆　名	台　湾　名	英　文　名
弓形孔口板	弓形孔口板	segmental orifice plate
公称尺寸	標稱尺寸	nominal size
公称管径	標稱管徑	nominal pipe diameter
公共设施	公用設施	common utility
公共设施检查报表	公共設施檢查報表	utility check list
公共设施流程图	公用設施流程圖	utility flowsheet
公式	[公]式	formula
公用服务事业	公用設施	utility service
公用设施	公用設施	utility
公制	公制,米制	metric system
功函数	功函數	work function
功率	功率,能力	power
功率数	功率數	power number
功率损耗	功率損失	power loss
功率消耗	功率消耗[量]	power consumption
功率效率	功率效率	power efficiency
功率需要量	功率需要量	power requirement
功率因数	功率因數	power factor
功能高分子	功能性聚合物	functional polymer
功需要量	功需要量	work requirement
功指数	功指數	work index
供给压力	供給壓力	supply pressure
供酸罐	飼酸箱	acid feed box
供氧	供氧	oxygen supply
供应	供應	supply
汞电池	汞電池,水銀電池	mercury cell
汞分解器	汞分解器	mercury decomposer
汞浮子压力计	汞浮子壓力計	mercury float manometer
汞扩散泵	汞擴散泵	mercury diffusion pump
汞齐	汞齊	amalgam
汞污染	汞污染	mercury pollution
汞阴极电解槽	汞陰極電解槽	mercury cathode cell
拱砖	拱磚	arch brick
共萃取	共萃取	coextraction
共存方程	共存方程式	coexistence equation
共轭对	共軛對	conjugate pair
共轭溶液	共軛溶液	conjugate solution
共轭深度	共軛深度	conjugate depth

大　陆　名	台　湾　名	英　文　名
共轭相	共軛相	conjugate phase
共反应剂	共反應物	coreactant
共沸参数	共沸參數	azeotropic parameter
共沸点	共沸點	azeotropic point
共沸范围	共沸範圍	azeotropic range
共沸干燥	共沸乾燥	azeotropic drying
共沸混合物	共沸混合物	azeotropic mixture
共沸塔	共沸塔	azeotrope tower
共沸温度	共沸溫度	azeotropic temperature
共沸物	共沸液	azeotrope
共沸系统	共沸系統	azeotropic system
共沸现象	共沸現象	azeotropic phenomenon
共沸线	共沸線	azeotropic line
共沸[性]	共沸[現象]	azeotropy
共沸性质	共沸性質	azeotropic property
共沸蒸馏	共沸蒸餾	azeotropic distillation
共沸[蒸馏]过程	共沸程序	azeotropic process
共沸状态	共沸狀態	azeotropic state
共沸组成	共沸組成	azeotropic composition
共混	摻配	blending
共混聚合物	高分子摻合物	polymer blend
共混物	摻合物	blend
共价电子对	共用電子對	shared electron pair
共聚合	共聚合[反應]	copolymerization
共聚物	共聚物	copolymer
共离子	共離子	co-ion
共栖	共生,共棲	commensalism
共熔点	共熔點	eutectic point
共熔合金	共熔合金	eutectic alloy
共熔平衡	共熔平衡	eutectic equilibrium
共熔温度	共熔溫度	eutectic temperature
共熔物	共熔物,共熔混合物	eutectics, eutectic mixture
共熔系统	共熔系統	eutectic system
共熔状态	共熔狀態	eutectic state
共生	共生	symbiosis
共形映射	保角映射	conformal mapping
共振	共振	resonance
共振峰	共振峰	resonance peak, resonant peak

大　陆　名	台　湾　名	英　文　名
共振积分	共振積分	resonance integral
共振频率	共振頻率	resonance frequency, resonant frequency
构象	構形	conformation
构型配分函数(= 位形 配分函数)		
估计方差	估計變異數	estimated variance
估计器	估計器	estimator
孤立系统(= 隔离系统)		
箍缩	狹點	pinch
箍缩效应	狹點效應	pinch effect
古伊-斯托多拉定理	古依-斯託多拉定理	Gouy-Stodola theorem
骨架振动	骨架振動	skeletal vibration
骨炭	骨炭, 骨黑	bone black, bone char
鼓风	鼓風	blast, air blasting
鼓风机	鼓風機, 鼓風扇	air blower, forced draft fan, blast blower
鼓风炉	鼓風爐	blast furnace
鼓风通气机	鼓風通氣機	forced draft aerator
鼓泡	鼓泡作用	bubbling
鼓泡反应器	鼓泡反應器	sparging reactor
鼓泡流化床	鼓泡流[體]化床	bubbling fluidized bed
鼓泡流态化	鼓泡流體化	bubbling fluidization
鼓泡气	鼓泡氣	sparging gas
鼓泡器	鼓泡器	sparger
鼓泡式吸收器	鼓泡吸收器	bubbling absorber
鼓泡塔	氣泡塔	bubble column
鼓泡塔盘	泡罩板	bubble tray
鼓式标度指示计	鼓式標[度指]示計	drum type scale indicator
鼓式分离器	桶式分離器	drum separator
鼓式干燥器	鼓式乾燥器, 桶式乾燥 器	drum drier
鼓式计数器	鼓式計數器	drum type counter
鼓式记录器	鼓式記錄器	drum type recorder
鼓型过滤机	桶式過濾器	drum filter
固醇(= 甾醇)		
固氮菌	固氮[細]菌	azotobacteria
固氮[作用]	固氮作用	nitrogen fixation
固定	固定	fixation
固定成本	固定成本	fixed cost

大 陆 名	台 湾 名	英 文 名
固定成长系统	固著生長系統	fixed-growth system
固定床	固定床	fixed bed
固定床反应器	固定床反應器	fixed-bed reactor
固定床过程	固定床法	fixed-bed process
固定催化剂床	固定觸媒床	fixed catalyst bed
固定氮	固定氮	fixed nitrogen
固定费用	固定費用	fixed charge
固定化技术	固定化技術	immobilization technology
固定化酶	固定化酶	immobilized enzyme
固定化细胞反应器	固定化細胞反應器	immobilized cell reactor
固定剂	固定劑	fixing agent
固定台	固定平台	stationary platen
固定消耗量	固定費用	fixed charge
固定浴	固定浴	fixation bath
固定资本	固定資本	fixed capital
固定资本投资额	固定資本投資額	fixed capital investment
固–固相反应	固固相反應	solid-solid reaction
固–固相转变	固固相轉變	solid-solid transition
固化酒精	固化酒精	solidified alcohol
固化汽油	固化汽油	solidified gasoline
固化热	固化熱	heat of solidification, solidification heat
固化时间	固化時間	curing time
固化速率	固化速率	solidification rate
固化速率参数	固化速率參數	solidification rate parameter
固溶体	固溶體	solid solution
固溶体合金	固溶體合金	solid solution alloy
固溶线	固溶線	solvus
固态	固態	solid state
固态高分子电解质	固態聚[合物]電解質	solid state polyelectrolyte
固态化学	固態化學	solid state chemistry
固态扩散	固態擴散	solid state diffusion
固态物理学	固態物理學	solid state physics
固体	固體	solid
固体超载	固體超載	solid overload
固体催化反应	固體催化反應	solid-catalyzed reaction
固体废物处理	固體廢棄物處理	solid waste disposal
固体废物掩埋场	固體廢棄物掩埋場	solid waste landfill
固体废物	固體廢棄物	solid waste

大　陆　名	台　湾　名	英　文　名
固体聚合	固態聚合[反應]	solid polymerization
固体力学	固體力學	solid mechanics
固体粒子	固體粒子	solid particle
固体浓度	固體濃度	solid concentration
固体燃料	固體燃料	solid fuel
固体润滑剂	固體潤滑劑	solid lubricant
固体输送	固體輸送	solid transport, solid handling
固体填充塑料	固體填充塑料	solid filled plastic
固体通量	固體通量	solid flux
固体通量理论	固體通量理論	solid flux theory
固体推进剂	固體推進劑	solid propellant
固体悬浮体	固體懸浮物	solid suspension
固相	固相	solid phase
固相反应	固相反應	solid phase reaction
固相含量	固[體]含量	solid content
固相合成	固相合成	solid phase synthesis
固相化	固定化	immobilization
固相聚合	固相聚合	solid phase polymerization, solid state polymerization
固相缩聚	固相縮聚	solid phase polycondensation
固相线	固相線	solidus
固液分离器	固液分離器	solid-liquid separator
固有活化能	固有活化能	intrinsic activation energy
固有能	固有能	intrinsic energy
固有速率	固有速率	intrinsic rate
故障形式和影响分析	失效模式及效應分析	failure mode and effect analysis
故障诊断	故障診斷	failure diagnosis, fault diagnosis
刮板	刮刀	scraper
刮板式换热器	刮刀熱交換器	scraper heat exchanger
刮板输送机	刮運機,梯板運送機,梯運機	scraper conveyor, flight conveyor
刮铲角	刮勺角	angle of spatula
刮刀式冷却器	刮刀冷凍器	scraper chiller
刮痕	刮痕	scratch
刮痕硬度	刮痕硬度	scratch hardness
刮涂机	刮塗機	spread coater
寡糖	寡醣	oligosaccharide
拐点	拐點,反曲點	inflection point

大 陆 名	台 湾 名	英 文 名
拐角流	繞角流	corner flow
拐折空气加热器	交錯空氣加熱器	staggered air heater
关闭高差	關閉高差	shut-off head
关键字	關鍵字	key word
关键组分	關鍵成分	key component
关联	關聯,相關[性]	correlation
关联矩阵	關聯矩陣	incidence matrix
关税	關稅	tariff
观测次序	測得階	observed order
观察孔	視孔	peep hole
官能团	官能基	functional group
管	管	pipe
管板	管板	tube sheet
管壁厚度系列号	管壁厚度系列號	pipe schedule number, schedule number
管程	管程	tube pass, tube side pass
管道流程图	配管流程圖	piping flowsheet
管过滤器	管過濾器	pipe filter
管汇	歧管	manifold
管件	管接頭	pipe fitting
管接头	管接頭	pipe joint
管接头密封	管密封	pipe sealing
管径	管徑	pipe diameter
管理成本	行政成本	administrative cost
管理费用	管理費	overhead cost
管流	管流	pipe flow
管路(＝配管)		
管路混合器	管内混合器,沿線混合器	in-line mixer
管路摩擦	管路摩擦	pipe friction
管路网络	管線網路	pipeline network
管路系统	管路系統,配管系統	piping system
管路-仪表流程图	管線及儀器圖	piping and instrument diagram, PID
管磨机	管磨機	tube mill
管排	管排	tube bank
管式反应器	管式反應器,明火管式反應器	tubular reactor, fired-tubular reactor
管式加热器	管式加熱器	tubular heater
管式冷凝器	管式冷凝器	tubular condenser

大　陆　名	台　湾　名	英　文　名
管式炉	管形爐	tube furnace
管式气压计	管式壓力計	tube gage
管式蒸发器	管式蒸發器	tubular evaporator
管式蒸馏釜	管餾器	tube still, tubular still
管式蒸馏器	管餾器	pipe still
管式组件	管式組件	tubular module
管束	管束	tube bundle
管束干燥器	管式乾燥器	tube drier
管线表	管線表	pipe line list
管线过滤器	管線過濾器	in-line filter
管线应力	管線應力	piping stress
管肘	管肘	pipe bend
管子尺寸	管子尺寸	pipe size
管子粗糙度	管粗糙度	pipe roughness
管子相对粗糙度	管子相對粗糙度	pipe relative roughness
惯性	慣性	inertia
惯性沉降	慣性沈降	inertial settling
惯性离心分离器	慣性離心分離器	inertial centrifugal separator
惯性力	慣性力	inertial force
惯性装置	慣性裝置	inertial device
灌注培养	灌注培養	perfusion culture
罐头厂废水	罐頭廠廢水	cannery waste water
光比色法	光比色法	photocolorimetry
光催化	光催化[作用]	photocatalysis
光催化剂	光觸媒	photocatalyst
光导材料	光導電材料	photoconductive materials
光电比色	光電比色計	photoelectric colorimeter
光电池	光電池	photocell
光电池[管]	光電池,光電管	photoelectric cell
光电导性	光導電性	photoconductivity
光电分光光度计	光電光譜儀	photoelectric spectrophotometer
光电管	光電管	phototube
光电计	光電計	photoelectrometer
光电检测器	光偵檢計	photodetector
光电效应	光電效應	photoelectric effect
光电旋光计	光電旋光計	photoelectric polarimeter
光电学	光電學	photoelectricity
光碟	光碟	optical disc

大　陆　名	台　湾　名	英　文　名
光度测定法	光度測定法	photometric method
光度分析	光度分析	photometric analysis
光度计	光度計	photometer
光度学	光度學	photometry
光反应	光反應	photoreaction
光感受体	光受器,光受體	photoreceptor
光合作用	光合作用	photosynthesis
光呼吸	光呼吸	photorespiration
光化反应器	光化學反應器	photochemical reactor
光化学	光化學	photochemistry
光化学产量	光化學產率	photochemical yield
光化学臭氧化	光化[學]臭氧化	photochemical ozonization
光化学当量	光化[學]當量	photochemical equivalent
光化学低限	光化[學]低限	photochemical threshold
光化学动力学	光化學動力學	photochemical kinetics
光化学反应	光化學反應	photochemical reaction
光化学反应性	光化學反應性	photochemical reactivity
光化学分解	光化[學]分解	photochemical decomposition
光化学过程	光化學程序	photochemical process
光化学活性	光化[學]活性	photochemical activity
光化学激化	光化[學]激化	photochemical excitation
光化学降解	光化[學]降解	photochemical degradation
光化学氯化	光氯化	photochemical chlorination
光化学吸收律	光化學吸收律	photochemical absorption law
光化学效率	光化[學]效率	photochemical efficiency
光化学诱导	光化[學]誘發	photochemical induction
光黄化	光黃化	photoyellowing
光活化	光活化[作用]	photoactivation
光降解	光降解	photodegradation
光解	光解[作用],光分解	photolysis, photodecomposition
光刻	光蝕刻法,微影術	photolithography
光刻蚀过程	光蝕刻程序	photoetching process
光卤石	光鹵石	carnallite
光氯化	光氯化	photochlorination
光密度计	光密度計	photodensitometer
光敏剂	光敏劑	photosensitizer
光敏性	光敏性	photosensitivity
光敏氧化	光敏氧化[反應]	photosensitized oxidation

大　陆　名	台　湾　名	英　文　名
光谱测定法	光譜測定法	spectrometry
光谱分析	光譜分析	spectroscopic analysis, spectral analysis
光谱化学	光譜化學	spectrochemistry
光谱图	光譜圖	spectrogram
光谱学	光譜學	spectroscopy
光谱仪	光譜儀	spectrometer
光气	光氣, 二氯化羰	phosgene
光气法	光氣法	phosgenation process
光强测定仪	光量計	actinometer
光热分析	光熱分析[法]	photothermal analysis
光散射	光散射	scattering of light
光弹性	光彈性	photoelasticity
光纤	光纖[維]	optical fiber
光学玻璃	光學玻璃	optical glass
光学玻璃板	[光學]玻璃板	slab glass
光学补偿器	光學補償器	optical compensator
光学高温计	光學高溫計	optical pyrometer
光学密度	光學密度	optical density
光氧化	光氧化	photooxidation
光氧化老化	光氧化老化	photooxidative aging
光致发光	光致發光	photoluminescence
光[致]核反应	光[致]核反應	photonuclear reaction
光致聚合	光聚合[反應]	photopolymerization
光[致]聚合物	光聚合物	photopolymer
光致抗蚀剂	光阻劑	photoresist
光致离解	光解離	photodissociation
光中子	光中子	photoneutron
光子	光子	photon
光子材料	光子材料	photonic materials
广度性质	廣度性質, 外延性質	extensive property
广义牛顿流体	廣義牛頓流體	generalized Newtonian fluid
广义坐标	廣義坐標	generalized coordinates
归一化	正規化	normalization
规格	規格	specification
规号	規號	gage number
硅氮聚合物	矽氮聚合物	silicon-nitrogen polymer
硅氮烷聚合物	矽氮烷聚合物	silazane polymer
硅灰石	矽灰石	wollastonite

大 陆 名	台 湾 名	英 文 名
硅胶	矽膠	silica gel
硅锰铁合金	矽錳鐵合金	ferro-silico-manganese
硅片	矽晶片	silicon wafer
硅气凝胶	矽氣凝膠	silica aerogel
硅[润滑]脂	聚矽氧潤滑脂	silicone grease
硅石灰石	矽質石灰石	siliceous limestone
硅酸钍矿	矽酸釷礦	thorite
硅酸盐类黏合剂	矽酸鹽黏合劑	silicate adhesive
硅铁合金	矽鐵合金,矽鐵[齊]	ferrosilicon alloy
硅烷	矽烷	silane
硅烷类增黏剂	矽烷助黏著劑	silane adhesion promoter
硅烷偶联剂	矽烷偶合劑	silane coupling agent
硅线石	矽線石	sillimanite
硅橡胶	矽氧橡膠	silicone rubber, silicon rubber
硅橡胶混炼胶	矽氧橡膠聚合物	silicone rubber compound
硅橡胶黏合剂	矽氧橡膠黏著劑	silicone rubber adhesive
硅锌矿	矽鋅礦	willemite
硅岩	矽岩	silica rock
硅氧烷	矽氧烷	siloxane
硅油	[聚]矽[氧]油	silicone oil
硅藻土	矽藻土	diatomaceous earth, diatomite, kieselguhr, siliceous earth
硅藻土过滤器	矽藻土過濾器	diatomaceous-earth filter
硅藻土炸药	矽藻土炸藥	kieselguhr dynamite
硅整流器	矽整流器	silicon rectifier
硅质耐火材料	矽石耐火物	silica refractory materials
硅质耐火黏土	矽質耐火黏土	siliceous fireclay
硅质耐火黏土砖	矽質耐火黏土磚	siliceous fireclay brick
硅质黏土	矽質黏土	siliceous clay
硅砖	矽磚	silica brick
鲑鱼油	鮭魚油	salmon oil
轨迹	軌跡	locus
癸二酸	癸二酸	sebacic acid
柜式干燥机	箱形乾燥器	shelf dryer
辊式破碎机	輥[壓]碎機	roll crusher
辊[式研]磨机	輥磨機	roller mill
辊压机	輥壓機	roller mill press
[辊]轧机	輥軋機	rolling mill

大　陆　名	台　湾　名	英　文　名
滚磨机	滚磨機	tumbling mill
滚筒干燥器	旋桶乾燥器	rotating drum dryer
滚筒印花机	滚筒印花機	roller printing machine
滚轴支座	滚筒軸承,輥軸承	roller bearing
滚珠轴承	球軸承	ball bearing
滚柱	輥,滚筒	roller
锅	鍋	kettle
锅炉	鍋爐	boiler
锅炉给水	鍋爐給水	boil feed water
锅盐	鍋鹽	pan salt
国际[单位]系统	國際單位系統	system international
国际实用温标	國際實用溫標	international practical temperature scale
果胶	果膠	pectin
果胶酶	果膠酶	pectinase
果胶纤维素	果膠纖維素	pecto-cellulose
果糖	果糖	fructose, levulose
过饱和	過飽和	oversaturation
过饱和度	過飽和度	degree of supersaturation
过饱和溶液	過飽和溶液	supersaturated solution
过程变量	程序變數	process variable
过程变异性	程序變異性	process variability
过程辨识	程序辨識	process identification
过程超载	程序超載	process overload
过程动态[学]	程序動態學	process dynamics
过程反弹	程序韌性	process resilience
过程反应曲线	程序反應曲線	process reaction curve
过程非定态性	程序不穩定性	process instability
过程分解	程序分解	process decomposition
过程分析	程序分析	process analysis
过程工程	程序工程,方法工程	process engineering
过程工程核查表	程序工程查核表	process engineering check list
过程工业	程序工業	process industry
过程集成	程序整合	process integration
过程进料	程序進料	process feed
过程开发	程序開發	process development
过程可靠性	程序可靠性	process reliability
过程控制	程序控制	process control
过程控制模式	程序控制模式	process control pattern

大　陆　名	台　湾　名	英　文　名
过程灵敏度	程序靈敏度	process sensitivity
过程模拟	程序模擬	process simulation
过程模型化	程序模式化	process modeling
过程能力	製程能力指數	process capacity
过程评价	程序評估	process evaluation
过程热力学分析	程序熱力學分析	thermodynamic analysis of process
过程失稳	程序失穩	process upset
过程时间滞后	程序時延	process time lag
过程特性	程序特性	process characteristics
过程稳定性	程序穩定性	process stability
过程系统分析	程序系統分析	process system analysis
过程系统工程	程序系統工程	process system engineering
过程响应	程序應答	process response
过程响应曲线	程序應答曲線	process response curve
过程行为特性	程序行為	process behavior
过程优化	程序最適化	process optimization
过程增益	程序增益	process gain
过程自动化	程序自動化	process automation
过程组成元件	程序元件	process element
过电位	過電位	overpotential
过渡长度	過渡長度	transition length
过渡金属	過渡金屬	transition metal
过渡流	過渡流	transition flow
过渡区	過渡區	transition region
过渡态	過渡狀態	transition state
过渡态理论	過渡狀態理論	transition-state theory
过渡元素	過渡元素	transition element
过冷	過冷	subcooling, supercooling
过冷沸腾	過冷沸騰	subcooled boiling
过冷液	過冷液體	supercooled liquid
过冷液体	過冷液體	subcooled liquid
过冷蒸气	過冷蒸氣	supercooled vapor
过冷状态	過冷狀態	subcooled state
过量法	過量法	method of excess
过量空气	過量空氣	excess air
过量水	過量水	excess water
过磷酸钙	過磷酸鹽	superphosphate
过滤	過濾	filtration

大　陆　名	台　湾　名	英　文　名
过滤介质	過濾介質	filtration medium
过滤流程	過濾流程圖	filter flowsheet
过滤面积	過濾面積	filter area, filtration area
过滤灭菌	過濾滅菌	filtration sterilization
过滤器堵塞	過濾器阻塞	filter clogging
过滤器容量	過濾器容量	filter capacity
过滤速率	過濾速率	filtration rate
过滤网调换装置	換濾網裝置	screen changer
过滤网组[合]	篩組件	screen pack
过滤效率	過濾效率	filter efficiency
过滤性能	過濾性能	filter performance
过滤周期	過濾週期	filtration cycle
过膨胀现象	過膨脹現象	overexpansion phenomenon
过热	過熱	overheating, superheating
过热[量]	過熱[量]	superheat
过热器	過熱器	superheater
过热水	過熱水	superheated water
过热状态	過熱狀態	superheated state
过失误差	總誤差, 毛誤差	gross error
过失误差检出	總誤差鑑定	gross error identification
过调节	超越, 超調	overshooting
过氧化苯甲酰	過氧化苯甲醯[基]	benzoyl peroxide
过氧化氢异丙苯	異丙苯氫過氧化物	cumene hydroperoxide
过氧化物	過氧化物	peroxide
过氧化物分解剂	過氧化物分解劑	peroxide decomposer
过乙酸	過醋酸	peracetic acid
过载	超載, 超負荷	overload
过载磨石机	上動石磨機	over-driven buhrstone mill
过增益	過增益	over gain
过阻尼	過阻尼	overdamping
过阻尼系统	過阻尼系統	overdamped system
过阻尼响应	過阻尼應答	overdamped response

H

大 陆 名	台 湾 名	英 文 名
哈斯特镍基合金	赫史特合金	hastelloy
海岸污染	海岸污染	coastal pollution
海豹油	海豹油	seal oil
海岛型棉	海島棉	sea island cotton
海沟	溝	trench
海绵	海綿	sponge
海绵铂	海綿鉑	sponge platinum
海绵胶	海綿膠	sponge gum
海绵镍	海綿鎳	sponge nickel
海绵钛	海綿鈦	titanium sponge
海绵铁	海綿鐵	sponge iron
海鸟粪	[海]鳥糞[石]	guano
海洋污染	海洋污染	ocean pollution
海藻纤维	海藻纖維	seaweed fiber
[亥姆霍兹]自由能	亥姆霍兹自由能	Helmholtz free energy
含铵过磷酸钙	氨化過磷酸鈣	ammoniated superphosphate
含残油软石蜡	含油石蠟,鬆蠟	slack wax
含尘量	含塵量	dust content
含尘气体	含塵氣體	dust-laden gas
含钾过磷酸钙	含鉀過磷酸鈣	potash superphosphate
含蜡废油	含蠟污油	paraffin slop
含蜡分馏	蠟分馏	wax fractionation
含蜡油	含蠟油,蠟油	waxy oil, wax oil
含蜡原油	含蠟原油	waxy crude
含蜡渣油	蒸餘蠟	wax residuum
含量	含量	content
含硫聚合物	含硫聚合物	sulfur-containing polymer
含铅汽油	加鉛汽油	leaded gasoline
含铅氧化锌	含鉛氧化鋅	leaded zinc
含水量	含水量	water content
含水皂	含水皂	semi-boiled soap
含碳燃料	含碳燃料	carbonaceous fuel
含皂量	皂含量	soap content

大　陆　名	台　湾　名	英　文　名
函数发生器	函數波產生器	function generator
函数方程	泛函方程式	functional equation
函数关系	函數關係[式]	functional relationship
函数平移	函數平移	translation of function
焓	焓	enthalpy
焓浓图	焓-濃度圖	enthalpy-concentration diagram
焓-熵图	焓-熵圖	enthalpy-entropy diagram
焓-湿图	焓-濕度圖	enthalpy-humidity chart
焓-组成图	焓-組成圖	enthalpy-composition chart, enthalpy-composition diagram
焊接	焊接	welding
焊接变位机	定位器	positioner
夯锤法	夯鎚法	rammer process
行列式	行列式	determinant
航空汽油	航空汽油	aviation gasoline
航空燃料	航空燃料	aviation fuel
毫伏表	毫伏特計	millivoltmeter
毫克	毫克	milligram
毫米	毫米	millimeter
好气菌(=好氧菌)		
好氧处理	好氧處理	aerobic treatment
好氧发酵	需氧發酵[法]	aerobic fermentation
好氧分解	需氧分解	aerobic decomposition
好氧菌,好气菌	好氧菌,嗜氧菌	aerobic bacteria
好氧培养	好氧培養,嗜氧培養	aerobic culture
好氧生物氧化	好氣生物氧化[法]	aerobic biological oxidation
好氧污泥消化	需氧污泥消化	aerobic sludge digestion
好氧消化	需氧消化	aerobic digestion
好氧氧化	需氣氧化[反應]	aerobic oxidation
耗尽	耗盡	depletion
耗尽容许度	耗盡容許度	depletion allowance
耗散	散逸	dissipation
耗散函数	散逸函數	dissipation function
耗散能	散逸能	dissipated energy
耗氧速率	耗氧速率	oxygen consumption rate
合成	合成	synthesis
合成氨	合成氨	synthetic ammonia
合成宝石	合成寶石	synthetic gem

大　陆　名	台　湾　名	英　文　名
合成单宁	合成單寧	synthetic tannin
合成氮肥	合成氮肥	synthetic nitrogenous fertilizer
合成靛	合成靛	synthetic indigo
合成多肽	合成多肽	synthetic polypeptide
合成反应	合成反應	synthesis reaction, synthetic reaction
合成反应器	合成反應器	synthesis reactor
合成干性油	合成乾性油	synthetic drying oil
合成甘油	合成甘油	synthetic glycerol
合成高分子	合成高分子,合成大分子	synthetic high polymer, synthetic macro-molecule
合成革	合成皮革	synthetic leather
合成过程	合成程序	synthesis process
合成化学	合成化學	synthetic chemistry
合成胶乳	合成乳膠	synthetic latex
合成聚合物	合成聚合物	synthetic polymer
合成木材	合成木材	synthetic wood
合成泡沫塑料	合成發泡材	synthetic foam
合成气	合成氣	syngas, synthesis gas, synthetic gas
合成汽油	合成汽油	synthetic gasoline
合成器	合成器	synthesizer
合成燃料	合成燃料	synfuel, synthetic fuel
合成鞣剂	合成單寧	syntan
合成鞣料	合成鞣料	synthetic tanning materials
合成色素	合成色素	synthetic coloring agent
合成石膏	合成石膏	synthetic gypsum
合成树脂	合成樹脂	synthetic resin
合成树脂胶泥	合成樹脂膠合劑	synthetic resin cement
合成树脂黏合剂	合成樹脂黏合劑	synthetic resin adhesive
合成树脂涂料	合成樹脂清漆	synthetic resin varnish
合成塑料	合成塑膠	synthetic plastic
合成塔	轉化器	converter
合成弹性体	合成彈性體	synthetic elastomer
合成天然气	合成天然氣	synthetic natural gas
合成天然橡胶	合成天然橡膠	synthetic natural rubber
合成涂布材料	合成塗料	synthetic coating materials
合成洗涤剂	合成清潔劑	synthetic detergent
合成纤维	合成纖維	synthetic fiber
合成纤维纸	合成纖維紙	synthetic fiber paper

大　陆　名	台　湾　名	英　文　名
合成橡胶	合成橡膠	synthetic rubber
合成橡胶胶乳	合成橡膠乳膠	synthetic rubber latex
合成橡胶黏合剂	合成橡膠黏合劑	synthetic rubber adhesive
合成羊毛	合成羊毛	synthetic wool
合成油	合成油	synthetic oil
合成元,合成子	合成組元	synthon
合成原油	合成原油	synthetic crude oil, synthetic crude
合成增塑剂	合成塑化劑	synthetic plasticizer
合成脂肪	合成脂肪	synthetic fat
合成脂肪酸	合成脂肪酸	synthetic fatty acid
合成纸	合成紙	synthetic paper
合成子(=合成元)		
合金	合金	alloy
合金钢	合金鋼	alloy steel
合作吸附	合作吸附	cooperative adsorption
核弹	核彈	nuclear bomb
核反应堆	核反應器	nuclear reactor
核辐射	核輻射	nuclear radiation
核苷	核苷	nucleoside
核苷酸	核苷酸	nucleotide
核化学	核化學	nuclear chemistry
核聚变	核熔合	nuclear fusion
核能	核能	nuclear energy
核燃料	核燃料	nuclear fuel
[核燃料]后处理工厂	再處理廠	reprocessing plant
核素	核種	nuclide
核酸	核酸	nucleic acid
核糖	核糖	ribose
核糖核酸	核糖核酸	ribonucleic acid, RNA
核物理学	核子物理學	nuclear physics
荷电膜	荷電膜	charged membrane
褐煤	褐煤	lignite
褐铁矿	褐鐵礦	limonite
褐烟煤	褐煙煤	lignitic bituminite
褐藻酸钠	藻酸鈉	sodium alginate
赫斯定律	赫斯定律	Hess law
黑度	光學密度	optical density
黑灰	黑灰	black ash

大 陆 名	台 湾 名	英 文 名
黑漆	黑漆	japan black
黑素原染料	硫化染料	melanogen dye
黑体	黑體	black body
黑体辐射	黑體輻射	black body radiation
黑体系数	黑體係數	black body coefficient
黑铜矿	黑銅礦	tenorite
黑箱	黑箱	black box
黑箱法	黑箱法	black box method
黑箱模型	黑箱模型	black-box model
黑液	黑液	black liquor
痕量元素	痕量元素,微量元素	trace element
亨利定律	亨利定律	Henry law
恒沸点	共沸點	constant boiling point
恒化器	恆化器	chemostat
恒剪切力黏度计	恆剪力黏度計	constant shear viscometer
恒流率过滤	恆流速過濾	constant flow rate filtration
恒摩尔溢流	恆莫耳溢流	constant molar overflow
恒速泵	恆速泵	constant speed pump
恒速干燥[阶]段	恆速乾燥期	constant rate drying period
恒温烘箱	恆溫烘箱	constant temperature oven
恒温阱	恆溫阱	thermostatic trap
恒温器	恆溫器	thermostat
恒温箱	培養器,孵化器,保溫箱	incubator
恒温浴	恆溫浴,恆溫槽	constant temperature bath
恒压	恆壓	constant pressure
恒压过滤	恆壓過濾	constant pressure filtration
恒组成溶液	恆組成溶液	constant composition solution
恒组分共聚	恆組成共聚[作用]	azeotropic copolymerization
恒组分共聚物	恆組成共聚物	azeotropic copolymer
横晶	橫穿結晶	transcrystallization
横式喷雾室	橫式噴霧室	horizontal spray chamber
横向翅片	橫向鰭片	transverse fin
烘干	烘乾	stoving
烘炉	烘爐	baking furnace
烘箱	烘箱,爐	oven, baking oven
红霉素	紅黴素	erythromycin
红铅	紅鉛,紅丹	red lead
红外分光光度计	紅外線光譜儀	infrared spectrophotometer

大　陆　名	台　湾　名	英　文　名
红外分析	紅外線分析	infrared analysis
红外干燥	紅外線乾燥	infrared drying
红外光谱	紅外線光譜法	infrared spectrophotometry
红外光谱法	紅外線光譜法	infrared spectroscopy
红外线分析仪	紅外線分析儀	infrared analyzer
红外线干燥器	紅外線乾燥器	infrared drier
红锌矿	紅鋅礦	zincite
红柱石	紅柱石	andalusite
宏观动力学	宏觀動力學	macrokinetics
宏观规模行为	巨觀行為	macroscale behavior
宏观混合	巨觀混合	macromixing
宏观可逆性	巨觀可逆性	macroscopic reversibility
宏观流体	巨觀流體	macrofluid
宏观平衡	巨觀均衡	macroscopic balance
虹吸	虹吸	syphon, siphon
虹吸管	虹吸管	siphon
虹吸加油器	虹吸加油器	siphon oiler
虹吸气压表	虹吸氣壓計	siphon barometer
虹吸式密封	虹吸式密封	syphon seal
虹吸压力计	虹吸表	siphon gauge
喉道流速	喉道流速	throat velocity
喉管	喉道	throat
后产物	後產物	after product
后处理	後處理,再處理,再加工	after treatment, reprocessing
后发酵	後發酵	after fermentation
后过滤	後過濾	after filtration
后过滤器	後[過]濾器	after filter
后烘	後焙	postbaking
后净化	後純化	after purification
后硫化	後硫化	after vulcanization, post vulcanization
后期结晶	二次結晶	secondary crystallization
后期燃烧	後燃燒	after burning
后燃装置	後燃裝置	after burner
后收缩	後收縮	after contraction, after shrinkage
后硬化	後硬化	after hardening
后置冷却器	後冷卻器	after cooler
后轴承	後軸承	rear bearing
厚漆	厚漆	paste paint

大 陆 名	台 湾 名	英 文 名
呼吸	呼吸	respiration
呼吸链	呼吸鏈	respiratory chain
呼吸式油罐	通氣槽	breathing tank
弧鞍填料	弧鞍[形]填料	Berl saddle
胡桃油	胡桃油	pecan oil
糊剂	糊	paste
糊精	糊精	dextrin
糊状树脂	糊狀樹脂	paste resin
琥珀	琥珀	amber
互变异构体	互變異構物	tautomer
互变异构[现象]	互變異構現象	tautomerism
互补反馈	互補回饋	complementary feedback
互混	互混	intermixing
互混度	互混度	degree of intermixing
互溶度	互溶度	mutual solubility
互溶泛滥	互溶氾流[法]	miscible flooding
互溶系统	互溶系統	miscible system
互溶性	互溶性	miscibility
护床	護床	guard bed
护罩	護罩	casing
花岗岩	花崗岩	granite
花生油	花生油	peanut oil
滑板式泵	滑葉泵	slide vane pump
滑槽	滑槽	chute
滑动	滑動	slip
滑动角	滑動角	angle of slide
滑动摩擦	滑動摩擦	slip friction
滑动型芯	滑動芯	slide core
滑动因子	滑動因子	slippage factor
滑阀	滑閥	slide valve
滑石	滑石	talc
滑线	滑線	slide wire
滑移带	滑移帶	slip band
滑移速度	滑動速度	slip velocity
滑移系数	滑動因數	slip factor
猾子皮	羔羊皮	kid-skin
化工动力学	化工動力學	chemical engineering kinetics
化工分析	化工分析	chemical engineering analysis

大　陆　名	台　湾　名	英　文　名
化工过程分析	化工程序分析	chemical engineering process analysis
化工技术	化工技術	chemical technology
化工热力学	化工熱力學	chemical engineering thermodynamics
化合热	化合熱	heat of combination
化合物	化合物	compound
化石	化石	fossil
化石燃料	化石燃料	fossil fuel
化学变化	化學變化	chemical change
化学参数	化學參數	chemical parameter
化学操作	化學操作	chemical operation
化学沉淀	化學沈澱[法]	chemical precipitation
化学成键	化學鍵結	chemical bonding
化学澄清器	化學澄清器	chemical clarifier
化学处理	化學處理	chemical treatment
化学处理法	化學處理法	chemical treatment process
化学促进剂	化學促進劑	chemical promoter
化学动力过程	機械化學製漿法	mechanochemical pulping process
化学动力学	化學動力學	chemical kinetics
化学动态学	化學動態學	chemical dynamics
化学镀	無電[電]鍍	electroless plating
化学发光	化學發光	chemiluminescence
化学反应	化學反應	chemical reaction
化学反应工程	化學反應工程	chemical reaction engineering
化学反应平衡	化學反應平衡	chemical reaction equilibrium
化学反应平衡判据	化學反應平衡準則	criterion for chemical reaction equilibrium
化学反应器	化學反應器	chemical reactor
化学反应器稳定性	化學反應器穩定性	chemical reactor stability
化学反应稳定性	化學反應穩定性	chemical reaction stability
化学方程式	化學方程式,化學反應式	chemical equation
化学分离	化學分離	chemical separation
化[学]工厂	化學工廠	chemical plant
化学工程	化學工程	chemical engineering
化学工程建造成本指数	化學工程營建成本指數	chemical engineering construction cost index
化学工程师	化學工程師	chemical engineer
化学工程学	化工科學	chemical engineering science
化学过程	化學程序	chemical process

大　陆　名	台　湾　名	英　文　名
化学[过程]工业	化學程序工業	chemical process industry
化学机械抛光	化學機械研磨	chemical mechanical polishing
化学极化	化學極化	chemical polarization
化学计量	化學計量學	stoichiometry
化学计量混合物	化學計量混合物	stoichiometric mixture
化学计量计算	化學計量計算	stoichiometric calculation
化学计量兼容性	化學計量相容性	stoichiometric compatibility
化学计量数	化學計量數	stoichiometric number
化学计量系数	化學計量係數	stoichiometric coefficient
化学键	化學鍵	chemical bond
化学扩散系数	化學擴散係數	chemical diffusivity
化学密封	化學密封	chemical seal
化学能	化學能	chemical energy
化学凝聚	化學凝聚[作用]	chemical coagulation
化学品	化學品	chemical
化学平衡	化學平衡	chemical equilibrium
化学气相沉积	化學氣相沈積	chemical vapor deposition, CVD
[化学]式	化學式	formula
化学势	化學勢	chemical potential
化学输送	化學輸送	chemical transport
化学调理	化學調理[法]	chemical conditioning
化学文摘	化學摘要	Chemical Abstracts
化学稳定性	化學穩定性,化學安定性	chemical stability
化学物理[学]	化學物理學	chemical physics
化学物种	化學物種	chemical species
化学吸附	化學吸附	chemical adsorption, chemisorption
化学吸收	化學吸收	chemical absorption
化学吸收剂	化學吸收劑	chemical absorbent
化学性质	化學性質	chemical property
化学需氧量	化學需氧量	chemical oxygen demand, COD
化学烟雾	化學煙霧	chemical smoke
化学液泛	化學氾流[法]	chemical flooding
化学原料	化學原料	chemical feedstock
化学振荡	化學振盪	chemical oscillation
化学振荡器	化學振盪器	chemical oscillator
化学蒸汽	化學蒸氣	chemical vapor
化学制浆法	化學製漿法	chemical pulping process

大　陆　名	台　湾　名	英　文　名
槐子油	槐子油	ash-seed oil
还原	還原[反應]	reduction
还原电位	還原電位	reduction potential
还原反应	還原反應	reduction reaction
还原剂	還原劑	reducing agent, reductant
还原区	還原帶	reduction zone
还原染料	甕染料	vat dye
还原性漂白	還原漂白	reductive bleaching
还原性脱卤	還原性脱鹵[反應]	reductive dehalogenation
还原值	還原值	reduction value
环滚粉碎机	環輥粉碎機	ring roll pulverizer
环滚磨机	環輥磨	ring roll mill
环化	環化	cyclization
环化反应	環化反應	cyclization reaction
环化橡胶	環化橡膠	cyclize rubber
环境	環境	environment
环境保护	環境保護	environmental protection
环境成本	環境成本	environmental cost
环境冲击	環境衝擊	environmental impact
环境工程	環境工程	environmental engineering
环境化学	環境化學	environmental chemistry
环境技术	環境技術	environmental technology
环境科学	環境科學	environmental science
环境态	環境態	environmental state
环境温度	環境溫度	ambient temperature, surrounding temperature
环境污染	環境污染	environmental contamination, environmental pollution
环境压力	環境壓力	ambient pressure
环境质量	環境品質	environmental quality
环流反应器	環流反應器,循環反應器	loop reactor, circulating reactor
环流空间供暖	空間加熱	space heating
环路构型	環形組態	loop configuration
环路,回路	環路,回圈	loop
环路应答	環路應答	loop response
环路转移函数	環路轉移函數	loop transfer function
环丝氨酸	環絲胺酸	cycloserine

大　陆　名	台　湾　名	英　文　名
环烷	環烷	naphthene
环隙	環隙	annular space
环向应力	環應力	hoop stress
环形接头	環形接頭	ring joint
环形流压计	測壓環	piezometer ring
环形喷雾器	環狀噴霧器	ring sprayer
环形压强计	環形測壓計,環形壓力計	ring piezometer, ring type manometer
环形研磨机	環形研磨機	ring type grinder
环氧化	環氧化[作用]	epoxidation
环氧氯丙烷	環氧氯丙烷	epichlorohydrin
环氧树脂	環氧樹脂	epoxy resin
环状反应器	環狀反應器	annular reactor
环状管路系统	環狀管路系統	looped pipe system
环状流	環狀流	annular flow
缓冲板	緩衝板	cushion plate
缓冲层	緩衝層	buffer layer
缓冲罐	緩衝槽	buffer tank, surge tank
缓冲辊	緩衝輥	snubber roll
缓冲器	緩衝劑	buffer
缓冲区	緩衝區[域]	buffer region, buffer zone
缓冲容量	緩衝容量	buffer capacity
缓冲溶液	緩衝[溶]液	buffer solution
缓冲指数	緩衝指數	buffer index
缓冲作用	緩衝作用	buffer action
缓慢反应	慢反應	slow reaction
缓蚀剂	抑蝕劑,腐蝕抑制劑	corrosion inhibitor
缓释	緩釋	slow release
缓释胶囊	緩釋膠囊	slow release capsule
缓释控制	緩釋控制	slow release control
缓释器件	緩釋裝置	slow release device
缓释药物	緩釋藥物	slow release drug
换罐混合器	換罐式混合器	change-can mixer
换热	熱交換	heat exchange
换热器,热交换器	熱交換器	heat exchanger
换热器网路	熱交換器網路	heat exchanger network
换算表	換算表	conversion table
换算常数	換算常數	conversion constant

大　陆　名	台　湾　名	英　文　名
换算方程式	換算方程式	conversion equation
换算曲线	轉化曲線	conversion curve
换算因子	換算因數	conversion factor
黄饼	黄餅	yellow cake
黄丹	黄丹	yellow lead
黄金分割法	黄金分割搜尋法	golden section searching method
黄芪胶	黄芪膠	tragacanth gum
黄铁矿	黄鐵礦	pyrite
黄铜矿	黄銅礦	chalcopyrite
黄原酸化[反应]	黄酸化[作用]	xanthation
黄原酸纤维素	黄原酸纖維素	cellulose xanthate
磺化	磺[酸]化[反應]	sulfonation
磺化剂	磺[酸]化劑	sulfonating agent
磺化器	磺[酸]化器	sulfonator
磺化油	磺[酸]化油	sulfonated oil
磺酸型离子交联聚合物	磺酸離子聚合物	sulphonic acid ionomer
磺酸盐	磺酸鹽	sulfonate
灰分	灰分	ash
灰分分析	灰分分析	ash analysis
灰分扩散控制	灰分擴散控制	ash diffusion control
灰面	灰面	gray surface
灰区	灰區	ash zone
灰色系统	灰色系統	gray system
灰体	灰體	grey body
恢复时间	回收時間	recovery time
挥发	揮發[作用]	volatilization
挥发度	揮發度	volatility
挥发性悬浮固体	揮發性懸浮固體	volatile suspended solid
辉铜矿(=铜铀云母)	銅鈾雲母	chalcolite
回归	回歸	regression
回归方程	回歸方程式	regression equation
回归分析	回歸分析	regression analysis
回归系数	回歸係數	regression coefficient
回归线	回歸線	regression line
回火	回火,逆火	back fire, tempering
回火空气	回火空氣	tempering air
回流	回流,逆流	backflow, reflux
回流比	回流比	reflux ratio

大　陆　名	台　湾　名	英　文　名
回流操作	回流操作	reflux operation
回流槽	回流槽	reflux drum
回流管	回流管	reflux tube
回流罐	回流貯槽	reflux accumulator
回流阱	回流阱	return trap
回流冷凝器	回流冷凝器	reflux condenser
回流塔	回流塔	reflux column
回流蒸馏	回流蒸餾	reflux distillation
回路(=环路)		
回收	回收	reclamation, regain
回收槽	回收槽	reclaiming tank
回收[率]	回收	recovery
回收率曲线	回收曲線	recovery curve
回收石膏	再生石膏	reclaimed gypsum
回收塔	回收塔	recovery tower
回收系数	回收因數	recovery factor
回收系统	回收系統	recovery system
回收装置	回收裝置	reclaiming unit
回填机	回填機	backfiller
回填(=再装满)		
回弯头	回彎管	return bend
回转焙烧炉	旋轉焙燒爐	rotary roaster
回转泵	旋轉泵	rotary pump
回转干燥器	旋轉[式]乾燥器	rotary dryer
回转鼓风机	旋轉鼓風機	rotary blower
回转流量计	回轉流量計	gyroscopic flowmeter
回转破碎机	偏旋壓碎機	gyratory crusher
回转筛	迴轉篩	revolving screen
回转压缩机	旋轉壓縮機	rotary compressor
回转窑	旋[轉]窯	rotary kiln
回转轧碎机	偏旋破碎機	gyratory break
回转振动筛	偏旋篩	gyrating screen
汇	匯座	sink
汇点	匯座	sink
汇壑点	匯座點	sink point
[绘]图	圖	plot
绘图仪	製圖儀,繪圖儀	drawing apparatus
绘图纸	繪圖紙	drawing paper

大　陆　名	台　湾　名	英　文　名
桧树胶	檜樹膠	sandarac
混合	混合	mixing
混合杯温	混合杯溫	mixing cup temperature
混合槽	混合槽	mixing tank
混合长	混合長度	mixing length
混合澄清槽	混合沈降槽	mixer-settler
混合催化剂	混合觸媒	mixed catalyst
混合度	混合度	degree of mixing
混合反应器	混合反應器	mixed reactor
混合肥料	混合肥料	mixed fertilizer
混合罐	混合桶	mixing drum
混合规则	混合法則	mixing rule
混合计	混合汁	mixed juice
混合计算机	併合計算機	hybrid computer
混合桨叶	混合槳葉	mixing paddle
混合搅拌器	混合攪拌機	mixing agitator
混合节点	混合節點	mixed node
混合进料	混合進料	mixed feeding
混合流动反应器	混合流動反應器	mixed flow reactor
混合能力	混合能力	mixing capacity
混合器功率	混合器功率	mixer power
混合强度	混合強度	mixing intensity
混合热	混合熱	heat of mixing
混合深度	混合深度	mixing depth
混合生产	混合生產	mixed production
混合时间	混合時間,攪拌時間	mixing time
混合室	混合室	mixing chamber
混合速率	混合速率	mixing rate
混合酸	混合酸	mixed acid
混合温度	混合溫度	mixing temperature
混合物	混合物	mixture
混合雾沫	混合霧沫	mixing entrainment
混合型式	混合型式	mixing pattern
混合序列	混合階	mixed order
混合悬浮	混合懸浮	mixed suspension
混合液	混合液	mixed liquor
混合液挥发性悬浮固体	混合液揮發性懸浮固體	mixed-liquor volatile suspended solid
混合液悬浮固体	混合液懸浮固體	mixed-liquor suspended solid

大　陆　名	台　湾　名	英　文　名
混合引管	混合導管	mixing draft tube
混合用挡板	混合用擋板	mixing baffle
混合整数非线性规划	混合整數非線性規劃	mixed integer nonlinear programming, MINLP
混合指数	混合指數	mixing index
混合柱	混合[管]柱	mixing column
混合装置	混合裝置	mixing device
混晶	混晶	mixed crystal
混流	混合流動	mixed flow
混流泵	混流泵	mixed flow pump
混流器	混合器,混合機	mixer
混凝土	混凝土	concrete
混糖	黃砂糖	muscovado sugar
活度	活度	activity
活度系数	活性係數	activity coefficient
活度系数通用准化功能团贡献法	UNIFAC 法	universal quasi-chemical functional group activity coefficient method, UNIFAC method
活度系数通用准化学关联法	UNIQUAC 法	universal quasi-chemical activity coefficient method, UNIQUAC method
活[管]接头	活[管]接頭	union
活化分析	活化分析	activation analysis
活化分子	活化分子	activated molecule
活化硅石	活性矽石	activated silica
活化焓	活化焓	enthalpy of activation
活化化学吸附	活化化學吸附	activated chemisorption
活化剂	活化劑	activator
活化截面	活化截面	activation cross section
活化络合物	活化錯合物,活化複合體	activated complex
活化能	活化能	activation energy
活化熵	活化熵	entropy of activation
活化体积	活化體積	activation volume
活化[作用]	活化[作用]	activation
活塞	活塞	piston
活塞泵	活塞泵	piston pump
活塞阀	活塞閥	piston valve
活塞流反应器	塞流反應器	piston flow reactor, plug flow reactor

大　陆　名	台　湾　名	英　文　名
活塞流量计	活塞流量計	piston meter
活塞流(=平推流)		
活塞式压力计	活塞壓力計	piston gauge
活塞执行机构	活塞致動器	piston actuator
活性	活性	activity
活性表面积	活性表面積	active surface area
活性[部]位	活性[部]位	active site
活性部位理论	活性[部]位理論	active-site theory
活性催化剂	活性觸媒	active catalyst
活性催化中心	觸媒活性中心	active catalytic center
活性点	活性點	active point
活性矾土	活性礬土	activated alumina, active alumina
活性分布	活性分佈	activity distribution
活性干酵母	活性乾酵母	active dry yeast
活性固体	活性固體	active solid
活性剂	活性劑	active agent
活性碱	活性鹼	active alkali
活性焦	活化[焦]炭	activated char
活性校正因子	活性校正因子	activity correction factor
活性硫	活性硫	active sulfur
活性络合物	活性錯合物	active complex
活性黏土	活性黏土	active clay
活性生物过滤	活性生物過濾	activated biological filtration
活性衰变	活性衰變,活性衰退	activity decay
活性炭	活性碳	activated carbon, active carbon
[活性]炭过滤器	碳濾器	carbon filter
活性炭塔	活性碳塔	activated carbon column
活性污泥	活性污泥	activated sludge
活性污泥处理场	活性污泥處理場	activated sludge plant
活性污泥法	活性污泥法	activated sludge process
活性吸附	活化吸附	activated adsorption
活性氧	活性氧	active oxygen
活性氧化铝	活性氧化鋁	activated aluminum oxide
活性因子	活性因子,活性因數	activity factor
活性指数	活性指數	activity index
活性中心	活性中心	active center
活性中心络合物	活性中心錯合物	active-center complex
火成石英	火成石英	igneous quartz

大　陆　名	台　湾　名	英　文　名
火成岩	火成岩	igneous rock
火法冶金过程	火法冶金法, 高溫冶金法	pyrometallurgical process
火花电蚀	電火花腐蝕	spark erosion
火花放电	火花放電	spark discharge
火花放电检测器	火花放電偵檢器	spark discharge detector
火箭	火箭	rocket
火箭发动机	火箭引擎	rocket engine
火箭燃料	火箭燃料	rocket fuel
火漆	火漆, 封蠟	sealing wax
火山灰	火山灰	pozzolana
火山灰水泥	火山灰水泥	puzzolana cement
火石玻璃	燧石玻璃	flint glass
火焰	火焰	flame
火焰反应器	火焰反應器	flame reactor
火焰分光仪	火焰分光計	flame spectrometer
火焰温度	火焰溫度	flame temperature
火焰稳定性	火焰穩定性	flame stability
火灾及爆炸指数	火災及爆炸指數	fire and explosion index
和面机	和麵機	dough mixer

J

大　陆　名	台　湾　名	英　文　名
机壳	護罩	casing
机理结构	機制結構	mechanism structure
机理缺陷	機構缺陷	mechanism deficiency
机器人	機器人	robot
机械镀	機械鍍	mechanical plating
机械分离	機械分離	mechanical separation
机械分离器	機械分離器	mechanical separator
机械浮选池	機械浮選池	mechanical flotation cell
机械功	機械功	mechanical work
机械过滤器	機械過濾器	mechanical filter
机械搅拌	機械攪拌	mechanical agitation
机械离心分离器	機械離心分離器	mechanical centrifugal separator
机械密封	機械密封	mechanical seal
机械能	機械能	mechanical energy

大　陆　名	台　湾　名	英　文　名
机械能方程式	機械能方程式	mechanical energy equation
机械能量平衡	機械能均衡	mechanical energy balance
机械破碎	機械粉碎	mechanical disintegration
机械设计	機械設計	mechanical design
机械式控制器	機械式控制器	mechanical controller
机械损失	機械損失	mechanical loss
机械通风	機械通風	mechanical draft
机械通风冷却塔	機械通風冷卻塔	mechanical draft cooling tower
机械-物理分离过程	機械-物理分離程序	mechanical-physical separation process
机械学方程式	機理方程式	mechanistic equation
机械造型	機械成型	machine molding
机械自动化	機械自動化	mechanical automation
奇点	奇點	singular point
奇偶性	奇偶性	parity
积分	積分[式]	integral
积分饱卷	積分飽和	integral windup
积分动量方程	積分動量方程式	integral momentum equation
积分法	積分法	integral method
积分反馈	積分回饋	integral feedback
积分反应器	整體反應器	integral reactor
积分分布函数	累積分佈曲線	integral distribution curve
积分检验[法]	積分檢驗[法]	integral test
积分控制	積分控制	integral control, I control
积分控制作用	積分控制作用	integral control action
积分能量方程	積分能量方程式	integral energy equation
积分溶解热	積分溶解熱	integral heat of solution
积分时间	積分時間	integral time
积分时间常数	積分時間常數	integral time constant
积分仪	積分器	integrator
积分作用	積分作用	integral action
基本单位	基本單位	fundamental unit
基本性质关系式	基本性質關係[式]	fundamental property relation
基本原理	基本原理	fundamental principle
基尔霍夫定律	克希何夫定律	Kirchhoff law
基体	基體	matrix
基团分率	基團分率	group fraction
基团[贡献]解析法	基團[貢獻]解析法	analytical solution of group contribution method

大　陆　名	台　湾　名	英　文　名
基团活度系数	基團活性係數	group activity coefficient
基团专一性	基特異性	group specificity
基因	基因	gene
基因改造细胞	基因改造細胞	genetically modified cell
基因工程	基因工程	genetic engineering
基因工程细胞	基因[工程]改造細胞	genetically engineered cell
基元反应	基本反應	elementary reaction
基质	鹼基	base
基质专一性	受質專一性	substrate specificity
基准	基準	datum
基准位超越	基準位超越	base-level override
基准位共振	基準位共振	base-level resonance
基准温度	基準溫度	datum temperature
基准物	標準物質	standard substance
激发态	激態	excited state
激酶	激酶	kinase
吉布斯-杜安方程	吉布斯-杜安方程	Gibbs-Duhem equation
吉布斯自由能,自由焓	吉布斯自由能,自由焓	Gibbs free energy
级间加热	階間加熱	interstage heating
级间冷却	階間冷卻	interstage cooling
级间循环	階間循環	interstage recirculation
级间注射进料	階間[注射]進料	interstage feed injection
级效率	階效率	stage efficiency
极大值原理	極大原理	maximum principle
极化	極化	polarization
极化率	極化率,極化性	polarizability
极化因子	偏光因子	polarization factor
极谱法	極譜法	polarographic method, polarography
极谱仪	極譜儀	polarograph
极限电流	極限電流	limit current
极限环	極限循環	limit cycle
极限回流条件	極限回流條件	limiting reflux condition
极限灵敏度	極限靈敏度	ultimate sensitivity
极限黏度	極限黏度	limiting viscosity
极限批量	極限批式產能	limiting batch size
极限因子	限制因素	limiting factor
极限应力	極限應力	ultimate stress
极限周期	極限周期	limiting cycle time

大 陆 名	台 湾 名	英 文 名
极性	極性	polarity
极性溶剂	極性溶劑	polar solvent
极坐标	極坐標	polar coordinates
极坐标图	極坐標圖	polar plot
急冷系统	［已］抑止系統	quenched system
急骤干燥器	驟乾器	flash drier
集尘	集塵	dust collection
集尘极	收集電極	collector electrode
集尘器	集塵器	dust collector
集成	積分	integration
集成过程	整合程序	integrated process
集管	集管［箱］	header
集散控制系统	分散控制系統	distributed control system
集雾器	集霧器	mist collector
集总参数模型	集總參數模式	lumped parameter model
集总参数系统	集總參數系統	lumped parameter system
集总动力学	集總動力學	lumping kinetics
集总系统	集總系統	lumped system
几何表面积	幾何表面積	geometric surface area
几何规划	幾何規畫	geometric programming
几何平均	幾何平均	geometric mean
几何平均面积	幾何平均面積	geometric mean area
几何平均温差	幾何平均溫差	geometric mean temperature difference
几何相似	幾何相似性	geometric similarity
几何异构物	幾何異構物	geometric isomer
几何因数	幾何因數	geometric factor
己内酰胺	己內醯胺	caprolactam
挤出	擠壓	extrusion
挤出机	擠壓機	extruder
挤塑机	擠壓機	extruder
挤压泵	擠壓泵	squeeze pump
给水	給水	feed water
给水槽	給水井,進料井	feed well
给水处理	給水處理	feed water treatment
麂皮	麂皮	chamber leather
计	計	gauge
pH 计	pH 計,酸度計	pH meter
计划评审法	計畫評審法	program evaluation and review technique

大　陆　名	台　湾　名	英　文　名
计量	計量	metering
计量泵	計量泵	metering pump
计时器	計時器,定時器	timer
计数管	計數[器]管	counter tube
计数器	計數器	counter
计数器范围	計數器範圍	counter range
计算	計算	calculation, computation
计算机程序	電腦程式	computer program
计算机辅助工程	電腦輔助工程	computer aided engineering, CAE
计算机辅助过程设计	電腦輔助程序設計	computer-aided process design, CAPD
计算机辅助教学	電腦輔助教學	computer aided instruction, CAI
计算机辅助设计	電腦輔助設計	computer aided design, CAD
计算机辅助制造	電腦輔助製造	computer aided manufacturing, CAM
计算机控制	計算機控制	computer control
计算机模拟	電腦模擬	computer simulation
计算器	計算器	calculator
记号	記號,記法	notation
记录材料	記錄材料	recording materials
记录控制器	記錄控制器	recorder-controller
记录设备	記錄裝置	recording equipment
记录式气压计	記錄氣壓計	recording barometer
记录图表	記錄圖表	recording chart
记录仪	記錄器	recorder
记忆函数	記憶函數	memory function
记忆流体	記憶流體	memory fluid
记忆衰退材料	記憶衰退材料	fading memory materials
记忆细胞	記憶細胞	memory cell
技术	技術	technology
技术服务	技術服務	technical service
技术评估	技術評估	technical evaluation
技术转移	技術轉移	technology transfer
剂量	劑量	dosage, dose
迹线	徑線,路線	path line
继电控制	開關式控制	bang-bang control
继电器	繼電器	relay
继动阀	繼動閥	relay valve
寄生物	寄生物,寄生蟲	parasite
加成定律	加成律	additive law

大　陆　名	台　湾　名	英　文　名
加成反应	加成反應	addition reaction
加成聚合	加成聚合[作用]	addition polymerization
加成脱色	加成脱色	additive decolorization
加成性	加成性	additive property
加工	加工, 處理	processing
加工变量	加工變數	processing variable
加工工程	加工工程	processing engineering
加工工业	加工業	processing industry
加工技术	加工技術	processing technology
加合结晶	締合結晶	adductive crystallization
加合物	加成物	adduct
加酵母槽	種母槽[釀酒]	pitching tank
加晶种	播晶種	seeding
加料槽	進料槽	feed tank
加料斗	加料斗	charge hopper
加料机	加料機	charger
加料盘	加料盤	pan feeder
加料升降机	加料升降機	charge elevator
加料速度	進料速率	feed rate
加料运输机	加料運送機	charge conveyor
加氢	氫化	hydrogenation
加氢重整	加氢重組	hydroreforming
加氢重整催化剂	加氢重組觸媒	hydroreforming catalyst
加氢处理	加氫處理	hydrotreating, hydrotreatment
加氢处理装置	加氫處理器	hydrotreater
加氢裂化	加氫裂解[反應]	hydrocracking
加氢裂化单元	加氫裂解單元, 加氫裂解工場	hydrocracking unit
加氢裂化反应器	加氫裂解工場, 加氫裂解爐	hydrocracker
加氢气化	加氫氣化	hydrogasification
加氢气化器	加氫氣化器	hydrogasifier
加氢热解	加氫高溫裂解	hydropyrolysis
加氢脱氮	加氫脱氮	hydrodenitrogenation
加氢脱金属催化剂	加氫脱金屬觸媒	hydrodemetallization catalyst
加氢脱硫	加氫脱硫	hydrodesulfurization
加氢脱烷基化	加氫脱烷[反應]	hydrodealkylation
加氢液化	加氫液化	hydroliquefaction

大　陆　名	台　湾　名	英　文　名
加权残差	加權殘差	weighted residual
加权函数	加權函數	weighting function
加权平均[值]	加權平均[值]	weighted mean
加权温差	加權溫差	weighted temperature difference
加权因子	加權因數	weighting factor
加热	加熱	heating
加热成本	加熱成本	heating cost
加热成型	加熱成型	hot molding
加热单元	加熱單元,加熱裝置	heating unit
加热面积	加熱面積	heating area
加热盘管	加熱旋管	heating coil
加热器	加熱器,加熱爐	heater
加热时间	加熱時間	heating time
加热再生	熱再生	heat regeneration
加水裂化过程	加水裂化過程	aquolization process
加速度计	加速計	accelerometer
加速度误差系数	加速誤差係數	acceleration error coefficient
加速剂	加速劑	accelerator
加速老化	加速老化	accelerated aging
加速器	加速器	accelerator
加速速率	加熱速率	heating rate
加压分馏	加壓分餾	pressure fractionation
加压浮选	加壓浮選[法]	pressure flotation
加压过滤	加壓過濾	pressure filtration
加压馏出物	加壓餾出物	pressure distillate
加压砂滤器	加壓砂濾器	pressure sand filter
加压筛	壓力篩	pressure screen
加压氧化	加壓氧化[反應]	pressure oxidation
加压叶滤机	加壓葉濾機	pressure leaf filter
加压蒸馏	加壓蒸餾	pressure distillation
加压蒸煮器	加壓蒸煮器,壓力鍋	pressure cooker
夹层安全玻璃	積層安全玻璃	laminated safety glass
夹层结构	夾層結構	sandwich structure
夹带床反应器	挾帶床反應器	entrained bed reactor
夹带剂	挾帶劑	entrainer
夹带剂泵	挾帶劑泵	entrainer pump
夹带速度	[霧沫]挾帶速度	entrainment velocity
夹带物分离器	[霧沫]挾帶阱	entrainment trap

大　陆　名	台　湾　名	英　文　名
夹点	夾點	pinch point
夹点技术	夾點技術	pinch technology
夹角	挾角	angle of nip
夹丝玻璃	夾網玻璃	wire glass
夹套	夾套	jacket
夹套壁	夾套壁	jacketed wall
夹套锅	[夾]套鍋	jacketed kettle
夹套加热	夾套加熱	jacket heating
夹套加压釜	夾套高壓釜	jacketed autoclave
夹套结晶器	夾套結晶器	jacketed crystallizer
夹套蒸发器	夾套蒸發器	jacketed evaporator
夹套蒸馏器	夾套蒸餾器	jacketed still
甲醇	甲醇	methanol, methyl alcohol
甲酚	甲酚	cresol
甲硅烷醇	矽醇	silanol
甲基丙烯酸甲酯-丙烯腈-丁二烯-苯乙烯共聚物	甲基丙烯酸甲酯-丙烯腈-丁二烯-苯乙烯共聚物	methylmethacrylate-acrylonitrile-butadiene-styrene copolymer, MABS
甲基橙	甲基橙	methyl orange
甲基对硫磷	甲[基]巴拉松	methyl parathion
甲基红	甲基紅	methyl red
甲基化剂	甲基化劑	methylating agent
甲基化作用	甲基化[作用]	methylation
甲基黄	甲基黃	methyl yellow
甲基纤维素	甲基纖維素	methyl cellulose
甲基橡胶	甲基橡膠	methyl rubber
甲基转移作用	轉甲基化[作用]	transmethylation
甲基紫	甲基紫	methyl violet
甲烷富气	甲烷富氣	methane rich gas
甲烷化	甲烷化[法]	methanation
甲烷化反应	甲烷化反應	methanation reaction
甲烷化反应器	甲烷化器	methanator
甲烷菌	甲烷菌	methane bacteria
甲烷贫气	甲烷貧氣	methane lean gas
甲烷水合物	甲烷水合物	methane hydrate
甲烷细菌	甲烷菌	methane-forming bacteria
甲酰胺	甲醯胺	formamide
甲酰胺法	甲醯胺法	formamide process

大　陆　名	台　湾　名	英　文　名
钾玻璃	鉀玻璃	potash glass
钾长石	鉀長石	potash feldspar
钾肥	鉀肥	potash fertilizer
钾钙玻璃	鉀鈣玻璃	potash-lime glass
钾铅玻璃	鉀鉛玻璃	potash-lead glass
钾盐	鉀鹽	kali salt
钾盐镁矾	鉀鎂礬,鉀瀉鹽	kainite
钾皂	鉀皂	potash soap
假动力学	擬動力學	pseudokinetics
假临界常数	擬臨界常數	pseudocritical constant
假临界点法	擬臨界點法	pseudocritical method
假临界温度	擬臨界溫度	pseudocritical temperature
假临界压力	擬臨界壓[力]	pseudocritical pressure
假说	假說	hypothesis
假塑性	假塑性,擬塑性	pseudo plasticity
假塑性流体	擬塑性流體	pseudo plastic fluid
假稳态	擬穩態	pseudo steady state
兼性呼吸	兼性呼吸	facultative respiration
兼性污水塘	兼性污水塘	facultative lagoon
兼性细菌	兼性菌	facultative bacteria
兼性需氧细菌	兼性需氣菌	facultative aerobic bacteria
监测站	監測站	monitoring station
监督	監督	supervision
监督控制	監督控制	supervisory control
监视	監測	monitoring
监视器	監測器	monitor
减活化能	去活化能	deactivation energy
减量	①减量 ②降低	abatement
减黏剂,稀释剂	減黏劑,稀釋劑	thinner
减黏裂化	減黏裂煉	visbreaking
减湿	除濕	dehumidification
减速齿轮	減速齒輪	reducing gear
减速力	減速力	retarding force
减速扭矩	減速扭矩	retarding torque
减压阀	減壓閥	pressure reducing valve, reducing valve
减压器	減壓器	decompressor
减振器	緩衝輥	snubber roll
减阻	拖曳效應	drag effect

大　陆　名	台　湾　名	英　文　名
剪切	剪切,切變	shear, shearing
剪切变形	剪切變形	shear deformation
剪切储能模量	剪切儲存模數	shear storage modulus
剪切带	剪切帶	shear band, shear banding
剪切刚度	剪切勁度	shearing stiffness
剪切力	剪[切]力	shear force
剪切率	剪[切速]率	shear rate
剪切模量	剪切模數,切變模數	shear modulus
剪切挠曲	剪切撓曲	shear flexure
剪切黏度	剪切黏度	shear viscosity
剪切黏滞系数	切變黏滯係數	shear viscosity coefficient
剪切强度	剪切強度	shearing strength, shear strength
剪切柔量	剪切柔量,剪力順性	shear compliance
剪切蠕变	剪切潛變	shear creep
剪切试验	剪切試驗	shear test
剪切松弛	剪切鬆弛	shear relaxation
剪切速度	剪速度	shear velocity
剪切稀化流体	剪切減黏流體	shear-thinning fluid
剪切应变	剪應變	shear strain
剪切圆盘式黏度计	剪切盤式黏度計	shearing disc viscometer
剪切增稠流体	剪切增黏流體	shear-thickening fluid
剪切致稠	剪切增黏	shear thickening
剪切致稀	剪切減黏	shear thinning
剪应力	剪[切]應力	shear stress
检测器	偵檢器	detector
检错	除錯	debug
检流计	電流計	galvanometer
检漏器	測漏器	leak detector
简单反应	簡單反應	simple reaction
简单剪流动	簡單剪切流	simple shear flow
简单剪切	簡單剪切	simple shear
简单剪切形变	簡單剪切變形	simple shear deformation
简单[壳式]蒸馏釜	[殼式]蒸餾釜	shell still
简单立方晶格	簡單立方晶格	simple cubic lattice
简单流体	簡單流體	simple fluid
简单伸长	簡單伸長	simple elongation
简捷法	簡捷法	shortcut method
简捷设计法	簡捷設計法	shortcut design method

大 陆 名	台 湾 名	英 文 名
简式黄原酸化机	簡式黃[原酸]化機	simple xanthating machine
碱	鹼	alkali, base
碱测定法	鹼定量法	alkallimetry
碱催化	鹼催化	base catalysis
碱度	鹼度	alkalinity
碱法	鹼法	alkali process
碱法制浆	鹼式製漿	alkaline pulping
碱量计	鹼計	alkalimeter
碱木质素	鹼[溶]木質素	alkali lignin
碱石灰	鹼石灰	soda lime
碱纤维素	鹼纖維素, 鹼化纖維素	alkali cellulose, soda cellulose
碱性催化剂	鹼[性]觸媒	basic catalyst
碱性固体	鹼性固體	basic solid
碱性硅酸盐	鹼矽酸鹽	alkali silicate
碱性耐火砖	鹼性耐火磚	basic firebrick
碱性染料	鹼性染料	basic dye
碱液	鹼水	lye
碱液泛	鹼液氾流[法]	alkaline flooding
碱釉	鹼性釉	alkaline glaze
间隔	間距	spacing
间接成本	間接成本	indirect cost
间接法	間接法	indirect method
间接隔膜阀	間接隔膜閥	indirect diaphragm valve
间接加热	間接加熱	indirect heating
间接氢化	間接氫化	indirect hydrogenation
间接热交换	間接熱交換	indirect heat exchange
间接式冷凝器	間接[式]冷凝器	indirect condenser
间接液化	間接液化	indirect liquefaction
间接液位测量	間接液位量測	indirect liquid level measurement
间接蒸发	間接蒸發	indirect evaporation
间同立构单元	對排立構單元	syndiotactic unit
间同立构规整度	對排立構度	syndiotacticity
间同立构键接	對排立構鍵接	syndiotactic placement
间同立构聚合	對排立構聚合[作用]	syndiotactic polymerization
间同立构聚合物	對排立構聚合物	syndiotactic polymer
间同立构序列	對排立構序列	syndiotactic sequence
间隙	間隙	clearance
间隙测量计	間隙計	clearance meter

大　陆　名	台　湾　名	英　文　名
间隙体积	間隙體積	clearance volume
间歇操作	批式操作	batch operation
间歇沉降	批式沈降	batch settling
间歇萃取	批式萃取	batch extraction
间歇萃取器	批式萃取器	batch extractor
间歇反应器	批式反應器	batch reactor
间歇分馏	批式分餾	batch fractionation
间歇釜式蒸馏	批式蒸餾	batch still distillation
间歇干燥器	批式乾燥器	batch drier
间歇干燥室	間歇乾燥室	intermittent drier
间歇过程	批式程序	batch process
间歇过滤	批式過濾	batch filtration
间歇过滤器	批式過濾機	batch filter
间歇混合器	批式混合器	batch mixer
间歇加工	批式加工[法]	batch processing
间歇加料机	批式加料機	batch charger
间歇焦化蒸馏器	批式焦化蒸餾器	batch coke still
间歇搅拌器	批式攪拌器	batch agitator
间歇结晶	批式結晶	batch crystallization
间歇结晶器	批式結晶器	batch crystallizer
间歇进料器	批式進料器	batch feeder
间歇精馏	批式精餾	batch rectification
间歇净化	批式純化	batch purification
间歇控制	批式控制	batch control
间歇离心机	批式離心機	batch centrifuge
间歇汽化	批式汽化	batch vaporization
间歇生产	批式生產	batch production
间歇系统	批式系統	batch system
间歇窑	間歇窯,週期窯	periodic kiln
间歇蒸馏	批式蒸餾	batch distillation
间歇蒸馏器	批式蒸餾器	batch still
间周立构加成	對排立構加成[反應]	syndiotactic addition
建模	建模	modeling
建筑材料	營建材料	materials of construction
渐减曝气	漸減通氣,漸減曝氣	tapered aeration
渐近解	漸近解	asymptotic solution
渐近稳定区	漸近穩定區	region of asymptotic stability
渐近稳定性	漸近穩定性	asymptotic stability

大　陆　名	台　湾　名	英　文　名
渐近稳定性联合	漸近穩定聯合域	union of asymptotic stability
渐近线	漸近線	asymptote
渐近线图	漸近線圖	asymptotic plot
渐近线中心	漸近線中心	center of asymptote
渐进转化模型	漸進轉化模型	progressive conversion model
渐缩管	漸縮管	convergent tube
溅射	濺射,濺鍍	sputtering
溅射沉积	濺鍍[法]	sputtering deposition
溅射蚀刻	濺射蝕刻[法]	sputtering etching
鉴定	鑑定,鑑別	identification
键	鍵	bond
键角	鍵角	bond angle
键结力	鍵結力	bonding force
键能	鍵能	bond energy
箭头[流程]图	箭頭[流程]圖	arrow diagram
浆糊	漿糊,澱粉糊	starch paste
浆尖速度	梢速	tip speed
浆料	漿料,上漿劑	slurry, sizing agent
浆料反应	漿料反應	slurry reaction
浆料反应动力学	漿料反應動力學	slurry reaction kinetics
浆料反应器	漿料反應器	slurry reactor
浆料配管系统	漿料配管系統	slurry piping system
桨式混合器	槳式混合器	paddle mixer
桨式搅拌器	槳式攪拌器	paddle agitator, paddle stirrer
桨式絮凝器	槳式絮凝器	paddle flocculator
桨式叶轮	槳式葉輪	paddle impeller
降阶模型	降階模式	reduced model
降解	降解,裂解	degradation
降膜	降膜,落膜	falling film
降膜吸收器	降膜式吸收器	falling film type absorber
降膜蒸发器	降膜[式]蒸發器	falling-film evaporator
降速干燥[阶]段	減速乾燥期	falling rate drying period
降速期	減速期	falling rate period
降液	降流	downflow
降液管	降流管	downcomer, down-pipe
降液管液柱高度	降液管液柱高度	downcomer backup
交叉式构象	相錯式	staggered form
交错管排	交錯管	staggered tubes

大　陆　名	台　湾　名	英　文　名
交迭	交越	crossover
交迭频率	交越頻率	crossover frequency
交迭增益	交越增益	crossover gain
交换剂	交換劑	exchanger
交换能	互換能	exchange energy
交换器	交換器	exchanger
交换吸附	交換吸附	exchange adsorption
交联	交聯	cross linkage, cross-linking
交联度	交聯度	degree of cross-linking
交联剂	交聯劑	cross-linking agent
交联聚合物	交聯聚合物	cross-linked polymer
交联密度	交聯密度	cross-linking density
交联葡聚糖[凝胶]	交聯葡凝膠	sephadex gel
交替	交替[现象]	alternation
交替操作	交替操作	alternative operation
交替操作塔	交替操作塔	alternate operating column
交替共聚合	交替共聚合[反應]	alternating copolymerization
交替共聚物	交替共聚物	alternative copolymer
交替投资	交替投資	alternative investment
交替系统	交替系統	alternating system
浇道	澆道	runner
浇道套	[壓鑄]澆口套筒	sprue bush
浇注釉	澆釉	pouring glazing
浇铸	鑄造	casting
胶	膠	glue
胶版印刷纸	平板印紙	offset paper
胶带	膠帶	adhesive tape
胶管	橡皮軟管	rubber hose
胶合板	合板	plywood
胶化	膠體化	colloidization
胶结	膠接[作用]	cementation
胶结过程	膠接程序	cementation process
胶囊	膠囊	capsule
胶囊化	膠囊封裝	encapsulation
胶黏剂	上漿劑	sizing agent
胶凝剂	膠凝劑	gelatinizing agent
胶凝作用	膠凝[作用],膠化[作用]	gelatination, gelation

大　陆　名	台　湾　名	英　文　名
胶清橡胶	膠清橡膠	skim rubber
胶溶	解膠[作用]	peptization
胶溶剂	解膠劑	peptizing agent
胶乳	乳膠	latex
胶束	微胞	micelle
胶束催化	微胞催化[作用]	micellar catalysis
胶束化	微胞化	micellization
胶束驱油	微胞氾流[法]	micellar flooding
胶态白土	膠質黏土	colloidal clay
胶态分散	膠態分散	colloidal dispersion
胶体	膠體	colloid
胶体包埋	膠體包埋	gel entrapment
胶体铂	膠態鉑	colloidal platinum
胶体沉淀	膠體沈澱	colloidal precipitation
胶体化学	膠體化學	colloid chemistry
胶体粒子	膠體粒子	colloidal particle
胶体磨	膠體磨機	colloidal mill
胶体溶胶	膠體溶液	colloid solution
胶体溶液	膠體溶液	colloidal solution
胶体稳定化	膠體穩定化[法]	colloid stabilization
胶体现象	膠體現象	colloidal phenomenon
胶体悬浮液	膠體懸浮液	colloidal suspension
胶体状态	膠態	colloidal state
胶原	膠原蛋白	collagen
胶质炸药	膠質炸藥	gelatin dynamite
焦点	焦點	focus
焦[耳]	焦耳	joule
焦耳-汤姆孙系数	焦[耳]–湯[姆森]係數	Joule-Thomson coefficient
焦耳-汤姆孙效应	焦[耳]–湯[姆森]效應	Joule-Thomson effect
焦化反应	焦化反應	coking reaction
焦化器	煉焦器	coker
焦化蒸馏器	焦化蒸餾器	coke still
焦炉煤气	焦爐氣	coke oven gas
焦炉油	焦爐溚	coke oven tar
焦烧	焦化[現象]	scorching
焦烧试验	焦化試驗	scorching test
焦炭	焦炭	coke
焦炭孔隙度	焦炭孔隙度	coke porosity

大 陆 名	台 湾 名	英 文 名
焦糖	焦糖	caramel
焦油	焦油,潜,潜油	tar, tar oil
焦油馏出液	潜馏出物,焦油馏出物	tar distillate
焦油砂(=沥青砂)		
焦油酸	焦油酸,潜酸	tar acid
焦值	焦值	coke number
角冲量	角衝量	angular impulse
角蛋白	角蛋白	keratin
角动量	角動量	angular momentum
角动量守恒定律	角動量守恆定律	conservation law for angular momentum
角阀	角閥	angle valve
角畸变	角畸變	angular distortion
角频率	角頻率	angular frequency
角铅矿	角鉛礦	phosgenite
角散射	角散射	angular scattering pattern
角速度	角速度	angular velocity
角系数	角度因數	angle factor
角[向]变形	角[度]形變	angular deformation
角锥[沉淀]池	錐形選粒器	spitzkasten
绞丝	絞紗	skein
搅拌	攪拌	agitating, stirring
搅拌槽	攪拌槽	agitating tank
搅拌槽接触器	攪拌槽接觸器	stirred-tank contactor
搅拌反应器	攪拌反應器	agitated reactor, stirred reactor
搅拌釜	攪拌槽	agitated vessel
搅拌干燥器	攪拌乾燥器	agitated drier, agitator drier
搅拌罐	攪拌槽	stirred tank
搅拌罐式反应器	攪拌反應槽	stirred tank reactor
搅拌结晶器	攪拌結晶器	agitated crystallizer
搅拌流反应器	攪拌流動反應器	stirred flow reactor
搅拌流化床	攪拌流[體]化床	stirred fluidized bed
[搅拌]流量数	[攪拌]流量數	flow number
搅拌器	攪拌器	agitator, stirrer
校正曲线	校準曲線	calibration curve
校正作用	校正作用	corrective action
校准	校準	calibration
校准图	校準圖	calibration chart
酵母	酵母	yeast

大 陆 名	台 湾 名	英 文 名
酵母菌属	酵母菌	saccharomyces
阶梯分析	階梯分析	step analysis
阶梯函数	階梯函數	step function
阶跃干扰	階梯擾動	step disturbance
阶跃函数响应	階梯函數應答	step function response
阶跃输入	階梯輸入	step input
阶跃响应	階梯應答	step response
接触成核	接觸成核[作用]	contact nucleation
接触催化	接觸催化[作用]	contact catalysis
接触点	接觸點	point of contact
接触电阻	接觸電阻,接觸阻力	contact resistance
接触方式	接觸方式	contacting pattern
接触过滤	接觸過濾	contact filtration
接触角	接觸角	contact angle
接触冷凝器	接觸冷凝器	contact condenser
接触器	接觸器	contactor
接触器效率	接觸器效率	contactor efficiency
接触热阻	接觸熱阻	thermal contact resistance
接触室	接觸室	contact chamber
接触稳定过程	接觸穩定程序	contact stabilization process
接触相	接觸相	contacting phase
接点	接點	junction point
接点电位	接合電位	junction potential
接合	接點,接面	junction
接合效率	接合效率	joint efficiency
接取点	接取點	take off point
接头	接頭	coupling joint, joint
接线板	接線板	patch board
接线板模型	插線板模型	switchboard model
接线盒	接線盒	junction box
接枝共聚物	接枝共聚物	graft copolymer
接枝聚合	接枝聚合[反應]	graft polymerization
接枝聚合物	接枝聚合物	graft polymer
接种	接種	inoculation
接种物	接種液	inoculum
节点	節點	node
节距	節距	pitch
节流	節流	throttling

大　陆　名	台　湾　名	英　文　名
节流带	節流帶	throttling band
节流阀	節流閥	throttle valve, throttling valve
节流过程	節流程序	throttling process
节流量热计	節流卡計	throttling calorimeter
节流装置	節流裝置	throttling device
节能器	節能器	economizer
节涌流	氣包流動	slug flow
节涌流反应器	氣包流反應器	slug flow reactor
结构单元	結構單元	structural unit
结构分析	結構分析	structure analysis
结构复合材料	結構複合材料	structural composite
结构化模型	結構化模式	structured model
结构黏度	結構黏性	structural viscosity
结构黏合剂	結構性接著劑	structural adhesive
结构泡沫塑料	結構泡沫塑膠	structural foam plastic
结构疲劳	結構疲勞	structural fatigue
结构缺陷	結構缺陷	structural defect
结构双折射	結構雙折射	structural birefringence
结构稳定性	結構穩定性	structural stability
结构无序	結構失序	structural disorder
结垢	積垢	fouling
结垢形成	積垢	scale formation
结合力	結合力	binding force
结合能	結合能	binding energy
结合能力	結合能力	binder power
结合水分	結合水分	bound moisture
结焦	結焦,焦化,煉焦	coking
结焦槽	結焦槽	coke drum
结晶	結晶	crystallization
结晶度	結晶度	degree of crystallization
结晶聚合物	晶質聚合物	crystalline polymer
结晶器	結晶器	crystallizer
结晶热	結晶熱	heat of crystallization
结晶水	結晶水	crystal water, water of crystallization
结晶温度	結晶點	crystallizing point
结晶学	結晶學	crystallography
结块	結塊	caking
结块能力	結塊能力	caking power

大 陆 名	台 湾 名	英 文 名
结块性	結塊性	caking property
结皮[现象]	結皮	skinning
结筛	結篩	knot screen
结线	連結線	tie line
截留,保留	滯留	retention
截面	截面	cross section
截面积	截面積	cross sectional area
截止阀	截流閥	stop valve
截止频率	截止頻率	cut-off frequency
解冻	解凍	thawing
解毒	解毒[作用]	detoxification
解聚	解聚合[作用]	depolymerization
解聚集	解聚集[作用]	deaggregation
解离温度	解離溫度	dissociation temperature
解离吸附	解離吸附	dissociative adsorption
解偶联	解偶	uncoupling
解耦环路	解偶環路,去偶環路	decoupled loops
解耦器	解偶器,去偶器	decoupler
解耦系统	解偶系統,去偶系統	decoupled system
解吸	脫附	desorption
解吸剂	脫色劑	stripping agent
解吸塔	汽提塔	stripper
解吸因子	脫附因子,汽提因數	desorption factor, stripping factor
解析法	解析法	analytical method
解絮凝,反絮凝	解絮凝[作用]	deflocculation
介标行为	介觀行為	mesoscale behavior
介电常数	介電常數,比電容量	specific inductive capacity, dielectric constant
介电加热	介電加熱	dielectric heating
介电体	介電質	dielectric
介电吸收	介電吸收	dielectric absorption
介晶态	介[晶]態	mesomorphic state
介孔	中孔	mesopore
介质	介質,培養基	medium
介质过滤器	介質過濾器	medium filter
芥子油	芥子油	mustard oil
界面	界面	interface
界面薄膜	界面薄膜	interfacial film

大　陆　名	台　湾　名	英　文　名
界面工程	界面工程	interfacial engineering
界面活性剂	界面活性劑	interfacial agent
界面聚合	界面聚合[法]	surface polymerization
界面控制	界面控制	interface control
界面面积	界面面積	interfacial area
界面能	界面能	interfacial energy
界面浓度	界面濃度	interfacial concentration
界面势	界面勢,界面電位	interfacial potential
界面湍流	界面紊流	interfacial turbulence
界面温度	界面溫度	interfacial temperature
界面现象	界面現象	interfacial phenomenon
界面相(＝分界表面)		
界面形状	界面形狀	interfacial shape
界面性质	界面性質	interfacial property
界面张力	界面張力	interfacial tension
界面张力计	界面張力計	interfacial tensimeter
界面阻力	界面阻力	interfacial resistance
界面组成	界面組成	interface composition
金刚砂	剛砂	emery
金红石	金紅石	rutile
金绿宝石	金綠[寶]石	chrysoberyl
金霉素	金黴素	aureomycin
金相学	金相學	metallography
金属	金屬	metal
金属表面处理	金屬表面處理	metal finishing
金属玻璃	金屬玻璃	metallic glass
金属催化剂	金屬觸媒	metallic catalyst
金属电镀	金屬電鍍	metal plating
金属间化合物	介金屬[化合物]	intermetallic compound
金属离子惰化剂	金屬離子惰化劑	metal ion inactivator
金属络合酸性染料	金屬錯合物酸性染料	metal complex acid dye
金属媒染剂	金屬媒染劑	metallic mordant
金属配位染料	金屬化染料	metallized dye
金属喷镀	金屬噴霧	metal spraying
金属陶瓷	陶瓷金屬,陶金	cermet
金属陶瓷电阻器	陶瓷金屬電阻器	cermet resistor
金属有机化学蒸气沉积	金屬有機化學氣相沈積法	metallorganic chemical vapor deposition

大　陆　名	台　湾　名	英　文　名
金属皂	金屬皂	metallic soap
紧凑型换热器	緊緻熱交換器	compact heat exchanger
紧急操作	緊急操作	emergency operation
紧密度	緊密度	compactness
紧束效应	有效填充	effective packing
近程	近程,短距離	short range
近程分子内相互作用	分子内短程[相互]作用	short-range intramolecular interaction
近程结构	短程結構	short-range structure
近程链内曲柄运动	近程鏈内曲柄運動	short-range intrachain crankshaft movement
近程相互作用	短程交互作用	short-range interaction
近程有序	短程有序	short-range order
近晶型结构	[液晶]層列狀結構	smectic structure
近似	近似	approximation
近似法	近似法	approximate method
近似解	近似解	approximate solution
近似式	近似式	approximation formula
进口长度	進口長度	entry length
进口区	進口區[域]	entrance region
进口效应	進口效應	entrance effect
进料	進料,饋入	feed
进料板	進料板	feed tray
进料板位置	進料板位置	feed tray location
进料比	進料比	feed ratio
进料传送机	進料輸送機	feed conveyor
进料带	進料帶	feed belt
进料阀	進料閥	feed valve
进料管	進料管	feed pipe
进料滚柱	進料輥	feed roller
进料孔	進料孔	feed hole
进料流	進料流	feed stream
进料滤网	進料篩網	intake screen
进料器	進料器,飼機	feeder, bell and hopper
进料调节器	進料調節器	feed regulator
进料温度	進料溫度	feed temperature
进料系统	進料系統	feed system
进料质量	進料品質	feed quality

大　陆　名	台　湾　名	英　文　名
进料状态线	進料狀態線, q線	q-line
进料组成	進料組成	feed composition
进气行程	進氣衝程	intake stroke
进气压力	進氣壓[力]	intake pressure
浸管液位计	浸管液位計	dip pipe level gage
浸灰	浸灰法	liming
浸灰法	浸灰法	liming process
浸滤	瀝取	lixiviation
浸没表面	浸沒表面	immersed surface
浸没燃烧	沈沒燃燒	submerged combustion
浸没燃烧蒸发器	浸沒燃燒蒸發器	submerged combustion evaporator
浸没式冷凝器	沈式冷凝器	submerged condenser
浸没误差	浸沒誤差	immersion error
浸泡试验	浸泡試驗	soak test
浸漂	浸漂	dip bleaching
浸取	瀝取	leaching
浸取槽	瀝取槽	leaching tank
浸取法	瀝取法	leaching process
浸取液	瀝取液	leachate
浸染	浸染, 襯墊料	dip dyeing, padding, steeping
浸染机	浸染機	padding machine
浸染液	浸染液	padding liquor
浸湿法	浸漬[液]	soakage
浸蚀	蝕刻	etching
浸提物	浸提物	educt
浸渍	浸解[作用]	maceration, impregnation, dipping
浸渍槽	浸漬槽	dipping tank
浸渍镀锡	浸漬鍍錫	dip tinning
浸渍树脂	溶劑浸漬樹脂	solvent impregnated resin
浸渍压榨机	浸[漬]壓[榨]機	steeping press
浸渍釉	浸釉	dip glazing
浸渍纸	吸油紙	saturating paper
经济冲击	經濟衝擊	economic impact
经济回流比	經濟回流比	economic reflux ratio
经济可行性	經濟可行性	economic feasibility
经济评价	經濟評估	economic evaluation
经济寿命	經濟壽命	economic life
经济压力降	經濟壓力降	economic pressure drop

大　陆　名	台　湾　名	英　文　名
经验法	經驗法	empirical method
经验法则	經驗法則	empirical rule
经验方程式	經驗方程式,實驗方程式	empirical equation
经验模型	經驗模式	empirical model
经验式	實驗式	empirical formula
晶格	晶格	crystal lattice, lattice
晶硅石	晶矽石	crystalline silica
晶核	晶核	nucleus
晶核形成	晶核形成	nucleus formation
晶间腐蚀	粒間腐蝕	intergranular corrosion
晶粒	晶粒	grain
晶粒长大	晶粒生長	grain growth
晶粒大小	晶粒大小,粒徑	grain size
[晶]粒形[状]	粒形	grain shape
晶面	晶面	crystal face
晶片	晶片,晶圓	wafer
晶石	晶石	spar
晶体	晶體	crystal
晶体成核	晶體成核	crystal nucleation
晶体大小	晶體大小,晶體尺寸	crystal size
晶体大小分布	晶體粒徑分佈	crystal size distribution
晶体管	電晶體	transistor
晶体生长	晶體成長	crystal growth
晶体习性	晶體習性,晶癖	crystal habit
晶系	晶系	crystal system
晶形	晶形	crystal form, crystal shape
晶形结构	晶體結構	crystalline structure
晶形转变	晶形轉變,晶相轉變	crystalline transition
晶种	晶種	crystal seed, seed crystal
晶种混合槽	晶種混合槽	seed mixer
晶种结晶	晶種結晶	seeded crystallization
晶种粒	晶種粒	seed grain
精炼	精煉	refinement, refining
精炼度	精煉度	degree of refining
精馏	精餾	rectification
精馏板	精餾板	rectifying tray
精馏段	精餾段	rectification section, rectifying section

大 陆 名	台 湾 名	英 文 名
精馏釜	精餾釜	rectifying still
精馏器	精餾器	rectifier
精馏塔	精餾塔	rectifier column, rectifying column, rectifying tower
精馏塔板	精餾板	rectifying plate
精[密]度	精密度,精確度	precision
精密分馏	精密分餾	precise fractionation
精密分馏分离	精密分離	close cut separation
精密分馏馏出物	精密餾出物	close cut distillate
精密分馏馏分	精密餾分	close cut fraction
精密蒸馏	精密分餾	precision fractional distillation
精确解	精確解	exact solution
精筛	精篩	fine screen
精细分散	精細分散	fine dispersion
精细平衡原理	細部平衡原理	principle of detailed balance
精油	精油	essential oil
精制机	精製機	refiner
精制棉短绒	棉絨	linter
鲸蜡	鯨蠟	spermaceti wax
鲸蜡油	鯨蠟油,抹香鯨油	sperm oil, spermaceti oil
鲸油	鯨油	whale oil
鲸脂	鯨脂	whale tallow
井口气	井口天然氣	casing head gas
井式喷嘴	井型噴嘴	well type nozzle
阱	阱	trap
肼	肼,聯胺	hydrazine
净功	淨功	net work
净化过程	純化程序	purification process
净化剂	去除劑	scavenger
净化水	純化水	purified water
净利润	淨利潤	net profit
净热值	淨熱值	net heating value
净速率	淨速率	net rate
净通量	淨通量	net flux
净现值	淨現值	net present value, net present worth
净重	淨重	net weight
径向反应器	徑向流反應器	radial flow reactor
径向分布函数	徑向分佈函數	radial distribution function

大　陆　名	台　湾　名	英　文　名
径向分散	徑向分散	radial dispersion
径向分散系数	徑向分散係數	radial dispersion coefficient
径向混合	徑向混合	radial mixing
径向速度	徑向速度	radial velocity
径向应力	徑向應力	radial stress
径向轴承	徑向軸承	radial bearing
竞争反应	競爭反應	competitive reaction
竞争吸附	競爭吸附	competitive adsorption
竞争性	競爭性	competitiveness
竞争性抑制	競爭性抑制[作用]	competitive inhibition
静电沉降器,电除尘器	靜電集塵器,靜電沈積器	electrostatic precipitator
静电除尘	靜電沈積[法]	electrostatic precipitation
[静电]除尘器	集塵器	precipitator
静电除尘器	靜電集塵器,電集塵器	electrostatic dust collector, electric dust precipitator
静电分离	靜電分離	electrostatic separation
静电分离器	靜電分離器	electrostatic separator
静电介电常数	靜介電常數	static dielectric constant
静电吸引	靜電吸引	electrostatic attraction
静电消除器	靜電消除器	static eliminator
静电效应	靜電效應	electrostatic effect
静力学	靜力學	statics
静摩擦	靜摩擦	static friction
静态	靜態	static state
静态混合器	靜態混合器	static mixer
静态模量	靜模數	static modulus
静态疲劳	靜態疲勞	static fatigue
静态热重分析法	靜態熱重量法	static thermogravimetry
静态试验	靜態試驗,靜力試驗	static test
静态误差	靜態誤差	static error
静态误差系数	靜態誤差係數	static error coefficient
静态优化	靜態最適化	static optimization
静压	靜壓[力]	static pressure
静压头	靜高差	static head
静止边界层	[停]滯邊界層	stagnant boundary layer
静止床	靜止床	quiescent bed
静止期	靜止期	stationary phase

大　陆　名	台　湾　名	英　文　名
静滞区	静滞區	dead space
静重活塞压力计	静重活塞壓力計	dead weight piston gage
静重仪	静重計	dead weight gage
镜面磨光	鏡面磨光	mirror polishing
镜式比色计	鏡式比色計	mirror colorimeter
镜式光谱仪	鏡式光譜儀	mirror spectrophotometer
酒精	酒精	spirit
酒精汽油	酒精汽油	gasohol
酒石	酒石	tartar
酒石酸	酒石酸	tartaric acid
酒石酸铁钠	酒石酸鐵鈉	sodium iron tartrate
就地回收	就地回收	*in-situ* recovery
就地加工	就地加工	*in-situ* processing
就地燃烧	就地燃燒	*in-situ* combustion
就地再生	就地再生	*in-situ* regeneration
局部加速度	局部加速度	local acceleration
局部平衡	局部平衡	local equilibrium
局部稳定性	局部穩定性	local stability
局部组成	局部組成	local composition
局部最优[值]	局部最適[值]	local optimum
矩	矩	moment
矩鞍形填料	矩鞍形填料	intalox saddle
矩形脉冲	矩形脈衝	rectangular pulse
矩形堰	矩形堰	rectangular weir
矩阵	矩陣	matrix
矩阵求逆	矩陣求逆	matrix inversion
矩阵算术	矩陣算術	matrix arithmetics
巨正则配分函数	大正則分配函數	grand canonical partition function
巨正则系综	大正則系綜	grand canonical ensemble
距离因数	距離因數	distance factor
距速滞后	距速滯延	distance velocity lag
锯齿形	鋸齒形	serration
聚氨基甲酸酯	聚胺甲酸酯	polyurethane
聚氨酯橡胶	聚胺[甲酸]酯橡膠	polyurethane rubber
聚苯乙烯	聚苯乙烯	polystyrene
聚丙烯	聚丙烯	polypropylene
聚丙烯腈	聚丙烯腈	polyacrylonitrile
聚丙烯腈纤维	聚丙烯腈纖維	polyacrylonitrile fiber

大　陆　名	台　湾　名	英　文　名
聚丙烯酸钠	聚丙烯酸鈉	sodium polyacrylate
聚丙烯酸酯	聚丙烯酸酯	polyacrylate
聚丙烯酸酯树脂	聚丙烯酸酯樹脂	polyacrylate resin
聚丙烯纤维	聚丙烯纖維	polypropylene fiber
聚并	凝聚,聚結	coalescence
聚并模式	聚結模式	coalescence model
聚并器	凝聚器,聚結器	coalescer
聚砜	聚碸	polysulfone
聚合动力学	聚合[反應]動力學	polymerization kinetics
聚合度	聚合度	degree of polymerization
聚合[反应]	聚合[反應]	polymerization
聚合反应	聚合反應	polymerization reaction
聚合反应工程	聚合反應工程	polymerization reaction engineering
聚合反应器	聚合反應器	polymerization reactor
聚合机理	聚合反應機構	polymerization mechanism
聚合物	聚合物,高分子,聚合體	polymer
聚合物加工	聚合物加工,高分子加工	polymer processing
聚合物加工工程	聚合物加工工程	polymer processing engineering
聚合物加工技术	聚合物加工技術	polymer processing technology
聚合物降解	聚合物降解	polymer degradation
聚合物驱	聚合物汜流[法]	polymer flooding
聚合物稳定剂	聚合物穩定化	polymer stabilization
聚合油	聚合油	polymerized oil
聚集	聚集[作用]	aggregation
聚集结晶	聚集結晶	accumulative crystallization
聚集数	聚集數	aggregation number
聚集体	聚集體	aggregate
聚己内酰胺	聚己內醯胺	polycaprolactam
聚甲醛	聚甲醛	polyformaldehyde
聚硫聚合物	多硫聚合物	thiokol polymer
聚硫橡胶	多硫橡膠	polysulfide rubber, thiokol
聚氯乙烯	聚氯乙烯	polyvinyl chloride
聚氯乙烯管	聚氯乙烯管	vinyl pipe
聚氯乙烯树脂	聚氯乙烯樹脂	polyvinyl chloride resin
聚醚	聚醚	polyether
聚偏氟乙烯	聚偏二氟乙烯	polyvinylidene fluoride
聚偏氯乙烯	聚偏二氯乙烯	polyvinylidene chloride

大　陆　名	台　湾　名	英　文　名
聚三氟氯乙烯	聚三氟氯乙烯	polytrifluorochloroethylene
聚式流态化	聚集流體化	aggregative fluidization
聚四氟乙烯	聚四氟乙烯	polytetrafluoroethylene
聚碳酸酯	聚碳酸酯	polycarbonate
聚碳酸酯树脂	聚碳酸酯樹脂	polycarbonate resin
聚烯烃树脂	聚烯烴樹脂	polyolefin resin
聚酰胺	聚醯胺	polyamide
聚酰胺树脂	聚醯胺樹脂	polyamide resin
聚酰胺纤维	聚醯胺纖維	polyamide fiber
聚乙酸乙烯酯	聚乙酸乙烯酯	polyvinyl acetate
聚乙烯	聚乙烯	polyethylene
聚乙烯吡咯烷酮	聚乙烯氫吡咯酮	polyvinyl pyrrolidone
聚乙烯醇	聚乙烯醇	polyvinyl alcohol
聚乙烯基醚	聚乙烯醚	polyvinyl ether
聚乙烯醚黏合剂	聚乙烯醚黏合劑	polyvinyl ether adhesive
聚乙烯酯缩丁醛树脂	聚乙烯醇縮丁醛樹脂	polyvinyl butyral resin
聚异丁烯	聚異丁烯	polyisobutylene
聚异戊二烯橡胶	聚異戊二烯橡膠,異平橡膠	polyisoprene rubber
聚酯	聚酯	polyester
聚酯类树脂	聚酯樹脂	polyester resin
聚酯类纤维	聚酯纖維	polyester fiber
聚酯类橡胶	聚酯橡膠	polyester rubber
卷积	捲積,摺積,疊積	convolution
卷绕	繞緊	windup
卷缩	扭結	snarl
卷筒	有邊筒子,線軸	spool
绢丝橡胶	絲狀橡膠	silk rubber
决策	決策	decision making
决策变量	決策變數	decision variable
决策树	決策樹	decision tree
绝对反应速率	絕對反應速率	absolute reaction rate
绝对沸点	絕對沸點	absolute boiling point
绝对干纤维	絕對乾纖維	absolute dry fiber
绝对计数	絕對計數	absolute counting
绝对零度	絕對零度	absolute zero
绝对黏度	絕對黏度	absolute viscosity
绝对偏差	絕對偏差	absolute deviation

大　陆　名	台　湾　名	英　文　名
绝对热力学温标	絕對熱力溫標	absolute thermodynamic temperature scale
绝对色值	絕對色值	absolute color value
绝对熵	絕對熵	absolute entropy
绝对湿度	絕對濕度	absolute humidity
绝对湿含量	絕對含水量	absolute moisture content
绝对温度	絕對溫度	absolute temperature
绝对稳定性	絕對穩定性	absolute stability
绝对误差积分	絕對誤差積分	integral of absolute error
绝对压力	絕對壓力	absolute pressure
绝对压力计	絕對壓力計	absolute pressure gauge
绝对值	絕對值	absolute value
绝对专一性	絕對專一性	absolute specificity
绝对浊度	絕對濁度	absolute turbidity
绝热	隔熱	heat insulation
绝热饱和	絕熱飽和	adiabatic saturation
绝热饱和温度	絕熱飽和溫度	adiabatic saturation temperature
绝热壁	絕熱壁	adiabatic wall, diathermal wall
绝热材料,保温材料	隔熱材料,保溫材料	heat insulating materials
绝热操作	絕熱操作	adiabatic operation
绝热操作线	絕熱操作線	adiabatic operating line
绝热反应	絕熱反應	adiabatic reaction
绝热反应器	絕熱反應器	adiabatic reactor
绝热干燥	絕熱乾燥	adiabatic drying
绝热过程	絕熱程序	adiabatic process
绝热化学反应	絕熱化學反應	adiabatic chemical reaction
绝热节流过程	絕熱節流程序	adiabatic throttling process
绝热节流装置	絕熱節流裝置	adiabatic throttling device
绝热精馏	絕熱精餾	adiabatic rectification
绝热冷却	絕熱冷卻	adiabatic cooling
绝热冷却曲线	絕熱冷卻曲線	adiabatic cooling curve
绝热冷却线	絕熱冷卻線	adiabatic cooling line
绝热[沥青]纸	絕熱紙	sheathing paper
绝热流	絕熱流	adiabatic flow
绝热膨胀	絕熱膨脹	adiabatic expansion
绝热式量热计	絕熱卡計	adiabatic calorimeter
绝热收缩	絕熱收縮	adiabatic contraction
绝热塔	絕熱塔	adiabatic column
绝热条件[状态]	絕熱狀況	adiabatic condition

大　陆　名	台　湾　名	英　文　名
绝热图	絕熱圖	adiabatic plot
绝热温升	絕熱溫升	adiabatic temperature rise
绝热系统	絕熱系統	adiabatic system
绝热压缩	絕熱壓縮	adiabatic compression
绝热仪器	絕熱裝置	adiabatic apparatus
绝热因子	絕熱因數	adiabatic factor
绝热增湿	絕熱增濕	adiabatic humidification
绝热蒸发	絕熱蒸發	adiabatic evaporation
绝热转化	絕熱轉化	adiabatic conversion
绝缘	絕緣	insulation
绝缘体	絕緣體	insulator
绝缘油	隔離油	isolating oil
绝缘纸	絕緣紙	insulating paper
均镀能力	均鍍力	throwing power
均方根偏差	均方根偏差	root-mean-square deviation
均方根误差	均方根誤差	root-mean-square error
均方根振荡速度	均方根波動速度	root-mean-square fluctuating velocity
均方误差	均方誤差	mean square error
均衡肥料	均衡肥料	balance fertilizer
均衡伸缩阀	均衡伸縮閥	balanced bellow valve
均衡生长	均衡生長	balanced growth
均衡值	均衡值	balanced value
均聚反应	均聚合[反應],同元聚合[反應]	homopolymerization
均聚物	均聚物,同元聚合物	homopolymer
均热因子	均熱因數	soaking factor
均相	均相	homogeneous phase
均相成核	均相成核	homogeneous nucleation
均相催化	均相催化[作用]	homogeneous catalysis
均相反应	均相反應	homogeneous reaction
均相反应器	均相反應器	homogeneous reactor
均相共沸混合物	均相共沸液	homoazeotrope
均相共沸液	均相共沸液	homogeneous azeotrope
均相平衡	均相平衡	homogeneous equilibrium
均相燃烧	均相燃燒	homogeneous combustion
均相系统	均匀系統	homogeneous system
均匀分布	均匀分佈	uniform distribution
均匀结焦	均匀結焦	uniform coking

大 陆 名	台 湾 名	英 文 名
均匀晶体	均匀晶體	uniform crystal
均匀流	均匀流動	uniform flow
均匀流动	均匀流	uniform stream
均匀流化床	均匀流[體]化床	homogeneously fluidized bed
均匀体	均匀體	homogeneous body
均匀湍流	均匀紊流	homogeneous turbulence
均匀系数	均匀性係數	uniformity coefficient
均匀性	均匀性	homogeneity, uniformity
均匀液位控制	平均[化]液位控制	averaging level control
均匀载荷	均匀負荷	uniform loading
均匀中毒	均匀中毒	uniform poisoning
均匀转化	均匀轉化	uniform conversion
均匀转化模型	均匀轉化模型	uniform conversion model
均值	平均[值]	mean

K

大 陆 名	台 湾 名	英 文 名
咔唑	咔唑	carbazole
卡必醇	卡必醇	carbitol
卡[路里]	卡	calorie
开敞喷嘴	開敞噴嘴	open nozzle
开发项目	發展專案	development project
开[放式]炼胶机	輥磨機	roller mill
开关	開關	switch
开关作用	開關作用	on-off action
开环	開環	open loop
开环测试	開環測試	open-loop test
开环极点	開環極點	open-loop pole
开环控制	開環控制	open-loop control
开环零点	開環零點	open-loop zero
开环系统	開環系統	open-loop system
开环转移函数	開環轉移函數	open-loop transfer function
开路	開路,斷路	open circuit
开路研磨	開路研磨	open circuit grinding
开始[硫化]效应	開始[硫化]效應	set-up effect
开式边界	開放邊界	open boundary
开式容器	開放式容器	open vessel

大　陆　名	台　湾　名	英　文　名
开式叶轮	敞動葉輪	open impeller
开斯米纶	開司米綸	cashmilon
抗爆[震]	抗[爆]震	antiknock
抗爆[震]化合物	抗[爆]震化合物	antiknock compound
抗爆[震]值	抗[爆]震值	antiknock value
抗剥落性	抗散裂性	spalling resistance
抗臭氧剂	抗臭氧[化]劑	antiozonant
抗冻混合物	防凍混合物	antifreeze mixture
抗刮性	抗刮性	scratch resistance
抗碱性	抗鹼性,耐鹼性	alkali resistance
抗焦处理	防焦處理	scorch-resisting treatment
抗静电剂	去靜電劑	antistatic agent
抗卷绕	抗繞緊	anti-windup
抗拉强度	抗拉強度	tensile strength
抗磨剂	抗磨劑	attrition resistant
抗磨损	抗磨損	attrition resistance
抗磨性	抗磨性	abrasiveness
抗凝剂	抗凝劑	anticoagulant agent
抗凝效应	抗凝效應	anticoagulant effect
抗日光龟裂剂	抗晒劑	sun-checking agent
抗生素	抗生素	antibiotic
抗水剂	撥水劑	water repellence agent
抗水[作用]	撥水劑	water repellent
抗体	抗體	antibody
抗微生物活性	抗微生物活性	antimicrobial activity
抗微生物药	抗微生物藥	antimicrobial drug
抗微生物作用	抗微生物作用	antimicrobial action
抗絮凝剂	解絮凝劑	deflocculation agent, deflocculant
抗氧[化]剂	抗氧化劑,氧化抑制劑	antioxidant, oxidation inhibitor
抗原	抗原	antigen
烤清漆	烤清漆	baking varnish
苛化	苛性[作用]	causticization
科尔莫戈罗夫尺度	科摩哥洛夫尺度	Kolmogorov scale
颗粒材料	粒狀物質	granular materials
颗粒层过滤器	顆粒層過濾器	granular bed filter
颗粒床	顆粒床	particulate bed
颗粒簇	粒子團簇	cluster of particle
颗粒过程	顆粒程序	particulate process

大 陆 名	台 湾 名	英 文 名
颗粒密度	顆粒密度	particle density
颗粒群	粒子群	particle swarm
颗粒特性	粒子特性	particle characteristics
颗粒体	顆粒	granule
颗粒形状	粒形	particle shape
颗粒形状因子	粒形因數	particle shape factor
颗粒学	顆粒學	particuology
可编程计算器	可程式計算器	programmable calculator
可编程逻辑控制	可程式邏輯控制	programmable logic control
可编程调制器	可程式控制器	programmable controller
可变孔板流量计	可變孔流量計	variable orifice meter
可剥性涂料	可剝塗層	strippable coating
可操作性	可操作性	operability
可测量性	可量測性	measurability
可测量性质	可量測性質	measurable property
可的松	可體松,皮質酮	cortisone
可纺性	可紡性	spinnability
可观测性	可觀測性	observability
可恢复应力	可復應力	recoverable stress
可获利程度	獲利能力	profitability
可及矩阵	可達性矩陣	reachability matrix
可降解塑料	可降解塑膠	degradable plastic
可接受的回报	可接受[的]報酬	acceptable return
可解性	可溶解性,溶劑合性	solvability
可靠性	可靠性	reliability
可控性	可控性	controllability
可扩展性	延伸性	extensibility
可裂变物质	可[核]分裂物質	fissionable materials
可裂变性	可[核]分裂性	fissionability
可流态化颗粒大小	可流體化粒子大小	fluidizable particle size
可磨性	可磨性	grindability
可磨性指数	可磨性指數	grindability index
可逆变化	可逆變化	reversible change
可逆沉淀	可逆沈澱	reversible precipitation
可逆电池	可逆電池	reversible cell
可逆度	可逆度	degree of reversibility
可逆反应	可逆反應	reversible reaction
可逆分解电位	可逆分解電位	reversible decomposition potential

大　陆　名	台　湾　名	英　文　名
可逆功	可逆功	reversible work
可逆过程	可逆程序	reversible process
可逆化学反应	可逆化學反應	reversible chemical reaction
可逆机理	可逆機構	reversible mechanism
可逆胶囊	逆微胞	reversed micelle
可逆绝热流动	可逆絕熱流動	reversible adiabatic flow
可逆扩散	可逆擴散	reversible diffusion
可逆性	①可逆性②可逆度	reversibility
可燃性	可燃性	flammability，inflammability
可燃液体规范	可燃液體規範	combustible liquid code
可染性	可染性	dyeability
可溶性	溶解度	solubility
可溶性淀粉	可溶澱粉	soluble starch
可溶性核糖核酸	可溶性核糖核酸,可溶性 RNA	soluble RNA，sRNA
可溶性胶	可溶[性]膠	soluble gum
可溶性聚合物	可溶性聚合物	soluble polymer
可溶性树脂	可溶性樹脂	soluble resin
可乳化性	可乳化性	emulsifiability
可湿性助剂	潤濕劑	wettable agent
可实现性	可實現性	realizability
可塑性	塑性	plasticity
可调参数	可調參數	adjustable parameter
可调节权重	可調節權重	adjustable weight
可调锐孔	可調孔口	adjustable orifice
可调轴承	可調軸承	adjustable bearing
可涂期	可塗期	spreadable life
可稳性	可穩[定]性	stabilizability
可洗软膏基	可洗軟膏基	washable ointment base
可行路径法	可行路徑法	feasible path method
可行性	可行性	feasibility
可行性调查	可行性調查	feasibility survey
可行性分析	可行性分析	feasibility analysis
可行性研究	可行性研究	feasibility study
可行域	可行區域	feasible region
可压缩流动	[可]壓縮流動	compressible flow
可压缩流体	可壓縮流體	compressible fluid
可压缩滤饼	可壓縮濾餅	compressible filter cake

大　陆　名	台　湾　名	英　文　名
可压缩性	壓縮性	compressibility
可再生能源	[可]再生能源	renewable energy source
可再生资源	[可]再生資源	renewable resources
克拉珀龙-克劳修斯方程	克[拉伯隆]-克[勞修斯]方程式	Clapeyron-Clausius equation
克劳修斯不等式	克勞修斯不等式	Clausius inequality
克隆	克隆	clone
克伦舍尔图	克倫舍爾圖	Kremser diagram
克努森扩散	努生擴散	Knudsen diffusion
克努森数	努生數	Knudsen number
刻度盘指示器	針盤指示器	dial indicator
刻蚀	蝕刻	etching
空管反应器	空管反應器	unpacked-tube reactor
空间点阵	空間晶格	space lattice
空间电荷	空間電荷	space charge
空间电荷限制电流	空間電荷限制電流	space charge limited current, SCLC
空间空调	空調	space conditioning
空间冷却	空間冷卻	space cooling
空间群	空間群	space group
空间速度	空間速度	spatial velocity
空间速率	空間速度	space velocity
空间系数	空間因數	space factor
空间效应	位阻效應,立體效應	steric effect
空间因子	位阻因素,立體因素	steric factor
空间张力	立體應變	steric strain
空间障碍	空間障礙	steric restriction
空间振荡	空間振盪	spatial oscillation
空冷换热器	氣冷式熱交換器	air-cooled heat exchanger
空气饱和器	空氣飽和器	air saturator
空气吹扫	空氣吹驅	air purge
空气动力学	空氣動力學,氣壓動態[學]	aerodynamics, air pressure dynamics
空气动力直径	空氣動力直徑	aerodynamic diameter
空气分离	空氣分離	air separation
空气分离罐	空氣分離槽	air separation tank
空气分离器	空氣分離器	air separator
空气分离塔	空氣分離塔	air separation tower
空气干燥器	空氣乾燥機	air dryer

大 陆 名	台 湾 名	英 文 名
空气干燥塔	空氣乾燥塔	air drying tower
空气供应	空氣供應	air supply
空气鼓泡器	空氣鼓泡器,空氣噴佈器	air sparger
空气过滤	空氣過濾	air filtration
空气过滤器	空氣過濾器	air filter
空气混合深度	空氣混合深度	air mixing depth
空气加热器	空氣加熱器	air heater
空气扩散器	空氣擴散器	air diffuser
空气冷凝器	空氣冷凝器	aerial condenser, atmospheric condenser
空气冷却	空氣冷卻	air cooling
空气冷却环	空氣冷卻環	air ring
空气冷却器	空氣冷卻器	air cooler
空气冷却塔	空氣冷卻塔	atmospheric cooling tower
空气流	空氣流動	air flow
空气流量计	空氣流量計	air flow meter
空气炉	空氣爐	air furnace
空气灭菌器	空氣滅菌器	air sterilizer
空气膜	空氣膜	air film
空气黏合	氣結	air binding
空气喷射器	空氣噴射器	air ejector
空气喷嘴	空氣噴嘴	air nozzle
空气膨胀机	空氣膨脹機	air expander
空气-燃料比	空氣-燃料比	air-fuel ratio
空气软管	空氣軟管	air hose
空气污染	空氣污染	aerial pollution, air pollution
空气污染减量	空氣污染減量	air pollution abatement
空气污染控制	空氣污染控制	air pollution control
空气污染物	空氣污染物	air pollutant
空气雾沫剂	空氣霧沫劑	air-entraining agent
空气吸滤器	空氣吸濾器	air suction filter
空气洗涤	空氣洗滌	air washing
空气洗涤器	空氣洗滌器	air scrubber
空气压力	空氣壓力	air pressure
空气压力降	空氣壓力降	air pressure drop
空气压缩机	空氣壓縮機	air compressor
空气扬析	空氣揚析,空氣淘析	air elutriation
空气预热器	空氣預熱器	air preheater

大 陆 名	台 湾 名	英 文 名
空气蒸发	空氣蒸發[法]	air evaporation
空蚀	空腔腐蝕	cavitation corrosion
空调机组	空調機組	air conditioning unit
空调器	空調[機]	air conditioner
空压调节器	空氣調節器	air regulator
空心叶轮搅拌器(=自吸搅拌器)		
空穴辐射	空腔輻射	cavity radiation
孔板	孔口板	orifice plate
孔板法兰	孔口凸緣,孔口法蘭	orifice flange
孔板计接头	孔口計接頭	orifice tap
孔板计系数	孔口計係數	orifice meter coefficient
孔板流量计	孔口[流量]計	orifice flowmeter, orifice meter
孔板塔	孔板塔	perforated plate column, perforated plate tower
孔板蒸馏塔	孔板蒸餾塔	perforated plate distillation column
孔半径	孔[隙]半徑	pore radius
孔径	孔徑,孔隙大小	pore size
孔径分布	孔徑分佈	pore size distribution
孔口	孔口	orifice
孔口环	孔口環	orifice ring
孔口排放	孔口排放	orifice discharge
孔口系数	孔口[計]係數	orifice coefficient
孔口中毒	孔口中毒	pore-mouth poisoning
孔流系数	排放係數	discharge coefficient
孔膜隔板	多孔隔板	porous barrier
孔雀绿	孔雀綠	malachite green
孔雀石	孔雀石	malachite
孔体积	孔隙體積	pore volume
孔隙	孔隙,[細]孔	pore
孔隙计	孔隙計	porosimeter
孔隙结构	孔隙[幾何]形狀,孔[隙]結構	pore geometry, pore structure
孔隙率	孔隙率	porosity
孔隙仪	孔隙儀	porosity apparatus
孔效应	孔隙效應	porosity effect
空隙	空隙	void
空隙分数	空隙分率	void fraction

大　陆　名	台　湾　名	英　文　名
空隙率	空隙[率]	voidage
空隙速度	間隙速度	interstitial velocity
空隙体积	空隙體積	void volume
控制	控制	control
pH 控制	pH 控制	pH control
控制变量	受控變數	controlled variable, control variable
控制表面	控制表面	control surface
控制策略	控制策略	control strategy
控制点	控制點	control point
控制阀	控制閥	control valve
控制阀量程	控制閥可調範圍	rangeability of control valve
控制构型	控制組態	control configuration
控制函数	控制函數	control function
控制回路	控制回路,控制環路	control loop
控制界限	控制界限	control limit
控制律	控制律	control law
控制器参数	控制器參數	controller parameter
控制器机理	控制器機構	controller mechanism
控制器校正	控制器校正	controller calibration
控制器匹配	控制器調適	controller adaptation
控制器设定	控制器設定	controller setting
控制器增益	控制器增益	controller gain
控制器振铃	控制器振鈴	controller ringing
控制区间	控制區間	control interval
控[制]释[放]	控制釋放	controlled release
控制算法	控制算則	control algorithm
控制体积	控制體積	control volume
控制图	控制圖	control chart
控制系统	控制系統	control system
控制系统相互作用	控制系統相互作用	interaction of control systems
控制线路	控制方案	control scheme
控制仪表板	控制儀表板	control panel
控制作用	控制作用	control action
枯草杆菌	枯草桿菌	*Bacillus subtilis*
苦艾	苦艾	*Absinthium*
库存控制	存量控制	inventory control
库尔特颗粒计数仪	庫爾特計數器	Coulter counter
库仑滴定	電量滴定[法]	coulometric titration

大 陆 名	台 湾 名	英 文 名
跨度	跨距	span
快干水泥,速凝水泥	快乾水泥,速凝水泥	accelerated cement
快速反应	快[速]反應	fast reaction
快速过滤器	快速過濾器	rapid filter
快速流化床	快速流[體]化床	fast fluidized bed
快速流态化	快速流體化	fast fluidization
快速砂滤速率	快速砂濾[法]	rapid sand filtration
宽床反应器	寬床反應器	massive bed reactor
矿蜡	礦蠟	mineral wax
矿石浮选	礦石浮選[法]	ore flotation
矿物肥料	礦物肥料	mineral fertilizer
矿物学	礦物學	mineralogy
矿物油	礦油	mineral oil
矿物资源	礦物資源	mineral resources
矿用炸药	礦用炸藥	mining explosive
矿渣水泥	熔渣水泥	slag cement
矿脂	石蠟脂	petrolatum
框架	框,架	frame
框式搅拌器	框式攪拌器	grid agitator
框式压滤机	框式壓濾機	frame filter press
框图	方塊圖,塊解圖	block diagram
窥箱	窺箱	sight box
葵花油	葵花油	heliotrope oil
扩大损失	擴大損失	enlargement loss
扩散	擴散	diffusion
扩散泵	擴散泵	diffusion pump
扩散薄膜	擴散薄膜	diffusion film
扩散层	擴散層	diffuse layer
扩散常数	擴散常數	diffusion constant
扩散传质	擴散質傳	diffusion mass transfer
扩散催化	擴散催化[作用]	diffusion catalysis
扩散电流	擴散電流	diffusion current
扩散电位	擴散勢	diffusion potential
扩散方程	擴散方程[式]	diffusion equation
扩散过程	擴散程序	diffusion process
扩散环	擴散環	diffusion ring
扩散控制	擴散控制	diffusion control
扩散控制反应	擴散控制反應	diffusion-controlled reaction

大　陆　名	台　湾　名	英　文　名
扩散膜	擴散障壁	diffusion barrier
扩散器	擴散器	diffuser
扩散器护罩	擴散器護罩	diffuser casing
扩散热效应	擴散熱效應	diffusion-thermo effect
扩散时间	擴散時間	diffusion time
扩散速率	擴散速率	diffusion rate
扩散通量	擴散通量	diffusion flux
扩散系数	擴散係數	diffusion coefficient, diffusivity
扩散压力	擴散壓	diffusion pressure
扩散真空泵	擴散真空泵	diffusion vacuum pump
扩散阻力	擴散阻力	diffusion resistance
扩展式表面管	延伸表面式管	extended surface tube

L

大　陆　名	台　湾　名	英　文　名
拉发线	引線	trip wire
拉杆	拉桿	draw bar
拉坯	拉坯	throwing
拉伸	拉伸	stretching
拉伸比	拉伸比	stretch ratio
拉伸成型	拉伸成形	stretch forming
拉伸纺丝	拉伸紡絲	stretch spinning
拉伸流动	拉伸流動	elongational flow
拉伸膜	拉伸膜	stretched membrane
拉伸黏度	拉伸黏度	elongational viscosity, extensional viscosity
拉伸取向	拉伸定向	stretch orientation
拉伸速率	拉伸速率,伸展速率	elongation rate, stretch rate
拉伸性	拉伸性	stretchability
拉乌尔定律	拉午耳定律	Raoult law
蜡	蠟	wax
蜡馏出物	蠟餾出物	wax distillate
蜡纸	蠟紙	wax paper
赖氨酸	離胺酸	lysine
兰金循环	郎肯循環	Rankine cycle
蓝宝石	藍寶石	sapphire
蓝矾	藍石,膽礬	blue vitriol

大　陆　名	台　湾　名	英　文　名
蓝铜矿	藍銅礦	azurite
篮式过滤器	籃式[過]濾器	basket filter
篮式离心机	籃式離心機	basket centrifuge
篮式升降机	籃式升降機	basket elevator
篮式提取器	籃式萃取器	basket extractor
老化	老化	aging
老化试验	老化試驗	aging test
老化试验机	老化試驗機	aging test machine
老化箱	熟化烘箱	aging oven
老化装置	老化裝置	aging equipment
酪蛋白	酪蛋白	casein
酪蛋白黏合剂	酪蛋白黏合劑,酪蛋白黏著劑	casein adhesive
雷蒙磨	雷蒙磨	Raymond mill
雷诺数	雷諾數	Reynolds number
类比	類比	analogy
类比理论	類比理論	analogy theory
类比系统	類比系統	analogy system
类黑素	類黑素	melanoid
累积产率	累積產率	integral yield
累积分布	累積分佈	cumulative distribution
累积概率	累積機率	cumulative probability
累积过细重量分率	累積過細重量分率	cumulative weight undersize fraction
累积筛析	累積篩析	cumulative screen analysis
累积速率	累積速率	accumulation rate
累积误差	累積誤差	integrated error
累积重量百分率过粗部分	累積過粗重量百分率	cumulative weight percent oversize
累积重量百分率过细部分	累積過細重量百分率	cumulative weight percent undersize
累积重量分布	累積重量分佈	cumulative weight distribution
[累计]总量表	總量表	quantity meter
冷冻	冷凍	refrigeration
冷冻吨	冷凍噸	refrigerating ton
冷冻干燥	冷凍乾燥	freeze drying, lyophilization
冷冻干燥机	冷凍乾燥機	freeze drier
冷冻过程	冷凍程序	freezing process
冷冻剂	冷凍劑,冷媒	refrigerant

大　陆　名	台　湾　名	英　文　名
冷冻能力	冷凍能力	refrigeration capacity
冷冻凝聚作用	冷凍凝聚[作用]	freeze coagulation
冷冻盘管	冷凍旋管	refrigerating coil
冷冻器	冷凍器	freezer
冷冻系数	性能係數	coefficient of performance, COP
冷冻液	冷凍流體	refrigerating fluid
冷光	冷光	luminescence
冷回流	冷回流	cold reflux
冷碱法	冷鹼法	cold-soda process
冷接点	冷接點	cold junction
冷浸灰法	冷汁加灰法	cool liming
冷硫化	冷硫化[法]	cold vulcanization
冷模试验	冷模試驗	cold-flow model experiment, mockup experiment
冷凝点	冷凝點	condensation point
冷凝管	冷凝管	condensation tube, condensing tube
冷凝器	冷凝器	condenser
冷凝热	冷凝熱,凝結熱	heat of condensation
冷凝室	冷凝室	condensing chamber
冷凝温度	冷凝溫度	condensing temperature
冷凝系数	冷凝係數	coefficient condensation
冷凝液	冷凝液	condensate
冷却段	冷卻區	cooling zone
冷却剂	冷卻劑,冷媒	coolant, cooling agent
冷却介质	冷卻介質,冷媒	cooling medium
冷却面	冷卻[表]面	cooling surface
冷却面积	冷卻面積	cooling area
冷却盘管	冷卻旋管	cooling coil
冷却器	冷卻器	cooler
冷却曲线	冷卻曲線	cooling curve
冷却时间	冷卻時間	cooling time
冷却水	冷卻水	cooling water
冷却速率	冷卻速率	cooling rate
冷却塔	冷卻塔	cooling tower
冷却系统	冷卻系統	cooling system
冷制橡胶	冷製橡膠	cold rubber
厘泊	厘泊	centipoise, cP
离析	偏析	segregation

大 陆 名	台 湾 名	英 文 名
离析度	離析度	degree of segregation
离线	離線	off-line
离心泵	離心泵	centrifugal pump
离心沉降	離心沈降	centrifugal settling
离心成型	離心成型	centrifugal molding
离心纯化	離心純化	centrifugal purification
离心萃取器	離心萃取機	centrifugal extractor
离心涤气机	離心洗氣器	centrifugal scrubber
离心[分离]	離心[分離],離心化	centrifugation, centrifugalization
离心分离机	離心分離器	centrifugal separator
离心干燥器	離心乾燥機	centrifugal dryer
离心过滤机	離心過濾機	centrifugal filter
离心机	離心機	centrifuge
离心[机]管	離心管	centrifuge tube
离心机盘	離心機籃	centrifuge basket
离心机叶片	離心機葉片	centrifuge blade
离心机转头	離心機轉頭	centrifuge head
[离心]精炼法	糖精煉法	affination
离心净化	離心淨化	centrifugal clarification
离心力	離心力	centrifugal force
离心碾磨机	離心磨機	centrifugal mill
离心喷淋塔	離心噴霧塔	centrifugal spray tower
离心喷雾	離心噴霧	centrifugal spray
离心筛	離心篩	centrifugal screen
离心式鼓风机	離心鼓風機	centrifugal blower
离心式固体分离器	離心式固體分離器	centrifugal solid separator
离心式螺旋桨	離心式螺旋槳	centrifugal propeller
离心式液体分离器	離心式液體分離器	centrifugal liquid separator
离心收集器	離心收集器	centrifugal collector
离心脱水器	離心脱水機	centrifugal dehydrator
离心旋风器	離心旋風器	centrifugal cyclone
离心压缩机	離心壓縮機	centrifugal compressor
离心扬程	離心揚程	centrifugal head
离子电位	離子電位	ionic potential
离子反应	離子反應	ionic reaction
离子浮选	離子浮選[法]	ion flotation
离子交换	離子交換	ion exchange
离子交换分离	離子交換分離	ion exchange separation

大　陆　名	台　湾　名	英　文　名
离子交换过程	離子交換法	ion exchange process
离子交换剂	離子交換劑	ion exchange agent, ion exchanger
离子交换膜	離子交換膜	ion exchange membrane
离子交换平衡	離子交換平衡	ion exchange equilibrium
离子交换容量	離子交換容量	ion exchange capacity
离子交换色谱[法]	離子交換層析法	ion exchange chromatography
离子交换树脂	離子交換樹脂	ion exchange resin
离子聚合	離子聚合[作用]	ionic polymerization
离子聚合物	離子聚合物	ionomer
离子扩散	離子擴散	ionic diffusion
离子排斥	離子排斥	ion exclusion
离子迁移	離子遷移	ionic migration
离子迁移率	離子移動率	ionic mobility
离子强度	離子強度	ionic strength
离子束沉积法	離子束沈積[法]	ion-beam deposition
离子束蚀刻法	離子束蝕刻[法]	ion-beam etching
离子型表面活性剂	離子[型]界面活性劑	ionic surfactant
离子注入	離子植入[法]	ion implantation
里迪尔机理	里迪爾機構	Rideal mechanism
理论	理論	theory
理论板当量高度(= 等 [理论]板高度)		
理论板数	理論板數	theoretical plate number, theoretical tray number
理论级	理論階	theoretical stage
理论空气量	理論空氣[需要]量	theoretical air
理论模型	理論模式	theoretical model
理论[塔]板	理論板	theoretical plate
理想板	理想板	ideal plate, ideal tray
理想反应器	理想反應器	ideal reactor
理想功	理想功	ideal work
理想管式反应器	理想管式反應器	ideal tubular reactor
理想混合物	理想混合物	ideal mixture
理想级	理想階	ideal stage
理想接触级	理想接觸階	ideal contact stage
理想流动	理想流動	ideal flow
理想流体	理想流體	ideal fluid, perfect fluid
理想气体	理想氣體	ideal gas, perfect gas

大　陆　名	台　湾　名	英　文　名
理想气体常数	理想氣體常數	ideal gas constant
理想气体定律	理想氣體定律	ideal gas law
理想气体焓	理想氣體焓	ideal gas enthalpy
理想气体温标	理想氣體溫標	ideal gas temperature scale
理想气体状态	理想氣體狀態	ideal gas state
理想溶液	理想溶液	ideal solution, perfect solution
理想循环	理想循環	ideal cycle
锂基润滑脂	鋰基潤滑脂	lithium-base grease
力-电流类比	力-電流類比	force-current analogy
力-电压类比	力-電壓類比	force-voltage analogy
力矩	力矩	moment of force
力矩法	矩法	method of moments
力平衡	力均衡	force balance
力势	力勢	force potential
力学	力學	mechanics
力-质量换算常数	力-質量換算常數	force-mass conversion constant
立方膨胀	體膨脹	cubic expansion
立方膨胀系数	體膨脹係數	cubical expansion coefficient
立方平均沸点	立方平均沸點	cubic average boiling point
立构重复单元	立構重複單元	stereorepeating unit
立构规整聚合	立體規則聚合[反應]，立體異構聚合	stereoregular polymerization, stereospecific polymerization, stereotactic polymerization
立构规整聚合物	立體特定聚合物，立體異構聚合物	stereospecific polymer, stereotactic polymer
立构规整橡胶	立構橡膠	stereo rubber
立构规整性	立體規律性	steric regularity
立构嵌段共聚物	立構嵌段共聚物	stereoblock copolymer
立构无规共聚物	隨機立構共聚物	stereorandom copolymer
立构显微照相法	立體顯微照相法	stereomicrography
立构有规性	立體規則性	stereoregularity
立构有择聚合	立構選擇聚合	stereoselective polymerization
立构杂化作用	立構混成作用	stereohybridization
立构中心	立構中心	stereocenter
立式蒸煮器	直立式消化器	vertical digester
立体定向反应	立體特定反應	stereospecific reaction
立体对称均聚物	立體對稱均聚物	stereosymmetrical homopolymer
立体化学	立體化學	stereochemistry

大　陆　名	台　湾　名	英　文　名
立体结构	空間結構	spatial structure
立体聚合物	立體聚合物	space polymer
立体排列	空間組態	spatial configuration
立体网状聚合物	立體網形聚合物	space network polymer
立体序列	立體序列	stereosequence
立体序列分布	立體序列分佈	stereosequence distribution
立体选择性	立體選擇性	stereoselectivity
立体异构体	立體異構物	stereoisomer
立体有规构型	立構特定組態	stereospecific configuration
立体有择催化剂	立構特定觸媒	stereospecific catalyst
立体专一性	立體特定性	stereospecificity
利润	利潤	profit
利润函数	利潤函數	profit function
利润限度	利潤率	profit margin
利润因子	利潤因數	profitability factor
利润指数	利潤指數	profitability index
利息	利息	interest
利息回收期	利息回收期	interest recovery period
沥青	瀝青,柏油	asphalt
沥青漆	瀝青漆,柏油漆	asphalt paint
沥青砂,焦油砂	溚砂,焦油砂	tar sand
砾磨机	卵石磨	pebble mill
粒度	粒度,粒子大小	particle size
粒度测量	粒度分析	granulometry
粒度分布	粒度分佈,粒子大小分佈,粒徑分佈	particle size distribution, size distribution
粒度分离	粒度分離,篩析	size separation
粒度分离器	顆粒分離器	particle size separator
粒度分析	粒度分析,粒徑分析	size analysis, particle size analysis, grain size analysis
粒度分析仪	晶粒分析儀,粒度計,粒徑析儀	grain size analyzer, granulometer, particle size analyzer
粒度增大	尺寸增大	size enlargement
粒间扩散	粒間擴散	interparticle diffusion
粒径	粒[子直]徑	particle diameter
粒内扩散	粒內擴散	intraparticle diffusion
粒内扩散系数	粒內擴散係數	intraparticle diffusivity
粒状催化剂	錠粒觸媒	pelleted catalyst

大　陆　名	台　湾　名	英　文　名
粒状固体过滤器	粒狀固體過濾機	granular solid filter
粒子	粒子	particle
粒子动力学	粒子動力學	particle kinetics
粒子旋风分离器	粒子旋風分離器	particle cyclone separator
粒子转化	粒子轉化	particle conversion
连串反应	逐次反應	consecutive reaction
连续操作	連續操作	continuous operation
连续萃取	連續萃取	continuous extraction
连续萃取器	連續萃取器	continuous extractor
连续分馏	連續分餾	continuous fractionation
连续干燥器	連續乾燥器	continuous drier
连续过程	連續程序	continuous process
连续过滤	連續過濾	continuous filtration
连续过滤器	連續過濾器	continuous filter
连续加工	連續加工[法]	continuous processing
连续搅拌槽反应釜设计	連續流動攪拌反應器設計	CSTR design
连续搅拌槽反应器	連續攪拌槽反應器	continuous stirred tank reactor, CSTR
连续搅拌反应器	連續攪拌反應器	continuous stirred reactor
连续搅拌反应器组	連續攪拌反應器組	battery of continuous stirred tank reactors, battery of CSTRs
连续结晶	連續結晶	continuous crystallization
连续结晶器	連續結晶器	continuous crystallizer
连续介质	連續介質	continuum
连续聚合	連續聚合[法]	continuous polymerization
连续离心机	連續離心機	continuous centrifuge
连续流动反应器	連續流動反應器	continuous flow reactor
连续流动式搅拌槽反应器	連續攪拌槽反應器	continuous flow stirred tank reactor
连续流动系统	連續流動系統	continuous flow system
连续培养	連續式培養	continuous culture
连续皮带鼓式过滤器	連續輸送鼓式過濾器	continuous belt drum filter
连续汽提器	連續汽提器	continuous stripping still
连续倾析	連續傾析[法]	continuous decantation
连续式反应器	連續式反應器	continuous reactor
连续式炭滤器	連續式碳濾器	continuous carbon filter
连续隧道干燥器	連續隧道式乾燥器	continuous tunnel drier
连续隧道窑	連續隧道窯	continuous tunnel kiln

大 陆 名	台 湾 名	英 文 名
连续炭滤法	連續碳濾[法]	continuous carbon filtration
连续细丝	連續細絲	continuous filament
连续现金流动	連續現金流	continuous cash flow
连续相	連續相	continuous phase
连续性定律	連續定律	continuity law
连续性方程	連續方程式	continuity equation, equation of continuity
连续循环法	連續環圈法	continuous-cycling method
连续研磨机	連續研磨機	continuous grinder
连续窑	連續窯	continuous kiln
连续蒸馏	連續蒸餾	continuous distillation
连续蒸煮器	連續消化槽,連續蒸煮器	continuous digester
联产品	聯產物	coproduct
联管节	接頭	coupling
联立方程法	聯立方程法	equation-oriented approach, equationsol-ving approach
炼厂气	煉油氣	refinery gas
炼焦过程	焦化程序,煉焦法	coking process
[炼]焦炉	煉焦爐	coke oven
[炼]焦煤	煉焦煤	coking coal
炼油厂	煉油廠	petroleum refinery, refinery
链段布朗运动	鏈段布朗運動	segmental Brownian motion
链段各向异性	鏈段各向異性	segment anisotropy
链段密度分布	鏈段密度分佈	segment-density distribution
链段摩擦因子	鏈段摩擦因子	segmental friction factor
链段相互作用参数	鏈段交互作用參數	segment-interaction parameter
链段旋转	鏈段旋轉	segment rotation
链段跃迁频率	鏈段躍遷頻率	segmental jump frequency
链段运动	鏈段運動	segmental motion
链反应	鏈反應	chain reaction
链节	[鏈]段	segment
链解聚[作用]	鏈解聚合[反應]	chain depolymerization
链聚合	鏈聚[合][作用]	chain polymerization
链霉素	鏈黴素	streptomycin
链式输送机	鏈式運送機,鏈運機	chain conveyor
链引发	鏈引發	chain initiation
链增长	鏈傳播[反應]	chain propagation
链增长反应	連鎖反應傳播,鏈[鎖]	propagation of chain reaction, propagation

大　陆　名	台　湾　名	英　文　名
	反應傳播,增長反應	reaction
链终止	鏈終止[作用]	chain termination
链终止反应	鏈終止反應	chain termination reaction
链转移	鏈轉移[作用],鏈傳遞[作用]	chain transfer
两段串级系统	二階串級系統	two-stage cascade system
两段蒸馏	二階蒸餾	two-stage distillation
两级导向阀	二階導引閥	two-stage pilot valve
两流体理论	雙流體說	two-fluid theory
两相传热	兩相熱傳	two-phase heat transfer
两相流	兩相流	two-phase flow
两相流动型式	兩相流動型式	two-phase flow pattern
两相模型	兩相模式	two-phase model
两相区	兩相區[域]	two-phase region
两相系统	兩相系統	two-phase system
两性霉素 B	雙性[殺]黴素 B	amphotericin
两液体理论	兩液體理論	two-liquid theory
亮度	亮度	brightness
量纲	尺寸	dimension
量纲不变性	因次不變性	dimensional invariance
量纲分析,因次分析	因次分析	dimensional analysis
量纲均匀性	因次均勻性	dimensional homogeneity
量纲一致性	因次一致性	dimensional consistency
量瓶	量瓶	measuring flask
量气管	量氣管	gas measuring tube
量热计	熱量計,卡計	calorimeter
量热学	熱量測定法,量熱法	calorimetry
量酸槽	量酸槽	acid measuring tank
量子效应	量子效應	quantum effect
料仓松动器	料倉鬆動器	bin activator
料仓卸料器	料倉卸料器	bin discharger
料槽	槽	trough
列管换热器	殼管熱交換器	shell-and-tube heat exchanger
列管式反应器	殼管反應器	shell-and-tube reactor
列线图	列線圖	alignment chart
裂变	分裂	fission
裂变产物	[核]分裂產物	fission product
裂变能	[核]分裂能	fission energy

大　陆　名	台　湾　名	英　文　名
裂缝	裂縫	fissure
裂化催化剂(＝裂解催化剂)		
裂化活性	裂解活性	cracking activity
裂化(＝裂解)		
裂化气(＝裂解气)		
裂解催化剂,裂化催化剂	裂解觸媒	cracking catalyst
裂解度	裂解度	cracking severity
裂解反应	裂解反應	cracking reaction
裂解分馏塔	裂解分餾塔	cracking fractionator
裂解活性	裂解活性	activity cracking
裂解,裂化	裂解,裂煉	cracking
裂解炉	裂解爐	cracking furnace
裂解气,裂化气	裂解氣	cracked gas
裂解汽油	裂解汽油	cracked gasoline
裂解室	裂解室	cracking chamber
裂解蒸馏	裂解蒸餾	cracking distillation
裂膜纱	[撕]裂[薄]膜紗	splitting yarn
林产化学品	木材化學品	silvichemical
临界比容	臨界比容	critical specific volume
临界常数	臨界常數	critical constant
临界尺寸	臨界尺寸	critical size
临界等温线	臨界等溫線	critical isotherm
临界底坡	臨界斜率	critical slope
临界点	臨界點	critical point
临界点轨迹	臨界點軌跡	locus of critical point
临界电位	臨界勢	critical potential
临界反压比	臨界反壓比,臨界背壓比	critical back pressure ratio
临界共溶温度	臨界互溶溫度	critical solution temperature, consolute temperature
临界喉道流速	臨界喉道流速	critical throat velocity
临界混合温度	臨界混合溫度	critical mixing temperature
临界胶束浓度	臨界微胞濃度	critical micelle concentration
临界晶核尺寸	臨界晶核尺寸	critical nucleus size
临界流	臨界流[動]	critical flow
临界流动量	臨界流動量	critical flow capacity

大 陆 名	台 湾 名	英 文 名
临界密度	臨界密度	critical density
临界摩尔体积	臨界莫耳體積	critical molar volume
临界凝析温度	臨界凝結溫度	cricondentherm
临界凝析压力	臨界凝結壓	cricondenbar
临界浓度	臨界濃度	critical concentration
临界频率	臨界頻率	critical frequency
临界溶解点	臨界溶解點	critical solution point
临界溶解度	臨界溶解度	critical solubility
临界深度	臨界深度	critical depth
临界湿度	臨界濕度	critical humidity
临界湿含量	臨界水分	critical moisture content
临界时间	臨界時間	critical time
临界速度	臨界速度	critical velocity
临界体积	臨界體積,臨界容積	critical volume
临界填料尺寸	臨界填料尺寸	critical packing size
临界途径	臨界途徑	critical path
临界温差	臨界溫差	critical temperature difference
临界温度	臨界溫度	critical temperature
临界现象	臨界現象	critical phenomenon
临界泄压器件	臨界洩壓裝置	critical pressure relief device
临界性质	臨界性質	critical property
临界压力	臨界壓[力]	critical pressure
临界压力比	臨界壓[力]比	critical pressure ratio
临界压缩比	臨界壓縮比	critical compressibility ratio
临界压缩因子	臨界壓縮因子	critical compressibility factor
临界音速性质	臨界音速性質	critical sonic property
临界增益	臨界增益	critical gain
临界直径	臨界直徑	critical diameter
临界值	臨界值	critical value
临界质量	臨界質量	critical mass
临界转速	臨界速率	critical speed
临界状态	臨界條件	critical state
临界阻尼	臨界阻尼	critical damping
临时费用因数	應急費用因數	contingency factor
淋粒反应器	淋粒反應器	raining solid reactor
磷肥	磷肥	phosphate fertilizer
磷光	磷光,燐光	phosphorescence
磷硅酸盐玻璃	磷矽酸鹽玻璃	phosphosilicate glass

大　陆　名	台　湾　名	英　文　名
磷灰石	磷灰石	apatite
磷酸化[作用]	磷酸化[反應]	phosphorylation
磷酸己糖旁路途径	磷酸己糖旁路途徑	hexose phosphate shunt pathway
磷酸盐玻璃	磷酸鹽玻璃	phosphate glass
磷酸盐缓冲剂	磷酸鹽緩衝劑	phosphate buffer
磷酸盐岩	磷鹽岩	phosphate rock
磷酸盐增效助剂	磷酸鹽增滌劑	phosphate builder
磷铁	磷鐵齊	ferrophosphorus
磷脂	磷脂[質]	phospholipid
灵敏度	靈敏度	sensitivity
灵敏度分析	靈敏度分析	sensitivity analysis
灵敏度估计	靈敏度估計	sensitivity estimate
菱镁矿	菱鎂礦	magnesite
菱铁矿	菱鐵礦	siderite
菱锌矿	菱鋅礦	zinc spar
零点平衡	零點平衡	null balance
零电位器	零點電位計	null potentiometer
零级反应	零階反應	zero-order reaction
零记忆系统	零記憶系統	zero memory system
零假设	虛無假設	null hypothesis
零频率增益	零頻率增益	zero frequency gain
零通量表面	零通量表面	zero-flux surface
零压状态	零壓狀態	zero-pressure state
流变测量学	流變測定法	rheometry
流变性质	流變性質	rheological property
流变学	流變學	rheology
流变仪	流變儀	rheometer
流程图	流程圖	flow diagram, flowsheet
流程图符号	流程圖符號	flow diagram symbol, flowsheet symbol
流程图设计	流程圖設計	flowsheet design
流程综合	流程組合	flowsheet synthesis
流出	流出	effluence
流出管	放流管,出料管	effluent pipe
流出式黏度计	流出黏度計	efflux viscometer
流出物	流出物	effluent
流动	流動	flow
流动比	流動比[率]	current ratio
流动法	流動法	flow method

大　陆　名	台　湾　名	英　文　名
流动反应器	流動反應器	flow reactor, mobile reactor
流动功	流動功	flow work
流动混合器	流動混合器	flow mixer
流动量热计	流動卡計	flow calorimeter
流动落后	流動滯延	flow lag
流动双折射	流動雙折射	streaming birefringence
流动网	流動網	flow net
流动系统	流動系統	flow system
流动显示	流動顯示	flow visualization
流动线	流動線	flow line
流动相	流動相	mobile phase
流动行为指数	流動行為指數	flow behavior index
流动资产	流動資產	current asset
流动资金	營運資金	working capital
流动阻力	流動阻力	flow resistance
流函数	流線函數	stream function
流化焙烧炉	流[體]化焙燒爐	fluidized roaster
流化床	流[體]化床	fluidized bed
流化床反应器	流[體]化床反應器	fluidized bed reactor
流化床干燥器	流[體]化床乾燥器	fluidized bed dryer
流化床锅炉	流[體]化床鍋爐	fluidized bed boiler
流化床气化器	流[體]化床氣化器	fluidized bed gasifier
流化床燃烧	流[體]化床燃燒	fluidized bed combustion
流化床燃烧器	流[體]化床燃燒器	fluidized bed combustor
流化床吸附器	流[體]化床吸附器	fluidized bed adsorber
流化床转化	流[體]化床轉化	fluidized-bed conversion
流化催化剂床	流[體]化觸媒床	fluidized catalyst bed
流化反应器	流[體]化反應器	fluidized reactor
流化混合	流[體]化混合	fluidized mixing
流化数	流[體]化數	fluidization number
流化速度	流[體]化速度	fluidizing velocity
流化蒸馏	流[體]化蒸餾	fluidized distillation
流颈接头	縮口分接頭	vena contracta tap
流颈,缩脉	收縮口	vena contracta
流控技术	流控技術	fluidics
流量比例调节	流量比控制	ratio-flow control
流量测量设备	流量測量裝置	flow measuring equipment
流量积分	流量積分	flow integration

大　陆　名	台　湾　名	英　文　名
流量积分器	流量積分器	flow integrator
流量计	流量計	flow gage, flowmeter
流量控制	流量控制	flow control
流量喷嘴	流體噴嘴	flow nozzle
流量系数	排放係數	discharge coefficient
流量指示计	流量指示器	flow indicator
流率	流率	flow rate
流能磨	流體能量研磨機	fluid energy mill
流入物	流入物	influent
流水作业	流動程序	flow process
流速计	流速計, 電流計	current meter
流态化	流體化	fluidization
流体	流體	fluid
流体动力边界层	流[體動]力邊界層	hydrodynamic boundary layer
流体动力稳定性	流[體動]力穩定性	hydrodynamic stability
流体动力相互作用	流[體動]力相互作用	hydrodynamic interaction
流体动力效应	流[體動]力效應	hydrodynamic effect
流体动力学	流體動力學	fluid dynamics
流体高差	流體高差	fluid head
流体-固体反应	流體-固體反應	fluid-solid reaction
流体静力	流體靜力	hydrostatic force
流体静力计	流體靜力計	hydrostatic force meter
流体静压条件	流體靜力狀態	hydrostatic condition
流体力学	流體力學	fluid mechanics, hydromechanics
流体力学平滑表面	流力平滑表面	hydrodynamically smooth surface
流体流动	流體流動	fluid flow
流体流动阻力	流體流動阻力	fluid flow resistance
流体-流体反应	流體-流體反應	fluid-fluid reaction
流体摩擦	流體摩擦	fluid friction
流体速度	流體速度	fluid velocity
流体相	流體相	fluid phase
流体运动学	流體運動學	fluid kinematics, hydrokinematics
流线	流線	streamline
流线截取	流線截取	flow-line interception
流线式过滤器	流線式濾器	streamlined filter
流线型阀	流線形閥	streamlined valve
流线型化	流線化	streamlining
流线型物体	流線形物體	streamlined body

大　陆　名	台　湾　名	英　文　名
流线运动	流線運動	streamline motion
流线坐标	流線坐標	streamline coordinates
流型	流動型式	flow pattern
硫	硫	sulfur
硫醇	硫醇,硫醇[類]	thiol, mercaptan
硫醇改性橡胶	硫醇改質橡膠	thiolmodified rubber
硫代水解	硫化氫解[反應]	thiohydrolysis
硫靛蓝	硫靛	thioindigo
硫化	硫化,交聯	vulcanization, sulfurization
硫化度	硫化度	sulfidity
硫化锅	硫化鍋	vulcanizing pan
硫化过程	硫化作用	sulfidation
硫化剂	硫化劑	vulcanizing agent
硫化加速剂	硫化加速劑	accelerator of vulcanization
硫化器	硫化器	vulcanizer
硫化染料	硫化染料	sulfur dye
硫化室	硫化室	vulcanizing chamber
硫化系数	硫化係數	vulcanization coefficient
硫化纤维纸	硬化紙	vulcanized fiber paper
硫化橡胶	硫化橡膠	vulcanizate
硫化仪	硬化計	cure meter
硫化油	硫化油	vulcanized oil
硫化状态	硬化程度	state of cure
硫[磺]硫化	[硫黃]硫化	sulfur vulcanization
硫交联	硫交聯	sulfur crosslinking
硫链丝菌肽	硫鏈絲菌肽	thiostrepton
硫脲树脂	硫脲樹脂	thiourea resin
硫漂白	硫[黃]漂白	sulfur bleach
硫氢解反应	硫氫解[反應]	thiohydrogenolysis
硫砷铜矿	硫砷銅礦	enargite
硫酸法	硫酸法	sulfuric acid process
硫酸化	硫酸化	sulfating
硫酸化油	硫酸化油	sulfated oil
硫酸盐化	硫酸化[反應]	sulfation
硫酸盐浆	牛皮紙漿	kraft pulp
硫酸盐纸浆	硫酸鹽紙漿	sulfate pulp
硫酸盐制纸浆法	硫酸鹽方法,硫酸鹽處理	sulfate process

大　陆　名	台　湾　名	英　文　名
硫酸酯	硫酸酯	sulfuric acid ester
馏程	蒸餾範圍	distillation range
馏出液	餾出液	distillate
六方最密堆积晶格	六方最密堆積	hexagonal closest packed lattice
六聚物	六聚物	sexamer
笼式水压机	籠式水壓機	cage hydraulic press
笼效应	籠[蔽]效應	cage effect
漏液	滴流	weeping
漏液孔	滴流孔	weeping hole
露点	露點	dew point
露点计	露點計	dew point meter
露点记录器	露點記錄器	dew point recorder
露点温度	露點溫度	dew point temperature
露点压力	露點壓力	dew point pressure
炉缸	爐爐,膛,爐床	hearth
炉黑	爐黑	furnace black
炉渣	熔渣	slag
卤化	鹵化[反應]	halogenation
鲁棒过程控制	韌性程序控制	robust process control
鲁棒稳定化	韌性穩定化	robust stabilization
鲁棒稳定性	韌性穩定化度	robust stabilizability
鲁棒性	韌性	robustness
滤板	濾板	filter plate
滤饼	濾餅	filter cake
滤饼运输机	濾餅運送機	filter cake conveyor
滤布	濾布	filter cloth
滤池深度	濾池深度	filter depth
滤袋	濾袋	filter bag
滤镜反光计	濾鏡反光計	mirror filter reflectometer
滤框	濾框	filter frame
滤膜	濾膜	filtration membrane
滤泥机	濾泥機	mud press
滤砂	濾砂	filter sand
滤桶	濾桶	filter drum
滤箱	濾箱	filter box
滤液	濾液	filter liquor
滤纸	濾紙	filter paper
路径函数	路徑函數	path function

大 陆 名	台 湾 名	英 文 名
路径追踪	路徑追蹤	path tracing
辘轳	簸選機	jigger
铝箔	鋁箔	aluminum foil
铝矾土	鋁礬土	bauxite
铝管	鋁管	aluminum pipe
铝媒染剂	鋁媒染劑	aluminum mordant
铝热剂	鋁熱劑	thermite
铝砂	鋁砂	aloxite
铝皂	鋁皂	aluminum soap
铝质黏土	鋁氧黏土	bauxitic clay
绿宝石	純綠寶石	emerald
绿脱石	綠脫石,矽鐵石	nontronite
绿液	綠液	green liquor
绿柱石	綠柱石,綠寶石	beryl
氯胺	氯胺	chloramine
氯丁橡胶	氯丁二烯橡膠,氯平橡膠,新平橡膠	chloroprene rubber, neoprene, neoprene rubber
氯丁橡胶黏合剂	新平黏合劑	neoprene adhesive
氯化	氯化[反應]	chlorination
氯化苯醌	氯化[苯]醌	chlorinated quinone
氯化酚	氯化酚	chlorinated phenol
氯化钾	氯化鉀	muriate of potash
氯化硫溶液硫化	氯化硫[溶液]硫化	sulfur chloride vulcanization
氯化石蜡	氯[化]蠟	chlorinated paraffin
氯化烃	氯[化]烴	chlorinated hydrocarbon
氯化物接触室	氯接觸槽	chloride contact chamber
氯化橡胶	氯化橡膠	chlorinated rubber
氯磺化聚乙烯	氯磺化聚乙烯	chlorosulfonated polyethylene, chlorosulfonic polyethylene
氯磺酰化	氯磺化	chlorosulfonation
氯霉素	氯黴素	chloromycetin
氯乙烯单体	氯乙烯單體	vinyl chloride monomer
卵磷脂	卵磷脂	lecithin
卵石	卵石	pebble
卵石层	卵石床	pebble bed
卵石加热器	卵石加熱器	pebble heater
轮碾机	輪輾機	edge runner
轮胎	輪胎	tire

大　陆　名	台　湾　名	英　文　名
轮胎帘线	輪胎簾布	tire cord
轮胎砂	輪胎紗	tire yarn
轮形动物	輪蟲	rotifer
罗茨鼓风机	魯氏鼓風機	Roots blower
螺带混合机	螺條混合機	ribbon mixer
螺带搅拌器	螺帶攪拌器	helical ribbon agitator
螺杆	螺桿	screw
螺杆泵	螺桿泵	screw pump
螺杆出料机	拔螺釘器	screw extractor
螺杆挤出反应器	螺桿擠出反應器	screw extruder reactor
螺杆式预塑化注压成型机	螺桿式預塑化射出成型機	screw preplasticizing type injection molding machine
螺杆式注塑成型机	螺桿式射出成型機	screw type injection molding machine
螺杆效率	螺桿效率	screw efficiency
螺杆芯孔销	螺桿蕊栓	screw core pin
螺杆压出机	螺桿擠出機	screw extruder
螺栓	螺栓	bolt
螺型位错	螺旋差排	screw dislocation
螺旋	螺旋	spiral
螺旋板换热器	螺旋板熱交換器	spiral plate heat exchanger
螺旋板式冷凝器	螺旋冷凝器	spiral condenser
螺旋齿轮泵	螺旋齒輪泵	spiral gear pump
螺旋翅片	螺旋鰭片	helical fin
螺旋管加热器	螺旋管加熱器	spiral coil heater
螺旋滑线	螺旋滑線	helical slidewire
螺旋环	螺旋環	spiral ring
螺旋挤出	螺桿擠出成形	screw extrusion
螺旋挤出机	鑽探機	auger machine
螺旋桨	螺旋槳,推進器	propeller
螺旋桨风扇	螺旋槳型風扇	propeller type fan
螺旋桨式搅拌器	螺旋槳攪拌器	propeller agitator
螺旋桨式叶轮	螺旋槳式葉輪	propeller impeller
螺旋搅拌器	螺旋攪拌器	helical agitator
螺旋结构	螺旋結構	screw structure
螺旋聚合物	螺旋聚合物	spiropolymer
螺旋流	螺旋流	spiral flow
螺旋线圈	螺旋管	helical coil
螺旋形分离器	螺旋分離器	spiral separator

大　陆　名	台　湾　名	英　文　名
螺旋形高分子	螺旋形聚合物	spiral polymer
螺旋形结构	螺旋結構	spiral structure
螺旋压力弹簧	螺旋壓力彈簧	helical pressure spring, spiral pressure spring
螺旋运输机	螺旋運送機	spiral conveyor
裸管	裸管	bare pipe, bare tube
裸面	裸面	bare surface
裸热电耦	裸熱電耦	bare thermocouple
络合催化剂	配位觸媒	coordination catalyst
络合滴定	錯合[法]滴定	complexometric titration
络合物平衡	錯合物均衡	complex balancing
落后元件	滯延元件	lag element
落角	落角	angle of fall
落球法	落球法	falling ball method
落球黏度计	落球黏度計	falling ball viscometer, falling sphere viscometer
落针黏度计	落針黏度計	falling needle viscometer

M

大　陆　名	台　湾　名	英　文　名
麻油	木麻油	hemp seed oil
麻醉剂	麻醉劑	anesthetic
马丁-侯[虞钧]方程	馬[丁]侯[虞鈞]方程	Martin-Hou equation
马丁-侯[虞钧]状态方程	馬[丁]侯[虞鈞]狀態方程式	Martin-Hou equation of state
马赫数	馬赫數	Mach number
马居尔方程	馬古利斯方程[式]	Margules equation [of state]
马力	馬力	horsepower, hp
麦克斯韦关系	馬克士威關係[式]	Maxwell relation
麦芽	麥芽	malt
麦芽糖	麥芽糖	maltose
麦芽糖酶	麥芽糖酶	maltase
麦芽汁	麥芽漿	wort
脉冲	脈衝	pulse
脉冲法	脈衝法	pulse method
脉冲反应器	脈衝反應器	pulse reactor
脉冲函数	脈衝函數	impulse function, pulse function

大 陆 名	台 湾 名	英 文 名
脉冲搅动	脈衝擾動	impulse disturbance
脉冲流	脈衝流動	pulsing flow
脉冲流技术	脈衝流動法	pulse flow technique
脉冲扰动	脈衝擾動	pulse disturbance
脉冲筛板塔	脈衝篩板塔	pulsed sieve plate column
脉冲响应	脈衝應答,脈衝響應	impulse response, pulse response
脉冲型	脈衝型	impulse type
脉冲转速计	脈衝轉速計	impulse tachometer
脉动	①脈動 ②突波 ③波動 ④衝擊	surging, fluctuation, pulsation
脉动流化床	脈動流[體]化床	pulsating fluidized bed
脉动流速	波動速度	fluctuating velocity
脉动输入	脈動輸入	pulse input
脉动速度	脈動速度	pulsation velocity
脉动阻尼器	脈動阻尼器	pulsation damper
莽草酸	莽草酸	shikimic acid
毛刺	溢料	spew
毛发湿度计	毛髮濕度計	hair hygrometer
毛利	毛利	gross earnings cost
毛细测液器	毛細計	capillarimeter
毛细穿透	毛細穿透	capillary penetration
毛细管	毛細管	capillary tube, capillary
毛细管法	毛細管法	capillary tube method
毛细管冷凝	毛細冷凝[作用]	capillary condensation
毛细管力	毛細[管]力	capillary force
毛细管膜组件	毛細管模組	capillary module
毛细管黏度计	毛細管黏度計	capillary viscometer
毛细管上升	毛細上升	capillary elevation, capillary ascent
毛细管吸收	毛細吸收	capillary absorption
毛细管吸引	毛細吸引,毛細引力	capillary attraction
毛细管型球	毛細管式球	capillary-type bulb
毛细管准数	毛細數	capillary number
毛细瓶	毛細瓶	capillary flask
毛细压力	毛細壓力	capillary pressure
毛细张力	毛細張力	capillary tension
毛纤维	毛纖維	wool fiber
锚式搅拌器	錨式攪拌器	anchor agitator
枚举法	列舉法	enumeration method

大 陆 名	台 湾 名	英 文 名
梅仁油	梅仁油	plum kernel oil
媒染法	媒染法	mordanting process
媒染剂	媒染劑	mordant
媒染偶氮染料	媒染偶氮染料	mordant azo dye
媒染染料	媒染染料	mordant dye
媒染色料	媒染色料	mordant color
媒染助剂	媒染助劑	mordant assistant
煤床	煤床	coal bed
煤脆性	煤脆性	coal friability
煤的脱挥发作用	煤去揮發物作用	coal devolatilization
煤矸石	脈石	gangue
煤加工技术	煤處理技術	coal processing technology
煤焦油	煤焦油,煤溚	coal tar
煤孔隙度	煤孔隙度	coal porosity
煤气	煤氣	coal gas
煤气表	氣量計	gas meter
煤气发生炉	煤氣發生爐,發生爐	gas producer, producer
煤气化	煤氣化[法]	coal gasification
煤气化过程	煤氣化程序	coal gasification process
煤气化炉	煤氣化器	coal gasifier
煤热解	煤熱解	coal pyrolysis
煤田	煤田	coal field
煤衍生气	煤衍生氣	coal-derived gas
煤衍生液[体]	煤衍生液[體]	coal-derived liquid
煤液化	煤液化[法]	coal liquefaction
煤液化过程	煤液化程序	coal liquefaction process
煤油	煤油	kerosene, kerosine
煤油裂解	煤油裂解	kerosene cracking
煤油燃料	煤油燃料	kerosene fuel
煤油乳剂	煤油乳劑	kerosene emulsion
煤油脱脂	煤油脱脂	kerosene degreasing
煤蒸馏	煤蒸餾	coal distillation
煤转化	煤轉化	coal conversion
酶	酶	enzyme
酶半衰期	酶半衰期,酶半生期	half life of enzyme
酶产物复合物	酶產物複合物	enzyme product complex
酶促电催化	酶電催化	enzymatic electrocatalysis
酶促反应	酶反應	enzymatic reaction

大　陆　名	台　湾　名	英　文　名
酶催化	酶催化［作用］	enzyme catalysis
酶催化反应	酶催化反應	enzyme catalyzed reaction
酶-底物反应	酶-受質反應	enzyme-substrate reaction
酶-底物复合物	酶-受質複合物	enzyme-substrate complex
酶电极	酶電極	enzyme electrode
酶发酵	酶發酵［法］	enzyme fermentation
酶法	酶試驗法	enzymatic method
酶法分析	酶分析法	enzymatic analysis
酶法水解	酶水解	enzymatic hydrolysis
酶反应动力学	酶動力學	enzyme reaction kinetics
酶活力	酶活性	enzyme activity
酶联免疫吸附测定	酶聯免疫吸附測定	enzyme-linked immunosorbent assay
酶免疫分析法	酶免疫分析法	enzyme immunoassay
酶膜	酶膜	enzyme membrane
酶失活	酶失活	enzyme deactivation
酶稳定	酶穩定化	enzyme stabilization
酶选择性	酶選擇性	enzyme selectivity
［酶］诱导契合学说	［酶］誘導契合學說	induced fit theory
酶专一性	酶特異性	enzyme specificity
霉菌	黴［菌］	mould, mold
霉菌淀粉酶	黴菌澱粉酶	mold amylase
每分钟转数	每分轉數,分鐘轉數	revolution per minute, rpm
镁橄榄石	矽酸鎂石,鎂橄欖石	forsterite
镁橄榄石瓷	鎂橄欖石瓷,矽酸鎂石瓷	forsterite porcelain
镁橄榄石砖	鎂橄欖石磚,矽酸鎂石磚	forsterite brick
镁铬砖	鎂鉻磚	magnesite-chrome brick
镁砂	氧化鎂	magnesia
镁质耐火材料	鎂氧耐火物	magnesite refractory
镁砖	鎂磚,鎂氧磚	magnesia brick, magnesite brick
蒙特卡罗模拟	蒙地卡羅模擬	Monte Carlo simulation
蒙脱石	蒙脫石,蒙脫土	montmorillonite
弥散流(=分散流)		
弥散模型	分散模型	dispersion model
弥散系数(=分散系数)		
醚	醚	ether
醚化	醚化［作用］	etherification

大 陆 名	台 湾 名	英 文 名
醚数	醚值	ether number
醚值	醚值	ether value
米糠油	米糠油	rice bran oil
米氏常数	米氏常數	Michaelis-Menten constant
米氏动力学	米氏動力學	Michaelis-Menten kinetics
米氏方程	米氏方程[式]	Michaelis-Menten equation
密闭反应器	密閉反應器	closed reactor
密度	密度	density
密度计	密度計	densimeter
密度梯度离心	密度梯度離心	density gradient centrifugation
密度指数	密度指數	density index
密封	密封	seal
密封革	墊皮	packing leather
密封管	密封管	sealed tube
密封罐	密封罐	seal pot
密封胶	密封劑,封口膠,密封劑	sealant, seal gum, sealing compound
密封胶合剂	密封膠合劑	seal cement
密封器	封口機	sealer
密封圈	墊圈	packing ring
密封性能	密封性質	sealing property
密封液	密封液	sealing liquid
密封油	密封油	sealing oil
密孔板	多孔板	porous plate
密炼机	密閉混合器	internal mixer
密媒分离	重媒分離	dense medium separation
密相	稠相	dense-phase
密相流动	重相流動	dense-phase flow
密相流化床	重相流[體]化床	dense-phase fluidized bed
幂函数型方程	冪函數型方程	power function type equation
幂律流体	冪次律流體	power-law fluid
棉纸	棉紙	cotton paper
棉状纱	棉狀紗	staple yarn
棉籽油	棉籽油	cotton-seed oil
免疫电泳	免疫電泳	immuno electrophoresis
免疫吸附	免疫吸附	immunoadsorption
面积计	面積計	area meter
面积流量计	面積流量計	area flowmeter, area type flowmeter
面积平均方程式	面積平均方程式	area-averaged equation

大　陆　名	台　湾　名	英　文　名
面筋	麵筋	gluten
面心晶格	面心晶格	free-centered lattice
描述函数	描述函數	describing function
描图纸	描圖紙	tracing paper
灭火器	滅火器	fire extinguisher
灭菌	滅菌	sterilization
灭菌器	滅菌器	sterilizer
pH 敏感电极	pH 敏感電極	pH-sensitive electrode
敏感元件	感測元件	sensing element
敏化剂	敏化劑	sensitizer
名义应力	標稱應力	nominal stress
明槽	明渠	open channel
明槽流	明渠流	open channel flow
明矾	明礬	alum
明火加热炉	明火爐	fired furnace
明火加热器	明火加熱器	fired heater
明胶	明膠	gelatin
明胶蛋白	明膠蛋白	glutin
明渠	明渠	open channel
命名法	命名[法]	nomenclature
模	模	mold
模[具]	模	die
模口膨胀	模頭膨脹	die swell
模块	模組,組件	module
模量	模數	modulus
模量比	模數比	modulus ratio
模拟	模擬	simulation
模拟仿真	類比模擬	analog simulation
模拟计算	類比計算[法]	analog computation
模拟计算机	類比計算機	analog computer
模拟模型	模擬模型,模擬模式	simulation model
模拟器	模擬器	simulator
模式	圖型,圖樣	pattern
模式识别	圖型辨識	pattern recognition
模式搜索	樣式搜尋[法]	pattern search
模塑	成型,模製	molding
模型	模型	model
模型比尺	模型比例	model scale

大　陆　名	台　湾　名	英　文　名
模型辨识	模型辨識	model identification
模型参数	模型参數	model parameter
模型建立	模式建立	model building
模型试验	模型試驗	model test
模造纸	模造纸	simile paper
膜	膜	membrane
膜泵	膜泵	membrane pump
膜传热系数	膜熱傳係數	film heat transfer coefficient
膜萃取	膜萃取	membrane extraction
膜反应器	[薄]膜反應器	membrane reactor
膜分离	[薄]膜分離	membrane separation
膜分离技术	[薄]膜分離技術	membrane separation technology
膜簧式驱动器	膜簧式致動器	diaphragm-spring actuator
膜技术	膜技術	membrane technology
膜酵母	膜酵母	film yeast
膜扩散	膜擴散	film diffusion
膜扩散控制	膜擴散控制	film diffusion control
膜理论	膜理論	film theory
膜裂纤维	[撕]裂[薄]膜纖維	split fibre
膜滤器	膜濾器	membrane filter
膜囊	膜囊泡	membrane vesicle
膜强度	膜強度	film strength, membrane strength
膜渗透	膜滲透	membrane permeation
膜生物反应器	膜生物回應器	membrane bioreactor
膜式冷凝	膜冷凝	film type condensation
膜式洗涤器	膜式洗滌器	film scrubber
膜式压力计	隔膜壓力計	diaphragm pressure gage
膜式蒸发器	膜蒸發器	film type evaporator
膜试验	膜試驗	film test
膜通透性	膜滲透性	membrane permeability
膜温度	膜溫度	film temperature
膜系数	膜係數	film coefficient
膜相	膜相	film phase
膜蒸馏	[薄]膜蒸餾	membrane distillation
膜支撑物	膜支撑物	membrane support
膜状沸腾	膜沸騰	film boiling
膜状冷凝	膜式冷凝	filmwise condensation
膜阻力	膜阻力	film resistance

大　陆　名	台　湾　名	英　文　名
膜组件	膜組件	membrane module
摩擦	摩擦[力]	friction
摩擦分离器	摩擦分離器	friction separator
摩擦腐蚀	摩擦腐蝕	friction corrosion
摩擦接头	摩擦接頭	friction coupling
摩擦试验	摩擦試驗	friction test
摩擦损失	摩擦損失	friction loss
摩擦损失因子	摩擦損失因數	friction loss factor
摩擦系数	摩擦係數	friction coefficient
摩擦压力降	摩擦壓力降	friction drop
摩擦压头	摩擦高差	friction head
摩擦因子	摩擦因數	friction factor
摩擦阻力	摩擦阻力	frictional resistance, friction drag
摩尔浓度	體積莫耳濃度	molarity, molar concentration
摩尔平衡	莫耳平衡	mole balance
摩尔平均沸点	莫耳平均沸點	molal average boiling point
摩尔平均扩散系数	莫耳平均擴散係數	molar average diffusivity
摩尔平均速度	莫耳平均速度	molar average velocity
摩尔热容	莫耳熱容[量]	molar heat capacity
摩尔溶液	重量莫耳溶液,莫耳溶液	molal solution, molar solution
摩尔湿度	莫耳濕度	molar humidity
摩尔体积	莫耳體積	molar volume
摩尔通量	莫耳通量	molar flux
磨床	磨床,研磨機	grinder, grinding machine
磨耗	磨耗	detrition
磨耗试验	磨耗試驗	abrasion test
磨耗指数	磨耗指數	abrasion index
磨料	[研]磨料	abrasive
磨木浆	磨木漿,細木漿,機械紙漿	groundwood pulp, mechanical pulp
磨木浆法	機械製漿法	mechanical pulping process
磨球	磨球	grinding ball
磨砂玻璃	磨砂玻璃,毛玻璃	frosted glass
磨蚀	侵蝕	erosion
磨损	磨損,磨耗	abrasion, attrition
末煤	碎煤	slack coal
莫来石	莫來石,富鋁紅柱石	mullite

大　陆　名	台　湾　名	英　文　名
莫来石瓷	富鋁紅柱石瓷	mullite porcelain
莫来石耐火材料	富鋁紅柱石耐火材	mullite refractory
莫利尔图	莫里爾圖	Mollier diagram
莫诺生长动力学	莫氏生長動力學	Monod growth kinetics
默弗里效率	莫氏效率	Murphree efficiency
模压	模[製]壓[縮]	molding compression
模压成分	模製成分	molding composition
母体核素	母核種	parent nuclide
母体元素	母元素	parent element
母细胞	母細胞,親細胞	parent cell, mother cell
母液	母液	mother liquor
木材防腐剂	木材防腐劑	wood preservative
木材化学品	木材衍生化學品	wood-derived chemical
木材热解	木材熱解	wood pyrolysis
木材糖化	木材糖化	wood saccharification
木材提取物	木材萃取物	wood extractive
木材蒸馏	木材蒸餾	wood distillation
木瓜蛋白酶	木瓜[蛋白]酶	papain
木浆除滓机	木漿除滓機	slab grating
木焦油	木溚	wood tar
木精	木精	wood alcohol, wood spirit
木煤气	木煤氣	wood gas
木棉油	木棉子油	kapok oil
木片干燥器	木片乾燥器	chip drier
木薯淀粉	樹薯粉	cassava starch
木松香	松香	wood rosin
木素纤维素	木質纖維素	lignocellulose
木炭	木炭	charcoal
木糖	木糖	wood sugar, xylose
木酮糖	木酮糖	xylulose
木杂酚油	木雜酚油	wood creosote
木质化	木質化	lignification
木质磺酸	木質磺酸	lignin sulphonic acid
木质酶	木質酶	ligninase
木[质]素	木質素	lignin
目标产物	目標產物	desired product
目标管理	目標管理	objective management
目标函数	目標函數	objective function

大　陆　名	台　湾　名	英　文　名
目标控制	目標控制	objective control
目标值	目標值	desired value
目录	存量	inventory
钼催化剂	鉬觸媒	molybdenum catalyst

N

大　陆　名	台　湾　名	英　文　名
纳米	奈米	nanometer
纳米材料	奈米材料	nanomaterials
纳米高分子材料	奈米高分子材料	nano-polymer materials
纳米过滤	奈米過濾	nanofiltration
纳米技术	奈米技術	nanotechnology
纳米结构	奈米結構	nanostructure
纳米科学[与]技术	奈米科技	nanoscale science and technology
纳米粒子	奈米粒子,奈米顆粒	nanoparticle
纳米微晶	奈米晶體	nano-crystal
纳米液滴	奈米液滴	nanodroplet
纳氏泵	納氏泵	Nash pump
钠玻璃	鈉玻璃	soda glass
钠长石	鈉長石	albite, soda feldspar
钠灯	鈉燈	sodium lamp
钠钙玻璃	鈉鈣玻璃	soda-lime glass
钠钙长石	鈉鈣長石	soda-lime feldspar
钠[引发]聚合作用	鈉聚合[作用]	sodium polymerization
钠[硬]皂	鈉皂	soda soap
耐电弧性	耐電弧性	arc resistance
耐光作用	耐光作用	photostabilization
耐候性	耐候性	weatherability
耐候性试验	耐候試驗	weathering test
耐火表面	耐火表面	refractory surface
耐火瓷	耐火瓷	refractory porcelain
耐火黏土	耐火黏土,燒黏土	fireclay, chamotte
耐火页岩	耐火頁岩	refractory shale
耐火砖	耐火磚,耐火[黏土]磚	firebrick, fireclay brick, refractory brick
[耐火砖]格子散裂试验	[耐火磚]屏列剝落試驗	panel spalling test
耐硫酸水泥	抗硫酸鹽水泥	sulfate resistant cement

大　陆　名	台　湾　名	英　文　名
耐磨圈	耐磨圈	wearing ring
耐磨性	①耐磨性 ②磨耗阻力	abrasion resistance
耐热性	耐熱性	heat resistance, thermal endurance
耐溶剂性	抗溶劑性	solvent resistance
耐水度	水容限	water tolerance
耐酸板	耐酸板	acid-proof slab
耐酸硅铁	高矽鐵	duriron
耐酸水泥	耐酸水泥	acid-proof cement
耐酸搪瓷	耐酸搪瓷	acid-proof enamel
耐酸物	耐酸	acid resistant
耐酸性	耐酸性	acid resistance
耐酸纸	耐酸紙	acid-proof paper
耐酸砖	耐酸磚	acid-proof brick
耐皂洗[色]牢度	耐皂洗性	soap fastness
耐皂性	耐皂性	soap resistance
萘酚染料	萘酚染料	naphthol dye
难流动[性]	僵硬流	stiff flow
脑磷脂	腦磷脂	cephalin, kephalin
内表面	內表面	internal surface
内[部]构件	內部零件	internals
内部加热	內部加熱	internal heating
内部收益率	內部收益率	internal rate of return
内插	內插法	interpolation
内阀	內閥	inner valve
内耗	內摩擦	internal friction
内回流	內回流	internal reflux
内回流控制	內回流控制	internal reflux control
内径	內徑	inside diameter
内聚功	內聚功	cohesion work
内聚力	內聚力,內聚性	cohesion
内聚能密度	內聚能密度	cohesive density
内扩散	內擴散	internal diffusion
内流	內部流動	internal flow
内酶	內酶	endoenzyme
内磨擦角	內摩擦角	angle of internal friction
内能	內能	internal energy
内燃	內燃	internal combustion
内燃机	內燃機,燃燒機	internal combustion engine, combustion

大　陆　名	台　湾　名	英　文　名
		engine
内扰	内在擾動	internal disturbance
内胎	内胎	inner tube
内吸附	内吸附	internal adsorption
内酰胺	内醯胺	lactam
内循环	内循環	internal recycle
内循环反应器	内循環反應器	internal recirculation reactor
内压力	内[部]壓[力]	internal pressure
内在容量	内在容量	internal capacity
内酯	内酯	lactone
内坐标	内坐標	internal coordinate
能汇	能量匯座	energy sink
能级	能階	energy level
能垒	能量障壁	energy barrier
能量	能量	energy
能量本构方程	能量本質方程式	energetic constitutive equation
能量储存	能量儲存	energy storage
能量传递	能量傳送,能量傳遞	energy transfer
能量当量	能[量]當量	energy equivalent
能量方程	能量方程式	energy equation
能量分布	能量分佈	energy distribution
能量分离剂	能量分離劑	energy separating agent
能量含量	能[量]含量	energy content
能量耗散	能量耗散	energy dissipation
能量级线	能量級線	energy grade line
能量集成	能量集成	energy integration
能量交换器	能量交換器	energy exchanger
能量密度	能量密度	energy density
能量平衡	能量均衡	energy balance
能量平衡方程式	能量均衡方程式	energy balance equation
能量生成曲线	能量生成曲線	energy generation curve
能量守恒	能量守恆	energy conservation
能量守恒原理	能量守恆原理	principle of conservation of energy
能量损耗	能量消耗[量]	energy consumption
能量损失	能量損失	energy loss
能量损失曲线	能量損失曲線	energy loss curve
能量效率	能量效率	energy efficiency
能量学	能量學	energetics

大 陆 名	台 湾 名	英 文 名
能量转化器	能量轉化器	energy converter
能量转换	能量轉換	energy conversion, energy transformation
能量转换工程	能量轉換工程	energy conversion engineering
能量状态方程	能量狀態方程式	energetic equation of state
能流	能[量]通量	energy flux
能谱	能譜	energy spectrum
能态	能態	energy state
能源	能源	energy sources
能源工程	能源工程	energy engineering
能源技术	能源技術	energy technology
能源科学	能源科學,能量科學	energy science
尼龙	尼龍,耐綸	nylon
泥浆沉降器	泥漿沈降器	mud settler
泥炭	泥煤	peat
泥炭沼地	泥煤田,泥煤沼	peat bog
泥釉	泥釉	slip glaze
拟均相模型	擬均相模型	pseudo-homogeneous model
拟稳态	似穩態	quasi-steady state
拟线性化	準線性化	quasi-linearization
逆变换	逆變換	inverse transformation
逆反冷凝	降壓冷凝	retrograde condensation
逆反应	逆反應	reverse reaction
逆反作用	[澱粉]老化,衰退[作用]	retrogradation
逆分馏法	逆相分餾[法]	reverse fractionation
逆扩散	逆[向]擴散	counter diffusion
逆流	對向流[動]	countercurrent flow, counterflow
逆流操作	對向流操作	countercurrent operation
逆流萃取	對向流萃取	countercurrent extraction
逆流反应器	對向流反應器	countercurrent flow reactor
逆流过程	對向流程序	countercurrent process
逆流喷射式冷凝器	對向流噴凝器	countercurrent jet condenser
逆流倾析	對向流傾析[法]	countercurrent decantation
逆流塔过程	對向流塔程序	countercurrent column process
逆流网络	對向流網路	countercurrent network
逆流吸收	逆流吸收,對向流吸收	absorption countercurrent, countercurrent absorption
逆流洗涤	對向流洗滌	countercurrent washing

大　陆　名	台　湾　名	英　文　名
逆溶度曲级	反轉溶解度曲線	inversed solubility curve
逆向反应	逆[向]反應	backward reaction
逆向进料	逆向進料	backward feed
逆压梯度	逆壓力梯度	adverse pressure gradient
年金表	年金表	annuity table
年龄分布	年齡分佈	age distribution
年龄分布函数	年齡分佈函數	age distribution function
黏度	黏度	viscosity
黏度测定法	黏度測定法	viscometry, viscosimetry
黏度掺和图	黏度摻配圖	viscosity blending chart
黏度计	黏度計	viscometer, viscosimeter
黏度平均分子量	黏度平均分子量	viscosity average molecular weight
黏度试验	黏度試驗	viscosity test
黏度数	黏度值	viscosity number
黏度系数	黏度係數	viscosity coefficient
黏度指数	黏度指數	viscosity index
黏附	黏著	adhesion
黏附功	黏附功	adhesion work
黏附力	黏著力	adhesion force
黏附能	黏著能	adhesion energy
黏附强度	黏著強度	adhesion strength, adhesive strength
黏附润湿	黏著潤濕	adhesional wetting
黏附试验	黏著試驗	adhesion test, adhesive test
黏附张力	黏著張力	adhesive tension
黏合剂	黏合劑,黏結劑,黏著劑	binder, binding agent, adhesive agent, adhesive
黏合剂配方	黏著劑配方	adhesive formulation
黏合强度	黏合強度,鍵結強度	bonding strength
黏合吸引	黏著吸引	adhesive attraction
黏胶人造短纤维	嫘縈棉	viscose staple
黏胶人造丝	嫘縈	viscose rayon
黏菌素	黏菌素	colistin
黏流态	可塑狀態	plastic state
黏泥	黏泥,黏液	slime
黏塑性	黏塑性	viscoplasticity
黏塑性流体	黏塑性流體	viscoplastic fluid
黏态	黏性階段	sticky stage
黏弹性	黏彈性	viscoelasticity

大　陆　名	台　湾　名	英　文　名
黏弹性流体	黏彈性流體	viscoelastic fluid
黏土	黏土	clay
黏性	黏性	stickiness
黏性底层	黏性次層	viscous sublayer
黏性耗散	黏性耗散	viscous dissipation
黏性减震器	黏性阻尼器	viscous damper
黏性力	黏滯力	viscous force
黏性流动	黏性流［動］	viscous flow
黏性应力	黏性應力	viscous stress
黏性运动	黏性運動	viscous motion
黏性阻力	黏性阻力	viscous resistance
黏滯发酵	黏液發酵	slimy fermentation
碾磨机	磨碎機	attrition mill
碾碎机	澆道	runner
酿酶	解醣酶	zymase
酿造	釀造	brewing
酿造厂	釀酒廠	brewery
鸟粪磷酸盐	［海］鳥糞磷礦	guano phosphate
鸟嘌呤	鳥嘌呤	guanine
鸟嘌呤酶	鳥嘌呤酶	guanase
尿素	尿素，脲	urea
尿烷	胺甲酸乙酯	urethane
脲酶	脲酶，尿素酶	urease
脲醛树脂	脲甲醛樹脂	urea-formaldehyde resin
捏合	捏合	kneading
捏合机	捏合機	kneader
镍电阻球	鎳電阻球	nickel resistance bulb
镍坩埚	鎳坩堝	nickel crucible
镍钢	鎳鋼	nickel steel
镍铬合金	鎳鉻合金	nichrome
镍铝氧催化剂	鎳鋁氧觸媒	nickel alumina catalyst
镍蓄电池	鎳蓄電池	nickel storage battery
柠檬草油	檸檬草油	lemon-grass oil
凝固	固化［作用］	solidification
凝固点	凝固點，固化點	freezing point, solidification point, solidifying point
凝固点降低	凝固點下降	freezing point depression
凝固过程	固化程序	solidification process

大　陆　名	台　湾　名	英　文　名
凝固时间	設定時間,凝結時間,固化時間	set time, setting time
凝固温度	凝固溫度	setting temperature
凝华作用	反升華[作用]	desublimation
凝胶	凝膠	gel
凝胶过滤	凝膠過濾	gel filtration
凝胶过滤色谱[法]	凝膠過濾層析法	gel filtration [chromatography]
凝胶免疫电泳	凝膠免疫電泳	gel immunoelectrophoresis
凝胶色谱[法]	凝膠層析法	gel chromatography
凝结热	凝聚熱	heat of coagulation
凝结相	凝相	condensed phase
凝聚槽	凝聚槽	coagulation tank
凝聚剂	凝聚劑	coagulant, coagulating agent
牛顿-拉弗森法	牛[頓]-拉[福森]法	Newton-Raphson method
牛顿流体	牛頓流體	Newtonian fluid
牛顿收敛法	牛頓[迭代]法	Newton method for convergence
牛皮纸	牛皮紙	kraft paper
牛皮纸浆法	牛皮紙漿法	kraft process
牛脂	牛脂	tallow
扭辫分析	扭辮分析	torsional braid analysis
扭矩式黏度计	扭矩式黏度計	drag torque type viscometer
扭力黏度计	扭力黏度計	torsion viscometer
扭力天平	扭力天平	torsion balance
扭曲度	偏斜度	skewness
扭转	扭轉,扭力	torsion
扭转流动	扭轉流動	torsional flow
扭转蠕变	扭轉潛變	torsional creep
扭转应力	扭轉應力	torsional stress
农业废物	農業廢料	agricultural waste
浓差电池	濃[度]差電池	concentration cell
浓差极化	濃度極化	concentration polarization
浓度	濃度,濃縮	concentration
浓度边界层	濃度邊界層	concentration boundary layer
浓度差	濃度差	concentration difference
浓度[分布]剖面[图]	濃度輪廓	concentration profile
浓度极限	閾濃度,低限濃度	threshold concentration
浓度扩散	濃度擴散	concentration diffusion
浓度速率曲线	濃度速率曲線	concentration-rate curve

大　陆　名	台　湾　名	英　文　名
浓度梯度	濃度梯度	concentration gradient
浓浆法	濃漿法	thick slurry process
浓硫酸	濃硫酸,礬油	oil of vitriol
浓溶液	濃溶液	rich solution
浓缩器	濃縮器	concentrator
浓缩液	濃縮物	concentrate
努塞特数	那塞數	Nusselt number

O

大　陆　名	台　湾　名	英　文　名
欧姆计	歐姆計,電阻計	ohmmeter
偶氮成分	偶氮成分	azo component
偶氮染料	偶氮染料,不溶性偶氮染料	azo dye, azoic dye
偶氮染色	不溶性偶氮染色	azoic dyeing
偶合	偶合	coupling
偶极	偶極	dipole
偶极矩	偶極矩	dipole moment
偶图	偶圖	bipartite graph
耦合	偶合	coupling
耦合方程式	偶合方程式	coupled equation
耦合系统	偶合系統	coupled system

P

大　陆　名	台　湾　名	英　文　名
帕[斯卡](压力单位)	帕[斯卡](壓力單位)	pascal
排出阀	排放閥	discharge valve
排出栓	排出拴	drawoff cock
排出压头	排放高差	discharge head
排放	排放,發射	emission
排放标准	排放標準	effluent standard, emission standard
排放集管	排放集管[箱]	discharge header
排放口	排放口	drain
排放时间	排放時間	discharge time
排放损失	排放損失	discharge loss
排放物清单	排放物清單	emission inventory

大 陆 名	台 湾 名	英 文 名
排放系数	排放係數,發射係數	emission factor
排放压力	排放壓力	discharge pressure
排风机	風扇	fan
排管	排管	calandria
排管式蒸发罐	排管式蒸發罐	calandria pan
排管型蒸发器	排管型蒸發器	calandria type evaporator
排空	通氣孔	vent
排料	洩料,排放	blowdown
排气泵	排氣泵	air displacement pump
排气阀	排氣閥,釋氣閥	air release valve, exhalation valve
排气管	排氣管,通風管	exhaust pipe, exit tube, vent pipe
排气管线	排放管線	vent line
排气机	排氣機	exhauster
排气及排泄口密封	排氣卸液口密封	vent and drain seal
排气孔	排氣孔	vent
排气锐孔	排氣銳孔	air discharge orifice
排气栓	排氣栓	blow cock
排气系统	排氣系統	air displacement system, gas venting system
排汽	排放蒸汽	exhaust steam
排水管道	排洩管線	drain line
排水台	排水檯	draining table
排泄泵	排洩泵	drain pump
排泄管	排洩管	drain pipe
排泄阱	排洩阱	drain trap
排序	排序	precedence ordering
盘管(=蛇管)		
盘管冷凝器(=蛇管冷凝器)		
盘管散热器(=蛇管散热器)		
盘管蒸发器(=蛇管蒸发器)		
盘滤机	盤濾機	disc filter
盘式过滤机	盤濾機	disk filter
盘式进料器	進料盤	feed disk
盘式离心机	盤式離心機	disc centrifuge
盘式流量计	盤式流量計	disc meter

大　陆　名	台　湾　名	英　文　名
盘式压碎机	盤碎機	disc crusher
盘式蒸发器	盤式蒸發器	disc evaporator
盘形膨胀管	盤形膨脹管	coil expansion pipe
盘柱吸收器	盤柱吸收器	disk column absorber
判据	準則	criterion
旁路	旁路	bypass
旁容量	旁容量	side capacitance
旁通比	旁通比	bypass ratio
旁通阀	旁通閥	bypass valve
旁通管	旁通管	bypass pipe
旁通过滤器	旁通過濾器	bypass filter
旁通控制	旁路控制	bypass control
旁通效应	旁通效應	bypassing effect
抛光	砑光,磨光	burnish, polishing
抛物线阀	抛物線閥	parabolic valve
抛物线溜槽	抛物線槽	parabolic flume
抛物线体聚光器	抛物線體聚光器	paraboloid condenser
泡点	泡點	bubble point
泡点温度	泡點溫度	bubble point temperature
泡点压力	泡點壓力	bubble point pressure
泡核沸腾	成核沸騰	nucleate boiling
泡沫	泡沫	foam, froth
泡沫分离	泡沫分餾	foam fractionation, foam separation
泡沫浮选	泡沫浮選[法]	froth flotation
泡沫硅橡胶	發泡矽氧橡膠	silicone rubber foam
泡沫混凝土	泡沫混凝土	foam concrete
泡沫深度	泡沫深度	froth depth
泡沫塑料	發泡塑膠	foam plastic
泡沫橡胶	泡沫橡膠,海綿橡膠	foam rubber, sponge rubber
泡罩	泡罩	bubble cap, bubbling cap
泡罩板	泡罩板	bubble cap plate, bubble cap tray, bubble plate
泡罩板蒸馏塔	泡罩板蒸餾塔	bubble cap plate distillation tower
泡罩接触器	泡罩接觸器	bubble cap contactor
泡罩塔	泡罩塔	bubble cap column, bubble cap tower, bubble plate column
泡罩塔顶馏分	泡罩塔頂餾分	bubble tower overhead
泡罩吸收塔	泡罩吸收塔	bubble cap absorption column

大　陆　名	台　湾　名	英　文　名
胚胎	胚胎	embryo
培养	培養,培養菌	culture, cultivation
培养器	培養器	cultivator
佩克莱数	佩克萊數	Peclet number
配电盘	儀表板	panel board
配方	①配方 ②公式化	formulation
配管标准	配管標準	piping standard
配管,管路	配管	piping
配件当量长度	配件相當長度	fitting equivalent length
配件当量管长	配件相當管長	fitting equivalent pipe length
配件流程图	[管件]接頭流程圖	fitting flowsheet
配件阻力	[管件]接頭阻力	fitting resistance
配矩法	配矩法	method of moments matching
配位反应	配位反應	coordination reaction
配位共价	配位共價	coordinates covalence
配位化合物	配位化合物	coordination compound
配位聚合	配位聚合[反應]	coordination polymerization
喷灯	噴燈	blast burner
喷动床	噴流床	spouted bed
喷发	噴發	eruption
喷粉塔	噴霧塔	spray tower
喷流速度	噴流速度	spouting velocity
喷漆	噴漆	spray paint
喷气燃料	噴射機燃料	jet fuel
喷气式发动机	噴射引擎	jet engine
喷砂	噴砂	sand-blast
喷射泵	噴射泵	jet pump
喷射出口压	噴射出口壓[力]	jet exit pressure
喷射干燥器	噴射乾燥器	jet drier
喷射冷凝器	噴射冷凝器	jet condenser
喷射器	噴射器,射出器	ejector, eductor
喷射器排列	噴射器排列,射出器排列	ejector arrangement
喷射器容量	噴射器容量,射出器容量	ejector capacity
喷射式洗涤器	噴射式洗滌器	jet type washer
喷射压缩	噴射壓縮	jet compression
喷射液泛	噴射溢流	jet flooding

大 陆 名	台 湾 名	英 文 名
喷水池	噴水池	spray pond
喷水器	噴水器,水刀	water jet
喷丝头	噴絲頭,紡絲頭,紡嘴	spinning jet, spinning head, spinning die, spinneret
喷涂	噴霧	spraying
喷涂法	噴塗法	spray coating process
喷雾	噴霧,噴霧劑	spray
喷雾弹	噴霧彈	aerosol bomb
喷雾阀	噴霧閥	aerosol valve
喷雾分离器	噴霧分離器	spray separator
喷雾干燥	噴霧乾燥	spray drying
喷雾干燥法	噴霧乾燥法	spray drying process
喷雾干燥器	噴霧乾燥器	spray drier, spraying drier
喷雾冷凝器	噴霧冷凝器	spray condenser
喷雾冷却	噴霧冷卻	spray cooling
喷雾冷却结晶器	噴霧冷卻結晶器	spray cooled crystallizer
喷雾器	噴霧器	sprayer
喷雾室	噴霧室	spray chamber
喷雾室接触器	噴霧室接觸器	spray chamber contactor
喷雾塔	噴霧塔	spray column
喷雾推进剂	噴霧[推進]劑,推噴劑	aerosol propellant
喷雾雾化器	噴霧[霧化]器	spray atomizer
喷釉	噴釉	spraying glazing
喷嘴	噴嘴	nozzle
弯头喷嘴	彎頭噴嘴	elbow nozzle
喷嘴压力	噴嘴壓力	nozzle pressure
硼玻璃	硼玻璃	borax glass
硼硅玻璃	硼矽酸玻璃	borosilicate glass
硼砂	硼砂	borax
硼酸盐	硼酸鹽	borate
膨润土	膨土,漿土,膠膨潤土	volclay, wilkinite
膨胀	膨脹,擴大	expansion, dilation, dilatation
膨胀比[挤塑]	膨潤比	swelling ratio
膨胀测定法	膨脹測量法	dilatometry
膨胀床	膨脹床	expanded bed
膨胀床接触器	膨脹床接觸器	expanded bed contactor
膨胀度	膨脹度	degree of expansion
膨胀功	膨脹功	expansion work

大　陆　名	台　湾　名	英　文　名
膨胀环	伸縮圈	expansion loop
膨胀计	膨脹計	dilatometer
膨胀节	脹縮接頭	expansion joint
膨胀器	膨脹器,膨脹機	expander
膨胀损失	膨脹損失	expansion loss
膨胀系数	膨脹係數	expansion coefficient
膨胀压力[模具]	膨潤壓[力]	swelling pressure
膨胀因子	膨脹因子	expansion factor
膨胀应力	膨脹應力	dilatational stress
碰撞理论	硬球理論	hard sphere theory
批号	批號	batch number
批量大小	批式產能	batch size
皮	皮,表層	skin
皮带输送机	帶[式]運[送]機	belt conveyor
皮带运输机	帶運機	band conveyor
皮革	皮革	leather
皮革态	皮革態	leathery state
皮托管	皮托管	Pitot tube
皮心结构	皮芯結構	skin-core structure
皮心效应	皮芯效應	skin and core effect
疲劳强度	疲勞強度	fatigue strength
疲劳试验	疲勞試驗	fatigue test
啤酒酵母	釀酒酵母	brewer yeast
片剂	片劑,錠	tablet
片料吹塑法	薄膜吹製法	sheet blowing method
片型聚合物	片型聚合物	sheet polymer
片状成型	片狀成型	sheet molding
片状成型料	片狀成型塑料	sheet molding compound
偏差	偏差	deviation
偏差变数	偏差變數	deviation variable
偏差形式	偏差形式	deviation form
偏光镜	偏光計	polariscope
偏离	偏差	offset
偏离电位计	偏轉電位計	deflection potentiometer
偏离函数	偏離函數	departure function
偏离角	偏離角,偏向角	angle of deflection
偏摩尔超额物性函数	部分莫耳過剩性質	partial molar excess property
偏摩尔焓	偏莫耳焓,部分莫耳焓	partial molar enthalpy

大　陆　名	台　湾　名	英　文　名
偏摩尔吉布斯自由能	偏莫耳吉布斯自由能, 部分莫耳吉布斯自由 能	partial molar Gibbs free energy
偏摩尔量	偏莫耳[數]量,部分 莫耳[數]量	partial molar quantity
偏摩尔内能	部分莫耳内能,偏莫耳 内能	partial molar internal energy
偏摩尔体积	部分莫耳體積,偏莫耳 體積	partial molar volume
偏摩尔物性函数	部分莫耳性質,偏莫耳 性質	partial molar property
偏微分方程	偏微分方程[式]	partial differential equation
偏相关	部分相關	partial correlation
偏心秤	偏心刻度	eccentric scale
偏心环	偏心環	eccentric ring
偏心渐缩管	偏心漸縮管	eccentric reducer
偏心孔口板	偏心孔口板	eccentric orifice plate
偏心率	偏心率	eccentricity
偏心因子	偏心因子	acentric factor
偏压	偏壓	bias
偏应力	偏向應力,軸差應力	deviatoric stress
偏振	偏光	polarization
偏振化因子	偏光因子	polarization factor
偏振片	偏光玻璃	polaroid
漂移	漂移	drift
漂白	漂白	bleaching
漂白粉	漂白粉	bleaching powder, bleach powder
漂白剂	漂白劑	bleaching agent
漂白土	漂白土	fuller's earth
漂白液	漂白液	bleaching liquor
漂白作用	漂白作用	bleaching action
撇去浮渣	[去]浮沫	skim
贫气	貧氣	lean gas
贫溶液	貧溶液	lean solution
贫油	貧油	lean oil
频带	頻帶	frequency band
频率	頻率	frequency
频率测试	頻率試驗[法]	frequency test

大　陆　名	台　湾　名	英　文　名
频率示踪器	頻率示蹤器	frequency tracer
频率响应	頻率應答	frequency response
频率响应分析	頻率應答分析	frequency response analysis
频率因子	頻率因數	frequency factor
频闪仪	頻閃轉速計	stroboscope
频移	頻移	frequency shift
频域	頻域	frequency domain
平板	平板	flat plate
平板架	棧板	pallet
平板桨	平板槳	flat paddle
平板近似法	平板近似法	flat-plate approximation
平板流动	平板流動	flat-plate flow
平板培养	平板培養	plate culture
平板式换热器	板式熱交換器	plate type heat exchanger
平版印刷术	微影術,蝕刻法	lithography
平槽滤板	平槽濾板	flush plate
平槽压滤板	平槽壓濾板	flush filter plate
平动配分函数	移動分配函數	translational partition function
平管排	［水］平管排	horizontal tube bank
平管式蒸发器	平管式蒸發器	horizontal tube evaporator
平衡	平衡	equilibrium
平衡槽	均化槽	equalization tank
平衡常数	平衡常數	equilibrium constant
平衡点	平衡點	equilibrium point
平衡电动机	平衡馬達	balancing motor
平衡阀	平衡閥	equalizing valve
平衡法	平衡法	equilibrium approach
平衡方程式	平衡方程式	balance equation
平衡分布	平衡分佈	equilibrium distribution
平衡分离过程	平衡分離程序	equilibrium separation process
平衡釜	平衡釜,平衡蒸餾器	equilibrium still
平衡化	平衡化	equilibration
平衡环	平衡環	balancing ring
平衡混合物	平衡混合物	equilibrium mixture
平衡级	平衡階	equilibrium stage
平衡近似	平衡近似［法］	equilibrium approximation
平衡冷凝	平衡冷凝	equilibrium condensation
平衡冷凝曲线	平衡冷凍曲線	equilibrium freezing curve

大　陆　名	台　湾　名	英　文　名
平衡浓度	平衡濃度	equilibrium concentration
平衡判据	平衡準則	equilibrium criterion
平衡汽化	平衡汽化	equilibrium vaporization
平衡曲线	平衡曲線	equilibrium curve
平衡闪蒸	平衡驟汽化	equilibrium flash vaporization
平衡闪蒸曲线	平衡驟汽化曲線	equilibrium flash vaporization curve
平衡湿含量	平衡含水量	equilibrium moisture content
平衡收率	平衡產率	equilibrium yield
平衡水分	平衡水分	equilibrium moisture
平衡速率	達平衡速率	equilibration rate
平衡图	平衡圖	equilibrium diagram, equilibrium plot
平衡系统	平衡系統	equilibrium system
平衡线	平衡線	equilibrium line
平衡蒸馏	平衡蒸餾	equilibrium distillation
平衡转化率	平衡轉化率	equilibrium conversion
平衡组成	平衡組成	equilibrium composition
平滑表面	平滑表面	smooth surface
平桨	平槳	paddle
平均半径	平均半徑	mean radius
平均沸点	平均沸點	average boiling point
平均分子量	平均分子量	average molecular weight
平均过滤比阻	平均過濾比阻	average specific filtration resistance
平均偏差	平均偏差	average deviation, mean deviation
平均速度	平均速度	average velocity
平均停留时间	平均滯留時間	average residence time, mean residence time
平均温差	平均溫差	average temperature difference, mean temperature difference
平均误差	平均誤差	average error, mean error
平均自由程	平均自由徑	mean free path
平面应力	平面應力	plane stress
平切	片	slice
平切薄膜	切片薄膜	sliced film
平台	平臺	platform
平坦效应	平台效應	plateau effect
平推流返混	塞流逆混	plug flow backmixing
平推流,活塞流	塞流,栓流	plug flow
平行处理	平行處理	parallel processing

大　陆　名	台　湾　名	英　文　名
平行反应	平行反應	parallel reaction
平行进料	平行進料	parallel feed
平行失活	平行去活化	parallel deactivation
平移位能	平移能,移動能	translational energy
平移因子	平移因子	shift factor
评估	評估	evaluation
苹果酸	蘋果酸	malic acid
苹果籽油	蘋果籽油	apple seed oil
屏蔽	屏蔽,屏護,影屏	shielding, shadowing
屏蔽长度	屏蔽長度	shielding length
屏蔽有效度	篩有效度	screen effectiveness
瓶颈	瓶頸	bottle neck
坡印亭校正	波印亭[校正]因子	Poynting correction
破裂	破裂	breakage
破乳	破乳化	breakup of emulsion
破碎	壓碎	crush
破碎比	磨碎比	reduction ratio
破碎机	壓碎機,崩解機	crusher, disintegrator
剖面[图]	剖面[圖]	profile
葡聚糖	聚葡[萄]糖,葡聚糖, 葡[萄]聚糖	dextran, glucan, glucosan
葡萄糖	葡萄糖,右旋糖	dextrose, glucose
普朗特数	卜朗特數	Prandtl number
普鲁士蓝	普魯士藍	Prussian blue
普适方程	通式	generalized equation
普适气体常量	通用氣體常數	universal gas constant
普通原料	普通原料	common stock
谱	光譜	spectrum
曝露界限	曝露界限	exposure limit
曝露时间	曝露時間	exposure time
曝露试验	曝露試驗	exposure test
曝露中心	曝露中心	exposed center
曝气	曝氣	aeration
曝气池	曝氣污水塘,通氣污水 塘	aerated lagoon
曝气浮选[法]	曝氣浮選[法]	dispersed-air flotation
曝气机	曝氣機	aerator
曝气水	氣溶水,汽水	aerated water

大　陆　名	台　湾　名	英　文　名
曝气稳定化池	曝氣穩定化池,通氣穩定化池	aerated stabilization pond
曝气系统	曝氣系統,通氣系統	aerated system

Q

大　陆　名	台　湾　名	英　文　名
期望值判据	期望值準則	expected value criterion
漆	漆	lacquer
漆革	漆革,黑漆皮	patent leather
漆料	媒液,展色劑	vehicle
歧点	歧點	bifurcation point
启发式方法	經驗推斷法	heuristic method
启发式规则	試探規則	heuristic rule
起爆	起爆	detonation
起动器	起動器,起動機	starter
起黏丝性	絲黏性	stringiness
起始流化速度	起始流體化速度	incipient fluidizing velocity
起始流化态	起始流體化	incipient fluidization
起酥油	酥脆油	shortening oil
起重机	起重機,吊車	hoist
气泵	氣泵	air pump
气吹磨	風掃磨	air-swept mill
气动测力仪	氣動測力計	pneumatic force meter
气动传动	氣動傳送	pneumatic transmission
气动传动接受器	氣動傳送接受器	pneumatic transmission receiver
气动传动器	氣動傳送器	pneumatic transmitter
气动传动线	氣動傳送線路	pneumatic transmission line
气动传感器	氣動感測器	pneumatic sensor
气动电流继电器	氣動繼電器	pneumatic electric relay
气动阀	氣動閥	air-operated valve, pneumatic valve
气动滑送	氣動滑送	air slide
气动机理	氣動機構	pneumatic mechanism
气动搅拌	氣動攪拌	pneumatic stirring
气动控制	氣動控制	pneumatic control
气动模拟	氣動類比計算機	pneumatic analog
气动设定控制器	氣動設定控制器	pneumatic set controller
气动式流动传送器	氣動式流動傳送器	pneumatic flow transmitter

大　陆　名	台　湾　名	英　文　名
气动式液面计	氣動液位計	pneumatic liquid level gage
气动式液体密度计	氣動液體密度計	pneumatic liquid density gage
气动输运	氣動運送	pneumatic transport
气动调节阀	氣動控制閥	pneumatic control valve
气动调节器	氣動控制器	pneumatic controller
气动系统	氣動系統	pneumatic system
气动信号	氣動信號	pneumatic signal
气动执行机构	氣動致動器	pneumatic actuator
气动装置	氣動裝置	pneumatic device
气固反应	氣固反應	gas-solid reaction
气固反应器	氣固反應器	gas-solid reactor
气固平衡	氣固平衡	gas-solid equilibrium
气化	氣化	gasification
气化反应	氣化反應	gasification reaction, gasifying reaction
气化过程	氣化程序	gasification process
气化技术	氣化技術	gasification technology
气化炉	氣化爐,氣化器	gasifier
气阱	氣阱	gas trap
气冷夹套	氣冷[夾]套	air cooled jacket
气力分级器	氣動類析器	pneumatic classifier
气力输送	氣動運送	pneumatic conveying
气力输送装置	氣動運送機	air conveyor
气量计	氣量計	air meter
气流	氣流	air stream
气流粉碎机	噴射磨機	jet mill
气流干燥器	氣流乾燥機	pneumatic dryer
气流输送器	氣動運送機	pneumatic conveyor
气膜	氣膜	gas film
气膜接触器	氣膜接觸器	gas film contactor
气膜控制	氣膜控制	gas film control
气囊	氣囊	air bag
气泡	氣泡	bubble
气泡反应器	氣泡反應器	bubble reactor
气泡分馏	氣泡分餾[法]	bubble fractionation
气泡计数器	氣泡計數器	bubble counter
气泡剂	發泡劑	bubble forming agent
气泡聚并	氣泡聚結	bubble coalescence
气泡流	氣泡流	bubble flow

大　陆　名	台　湾　名	英　文　名
[气]泡膜	氣泡膜	bubble film
气泡曝气	氣泡通氣	bubble aeration
气泡系统	氣泡[發生]系統	air bubble system
气泡型黏度计	泡式黏度計	bubble type viscometer
气泡云	氣泡雲	bubble cloud
气泡指示计	氣泡指示計	bubble gage
气溶胶	氣溶膠,霧劑	aerosol
气升	氣升	air-lift
气升泵	氣升泵	air-lift pump
气升搅拌	氣升攪拌	air-lift agitation
气升搅拌器	氣升攪拌器	air-lift agitator
气提	氣提	gas stripping
气[体]	氣	gas
气体比重计	量氣計	aerometer
气体常数	氣體常數	gas constant
气体重整	[烴]氣重組	gas reforming
气体定律	氣體定律	gas law
气体分布器	氣體分配器	gas distributor
气体分析仪	氣體分析儀	gas analyzer
气体浮选	氣體浮選[法]	gaseous floatation
气体干管	氣體幹管	gas main
气体量热计	氣體卡計	gas calorimeter
气体排放标准	氣體排放標準	gaseous emission standard
气体燃料	氣體燃料	gaseous fuel
气体燃烧器	氣體燃燒器	gas burner
气体溶液	氣體溶液	gaseous solution
气体渗透	氣體滲透	gas permeation
气体收集器	集氣器	gas receiver
气体调节器	氣體調節器	gas regulator
气体温度计	氣體溫度計	gas thermometer
气体洗涤器	洗氣器	gas scrubber
气田	天然氣田	gas field
气味	氣味	odor
气隙	空氣隙	air gap
气相	氣相	gas phase
气相沉积	氣相沈積[法]	vapor deposition
气相传质系数	氣相質傳係數	gas phase mass transfer coefficient
气相反应	氣相反應	gas phase reaction

大 陆 名	台 湾 名	英 文 名
气相控制	氣相控制	gas phase control
气相色谱法	氣相層析法	gas chromatography
气相色谱图	氣相層析圖	gas chromatogram
气相色谱仪	氣相層析儀	gas chromatograph
气压方程式	氣壓方程式	barometric equation
气压高度	氣壓高度	barometric height
气压管	氣壓管	barometer tube
气压计	氣壓計	air gage, air pressure gage, barometer
气压记录器	氣壓記錄器	barometrograph
气压系统	壓力系統	pressure system
气压效应	氣壓效應	barometric effect
气压柱泵	氣壓泵	barometric leg pump
气液传质设备	氣液質傳裝置	gas-liquid mass transfer equipment
气液反应	氣液反應	gas-liquid reaction
气液反应器	氣液反應器	gas-liquid reactor
气液分离	氣液分離	gas-liquid separation
气液固反应	氣液固反應	gas-liquid-solid reaction
气液固反应器	氣液固反應器	gas-liquid-solid reactor
气液接触	氣液接觸	gas-liquid contacting
气液接触器	氣液接觸器	gas-liquid contactor
气液平衡	氣液平衡	gas-liquid equilibrium
气液平衡过程	氣液平衡程序	gas-liquid equilibrium process
气液吸收	氣液吸收	gas-liquid absorption
气油比	氣-油比	gas-oil ratio
气闸	氣閘	air lock
气罩	氣罩	air cap
汽包	蒸汽鼓	steam drum
汽锤	蒸汽鎚	steam hammer
汽化	汽化	vaporization
汽化器	汽化器	vaporizer
汽化潜热	汽化潛熱	latent heat of vaporization
汽化曲线	汽化曲線	vaporization curve
汽化热	汽化熱	heat of vaporization
汽蚀	成腔現象	cavitation
汽蚀余量	淨正吸高差	net positive suction head, NPSH
汽提	蒸汽汽提	steam stripping
汽提剂	脫色劑	stripping agent
汽提器	汽提器	stripping still

大 陆 名	台 湾 名	英 文 名
汽提塔	汽提塔	stripping column, stripping tower, stripper
汽提用蒸汽	汽提蒸汽	stripping steam
汽相缔合	汽相缔合	vapor phase association
汽液分离器	汽液分離器	vapor-liquid separator
汽液平衡	汽液平衡	vapor-liquid equilibrium
汽液平衡比	汽液平衡比	vapor-liquid equilibrium ratio
汽油	汽油	gasoline
汽油机	汽油引擎	gasoline engine
汽油阱	汽油阱	gasoline trap
汽蒸	[汽]蒸	steaming
迁移电位	遷移電位	migration potential
迁移率	遷移率,移動率	mobility
钎焊	軟焊	soldering
钎焊膏	焊膏	soldering paste
铅	鉛	lead
铅白	鉛白	lead white
铅玻璃	鉛玻璃	lead glass
铅锤式浮标	錘式浮標	plumb bob float
铅丹	鉛丹,鉛紅	minium
铅管	鉛管	lead pipe
铅碱玻璃	鉛鹼玻璃	lead alkali glass
铅室	鉛室	lead chamber
铅室法	鉛室法	chamber process
铅蓄电池	鉛蓄電池	lead storage battery
铅皂	鉛皂	lead soap
前馈	前饋	feedforward
前馈控制	前饋控制	feedforward control
前馈系统	前饋系統	feedforward system
前体	前驅物	precursor
前沿	前緣	leading edge
潜伏期	培養期	incubation period
潜能	潛能	latent energy
潜热	潛熱	latent heat
潜望镜	潛望鏡	periscope
浅床反应器	淺床反應器	shallow-bed reactor
欠硫	低硫化	under cure
欠硫化橡胶	低硫化橡膠	under cured rubber
欠阻尼	欠阻尼	underdamping

大　陆　名	台　湾　名	英　文　名
欠阻尼系统	欠阻尼系統	underdamped system
欠阻尼响应	欠阻尼應答	underdamped response
茜草	茜草	madder
茜草根	茜草根	madder root
茜素染料	茜素染料	alizarine dyestuff
茜素色淀	茜素色澱	alizarine lake
嵌段共聚合	鏈段共聚合	segment copolymerization
嵌段共聚物	嵌段共聚物,團聯共聚物	block copolymer
嵌段聚合	嵌段聚合[反應],團聯聚合[反應]	block polymerization
嵌接	嵌接頭	scarf joint
强度	強度	intensity, strength
强度性质	強度性質	intensive property
强度状态变量	內含狀態變數	intensive state variable
强力玻璃	強力玻璃	steel glass
强力黏胶纤维	強黏液嫘縈	strong viscose rayon
强制对流	強制對流	forced convection
强制对流传热	強制對流熱傳	forced convection heat transfer
强制对流蒸发	強制對流蒸發	forced convection evaporation
强制函数	強制函數	forcing function
强制扩散	強制擴散	forced diffusion
强制通风	強制通風,鼓風	forced draft
强制通风凉水塔	鼓風冷卻塔	forced draft cooling tower
强制涡旋	強制漩渦	forced vortex
强制循环蒸发	強制循環蒸發	forced circulation evaporation
强制循环蒸发器	強制循環蒸發器	forced circulation evaporator
强制振荡	強制振盪	forced oscillation
羟基	羥基	hydroxyl group
羟醛	醛醇	aldol
羟值	羥值	hydroxyl value
翘曲	翹曲[變形]	warpe
壳程传热系数	殼側熱傳係數	shell-side heat transfer coefficient
壳模铸造	殼狀成型	shell molding
壳模铸造[用]树脂	殼狀成型樹脂	shell molding resin
壳平衡	殼均衡	shell balance
切段[定长]纤维	短纖維	staple fiber
切工组	切工組	draw gang

大　陆　名	台　湾　名	英　文　名
切换线	交換線	switching line
切口堰	V 形堰	notched weir
切块机	切塊機	slabber
切面直径	切面直徑	cut diameter
切片	切片,剖切	slicing, sectioning
切片机	切片機	slicer
切条机	切割機	slitter
切向力	切向力,切線力	tangential force
切向应力	切線應力	tangential stress
切削油	切削油	cutting oil
亲和标记	親和標記法	affinity labeling
亲和超滤	親和超[過]濾	affinity ultrafiltration
亲和沉淀	親和[性]沈澱	affinity precipitation
亲和膜	親和[薄]膜	affinity membrane
亲和色谱[法]	親和[力]層析術	affinity chromatography
亲和势	親和力	affinity
亲和吸附	親和吸附	affinity adsorption
亲和作用	親和[相互]作用	affinity interaction
亲水胶体	親水膠體	hydrophilic colloid
亲水亲油平衡	親水性-親油性均衡	hydrophilic-lipophilic balance, HLB
亲水物	親水物	hydrophile
亲水纤维	親水纖維	hydrophilic fiber
亲水性	親水性	hydrophilicity
亲水性颗粒	親水顆粒	hydrophilic particle
亲液胶体	親液膠體	lyophilic colloid
青霉素	青黴素,盤尼西林	penicillin
青霉素酶	青黴素酶	penicillinase
青铜	青銅	bronze
氢弹	氫彈	hydrogen bomb
氢电极	氫電極	hydrogen electrode
氢放电管	氫放電管	hydrogen discharge tube
氢化催化剂	氫化觸媒	hydrogenation catalyst
氢化反应	氫化反應	hydrogenation reaction
氢化酶	氫化酶	hydrogenase
氢化油	氫化油	hydrogenated oil
氢化脂	氫化脂肪	hydrogenated fat
氢甲酰化	氫甲醯化[反應]	hydroformylation
氢解	氫解	hydrogenolysis

大　陆　名	台　湾　名	英　文　名
氢离子比色计	氫離子比色計	hydrogen ion comparator
氢氯化橡胶	橡膠氫氯化物	hydrochloride rubber
氢能化学	氫能化學	chemistry of hydrogen energy
氢气爆炸	氫氣爆炸	hydrogen explosion
氢气泡法	氫氣泡法	hydrogen bubble technique
氢燃料	氫燃料	hydrogen fuel
氢燃料电池	氫燃料電池	hydrogen fuel cell
氢氧根离子	氫氧離子	hydroxide ion
氢氧化钠,烧碱	燒鹼,苛性鈉	caustic soda
轻垢洗涤剂	輕污清潔劑	light duty detergent
轻关键组分	輕關鍵成分	light key component
轻灰	輕鈉鹼灰	light soda ash
轻馏分	輕餾分	light ends
轻石脑油	輕石油腦	light naphtha
轻烃	輕烴	light hydrocarbon
轻直馏油	輕直餾油	light straight-run gas oil
轻质燃料油	輕燃料油	light fuel oil
轻质油	輕質油	light oil
倾角	傾[斜]角	angle of inclination
倾析	傾析[法]	decantation
倾析器	傾析器	decanter
倾斜管	傾斜管	dip tube
倾卸装置	跳動裝置	tripper
清漆	清漆	varnish
清晰分离	清晰分離	sharp separation
清洗过程	清洗程序	cleaning process
氰硅橡胶	[聚]矽[氧]腈橡膠	silicone nitrile rubber
琼脂电泳	瓊脂電泳	agar electrophoresis
琼脂过滤	瓊脂過濾	agar filtration
琼脂扩散	瓊脂擴散	agar diffusion
琼脂扩散技术	瓊脂擴散技術	agar diffusion technology
琼脂培养基	瓊脂培基	agar medium
琼脂糖	瓊脂糖	agarose
琼脂糖凝胶	瓊脂糖凝膠	agarose gel
琼脂,洋菜	瓊脂,洋菜	agar
求积仪	面積計	planimeter
球阀	球閥	globe valve
球晶	球晶	spherulite

大　陆　名	台　湾　名	英　文　名
球晶结构	球晶結構	spherulite structure
球面	球體	sphere
球面极坐标	球極坐標	spherical polar coordinates
球面坐标	球[形]坐標	spherical coordinate
球磨反应器	球磨反應器	ball mill reactor
球磨机	球磨機	ball mill
球磨精制机	球磨精製機	ball mill refiner
球土	球土	ball clay
球形度	球形度	sphericity
球形油罐	球形槽	spherical tank
球状团簇	球狀團簇	spherical cluster
区带电泳	帶域電泳	zone electrophoresis
区[域]	區[域]	region
区域沉降	帶域沈降	zone settling
区域精制	帶域精煉[法]	zone refining method
区域冷冻	帶域冷凍	zone freezing
区域熔炼	帶域熔煉	zone melting
区域熔炼法	區域熔融法	zone melting method
曲菌	麴菌	aspergillus
曲率	曲率	curvature
曲线	曲線	curve
曲线坐标	曲線坐標	curvilinear coordinates
曲折因子	扭曲度	tortuosity
驱动功率	驅動功率	driver horsepower
驱动函数	驅動函數	driving function
驱动器	驅動器,致動器	driver, actuator
驱动信号	致動信號	actuating signal
屈伏极限	降伏極限	yield limit
屈服点	降伏點	yield point
屈服强度	降伏強度	yield strength
屈服应力	降伏應力	yield stress
屈尼萃取塔	屈尼萃取塔	Kühni extractor
渠道流	渠流	channel flow
取代法则	代換法則	substitution rule
取代反应	取代反應	substitution reaction
取代基	取代基	substituent
取代基常数	取代基常數	substituent constant
取代基均匀度	取代基均匀度	substituent uniformity

大 陆 名	台 湾 名	英 文 名
取向	位向	orientation
取样	採樣	sampling
取样管	取樣管	sampling tube
取样误差	取樣誤差	sampling error
取样滞后	採樣延遲	sampling delay
去饱和作用	去飽和作用	desaturation
去垢力	清潔力	detergency
去过热	去過熱	desuperheat
去过热器	去過熱器	desuperheater
去灰分	去灰分	deashing
去灰分过程	去灰分程序	deashing process
去灰煤	去灰煤	deashed coal
去活化级	去活化階	order of deactivation
去矿化	去礦[物]質	demineralization
去离子水	去離子水	deionized water
去离子作用	去離子[作用]	deionization
去硫化	去硫化	devulcanization
去木素作用	去木質素[作用]	delignification
去溶剂化	去溶劑合[作用],去媒合[作用]	desolvation
去乳化	去乳化[作用]	demulsification
去湿器	濕氣分離器,除濕器	moisture separator
去稳定作用	去穩定作用	destabilization
去污剂	去污劑	decontaminant
去污指数	去污指數	decontamination factor
去氧化	去氧化[作用]	deoxidation
圈转电流计	圈轉電流計	moving coil galvanometer
全部循环过程	全部循環程序	total recycle process
全导数	全導數	total derivative
全分析	總體分析	bulk analysis
全回流	全回流	total reflux
全混	完全混合	complete mixing, perfect mixing
全混流	完全混合流	complete mixing flow
全同立构聚合物,等规聚合物	同排聚合物	isotactic polymer
全纤维素	全纖維素	holocellulose
权函数	權函數	weight function
权重分布	重量分佈	weight distribution

大　陆　名	台　湾　名	英　文　名
权[重]因子	權重[因數]	weight factor
缺胶层	缺膠層	starved line
缺省值	內定值,既定值	default value
缺陷固体	缺陷固體	defect solid
缺陷晶体	缺陷晶體	defect crystal
确定性模型	確定性模型	deterministic model
裙式传送机	裙式運送機	apron conveyor
裙式给料机	裙飼機	apron feeder
裙式输送机	裙式輸送機	apron carrier
群青	群青	ultramarine
群青黄	群青黃	ultramarine yellow
群青蓝	群青藍	ultramarine blue
群青绿	群青綠	ultramarine green
群青紫	群青紫	ultramarine violet
群青棕	群青棕	ultramarine brown

R

大　陆　名	台　湾　名	英　文　名
燃点	燃點,燃燒點,火點	ignition point, burning point
燃料	燃料	fuel
燃料比	燃料比	fuel ratio
燃料床	燃料床	fuel bed
燃料电池	燃料電池	fuel cell
燃料费	燃料成本	fuel cost
燃料工业	燃料工業	fuel industry
燃料酒精	燃料酒精	fuel alcohol
燃料-空气比	燃料-空氣比	fuel-air ratio
燃料油	燃料油	fuel oil
燃气轮发动机	燃氣渦輪引擎	gas turbine engine
燃气轮机	燃氣渦輪機	gas turbine
燃烧	燃燒	combustion
燃烧反应	燃燒反應	combustion reaction
燃烧反应器	燃燒反應器	combustion reactor
燃烧过程	燃燒程序	combustion process
燃烧机理	燃燒機構	combustion mechanism
燃烧率	燃燒速率	burning rate
燃烧面	燃燒鋒面	combustion front

大　陆　名	台　湾　名	英　文　名
燃烧器	燃燒器	burner
[燃烧器]调节比	[燃燒器]調節比	turndown ratio
燃烧区	燃燒區	combustion zone
燃烧曲线	燃燒曲線	combustion curve
燃烧热	燃燒熱	heat of combustion
燃烧室	燃燒室	combustion chamber
燃烧收缩	燃燒收縮	burning shrinkage
燃烧速率	燃燒速率	combustion rate
燃烧特性	燃燒特性	combustion characteristics
燃烧效率	燃燒效率	combustion efficiency
染料	染料,二苯乙烯染料	dye, stilbene dye
染料亲和色谱法	染料親和層析法	dye affinity chromatography
染料摄入法	染料攝入法	dye uptake method
染色	染色[法],著色	dyeing, staining
染色机	染色機	dyeing machine
扰动	擾動	disturbance
绕成绞	成絞	skeining
绕射	繞射	diffraction
热	熱	heat
热本构方程	熱本質方程式	thermal constitutive equation
热泵	熱泵	heat pump
热波	熱波	thermal wave
热沉降	熱沈澱[作用]	thermal precipitation
热成型	熱壓成型	thermoforming
热冲击	熱衝擊	thermal shock
热冲击试验	熱衝擊試驗	thermal shock test
热重整	熱重組[反應]	thermal reforming
热处理	熱處理	heat treatment
热传导	熱傳導	heat conduction, thermal conduction
热传导系数测定池	熱傳導係數測定槽	thermo-conductivity cell
热传声法	熱聲學	thermoacoustimetry
热磁分析	熱磁分析[法]	thermomagnetic analysis
热磁性	熱磁性	thermomagnetism
热催化重整工艺	熱觸媒重組法	thermofor catalytic reforming process
热催化裂化工艺	熱觸媒裂解法	thermofor catalytic cracking process
热单位	熱單位	thermal unit
热导检测器	熱傳導[係數]偵檢器	thermal conductivity detector
热导率	熱傳導係數	heat conductivity

大　陆　名	台　湾　名	英　文　名
热滴定[法]	溫度滴定[法]	thermometric titration
热点	熱點	hot spot
热电	熱電	thermoelectricity
热电池	熱電池	thermobattery
热电法	熱電法	thermoelectrometry
热电高温计	熱電高溫計	thermoelectric pyrometer
热电计	熱電計	thermoelectrometer
热电流	熱電流	thermocurrent
热电流计	熱電流計	thermogalvanometer
热电偶	熱[電]偶	thermocouple
热电偶参考接点	熱[電]偶參考接點	thermocouple reference junction
热电偶高温计	熱[電]偶高溫計	thermocouple pyrometer
热电偶接点	熱電[偶]接點	thermoelectric junction
热电偶式流速计	熱[電]偶式流速計	thermocouple type anemometer
热电式电位计	熱電[式]電位計	thermoelectric potentiometer
热电温度计	熱電溫度計	thermoelectric thermometer
热电效应	熱電效應	thermoelectric effect
热电泳	熱泳法	thermophoresis
热对流	熱對流	heat convection, thermal convection
热发射	熱發射	heat emission
热发声法	熱聲[分析]法	thermosonimetry
热反射法	熱反應[分析]法	thermoreflectometry
热方法	熱方法	thermal method, thermal technique
热分解	熱分解	thermal decomposition
热分析	熱分析[法],熱分析	thermal analysis, thermoanalysis
热分析法	熱分析法	thermoanalytical method
热分析显微镜	熱分析顯微術	thermoanalytical microscopy
热分析仪	熱分析儀	thermoanalyzer
热风炉	熱風爐	hot blast heater
热封黏合剂	熱封黏合劑	heat sealing adhesive
热辐射	熱輻射	radiation of heat, thermal radiation
热负载	熱負載	heat duty
热功当量	熱功當量	mechanical equivalent of heat
热功转化率	熱功轉化率	heat rate
热固定	熱固定	thermofixation
热固化	熱固性	thermosetting
热固性树脂	熱固性樹脂	thermosetting resin
热固性塑料	熱固物	thermoset

大　陆　名	台　湾　名	英　文　名
热管	熱管	heat pipe
热管换热器	熱管交換器	heat pipe exchanger
热光分析[法]	熱光分析[法]	thermophotometry
热含量	熱含量	heat content
热耗散	熱耗散,熱散逸	thermal dissipation, heat dissipation
热合	封口	sealing
热合成	熱合成	thermal synthesis
热合接头	密封接頭	sealing joint
热核反应	熱核反應	thermonuclear reaction
热虹吸	熱虹吸	thermo siphon
热虹吸式再沸器	熱虹吸再沸器	thermosiphon reboiler
热化学	熱化學	thermochemistry
热回流	熱回流	hot reflux
热机	熱機	heat engine
热机械分析	熱機械分析[法]	thermomechanical analysis
热集成	熱整合	heat integration
热降解	熱降解	thermal degradation
热交换器	熱交換器	heat interchanger
热交换器(＝换热器)		
热接点	熱接點	hot junction
热解	熱[裂]解	pyrolysis
热解过程	熱[裂]解程序	pyrolysis process
热解聚	熱解聚合[反應]	thermal depolymerization
热解离[反应]	熱解離[反應]	thermal dissociation
热解炉	熱[裂]解爐	pyrolysis furnace
热解油	熱解油	pyrolysis oil
热浸	熱浸[漬]	hot dipping
热阱	熱匯	heat sink
热静力学	熱靜力學	thermostatics
热聚合	熱聚合[反應]	thermal polymerization, thermopolymeriza-tion
热颗粒分析	熱顆粒分析[法]	thermoparticulate analysis
热坑干燥室	熱炕乾燥室	hot floor drier
热扩散	熱擴散	thermal diffusion, thermodiffusion
热扩散系数	熱擴散係數	thermal diffusion coefficient, thermal dif-fusivity
热力学	熱力學	thermodynamics
热力学本构方程	熱力學本質方程式	thermodynamic constitutive equation

大 陆 名	台 湾 名	英 文 名
热力学第三定律	熱力學第三定律	third law of thermodynamics
热力学第一定律	熱力學第一定律	first law of thermodynamics
热力学分析	熱力學分析	thermodynamic analysis
热力学概率	熱力學機率	thermodynamic probability
热力学关系	熱力學關係	thermodynamic relation
热力学函数	熱力學函數	thermodynamic function
[热力学]环境	環境	surroundings
热力学平衡	熱力學平衡	thermodynamic equilibrium
热力学特性函数	熱力學特徵函數	thermodynamic characteristic function
热力学通量	熱力學通量	thermodynamic flux
热力学图	熱力學圖	thermodynamic diagram
热力学温标	熱力學溫標	thermodynamic temperature scale
热力学温度	熱力學溫度	thermodynamic temperature
热力学系统	熱力學系統	thermodynamic system
热力学效率	熱力學效率	thermodynamic efficiency
热力学性质	熱力學性質	thermodynamic property
热力学性质表	熱力學性質表	thermodynamic property table
热力学一致性	熱力學一致性	thermodynamic consistency
热力学一致性检验	熱力學一致性試驗[法]	thermodynamic consistency test
热连续渗滤工艺	熱連續滲濾法	thermofor continuous percolation process
热量传递,传热	熱傳	heat transfer
热量衡算	熱量均衡	heat balance
热量回收	熱回收	heat recovery
热量平衡	熱量均衡	thermal balance
热量去除曲线	熱量去除曲線	heat removal curve, heat removal line
热量输送	熱輸送	heat transport
热裂变	熱[中子]分裂	thermal fission
热裂化	熱裂解	thermal cracking
热流	熱流動	heat flow
热流道	熱澆道	hot runner
热[流]通量	熱通量	heat flux
热硫化	熱硫化	hot vulcanization
热敏电阻器	熱[敏電]阻器	thermistor
热能	熱能	heat energy, thermal energy
热能方程式	熱能方程式	thermal energy equation
热偶安培计	熱偶安培計	thermoammeter
热膨胀	熱膨脹	thermal expansion

大　陆　名	台　湾　名	英　文　名
热膨胀系数	熱[膨]脹係數	thermal expansion coefficient
热平衡	熱平衡	thermal equilibrium
热气再循环过程	熱氣循環程序	hot gas recycle process
热汽	熱汽	hot vapor
热容	熱容[量]	heat capacity
热容量	熱容量	thermal capacity
热容流率	熱容流率	heat capacity flow rate
热渗透	熱滲透[作用]	thermoosmosis
热生成	熱生成	heat generation
热生成曲线	熱生成曲線	heat generation curve
热生成线	熱生成線	heat generation line
热失活	熱失活	thermal deactivation
热石灰法	熱石灰法	hot lime process
[热]收缩机	收縮機	shrinking machine
热水器	水加熱器	water heater
热塑性	熱塑性	thermoplasticity
热塑性聚合物	熱塑性聚合物	thermoplastic polymer
热塑性树脂	熱塑性樹脂	thermoplastic resin
热塑性塑料	熱塑性塑膠	thermoplastic plastics
热损失	熱損失	heat loss
热炭黑	熱碳黑	thermal black
热套管	熱套管	thermal well
热特性	熱示性[法]	thermal characterization
热天平	熱天平	thermobalance
热脱挥发物作用	熱脫揮發物作用	thermal devolatilization
热烷化	熱烷化[反應]	thermal alkylation
热稳定性	熱安定性	thermal stability
热稳性	熱安定性	thermostability
热污染	熱污染	thermal pollution, heat pollution
热物理性质	熱物理性質	thermophysical property
热效率	熱效率	thermal efficiency, heat efficiency
热效应	熱效應	heat effect
热行为	熱行為	thermal behavior
热循环	熱循環	thermal cycle
热压	熱壓	hot pressing
热压机	熱壓機	hot press
热氧化降解	熱氧化降解	thermooxidative degradation
热应变	熱應變	thermal strain

大 陆 名	台 湾 名	英 文 名
热应力	熱應力	thermal stress
热原子法	熱原子法	thermoatomic process
热源	熱源	thermal source, heat source
热运动学	熱運動學	thermokinematics
热汁加灰法	熱汁加灰法	hot liming
热值	熱值	heat value, thermal value, heating value
热值试验	熱值試驗	thermal value test
热[致]发光	熱[致]發光	thermoluminescence
热致液晶	熱致液晶	thermotropic liquid crystal
热重分析[法]	熱重分析[法],熱重分析術	thermogravimetric analysis, thermogravimetry
热重量分析仪	熱重量分析儀	thermogravimetric analyzer
热自氧化	熱自氧化	thermal autoxidation
热阻	熱阻	thermal resistance
人工操作	人工操作	manual operation
人工成本	人工成本	labor cost
人工肥料	人造肥料	artificial fertilizer
人工干燥器	人工乾燥器	artificial drier
人工老化	人工老化	artificial aging
人工器官	人造器官	artificial organ
[人工]神经网络	人工類神經網路,類神經網路	artificial neural network, ANN
人工肾脏	人造腎臟	artificial kidney
人工甜味剂	人工甜味料,人工甘味劑	artificial sweetener
人工智能	人工智慧	artificial intelligence, AI
人孔	人孔	manhole
人造短纤维	短纖維,棉狀纖維,絨[棉]	staple
人造沸石	人造沸石	permutite
人造革	人造皮	artificial leather
人造黄油	人造奶油	margarine
人造燃料	人造燃料	artificial fuel
人造丝	嫘縈	Rayon
人造丝短纤维	嫘縈棉	rayon staple
人造丝浆	嫘縈紙漿	rayon pulp
人造丝束	嫘縈束	rayon tow
人造纤维	人造纖維	artificial fiber, man-made fiber

大　陆　名	台　湾　名	英　文　名
认可概率	允收機率	probability of acceptance
韧度	韌度,抗拉強度	tenacity,toughness
韧皮纤维	莖纖維	stem fiber
韧性	韌性	toughness
韧性聚苯乙烯	韌化聚苯乙烯	toughened polystyrene
日本[天然]漆	日本[天然]漆	japan
日晒变色	日晒變色	sun-discoloration
容差	容許量	allowance
容积传氧系数,体积传氧系数	體積氧質傳係數	volumetric oxygen transfer coefficient
容积式泵	正排量泵	positive displacement pump
容积式流量计	正排量流量計	positive displacement flowmeter
容量滴定法	容量滴定[法]	volumetric titration
容量分析法	容量分析法	volumetry
容器	容器	container
容许粗糙度	容許粗糙度	admissible roughness
容许区间	容許區間	tolerance interval
溶度参数	溶解度參數	solubility parameter
溶度分级	溶解度分餾	solubility fractionation
溶度分析	溶解度分析	solubility analysis
溶度积	溶度積	solubility product
溶度极限	溶解極限	solubility limit
溶纺纤维	溶紡纖維	solvent spun fiber
溶剂	溶劑	solvent
溶剂萃取	溶劑萃取[法]	solvent extraction
溶剂法	溶劑法	solvent process
溶剂纺丝	溶劑紡絲	solvent spinning
溶剂分离器	溶劑分離器	solvent separator
溶剂锋面	溶劑鋒面	solvent front
溶剂复合	溶劑積層	solvent lamination
溶剂合物	溶劑合物	solvate
溶剂化	溶劑合[作用]	solvation
溶剂化反应	溶劑分解[作用]反應	solvolysis reaction
溶剂化能	溶劑合能	solvation energy
溶剂化能力	溶劑合能力	solvating power
溶剂化效应	溶劑合效應	solvating effect, solvation effect
溶剂化增塑剂	溶劑合塑化劑	solvating plasticizer
溶剂基黏合剂	溶劑基黏合劑	solvent based adhesive

大　陆　名	台　湾　名	英　文　名
溶剂基涂料	溶劑基塗料	solvent based coating
溶剂胶浆	溶劑膠合劑	solvent cement
溶剂解	溶劑解[作用]	solvolysis
溶剂精炼煤	溶劑精煉煤	solvent-refined coal
溶剂精制	溶劑精煉[法]	solvent refining
溶剂-链节相互作用	溶劑鏈節作用	solvent-segment interaction
溶剂裂化	溶劑裂解	solvent cracking
溶剂流铸法	溶液鑄造法	solvent cast process
溶剂滤油	溶劑油	solvent naphtha
溶剂黏接	溶劑熔接	solvent welding
溶剂浓度	溶劑強度	solvent strength
溶剂去灰分[法]	溶劑去灰分[法]	solvent deashing
溶剂去灰分过程	溶劑去灰分程序	solvent deashing process
溶剂染色	溶劑染色	solvent dyeing
溶剂蚀刻	溶劑蝕刻	solvent etching
溶剂试验	溶劑試驗	solvent test
溶剂梯度洗脱	溶劑梯度溶解	solvent-gradient elution
溶剂相互作用参数	溶劑相互作用參數	solvent interaction parameter
溶剂效应	溶劑效應	solvent effect
溶剂型黏合剂	溶劑型黏合劑	solvent type adhesive
溶剂型增塑剂	溶劑型塑化劑	solvent type plasticizer
溶剂银纹化	溶劑細紋	solvent crazing
溶剂滞留量	溶劑滯留量	solvent hold-up
溶剂助染	溶劑助染法	solvent assisted dyeing
溶剂最大容限	溶劑容許度	solvent tolerance
溶胶	溶膠	sol
溶胶部分	溶膠部分	sol fraction
溶胶化	溶膠化[作用]	solation
溶胶-凝胶转换	溶凝膠轉換	sol-gel transformation
溶胶橡胶	溶膠橡膠	sol rubber
溶解	溶解	dissolution
溶解度	溶解度	solubility
溶[解]度积常数	溶[解]度積常數	solubility product constant
溶解度曲线	溶解度曲線	solubility curve
溶解度势垒	溶解度障壁	solubility barrier
溶解度试验	溶[解]度測定	solubility test
溶解度-温度曲线	溶解度-溫度曲線	solubility-temperature curve
溶解固体	溶解固體	dissolved solid

大　陆　名	台　湾　名	英　文　名
溶解空气	溶解空氣	dissolved air
溶解器	溶解器	dissolver
溶解热	溶解熱	heat of solution
溶解压	溶解壓力	solution pressure
溶解氧	溶氧	dissolved oxygen
溶解氧分析仪	溶氧分析儀	dissolved oxygen analyzer
溶解氧探头	溶氧探針	dissolved oxygen probe
溶菌酶	溶菌酶	lysozyme
溶气浮选［法］	溶氣浮選［法］	dissolved-air flotation
溶液	溶液	solution
溶液的依数性	依數性	colligative property of solution
溶液丁苯橡胶	溶液［聚合］苯乙烯丁 二烯橡膠	solution styrene-butadiene rubber
溶液纺丝	溶液紡絲	solution spinning
溶液化学	溶液化學	solution chemistry
溶液浇铸	溶液澆鑄［法］	solution casting
溶液聚合	溶液聚合［作用］	solution polymerization
溶液双重折射	溶液雙折射	solution birefringence
溶胀	膨潤	swell
溶胀比	膨潤比	swelling ratio
溶胀度	膨潤度	degree of swelling
溶胀剂	膨潤劑	swelling agent
溶胀量	膨潤能力	swelling capacity
溶胀能力	膨潤能力	swelling power
溶胀压	膨潤壓［力］	swelling pressure
溶胀值	膨潤值	swelling value
溶质	溶質	solute
熔点	熔點	fusion point, melting point
熔点范围	熔點範圍	melting range
熔点下降	熔點下降	melting point depression
熔纺	熔紡［絲］	melt spinning
熔化	①熔體　②熔態	melt
熔化曲线	熔化曲線	fusion curve
熔化热	熔化熱,熔合熱	heat of fusion
熔化温度	熔化溫度,熔融溫度	melting temperature
熔化物	①熔體　②熔態	melt
熔炼	熔煉,冶煉	smelting
熔炼坩埚	熔煉［坩］堝	smelting pot

大 陆 名	台 湾 名	英 文 名
熔炼炉	熔煉爐	smelting furnace
熔融	熔化	fusion
熔融黏度	熔體黏度	melt viscosity
熔蚀作用	熔蝕作用	slag action
熔体流动指数	熔體[流動]指數	melt flow index
熔体指数	熔體指數	melt index
熔浴气化器	熔浴氣化器	molten-bath gasifier
冗余[度]	冗餘[度]	redundancy
柔软度	[柔]軟度	softness
柔软性	[柔]軟度	softness
揉捏机	捏泥機	pug mill
鞣革厂	製革廠	tannery
鞣酸,单宁酸	鞣酸,單寧酸	tannic acid
鞣制	鞣製	tannage
鞣质,单宁	鞣酯,單寧	tannin
肉桂皮油	桂皮油	cassia oil
肉桂油	肉桂油,桂皮油	cinnamon oil
蠕变	蠕變,潛變	creep
蠕变恒速区	二期潛變	secondary creep
蠕变试验	潛變試驗	creep test
蠕变运动	蠕[流運]動	creeping motion
蠕流	蠕流	creeping flow
乳白玻璃	乳白玻璃,乳光玻璃	milky glass, opalescent glass
乳二糖	乳糖	lactobiose
乳化剂	乳化劑	emulsifying agent, emulsifier
乳化胶体	乳化膠體	emulsifying colloid
乳化器	乳化器	emulsifier
乳化软膏	乳化軟膏	emulsifying ointment
乳化稳定性	乳液穩定性	emulsion stability
乳化液膜	乳液薄膜	emulsion liquid membrane
乳化[作用]	乳化[作用]	emulsification
乳胶	乳化膠體,類乳化液	emulsoid
乳胶漆	乳膠漆,乳化漆	emulsion paint, latex paint
乳清蛋白	乳白蛋白	lactalbumin
乳清[橡胶]	血清	serum
乳酸	乳酸	lactic acid
乳酸钙	乳酸鈣	calcium lactate
乳糖	乳糖	lactose

大　陆　名	台　湾　名	英　文　名
乳糖酶	乳酸酶	lactase
乳相	乳[化]相	emulsion phase
乳液聚合	乳化聚合[作用]	emulsion polymerization
乳油分离机	乳油分離機	skimming machine
乳浊玻璃	不透明玻璃	opaque glass
乳浊剂	失透劑	opacifier, opacifying agent
乳[浊]液	乳液,乳膠	emulsion
入口	入口	inlet
软玻璃	軟玻璃	soft glass
软钢	軟鋼	mild steel
软膏基质	軟膏基	ointment base
软管	軟管	hose
软化	軟化	softening
软化点	軟化點	softening point
软化剂	[水]軟化劑	softener, softening agent
软化器	去礦[物]質劑	demineralizer
软化水	軟化水,去礦[物]質水	softened water, demineral water, deminer-alized water
软化温度	軟化溫度	softening temperature
软钾镁矾	硫酸鉀鎂礦	schonite
软件	軟體	software
软件包	套裝軟體	package software
软聚氯乙烯薄膜	軟聚氯乙烯薄膜	soft poly(vinyl chloride) film
软[链]段	軟段	soft segment
软锰矿	軟錳礦	pyrolusite
软木材	軟木	softwood
软木粉	軟木粉	softwood flour
软泥法	軟泥法	soft-mud process
软润滑脂	軟潤滑脂	soft grease
软石蜡	軟石蠟	soft paraffin
软水	軟水	soft water
软水剂	水軟化劑	water softener
软酸	軟酸	soft acid
软纤维	軟纖維	soft fiber
软性 X 射线	弱[穿]X 射線	soft X-ray
软性洗涤剂	軟性清潔劑	soft detergent
软质聚合物	軟質聚合物	soft polymer
软质蜡	軟蠟	soft wax

大　陆　名	台　湾　名	英　文　名
软质橡胶	軟質橡膠	soft rubber
锐孔管	孔口管	orifice tube
锐缘孔口	銳緣孔口	sharp edged orifice
瑞利数	瑞立數	Rayleigh number
瑞斯托菌素	瑞斯特黴素	ristocetin
润滑	潤滑	lubrication
润滑剂	潤滑劑	lubricant
润滑力学	潤滑力學	lubrication mechanics
润滑性	潤滑性	lubricity
润滑油	潤滑油	lubricating oil
润滑脂	［潤］滑脂	grease, lubricating grease
润湿表面积	潤濕表面積	wetted surface area
润湿效应	潤濕效應	wettability effect
润湿周边	潤濕周邊	wetted perimeter
弱电解质	弱電解質	weak electrolyte
弱碱	弱鹼	weak base
弱可逆性	弱可逆性	weak reversibility
弱酸	弱酸	weak acid

S

大　陆　名	台　湾　名	英　文　名
洒水器	灑水器	sprinkler
赛璐玢	賽璐玢,玻璃紙	cellophane
赛璐珞	賽璐珞	celluloid
三次采油	三級採油	tertiary oil recovery
三对角矩阵	三對角矩陣	tridiagonal matrix
三分子反应	三分子反應	termolecular reaction, trimolecular reaction
三合盐	三合鹽	triple salt
三级处理	三級處理	tertiary treatment
三级化学澄清器	三級化學澄清器	tertiary chemical clarifier
三角缺口	三角凹槽	triangular notch
三角图	三角圖	triangular diagram
三角坐标	三角坐標	triangular coordinates
三聚丙烯	三聚丙烯	propylene trimer
三聚氰胺甲醛树脂	三聚氰胺甲醛樹脂	melamine formaldehyde resin
三聚氰胺黏合剂	三聚氰胺黏合劑,美耐	melamine adhesive

大 陆 名	台 湾 名	英 文 名
	皿黏合劑	
三聚体	三聚物	trimer
三氯乙醛	[三]氯[乙]醛	chloral
三式控制	三式控制	three-mode control
三式控制器	三式控制器	three-mode controller
三水铝石	水礬土,水鋁氧	gibbsite
三水铝石耐火土	水礬土耐火土	gibbsite fireclay
三羧酸循环	三羧酸循環	tricarboxylic acid cycle
三通(=T形件)		
三相点	三相點	triple point
三相反应器	三相反應器	three-phase reactor
三相流化床	三相流[體]化床	three-phase fluidized bed
三相流态化	三相流體化	three-phase fluidization
三元共沸物	三元共沸液	ternary azeotrope
三元系[统],三组分系统	三成分系[統],三元系[統]	ternary system
三轴应力	三軸應力	triaxial stress
三组分系统(=三元系[统])		
散堆填料	隨機填料	random packing
散热片	散熱片	cooling fin
散热器	散熱器	radiator
散射	散射	scattering
散射池	散射池	scattering cell
散射角	散射角	scattering angle
散射体积	散射體積	scattering volume
散射系数	散射係數	scattering coefficient
散射形式	散射圖樣	scattering pattern
散式流态化	顆粒流體化	particulate fluidization
散纤维染色	纖維染色	stock dyeing
散装储存	散裝儲存	bulk storage
散装填料	堆積填充	dumped packing
扫描电子显微镜	掃描電子顯微鏡	scanning electron microscope, SEM
扫描电子显微镜学	掃描電子顯微法	scanning electron microscopy
扫描量热法	掃描量熱法	scanning calorimetry
扫描速率	掃描速度	scanning rate
色淀	色澱,沈澱色料	lake
色淀颜料	色澱,沈澱色料	lake pigment

大 陆 名	台 湾 名	英 文 名
色谱电泳	層析電泳	chromatoelectrophoresis
色谱[法]	層析術	chromatography
色谱分离	層析分離	chromatographic separation
色谱聚焦	層析聚焦	chromatofocusing
色谱图	層析圖	chromatogram
色谱仪	層析儀,色譜儀	chromatograph
色谱柱	層析管柱	chromatographic column
色谱柱式脉冲反应堆	層析柱式脈動反應器	chromatographic column pulse reactor
色散力	分散力	dispersion force
色散曲线	分散曲線	dispersion curve
色原	色素原	chromogen
杀虫剂	殺蟲劑	insecticide, pesticide
杀稻瘟菌素 S	殺稻瘟菌素 S	blasticidin-S
杀菌剂	殺真菌劑,殺黴劑	fungicide, germicide
杀螨剂	殺蟎劑	acaricide, miticide
杀鼠剂	殺鼠劑	rodenticide
沙丁鱼油	沙丁魚油	sardine oil
纱	紗	yarn
纱染机	紗染機	yarn dyeing machine
砂布	砂布	emery cloth
砂浆	灰泥	mortar
砂滤法	砂濾[法]	sand filtration
砂滤器	砂濾器	sand-bed filter
砂轮	砂輪	emery wheel
砂磨	砂磨	sand mill
砂纸	砂紙	abrasive paper, sand paper
莎纶	紗綸,聚偏二氯乙烯	saran
莎纶管	紗綸管	saran pipe
鲨鱼皮	鯊魚皮	shark skin
筛	篩	sieve, sifter, bolt
筛板	篩板,篩盤	sieve tray, sieve plate, screen plate
筛板塔	篩板塔	sieve plate column, sieve plate tower
筛布	篩布	bolting cloth
筛分	篩[分]	sieving
筛分试验	篩分試驗	sieve test
筛孔	篩孔	screen aperture, screen opening
筛滤器	篩濾機	screen filter
筛磨	篩磨機	screen mill

大　陆　名	台　湾　名	英　文　名
筛目	篩目	mesh
筛目效率	篩網效率	mesh efficiency
筛上物分布曲线	篩上物分佈曲線	oversize distribution curve
筛网	篩網	screen
筛网分离器	篩網分離器	mesh separator
筛网过滤器	篩網過濾器	mesh filter
筛网输送机	篩運機	screen conveyor
筛析	篩析	mesh analysis, screen analysis, sieve analysis
筛选	篩選	screening
筛选测验	篩選試驗	screening test
筛选机	篩選機, 篩粉機	bolting mill, sifting machine
筛选效率	篩選效率	screening efficiency
筛选装置	篩選裝置	screening device
筛用丝	篩用絲	bolting silk
筛组	篩組	screen bank
晒裂	晒裂	sun crack, sun-cracking, sun crazing
闪点	閃[火]點	flash point
闪点测定器	閃[火]點測定器	flash point apparatus
闪烁火焰	閃爍火焰	flickering flame
闪速焙烧炉	急驟焙燒爐	flash roaster
闪速干燥	急驟乾燥	flash drying
闪速蒸馏	驟餾	flash distillation
闪锌矿	閃鋅礦	zinc blende
闪蒸	驟沸蒸發	flash vaporization, flash evaporation
闪蒸槽	驟沸桶	flash drum
闪蒸计算	驟沸計算	flash calculation
闪蒸浓缩器	驟沸濃縮器	flash concentrator
闪蒸气	驟沸蒸气	flash vapor
闪蒸曲线	驟沸蒸發曲線	flash vaporization curve
闪蒸室	驟沸室	flash chamber
闪蒸塔	驟餾塔	flash distillation column, flash tower
扇形排风机	扇式鼓風機	fan blower
商业发展	商業發展	commercial development
熵	熵	entropy
熵变化	熵變化	entropy change
熵产生	熵產生, 熵生成	entropy production, entropy generation
熵方程式	熵方程式	entropy equation

大　陆　名	台　湾　名	英　文　名
熵衡算	熵均衡	entropy balance
熵流	熵通量	entropy flux, entropy flow
熵通量	熵通量	entropy flux
熵增原理	熵增原理	principle of entropy increase
上光剂	亮光劑	lustering agent
上光盘	上光盤	polishing pan
上极限	上限	upper limit
上浆机	上漿機	sizing machine
上胶剂	上漿劑	sizing agent
上控制限	管制上限	upper control limit
上流塔	上向流塔,升流塔	upflow column
上清液	上澄液	supernatant
上升系数	上升係數	lift coefficient
上限	上限	upper bound
上行管	上升管	ascending pipe
上行色谱法	上升層析[法]	ascending chromatography
上行窑	上行窯,階梯窯	ascending kiln
上游	上游	upstream
上游压力	上游壓力	upstream pressure
上釉	上釉	glazing
上脂	加脂[法]	stuffing
烧碱法	鹼法	soda process
烧碱法浆	鈉鹼紙漿	soda pulp
烧碱(＝氢氧化钠)		
烧结	燒結	sinter, sintering
烧结玻璃	燒結玻璃	sintered glass
烧结玻璃过滤器	燒結玻璃濾器	fritted glass filter
烧结成型	燒結成型	sinter forming
烧结带	燒結區	sintering zone
烧结料	燒結材料	sintered materials
烧结膜	燒結膜	sinter membrane
烧瓶	燒瓶	flask
烧蚀	剝蝕	ablation
舌形板	噴射塔板	jet tray
蛇管冷凝器,盘管冷凝器	蛇管冷凝器,盤管冷凝器	coil condenser
蛇管,盘管	蛇管,盤管	coiled pipe
蛇管散热器,盘管散热	蛇管散熱器,盤管散熱	coiled radiator

大 陆 名	台 湾 名	英 文 名
器	器	
蛇管蒸发器,盘管蒸发器	蛇管蒸發器,盤管蒸發器	coil evaporator
蛇笼型聚电解质	蛇籠型聚電解質	snake-cage polyelectrolyte
设备安全	設備安全	equipment safety
设备成本	設備成本	equipment cost
设备可靠性	設備可靠性	equipment reliability
设备流程图	設備流程圖	equipment flowsheet
设定值	設定點	set point
设定值变动	設定點變動	set point variation
设定值跟踪	設定點追蹤	set point tracking
设定值控制	設定點控制	set point control
设定值扰动	設定點擾動	set point disturbance
设定值响应	設定點應答	set point response
设计	設計	design
设计变量	設計變數	design variable
设计标准	設計標準	design standard
设计参数	設計參數	design parameter
设计方程式	設計方程式	design equation
设计工程师	設計工程師	design engineer
设计工具	設計工具	design tool
设计模型	設計模式	design mode
设计数据	設計數據	design data
设计图	設計圖	design chart
设计项目	設計專案	design project
设计主管工程师	專案工程師	project engineer
设计准则	設計準則	design criterion
社会成本	社會成本	social cost
射流	射流,流出	efflux
射流穿透长度	噴射穿透長度	jet penetration length
射流反应器	噴射反應器	jet reactor
射流时间	射流時間	efflux time
射流速度	射流速度	efflux velocity
射流系数	射流係數	efflux coefficient
X 射线谱	X 射線譜	X-ray spectrum
X 射线衍射	X 射線繞射	X-ray diffraction
X 射线荧光	X 射線螢光	X-ray fluorescence
摄动理论(=微扰理论)		

大　陆　名	台　湾　名	英　文　名
摄谱仪	攝譜儀	spectrograph
摄取	攝取	uptake
摄取速率	攝取速率	uptake rate
摄氧	攝氧	oxygen uptake
摄氧速率	攝氧速率	oxygen uptake rate
伸长	延伸	extension
砷杀虫剂	砷殺蟲劑	arsenical insecticide
砷杀鼠剂	砷殺鼠劑	arsenical rodenticide
深度冷冻	[超]低溫程序	cryogenic process
深井注入[法]	深井注入[法]	deep-well injection
神经网络	類神經網路	neural network
神经网络训练	神經網絡訓練	neural network training
神经元	神經元	neuron
渗出	滲出	exudation
渗漏	漏洩	leakage
渗滤	滲濾,浸透	percolation, infiltration
渗滤器	滲漉[萃取]器	percolation extractor
渗滤液	滲漉液	percolate
渗入测试	針入試驗{煤}	penetration test
渗透萃取	透[過]萃[取]法	pertraction
渗透剂	滲透劑	penetrating agent, penetrant
渗透扩散	滲透擴散	osmotic diffusion
渗透率	滲透率,磁導率	permeability
渗透试验	滲透[性]試驗	permeability test
渗透试验器	滲壓測定器	osmoscope
渗透通量	滲透通量	permeation flux
渗透物	滲透物	permeate
渗透系数	滲透係數	osmotic coefficient
渗透性	滲透率,磁導率	permeability
渗透压	滲透壓[力]	osmotic pressure
渗透压计	滲[透]壓[力]計	osmometer
渗透仪	滲透儀	permeability apparatus
渗透蒸发	滲透蒸發	pervaporation
渗透[作用]	滲透	osmosis
渗余物	滲餘物	retentate
升华	升華[作用]	sublimation
升华干燥法	升華乾燥[法]	sublimate drying
升华器	升華器	sublimator, sublimer

大　陆　名	台　湾　名	英　文　名
升华热	升華熱	heat of sublimation
升华物	升華物	sublimate
升华装置	升華裝置	sublimation apparatus
升降色谱法	升降層析[法]	ascending-descending chromatography
升膜蒸发器	升膜蒸發器	climbing-film evaporator
升压泵	增壓泵	booster pump
升压喷射器	增壓噴射器	booster ejector
升压器	增壓器,助力器	booster
升压压力表	增壓壓力計	boost gage
生产	生產	production
生产成本	生產成本	production cost
生产过剩	過度生產	overproduction
生产进度表	生產時程表	production schedule
生产控制	生產管控	production control
生产力	生產力	productivity
生产率	生產速率	production rate
生产能力	生產能力	production capacity
生产能力过剩	生產能力過剩	overcapacity
生产数据	生產數據	production data
生产线	生產線	production line
生产用水	製程用水	process water
生成焓	生成焓	enthalpy of formation
生成热	生成熱	heat of formation
生成速率	生成[速]率	birth rate
生存理论	生存理論	survival theory
生存率	生存率	survival rate
生化反应	生化反應	biochemical reaction
生化反应工程	生化反應工程	biochemical reaction engineering
生化反应器	生化反應器	biochemical reactor
生化分离	生化分離	biochemical separation
生化工程	生化工程	biochemical engineering
生化热力学	生化熱力學	biochemical thermodynamics
生化需氧量	生化需氧量	biochemical oxygen demand
生活污水	生活污水	sanitary wastewater
生能反应	生能反應	energy producing reaction
生石灰	生石灰	quick lime
生丝精练	生絲精練,絲綢精練,絲洗練	silk degumming, silk scouring

大 陆 名	台 湾 名	英 文 名
生态系统	生態系統	ecosystem
生铁	生鐵	pig iron
生物	生物	organism
生物变质	生物變質	biodeterioration
生物材料	生物材料	biological materials, biomaterials
生物测定	生物檢定	bioassay
生物处理	生物處理	biological treatment
生物处理过程	生物處理程序	biological treatment process
生物传感器	生物感測器	biosensor
生物催化	生物催化[作用]	biocatalysis
生物催化反应	生物催化反應	biocatalytic reaction
生物催化剂	生物觸媒	biocatalyst
生物大分子	生物巨分子	biomacromolecule
生物电池	生物電池	biocell
生物毒性	生物毒性	biotoxicity
生物毒性试验	生物毒性試驗	biotoxicity test
生物发光	生物發光	bioluminescence
生物反应器	生物反應器	biological reactor, bioreactor
生物分离	生物分離	bioseparation
生物分散	生物分散	biological dispersion
生物腐蚀	生物腐蝕	biological corrosion
生物工程	生物工程	bioengineering
生物功能试剂	生物功能試劑	biofunctional reagent
生物固体	生物固體	biological solid
生物过程	生物程序	biological process, bioprocess
生物过滤器	生物[過]濾器	biological filter
生物合成	生物合成	biosynthesis
生物合成反应	生物合成反應	biosynthetic reaction
生物化学	生物化學	biochemistry
生物基聚合物	生物聚合物	biological polymer
生物技术	生物技術	biotechnology
生物碱	生物鹼	alkaloid
生物降解	生物降解	biodegradation
生物降解塑料	[可]生物降解塑膠	biodegradable plastic
生物降解洗涤剂	[可]生物降解清潔劑	biodegradable detergent
生物降解性	生物降解性	biodegradability
生物浸取	生物瀝取	bioleaching
生物聚合物	生物聚合物	biopolymer

大　陆　名	台　湾　名	英　文　名
生物可利用率	生體可用率	bioavailability
生物流体力学	生物流體力學	biofluid mechanics
生物膜	生物[薄]膜	biological film, biological membrane
生物能学	生物能量學	bioenergetics
生物群	生物相	biota
生物燃烧	生化燃燒	biochemical combustion
生物渗透	生物滲透	bio-osmosis
生物塔	生物塔	biological tower
生物絮凝[作用]	生物絮凝[作用]	biological flocculation
生物延时	生物時延	biological time lag
生物氧化	生物氧化[反應]	biological oxidation
生物医学	生物醫學	biomedicine
生物医学工程	生醫工程	biomedical engineering
生物医学技术	生醫技術	biomedical technology
生物医药器件	生醫裝置	biomedical device
生物医用聚合物	生醫聚合物	biomedical polymer
生物制剂	生物製劑	biological agent
生物质	生質	biomass
生物质能	生質能	biomass energy
生物质气化	生質氣化	biomass gasification
生物质燃料	生質燃料	biomass fuel
生物质热解	生質熱裂解	biomass pyrolysis
生物质转化	生質轉化	biomass conversion
生物质转化过程	生質轉化程序	biomass conversion process
生物专一性连结	生物專一性連結	biospecificity binding
生物转化	生物轉化	bioconversion, biotransformation
生物转盘法	[生物]旋轉盤法	rotating-disc process, bio-disc process
生长参数	生長參數	growth parameter
生长静止期	生長靜止期	stationary growth phase
生长率	生長速率	growth rate
生长曲线	生長曲線	growth curve
生长收率	生長收率	growth yield
生长收率系数,增殖收率系数	生長收率係數	growth yield coefficient
生长因子	生長因子	growth factor
声控	聲控	acoustic control
声速	音速	sonic velocity
声速流动	聲速流動,音速流動	acoustic flow, sonic flow

大　陆　名	台　湾　名	英　文　名
声学	聲學	acoustics
声致发光	聲致發光	sonoluminescence
失活	失活,去活化,去活[作用]	deactivation, inactivation
失活剂	去活化劑	deactivator
失控	失控	runaway
施肥	施肥	fertilization
施密特数	施密特數	Schmidt number
施陶丁格黏性定律	史陶丁格黏性定律	Staudinger viscosity law
湿壁塔	濕壁塔	wetted-wall column, wetted-wall tower
湿表面	濕面	wet surface
湿存水	吸濕[水]分	hygroscopic water
湿度	濕度	humidity
湿度比	濕度比	humidity ratio
湿度表	濕度表	humid chart, humidity chart
湿度测定法	濕度測定法	psychrometry
湿度测定器	水分測定器	moisture determination apparatus
湿度差	乾濕球溫度差	psychrometric difference
湿度计	濕度計	hygrometer, moisture meter
湿度图	濕度圖	psychrometric chart
湿度线	濕度線	psychrometric line
湿度因数	水分因數	moisture factor
湿法分离	濕法分離	wet separation
湿法过程	濕法	wet process
湿法冶金[学]	濕法冶金[學]	hydrometallurgy
湿纺	濕紡[絲]	wet spinning
湿含量	水分含量	moisture content
湿化	水化	slaking
湿井	濕井	wet well
湿面积	潤濕面積	wetted area
湿磨	濕磨	wet grinding
湿气	含油氣	wet gas
湿气计	濕氣計	wet gas meter
湿球温度表	濕球溫度計	wet-bulb thermometer
湿球下降	濕球下降	wet-bulb depression
湿热	濕[潤]比熱	humid heat
湿润剂	潤濕劑	wetting agent
湿筛选	濕篩選	wet screening

大　陆　名	台　湾　名	英　文　名
湿式密闭容器	濕式密封器	wet seal holder
湿式燃烧	濕式燃燒	wet combustion
湿式氧化	濕式氧化法	wet oxidation
湿体积	濕[潤]比容	humid volume
湿物料	濕物料	moist materials
湿辗机	濕輾機	wet pan mill
湿蒸汽	濕蒸汽	wet steam
十六烷值	十六烷值	cetane number
十六烷指数	十六烷指數	cetane index
石膏	石膏	gypsum
石膏模	石膏模	plaster mold
石灰	石灰	lime
石灰乳	石灰乳	lime milk
石灰石	石灰石	limestone
石灰苏打法	石灰蘇打法	lime-soda process
石灰窑	石灰窯	lime kiln
石灰皂	鈣皂	lime soap
石蜡	石蠟	paraffin, paraffin wax
石蜡馏分	石蠟餾出物	paraffin distillate
石蜡油	石蠟油	paraffin oil
石蜡皂	石蠟皂	paraffin soap
石蜡纸	[石]蠟紙	paraffin paper
石蜡转化	石蠟轉化	paraffin conversion
石棉	石棉,石綿	asbestos
石棉过滤器	石綿過濾器,石棉過濾器	asbestos filter
石墨	石墨	graphite
石墨电极	石墨電極	graphite electrode
石墨坩埚	石墨坩鍋	graphite crucible
石墨化	石墨化	graphitization
石墨耐火材料	石墨耐火物	graphite refractory
石墨润滑脂	石墨[潤]滑脂	graphite grease
石墨纤维	石墨纖維	graphitic fiber
石墨纤维强化塑料	石墨纖維強化塑膠	graphite fiber reinforced plastics
石脑油	石油腦,輕油	naphtha, petroleum naphtha
石脑油裂解	輕油裂解	naphtha cracking, naphtha pyrolysis
石蕊试纸	石蕊試紙	litmus paper
石盐	岩鹽	halite

大　陆　名	台　湾　名	英　文　名
石英	石英	quartz
石英玻璃	石英玻璃,矽玻璃	quartz glass, silica glass
石英管	石英管	quartz tube
石英细粉	矽砂粉	silica flour
石英岩	石英石	quartzite
石油	石油	petroleum
石油工程	石油工程	petroleum engineering
石油化工厂	石[油]化[學]工業區	petrochemical complex
石油化学	石油化學	petrochemistry
石油化学工业	石[油]化[學]工業	petrochemical industry
石油化学品	石[油]化[學]品	petrochemical
石油加工工程	石油煉製工程	petroleum processing engineering
石油加工技术	石油煉製技術	petroleum processing technology
石油焦	石油焦	petroleum coke
[石油]沥青	地瀝青	asphaltum
石油炼制	石油煉製	petroleum refining
石油醚	石油醚	petroleum ether
石油气	石油氣	petroleum gas
时变系统	時變系統	time variant system, time varying system
时程表	時程表	time schedule
时间比率法	時間比率法	time ratio method
时间标度	時間標度	time scaling
时间常量	時間常數	time constant
时间导数	時間導數	time derivative
时间关连	時間關連[性]	time dependence
时间控制	時間控制,時控	time control
时间线	時間線	time line
时间序列	時間序列	time series
时间序列模型	時間數列模型	time series model
时间最短控制	時間最短控制	time minimum control
时空	空間時間	space time
时效硬化	時效硬化	age hardening
时延	時[間遲]延	time delay
时域	時域	time domain
时滞	時[間遲]滯	time lag
实报酬率	實報酬率	true rate of return
实沸点蒸馏曲线	實沸點曲線	true boiling point curve
实际阶段	實際階	real stage

大　陆　名	台　湾　名	英　文　名
实际上升	實際上升	actual lift
实际塔板	實際塔板	actual plate
实际塔板数	實際[塔]板數	actual plate number
实际稳定区	實用穩定區	region of practical stability
实际稳定性	實用穩定性	practical stability
实平均	實平均	true mean
实验设计	實驗設計	experimental design
实验室反应器	實驗室反應器	laboratory reactor
实验室手册	實驗室手冊	laboratory manual
实验数据	實驗數據	experimental data
实验误差	實驗誤差	experimental error
实验用反应器	實驗反應器	experimental reactor
炻器	陶石器	stoneware
蚀刻剂	蝕刻劑	etchant
食品加工	食品加工	food processing
食物-微生物比	食物-微生物比	food-to-microorganism ratio
食盐	食鹽	table salt
食用牛脂	食用牛脂	edible tallow
食用脂肪	食用脂肪	edible fat
使用期	使用期	serviceable life
世代时间	世代時間	generation time
市场成本	市場成本	market cost
市场研究	市場研究	market research
市价	市價	market price
市值	市價	market value
示波器	示波器	oscillograph, oscilloscope
示差热膨胀测量术	微差膨脹測定法	differential dilatometry
示差扫描量热法	微差掃描熱量法	differential scanning calorimetry
示范装置	示範單元	demonstration unit
示意图	示意圖	schematic diagram
示踪化学	示蹤[劑]化學	tracer chemistry
示踪技术	示蹤劑法	tracer technique
示踪剂	示蹤劑	tracer
示踪曲线	示蹤劑曲線	tracer curve
示踪同位素	示蹤同位素	tracer isotope
示踪响应曲线	示蹤劑應答曲線	tracer response curve
示踪信息	示蹤劑資訊	tracer information
事件矩阵	事件矩陣	occurrence matrix

大　陆　名	台　湾　名	英　文　名
势	勢	potential
势差	勢差,[電]位差	potential difference
势垒	勢障,位能障壁	potential barrier
势流	勢流	potential flow
势能	位能	potential energy
势能面	位能面	potential energy surface
势偏差	勢偏差	potential deviation
势线	勢線	potential line
势修正	勢修正	potential correction
视镜	窺鏡	sight glass
视孔	視孔,窺孔	sight hole
视因数	視因數	view factor
试饼	試餅	pat
试管	試管	test tube
F 试验	F 檢定	F-test
适合性	適合性	suitability
适应调谐	適應調諧	adapt tuning
释放角	釋放角	angle of release
嗜冷菌	嗜冷菌	psychrophilic bacteria
嗜冷微生物	嗜冷菌	psychrophile
嗜热细菌	嗜熱菌	thermophilic bacteria
噬菌体	噬菌體	phage
收集环	收集環	collecting ring
收集器	收集器	collector
收敛加速	收斂加速	convergence acceleration
收敛判据	收斂準則	convergence criterion
收率	產率	yield
收率系数	產率係數	yield coefficient
收缩	收縮	constriction, shrink, shrinkage
收缩剂	收縮劑	shrinking agent
收缩裂纹	縮裂	shrinkage crack
收缩率	收縮率	shrinkage rate
收缩能力	收縮能力	shrinking power
收缩曲线	收縮曲線	shrinkage curve
收缩损失	收縮損失	contraction loss
收缩稳定性	收縮穩定性	contraction stability
收缩系数	收縮係數	coefficient contraction, contraction coefficient

大　陆　名	台　湾　名	英　文　名
收缩应力	收縮應力	shrinking stress
收缩张力试验机	收縮張力試驗機	shrinkage tension tester
收益率	收益率	earning rate
手动重设	手動重設[定]	manual reset
手性分离	手性分離,掌性分離	chiral separation
手摇干湿表	搖轉濕度計	sling psychrometer
守恒方程	守恆方程式	conservation equation
守恒律	守恆律	conservation law
受激分子	受激分子	energized molecule
受激响应技术	刺激應答法	stimulus-response technique
受控区域	受控區域	controlled field
受体	受體,受器	receptor, acceptor
受污染土壤	受污土壤	contaminated soil
受阻沉降	受阻沈降	hindered sedimentation, hindered settling
枢轴	樞軸	pivot
疏水胶体	疏水膠體	hydrophobic colloid
疏水器	①袪水器,蒸汽阱②閘	steam trap
疏水色谱[法]	疏水層析法	hydrophobic chromatography
疏水物	疏水物	hydrophobe
疏水纤维	疏水纖維	hydrophobic fiber
疏水性	疏水性	hydrophobicity
疏水性颗粒	疏水性粒子	hydrophobic particle
疏液胶体	疏液膠體	lyophobic colloid
舒尔茨-齐姆分布	舒爾茨-齊姆分佈	Schulz-Zimm distribution
输出	輸出	output
输出[变]量	輸出變數	output variable
输出层	輸出層	output layer
输出函数	輸出函數	output function
输出集	輸出集	output set
输出信号	輸出信號	output signal
输出压力	輸出壓力	output pressure
输入	輸入	input
输入变量	輸入變數	input variable
输入函数	輸入函數	input function
输入搅动	輸入擾動	input disturbance
输入节点	輸入節點	input node
输入信号	輸入信號,輸入訊號	input signal
熟化器	熟化器	aging machine

大　陆　名	台　湾　名	英　文　名
熟化试验	硬化試驗	curing test
熟料	燒粉	chamotte
熟石灰	熟石灰	hydrated lime
熟铁	熟鐵	wrought iron
熟油	熟[煉]油	boiled oil
曙红染料	曙紅染料	eosin dye
曙红色淀	曙紅色澱	eosin lake
树胶	膠	gum
树脂	樹脂	resin
ABS 树脂	丙烯腈,丁二烯,苯乙烯樹脂	acrylonitrile-butadiene-styrene resin, ABS resin
树脂鞣法	樹脂鞣法	resin tannage
树脂皂	樹脂皂	resin soap
竖管式蒸发器	豎管蒸發器	vertical tube evaporator
竖甑	豎甑	vertical retort
数据	數據	data, datum
数据不准性	數據不準性	data inaccuracy
数据处理	數據處理,資料處理	data processing
数据分析	數據分析	data analysis
数据校正	數據校正	data reconciliation
数据拟合	數據擬合	data fitting
数据日志	數據日誌	data log
数据筛选	數據篩選,資料篩選	data screening
数均分子量	數量平均分子量	number average molecular weight
数量级	[數]量級	order of magnitude
数学程序	數學規劃	mathematical programming
数学模拟	數學模式化	mathematical modeling
数学模型	數學模式	mathematical model
数值方法	數值方法	numerical method
数值分析	數值分析	numerical analysis
数值解	數值解	numerical solution
数字计算机	數位電腦,數位計算機	digital computer
数字记录器	數位記錄器	digital recorder
数字控制	數值控制,數位控制	numerical control, digital control
衰变	衰變,衰退	decay
衰变比	衰變比,衰退比	decay ratio
衰变常数	衰變常數	decay constant
衰变机理	崩解機制	disintegration mechanism

大　陆　名	台　湾　名	英　文　名
衰减	衰减[作用]	attenuation
衰减器	衰减器	attenuator
衰减效应	衰减效應	attenuation effect
衰减因子	衰减因數	attenuation factor
衰亡期	衰减期	decline phase
双倍定率递减折旧法	雙倍遞減均衡法	double declining balance method
双臂捏合机	雙臂捏合機	double arm kneading mixer
双部位机理	雙[部]位機構	dual-site mechanism
双层隔热玻璃板	雙層隔熱玻璃板	double-glazed insulating pane, thermopane
双层绝热量热器	絕熱雙卡計	adiabatic twin calorimeter
双层联立模块法	二階法	two tier approach
双层中空玻璃	雙層玻璃	pair glass
双重促进催化剂	雙重促進化觸媒	double promoted catalyst
双重上胶	雙重上膠	double sizing
双重塔	雙重塔	double column
双重循环	雙重循環	dual recirculation
双动泵	雙動泵	double acting pump
双动搅拌器	雙動攪拌器	double motion agitator
双分散	雙分散	bidispersion
双分子反应	雙分子反應,二分子反應	bimolecular reaction
双峰分布	雙峰分佈	bimodal distribution
双峰孔径分布	雙峰孔徑分佈	bimodal pore-size distribution
双缸往复泵	雙缸往復泵	duplex reciprocating pump
双功能催化剂	雙功能觸媒	dual-function catalyst
双辊轧碎机	雙輥軋碎機	double roll crusher
双滑线电桥	雙滑線電橋	double slidewire bridge
双极矩	雙極矩	bipolar moment
双极性电极	雙極性電極	bipolar electrode
双桨混合器	雙漿混合機	double paddle mixer
双结点溶度曲线	雙結點溶解度曲線	binodal solubility curve
双介质过滤器	雙介質過濾器	dual-medium filter
双金属簇	雙金屬團	bimetallic cluster
双金属催化剂	雙金屬催化劑,雙金屬觸媒	bimetallic catalyst
双金属片	雙金屬片,雙金屬	bimetallic strip, bimetal
双金属调温计	雙金屬調溫器	bimetallic thermoregulator
双金属温度计	雙金屬溫度計	bimetallic thermometer

大 陆 名	台 湾 名	英 文 名
双口阀	雙口閥	double ported valve
双连续结构	雙連續結構	bicontinuous structure
双螺带混合机	雙螺帶混合機	double helical ribbon mixer
双螺杆挤出机	雙螺桿擠出機	twin-screw extruder
双模式系统	雙[模]式系統	dual-mode system
双膜理论	雙膜理論	two-film theory
双曲线型[动力学]方程	雙曲線型[動力學]方程	hyperbolic type [kinetic] equation
双溶剂萃取	雙溶劑萃取	double solvent extraction
双式过滤器	雙式濾器	double filter
双式控制	雙式控制,雙[模]式控制	two-mode control, dual-mode control
双室稳定化池	雙室穩定化池	two-cell stabilization pond
双水相萃取	雙水相萃取,兩水相萃取	aqueous two-phase extraction
双水相体系(=双水相系统)		
双水相系统,双水相体系	雙水相系統,兩水相系統	aqueous two-phase system
双糖	雙醣	disaccharide
双头螺纹短接头	雙頭螺紋短接頭	short nipple
双位控制器	雙位控制器	two-position controller
双位驱动器	雙位致動器	two-position actuator
双稳定装置	雙穩定裝置	bistable device
双吸叶轮机	雙吸葉輪機	double suction impeller
双向狭缝	雙向狹縫	bilateral slit
双效蒸发	雙效蒸發	double effect evaporation
双效蒸发器	雙效蒸發器	double effect evaporator
双旋光	雙旋光[作用]	birotation
双循环	雙循環,二元循環	binary cycle
双用控制	雙工控制	duplex control
双折射	雙折射	birefraction, birefringence
双折射流体	雙折射流體	birefringent fluid
双锥掺合机	雙錐摻合機	double cone blender
双锥分级机	雙錐分級機	double cone classifier
双阻力理论	雙阻力理論	two-resistance theory
双座阀	雙座閥	double seat valve
爽身粉	滑石粉	talcum powder

大　陆　名	台　湾　名	英　文　名
水/油乳剂软膏基	水/油劑軟膏基	W/O emulsion ointment base
水包油乳状液	水中油乳液	oil in water emulsion
水玻璃	水玻璃	water glass
水簸机	水簸機	hydraulic jig
水处理	水處理	water treatment
水锤	水鎚	water hammer
水淬	水淬	shredding
水萃取物	水相萃取物	aqueous extract
水阀	水閥	water valve
水分	水分	moisture
水分离器	水分離器	water separator
水封	水[密]封	water seal
水合	水合	hydration
水合器	水合器	hydrator
水合热	水合熱	heat of hydration
水合物	水合物	hydrate
水合盐	水合鹽	hydrated salt
水化酶	水合酶	hydrase
水浆涂料	漿料塗布	slurry coating
水胶体	水膠體	hydrocolloid
水解	水解[作用]	hydrolysis
水解产物	水解產物	hydrolysate
水解酶	水解酶	hydrolase, hydrolytic enzyme
水解吸附	水解吸附	hydrolytic adsorption
水净化	水淨化	water purification
水冷结晶器	水冷結晶器	water cooled crystallizer
水冷却	水冷卻	water cooling
水冷却器	水冷卻器	water cooler
水力半径	水力半徑	hydraulic radius
水力超载量	水力超載量	hydraulic overload
水力传感器	水力感測器	hydraulic sensor
水力阀	水力閥	hydraulic valve
水力分类	水力類析[法]	hydraulic classification
水力分离	水力分離	hydraulic separation
水力分离器	水力分離器	hydraulic separator
水力负载量	水力負載量	hydraulic load
水力功率	水力功率	hydraulic power
水力共振	水力共振	hydraulic resonance

大 陆 名	台 湾 名	英 文 名
水力劈裂	液壓破裂[法]	hydraulic fracturing
水力平滑壁	水力平滑壁	hydraulically smooth wall
水力平滑管	水力平滑管	hydraulically smooth pipe
水力平均直径	水力平均直徑	hydraulic mean diameter
水力效率	水力效率,液壓效率	hydraulic efficiency
水力旋流分离器	液體旋流分離器	liquid cyclone separator
水力旋流器	液體旋流器	liquid cyclone
水力学	水力學	hydraulics
水力压头	水力高差	hydraulic head
水力直径	水力直徑	hydraulic diameter
水力转速计	水力轉速計,液壓轉速計	hydraulic tachometer
水力阻力	水力阻力	hydraulic resistance
水帘	水簾	water curtain
水淋冷凝器	滴水冷凝器	drip condenser
水铝石	水鋁石	boehmite
水轮机	水力渦輪機	hydraulic turbine
水煤浆	煤水漿	coal-water slurry
水煤气	水煤氣	blue gas, water gas
水煤气变换	水煤氣轉化	gas shift
水煤气变换反应	水煤氣[轉化]反應	gas shift reaction, water gas shift reaction
[水煤气]变换反应器	[水煤氣]轉化反應器	shift reactor
水煤气变换过程	水煤氣轉化程序	gas shift process
水镁矾	硫酸鎂石	kieserite
水锰矿	水錳礦	manganite
水磨光机	水純化器	water polisher
水泥	水泥	cement
水泥管	水泥管	concrete pipe
水泥输送泵	水泥泵	cement pump
[水泥]硬化指数	[水泥]硬化指數	cementation index
水喷头	灑水頭	sprinkler head
水平列管蒸发器	水平列管蒸發器	evaporator with horizontal tubes
水溶胶	水溶膠	hydrosol
水溶性软膏基	水溶性軟膏基	water-soluble ointment base
水溶液	水溶液	aqueous solution
水溶助长性	水溶助長性	hydrotropy
水溶助剂	水溶助劑	hydrotropic agent, hydrotrope
水软化	水軟化	water softening

大　陆　名	台　湾　名	英　文　名
水套	水[夾]套	water jacket
水调理	水調理	water conditioning
水头	水位高差	water head
水头损失	壓頭損失,高差損失	head loss
水污染	水污染	water pollution
水污染工程	水污染工程	water pollution engineering
水污染降低	水污染減量	water pollution abatement
水污染控制	水污染控制	water pollution control
水污染物	水污染物	water pollutant
水性漆	水性漆	water paint, water base paint
水压机	液壓機	hydraulic press
水压用水	液壓用水	hydraulic water
水杨酸	水楊酸,柳酸	salicylic acid
水银温度计	水銀溫度計,汞溫度計	mercury thermometer
水硬石灰	水硬石灰	hydraulic lime
水硬水泥	水硬水泥	hydraulic cement
水浴	水浴	water bath
水跃	水躍	hydraulic jump
水蒸气	水蒸氣	steam
水蒸气阀容量	蒸汽閥容量	steam valve capacity
水蒸气图表	蒸汽表	steam table
水蒸气蒸馏	蒸汽蒸餾	steam distillation
水质	水質	water quality
水质分析	水[質]分析	water quality analysis
水质判据	水質準則	water quality criterion
水质硬度	水質硬度	water hardness
顺磁体	順磁體	paramagnetic body
顺磁效应	順磁效應	paramagnetic effect
顺磁性	順磁性	paramagnetism
顺磁氧分析仪	順磁氧分析儀	paramagnetic oxygen analyzer
顺丁橡胶	順丁二烯橡膠	cis-butadiene rubber
顺式	順式	cis-form
顺式异戊二烯橡胶	順異戊二烯橡膠,順異平橡膠	cis-isoprene rubber
顺向进料	順向進料	forward feeding, forward feed
顺序操作	順序操作	sequential operation
顺序阀	順序閥	sequencing valve
顺序控制	順序控制	sequence control

大　陆　名	台　湾　名	英　文　名
瞬时产量分率	瞬間產量分率	instantaneous fractional yield
瞬时反应	瞬間反應	instantaneous reaction
瞬时反应速率	瞬間反應速率	instantaneous reaction rate
瞬时杀菌器	瞬間殺菌器	flash pasteurizer
瞬时系统	瞬間系統	instantaneous system
瞬态扩散	瞬態擴散,暫態擴散	transient diffusion
瞬态现象	瞬態現象,暫態現象	transient phenomenon
瞬态响应	瞬態應答,暫態應答	transient response
瞬态响应分析	瞬態應答分析,暫態應答分析	transient response analysis
瞬态行为	瞬態行為,暫態行為	transient behavior
朔佩尔张力试验机	肖珀張力試驗機	Schopper tensile tester
丝纺	絹紡	silk spinning
丝光处理	絲光處理	mercerizing, mercerization
丝光棉	絲光棉	mercerized cotton
丝光纱	絲光紗	mercerized yarn
丝胶蛋白	絲膠	sericin
丝鸣[绸料摩擦声]	絲鳴	scroop
丝纤维	絲纖維	silk fiber
丝状细菌	絲狀細菌	filamentous bacteria
斯科特挠度计	斯科特撓度計	Scott flexometer
斯氏黏度计	史托馬黏度計	Stormer viscometer
斯托克斯	斯托克(動黏度單位)	stokes
斯托克斯近似法	斯托克近似法	Stokes approximation
斯托克斯线	斯托克斯[譜]線	Stokes line
撕裂	裂痕	tearing
撕裂强度	撕裂強度	tearing strength
撕裂应力	撕裂應力	tearing stress
撕碎机	撕碎機	shredder
撕碎装置	撕碎裝置	shredding device
死带	死帶	dead band
死区	死區	dead zone
死水区	死水區[域]	dead water region
死亡率	死亡速率	death rate
死亡期	死亡期	death phase
四环素	四環[黴]素,鉑黴素	tetracycline, achromycin
四聚丙烯	四聚丙烯	propylene tetramer
四聚物	四聚物	tetramer

大　陆　名	台　湾　名	英　文　名
四氯苯醌	[四]氯[苯]醌	chloranil
四通阀	四通閥	four-way valve
四乙铅	四乙[基]鉛	tetraethyl lead
伺服机构	伺服機構	servo mechanism
伺服机构补偿	伺服機構補償	compensation servo-mechanism, servome-chanism compensation
伺服控制	伺服控制	servo control
伺服问题	伺服問題	servomechanism-type problem, servo pro-blem
松弛变量	鬆弛變量, 附加變數	slack variable
松弛法, 弛豫法	鬆弛法	relaxation method
松弛试验	鬆弛試驗	relaxation test
松弛丝光	鬆弛絲光加工	slack mercerization
松弛现象	鬆弛現象	relaxation phenomenon
松弛状态	鬆弛狀態	relaxed state
松节油	松節油	turpentine oil
松香	松香, 松脂	rosin
松香胶	松香膠料	rosin size
松香施胶	松香上膠	rosin sizing
松香油	松香油	rosin oil
松香皂	松香皂	rosined soap
松油	松油	pine oil
松子油	松子油	pine seed oil
苏打灰	蘇打灰, 鈉鹼灰	soda ash
速成胶质	速成膠	accelerated gum
速度	速度	velocity
速度边界层	速度邊界層	velocity boundary layer
速度边界层厚度	速度邊界層厚度	velocity boundary layer thickness
速度分布	速度分佈	velocity distribution
速度[分布]剖面[图]	速度剖面[圖], 速度分佈	velocity profile
速度计	速度計	velocity meter
速度距离滞后	速度距離滯延	velocity-distance lag
速度势	速度勢	velocity potential
速度梯度	速度梯度	velocity gradient
速度压头	速度高差	velocity head
速率	速率	speed
速凝水泥(=快干水泥)		

大 陆 名	台 湾 名	英 文 名
速启阀	速啟閥	quick-opening valve
速止剂	速止劑	short stopper, short stopping agent
速止聚合	速止聚合	short stopped polymerization
塑度计	塑性計	plastimeter, plastometer
塑解剂	解膠劑	peptizing agent
塑料	塑膠	plastic
塑料成形加工	塑膠加工	plastic processing
塑料工程	塑膠工程	plastic engineering
塑料计	塑性試驗計	plasticimeter
塑料黏合剂	塑膠黏結劑	plastic binder
塑料黏结剂	塑膠膠合劑	plastic cement
塑料泡沫	塑膠泡體,塑膠泡沫	plastic foam
塑料填料	塑膠填充物	plastic packing
塑性材料	塑性材料,塑膠材料	plastic materials
塑性计	塑性測定器,塑性儀	plastograph
塑性流体	塑性流體	plastic fluid
塑性凝胶	塑性凝膠	plastigel
塑性屈伏	塑性降伏	plastic yield
塑性屈伏点	塑性降伏點	plastic yield point
塑性指数	塑性指數	plastic index
酸败	酸化	souring
酸败发酵	酸敗發酵	sour fermentation
酸泵	酸泵	acid pump
酸槽车,酸罐车	酸槽車	acid tank car
酸催化	酸催化	acid catalysis
酸度	酸度	acidity
酸度试验	酸度試驗	acidity test
酸度指数	酸度指數	acidity index
酸法	酸法	acid process
酸罐车(=酸槽车)		
酸化	酸化,加酸	acidification, acidulation
酸化槽	加酸槽	acidulating tank
酸化剂	酸化劑	acidifier, acidulant
酸化器	加酸器	acidulator
酸冷却器	酸冷卻器	acid cooler
酸气	酸氣	sour gas
酸受体	酸受體	acid acceptor
酸水	酸水	sour water

大　陆　名	台　湾　名	英　文　名
酸水汽提	酸水汽提	sour-water stripping
酸雾	酸霧	acid mist
酸洗	酸洗	pickling
酸洗槽	酸洗槽	pickle bath, pickling bath
酸洗法	酸洗法	pickling process
酸洗液	酸洗液	pickle liquor
酸性	酸性	acidity
酸性成分	酸性成分	acid component
酸性催化	酸性催化[作用]	acidic catalysis
酸性催化剂	酸性觸媒	acidic catalyst
酸性肥料	酸性肥料	acid fertilizer
酸性铬媒染料	酸性鉻媒染料	acid chrome dye
酸性耐火材料	酸性耐火材	acid refractory
酸性耐火砖	酸性耐火磚	acid firebrick, acid refractory brick
酸性气体	酸性氣體	acid gas
酸性染料	酸性染料	acid color, acid dye
酸性污泥	酸性污泥	acid sludge
酸烟	酸煙	acid fume
酸雨	酸雨	acid rain
酸值	酸值	acid number
算法	演算[法]	algorithm
算法合成技术	算法合成技術	algorithmic synthesis technique
算术平均	算術平均	arithmetic mean
算术平均温差	算術平均溫差	arithmetic mean temperature difference
算术平均值	算術平均值	arithmetic mean value
随管加热蒸汽	隨管加熱蒸汽	tracing steam
随机变量	隨機變數	random variable
随机表面更新	隨機表面更新	random surface renewal
随机波动	隨機起伏	random fluctuation
随机抽样	隨機取樣	random sampling
随机分布	隨機分佈	random distribution
随机分析	隨機分析	stochastic analysis
随机负荷变化	隨機負載變化	random load change
随机过程	隨機程序	probabilistic process, random process, stochastic process
随机扰动	隨機擾動	random disturbance
随机数	隨機數	random number
随机搜索	隨機搜尋	random search

大　陆　名	台　湾　名	英　文　名
随机特性	隨機特質	stochastic feature
随机稳定性	隨機穩定性	stochastic stability
随机误差	隨機誤差	random error
随机信号	隨機信號	random signal
随机性	隨機性	randomness
碎玻璃	玻璃屑	cullet
碎料板	塑合板,碎屑膠合板	particle board
碎裂	碎斷[作用]	fragmentation
隧道干燥器	隧道乾燥器	tunnel drier
隧道窑	隧道窯	tunnel kiln
燧卵石	燧卵石	flint pebble
燧石	燧石	flint
损耗	損耗	loss
损耗角正切	損耗模數比,損耗正切	loss tangent
损失功	損失功	lost work
羧基化	羧化[作用]	carboxylation
羧基酶	羧基酶	carboxylase
羧甲基纤维素	羧甲纖維素	carboxymethyl cellulose
羧酸盐	羧酸鹽	carboxylate
缩二脲	縮二脲	biuret
缩合反应	縮合反應	condensation reaction
缩合系数	縮合係數	condensation coefficient
缩痕	縮陷處	shrink mark
缩径管接头	漸縮接頭	reducing coupling
缩聚[反应]	縮[合]聚合[作用]	condensation polymerization
缩聚反应	聚縮[作用]	polycondensation
缩扩喷嘴	縮擴噴嘴	converging-diverging nozzle
缩脉(＝流颈)		
缩醛	縮醛	acetal
缩醛树脂	縮醛樹脂	acetal resin
缩小	等比縮小	scale down
所得税	所得稅	income tax
索亥俄法	蘇黑澳法	Sohio process
锁钥学说	鎖鑰理論	lock-and-key theory

T

大　陆　名	台　湾　名	英　文　名
塔	塔［器］	tower
塔板	塔板	plate, tray
塔板间距	塔板間距	tray spacing
［塔］板效率	板效率	plate efficiency
塔顶冷凝器	塔頂冷凝器	overhead condenser
塔顶蒸汽	塔頂蒸氣	overhead vapor
塔内件	塔內件	column internals
塔填充物	塔填充物	tower packing
塔填料	塔填充物	column packing
塔贮留量	塔滯留量	tower hold-up
台架反应器	實驗室規模反應器	bench scale reactor
台架试验	實驗室規模試驗	bench scale test
台式天台	檯秤	platform balance
太古油	茜草油	alizarine oil
太阳常数	太陽常數	solar constant
太阳辐射	太陽輻射	solar radiation
太阳光谱	太陽光譜	solar spectrum
太阳能	太陽能	solar energy
太阳能电池	太陽能電池	solar cell
太阳能发动机	太陽能引擎	solar engine
太阳能集热器	太陽能收集器	solar collector
太阳能蒸馏	太陽能蒸餾	solar distillation
太阳［索拉］油	太陽油	solar oil
肽	［胜］肽	peptide
肽酶	肽酶	peptidase
钛白	鈦白	titanium white
钛白陶	鈦白陶［器］	titania whiteware
钛瓷	鈦瓷	titania porcelain
钛合金	鈦合金	titanium alloy
钛搪瓷	鈦搪瓷	titanium enamel
钛铁矿	鈦鐵礦	ilmenite
泰乐菌素	泰黴素	tylosin
泰勒标准筛	泰勒標準篩	Tyler standard sieve

大　陆　名	台　湾　名	英　文　名
酞菁蓝	酞[花]青藍	phthalocyanine blue
酞菁染料	酞[花]青染料	phthalocyanine dye
酞菁颜料	酞[花]青顏料	phthalocyanine pigment
酞青绿	酞[花]青綠	phthalocyanine green
弹簧常量	彈簧常數	spring constant
弹簧秤	彈簧秤	spring balance, spring scale
弹簧吊架	彈簧吊架	spring hanger
弹簧管压力计	彈簧管壓力計,波頓壓力計	Bourdon gauge
弹簧力	彈簧力	spring force
弹簧平衡型钟式压力计	鐘型彈簧壓力計	spring balanced type bell gauge
弹簧式安全阀	彈簧安全閥	spring safety valve
弹簧式硬度试验仪	彈簧式硬度試驗儀	spring type hardness tester
弹簧载荷减压阀	彈簧負載減壓閥	spring loaded reducing valve
弹簧载荷面积计	彈簧負載可變面積流量計	spring loaded area meter
弹簧载荷调压器	彈簧負載壓力調節器	spring loaded pressure regulator
弹簧执行机构	彈簧致動器	spring actuator
弹簧-质量-阻尼系统	彈簧-質量-阻尼系統	spring-mass-damper system
弹力纱	伸縮紗	stretch yarn
弹力纤维	彈性纖維	spandex fiber
弹力性	彈性	springness
弹内[加氧]陈化	彈內[加氧]老化	bomb aging
弹式反应器	彈[式]反應器	bomb reactor
弹式量热器	彈[式]卡計	bomb calorimeter
弹性	彈性,彈[性]能	elasticity, resilience
弹性胶囊	彈性膠囊	elastic capsule
弹性聚合物	彈性聚合物	elastopolymer
弹性硫	彈性硫	elastic sulphur, plastic sulfur
弹性模量	彈性模數	elastic modulus
弹性能	彈性能	elastic energy
弹性试验	彈性試驗	elasticity test
弹性数	彈性數	elasticity number
弹性塑料	彈性塑膠	elastoplastic
弹性体	彈性物,彈性體	elastomer
弹性体共混物	彈性體摻合物	elastomer blend
弹性纤维	彈性纖維	elastic fiber, snapback fiber
弹性液体	彈性液體	elastic liquid

大　陆　名	台　湾　名	英　文　名
檀香油	檀香油	sandal oil
炭	炭	char
炭电极	碳電極	carbon electrode
炭黑	碳黑	carbon black
炭滤	碳濾［法］	carbon filtration
炭刷	碳刷	carbon brush
炭砖	碳磚	carbon brick
探头	探頭	probe
碳钢	碳鋼	carbon steel
碳化钙,电石	碳化鈣,電石	calcium carbide
碳化硅	碳化矽	carbide silicon, silicon carbide
碳化硅纤维	碳化矽纖維	silicon carbide fiber
碳化硅砖	碳化矽磚	carbide brick
碳化硼	碳化硼	boron carbide
碳平衡	碳平衡	carbon balance
碳氢化合物	碳氫化合物	hydrocarbon
碳水化合物	碳水化合物,醣	carbohydrate
碳酸钙	碳酸鈣	calcium carbonate
碳酸化过程	碳酸法	carbonation process
碳吸附	碳吸附	carbon adsorption
碳纤维	碳纖維	carbon fiber
碳纤维强化塑料	碳纖維強化塑膠	carbon fiber reinforced plastic
碳循环	碳循環	carbon cycle
碳阳离子染料	碳陽離子染料	carbonium dyes
碳质耐火材料	碳質耐火物	carbon refractory
羰基合成	甲醯化	oxo process
羰基化	羰基化［作用］	carbonylation
羰基镍催化剂	鎳羰觸媒	nickel carbonyl catalyst
搪瓷	搪瓷	porcelain enamel
搪瓷钢	搪瓷鋼	porcelain enameled steel
搪塑	中空鑄型［法］	slush molding
糖	糖	sugar
糖化剂	糖化劑	saccharifying agent
糖化作用	糖化［作用］	saccharification
糖浆剂	糖漿	syrup
糖酵解	醣解［作用］	glycolysis
糖酵解途径	醣解途徑	glycolytic pathway
糖精	糖精	saccharin

大　陆　名	台　湾　名	英　文　名
糖类	醣	saccharide
糖量计	糖量計	saccharimeter
糖蜜	糖蜜	molasses
糖蜜酒精	糖蜜酒精	molasses alcohol
糖蜜起晶	糖蜜起晶	molasses graining
糖蜜塔养基	糖蜜培養基	molasses medium
糖原	肝醣	glycogen
淌流	淌流	sag flow
淌流板	淌流板	sag flow board
桃红釉	桃紅釉	peach bloom glaze
桃仁油	桃仁油	peach-kernel oil
陶瓷	陶瓷	ceramics
陶瓷管	陶瓷管	ceramic tube
陶瓷耐火材料	陶瓷耐火物	ceramic refractory
陶瓷器皿	陶瓷器	ceramic ware
陶瓷釉	陶瓷釉	ceramic glaze
陶瓷制品	陶瓷製品	ceramic
陶土	陶土	earthenware clay
淘矿机	淘洗器	vanner
淘析	淘析[作用]	elutriation
淘析器	淘析器	elutriator
淘洗	淘析[作用]	elutriation
套管反应器	[雙]套管反應器	double pipe reactor
套管接头	套筒接頭	sleeve joint
套管冷却结晶器	刮刀式熱交換器	votator heat exchanger
套管冷却器	[雙]套管冷卻器	double pipe cooler
套管热交换器	[雙]套管熱交換器	double-pipe heat exchanger
套筒联轴器	套筒連結器	sleeve coupling
特氟龙	鐵氟龍,聚四氟乙烯	teflon
特级耐火材料	特級耐火物,超強耐火物	super-duty refractory
特殊性能高分子	特用高分子	speciality polymer
特性常数	特性常數	characteristic constant
特性反应时间	特性反應時間	characteristic reaction time
特性方程	特性方程式,特徵方程式	characteristic equation
特性方程式	性能方程式	performance equation
特性根	特性根	characteristic root

大　陆　名	台　湾　名	英　文　名
特性函数	特性函數	characteristic function
特性化	特性分析,表徵分析	characterization
特性黏度	本質黏度,極限黏度	intrinsic viscosity
特性判据	性能準則	performance criterion
特性曲线	特性曲線	characteristic curve, performance curve
特性数	性能值	performance number
特性因数	特性因子,特性化因數	characteristic factor, characterization factor
特异性	特異性	specificity
特征长度	特性長度	characteristic length
特征时间	特性時間	characteristic time
特征值	特徵值,固有值	characteristic value, eigenvalue
特种黏合剂	特殊黏合劑	special adhesive
梯度	梯度	gradient
梯度法	梯度法	gradient method
梯形堰	梯形堰	trapezoidal weir
锑白	銻白	antimony white
锑电极	銻電極	antimony electrode
锑黄	銻黃	antimony yellow
锑皂	銻皂	antimonial soap
提纯	純化	purification
提斗浸取器	斗式萃取器	bucket-elevator extractor
提高石油采收率	提高石油採收[法]	enhanced oil recovery
提馏	汽提,脫除	stripping
提馏段	汽提段	stripping section
提浓段	增濃段	enriching section
提升	[品質]提升,升級	upgrading
提升管	上升管,豎板	riser
提升管反应器	上升管反應器	riser reactor
体积	體積,容積	volume
体积表面直径	體積表面[相當]直徑	volume surface diameter
体积传氧系数(=容积传氧系数)		
体积分率	體積分率	volume fraction
体积功	體積功	volume work
体积流率	體積流量	volumetric flow rate
体积模数	體積[彈性]模數	volume modulus
体[积]黏性	總體黏度	bulk viscosity

大　陆　名	台　湾　名	英　文　名
体积平均沸点	體積平均沸點	volumetric average boiling point
体积平均速度	體積平均速度	volume average velocity
体积平均直径	體積平均直徑	volume mean diameter
体积收缩	體積收縮[率]	volume shrinkage
体积效率	體積效率	volumetric efficiency
体积指数	體積指數	volume index
体扩散	體擴散	bulk diffusion
体内	[生]體內	*in vivo*
体内浓度	總體濃度	bulk concentration
体膨胀	體膨脹	cubical expansion
体膨胀率	體[積]膨脹係數	volume expansivity
体外	[生]體外	*in vitro*
体相	[總]體相	bulk phase
体心晶格	體心晶格	body-centered lattice
替代物	取代基	substituent
天然碱	天然鹼,碳酸鈉石	natural soda , trona
天然气	天然氣	natural gas
天然气重整	天然氣重組	natural gas reforming
天然气水合物	天然氣水合物	natural gas hydrate
天然气液化	天然氣液化	natural gas liquefaction
天然汽油	[井口]天然汽油,天然氣汽油,天然氣凝結油	casing head gasoline , natural gasoline
天然清漆	天然清漆	natural varnish
天然橡胶	天然橡膠	natural rubber
添加剂	添加劑	additive
甜菜糖	甜菜糖	beet sugar
甜味剂	甜味料	sweetener
填充	裝填	fill
填充床	填充床	packed bed
填充床反应器	填充床反應器	packed bed reactor
填充高度	填充高度	packed height
填充剂,辅料	助強劑,增容填料	builder , bulking filler
填充密度	填充密度	packed density , packing density
填充式喷淋塔	填料式噴霧塔	packed spray tower
填充式洗涤器	填料式洗滌塔	packed scrubber
填充物	填充物	packing materials
填充柱	填料柱	packed column

大 陆 名	台 湾 名	英 文 名
填料	①填料 ②增效劑 ③加脂[法]	filler, extender, stuffing
填料萃取塔	填料式萃取塔	packed extractor
填料函式密封	填料箱密封	stuffing box seal
填料环	墊圈	packing ring
填料塔	填料塔	packed tower, packed column
填料吸收器	填料式吸收塔	packed column absorber
填料箱	填料箱	stuffing box
填料因子	填料因子	packing factor
填料蒸馏塔	填充蒸餾塔	packed distillation column
填料支承板	填料支架	packing support
条件稳定性	條件穩定[性]	conditional stability
条纹	條紋,斑紋	stripe
条纹线	條紋線	streak line
调合漆	調和漆	ready-mixed paint
调和漆	調合漆	blended paint
调和汽油	摻配汽油	blended gasoline
调节剂	調節劑	conditioning agent, regulator
调节控制	調節控制	regulatory control
调节器	調節器	regulator
调节器参数整定	控制器調諧	controller tuning
调节箱	調節箱	regulating box
调味剂	調味料	flavoring materials
调温旋管	調溫旋管	tempering coil
调谐	調諧	tuning
调谐参数	調諧參數	tuning parameter
调压变压器	可調變壓器	variable transformer
调压阀	壓力調節閥,洩氣閥	pressure regulating valve, relief valve
调压器	壓力調節器	pressure regulator
调优操作	調優操作	evolutionary operation
调优法	調優法	evolutionary method
调整	調整	adjustment
调制	調變	modulation
跳开装置	跳動裝置	tripper
跳汰流化床	跳汰流[體]化床	jigged fluidized bed
贴壁细胞	貼壁細胞,固面依附細胞	anchorage-dependent cell
萜[烯]	[烯]萜[烯]類	terpene

大　陆　名	台　湾　名	英　文　名
萜[烯]油	萜油	terpene oil
铁氨催化剂	鐵氨觸媒	catalyst iron-ammonia
铁锭	熟鐵	ingot iron
铁合金	鐵合金	ferro-alloy
铁铝氧催化剂	鐵鋁氧觸媒	catalyst iron-alumina, iron-alumina cata-lyst
铁媒染剂	鐵媒染劑	iron mordant
铁锰合金	錳鐵合金,錳鐵[齊]	ferromanganese
铁鞣	鐵鞣	iron tannage
铁石棉	鐵石綿	amosite
铁细菌	鐵細菌	iron bacteria
烃发酵	烴發酵	hydrocarbon fermentation
烃黑	烴黑	hydrocarbon black
停车控制	停機控制	shutdown control
停堆检查	停工檢查	shutdown inspection
停工期	停工期	shutdown period
停工日程表	停機時程	shutdown schedule
停工时间	停工時間	shutdown time
停留时间	滯留時間	residence time
停留时间分布	滯留時間分佈	residence time distribution
停修时间	停工維護時間	maintenance downtime
停止流动法	停流法	stopped flow method
停滞期间	停滯期間	detention period
停滞时间	停滯時間,遲延時間	dead time, detention time
通道	通道	channel
通断控制	開關控制	on-off control
通风	通風	ventilation
通风比例	通風比率	vent ratio
通风机	通風器	ventilator
通风计	通風計	draft gage
通风扇	通風扇	draft fan
通风需求	通風需求	ventilation requirement
通量	通量	flux
通量密度	通量密度	flux density
通气管	通氣管	aeration candle
通用简约梯度法	通用簡約梯度法	general reduced gradient
通用冷凝器	通用冷凝器	universal condenser
通用清洁剂	通用清潔劑	general purpose detergent

大　陆　名	台　湾　名	英　文　名
通用压缩因子图	壓縮因數通用圖	generalized Z chart
通用指示剂	通用指示劑	universal indicator
同步传送器	同步傳送器	synchro-transmitter
同步互穿网络	同步[形成]互穿網結構	simultaneous interpenetrating network
同步机	同步器	synchro
同步加速器	同步加速器	synchrotron
同步控制变压器	同步控制變壓器	synchro-control transformer
同电子排列体	同電子排列體	isostere
同化	同化	assimilation
同时操作	同時操作	simultaneous operation
同时反应	併發反應	simultaneous reactions
同时聚合	同時聚合	simultaneous polymerization
同时吸收	同時吸收	simultaneous absorption
同位素	同位素	isotope
同位素法	同位素法	isotopic method
同位素分离	同位素分離	isotope separation
同位素丰度	同位素豐度	isotopic abundance
同位素示踪剂	同位素示蹤劑	isotopic tracer
同系化合物	同系物	homologous compound
同系物	同系物	homolog
同向流(=并流)		
同心变径管	同心漸縮管	concentric reducer
同心浮球	同心浮子	concentric float
同心管柱	同心管柱	concentric-tube column
同心环	同心環	concentric ring
同心刻度盘	同心儀	concentric dial
同心孔口板	同心孔口板	concentric orifice plate
同心柱黏度计	同軸圓柱黏度計	concentric cylinder viscometer
同型发酵菌	同型發酵菌	homofermentative bacteria
桐油	桐油	tung oil
铜铵人造纤维	銅銨嫘縈	cuprammonium rayon
铜铀云母,辉铜矿	銅鈾雲母	chalcolite
统筹法	計畫評審法	program evaluation and review technique
统计单元	統計單位	statistical unit
统计分析	統計分析	statistical analysis
统计估计	統計估計	statistical estimation
统计[结构]共聚物	統計[結構]共聚物	statistical copolymer

大　陆　名	台　湾　名	英　文　名
统计均匀性	統計均匀性	statistical homogeneity
统计理论	統計理論	statistical theory
统计力学	統計力學	statistical mechanics
统计链	統計鏈	statistical chain
统计链段	統計鏈段	statistical segment
统计权重矩阵	統計權重矩陣	statistical weight matrix
统计热力学	統計熱力學	statistical thermodynamics
统计设计	統計設計	statistical design
统计误差	統計誤差	statistical error
统计学	統計學	statistics
桶磨机	桶磨機	barrel mill
桶式锅炉	桶式鍋爐	drum boiler
桶式浸取	槽瀝取	vat leaching
桶式去皮机	桶式去皮機	barking drum, drum barker
筒	桶	drum
筒式进料机	進料鼓	feed drum
头馏分	塔頂餾出物	overhead distillate
投入产出	輸入輸出,投入產出	input-output
投资回收	投資回收	investment recovery
投资回收期	回收期	payback period
投资收益率	投資收益率,投資報酬率	rate of return on investment, return on investment
投资资本	投資資本	investment capital
透光率	透光率,透射係數	transmittance, spectral transmittance
透明度	透明度,透明性	transparency
透明釉	透明釉	transparent glaze
透明皂	透明皂	transparent soap
透气性	透氣性	air permeability, gas permeability
透射率	透射係數	transmissivity
透水性	透水性	water permeability
透析	透析[作用]	dialysis
透析槽	透析槽,滲析槽	dialysis cell
透析培养	透析培養	dialysis culture
透析器	透析器,滲析器	dialyzer, dialyzator
透析液	透析物,滲析物	dialyzate
透析蒸馏	透析蒸餾	perdistillation
突变	突變	mutation
突变论	劇變理論	catastrophe theory

大　陆　名	台　湾　名	英　文　名
突变模型	劇變模型	catastrophic model
突变频率	突變頻率	mutation frequency
M-T 图	M-T 圖	McCabe-Thiele diagram
P-S 图	焓-濃圖,P-S 圖	Ponchon-Savarit diagram
T-H 图（＝温湿图）		
Y-X 图	汽-液莫耳分率圖,Y-X 圖	Y-X diagram
图解	圖解	graphical solution
图解法	圖解法	graphical method
图解分析	圖解分析法	graphical analysis
图解积分	圖解積分法	graphical integration
图解计算	圖解計算	graphical calculation
图论	圖論	graph theory
图示板	圖示面板	graphic panel
图像分析仪	圖像分析儀	quantimet
图形流程图	圖形流程圖	pictorial flowsheet
涂胶压延机	塗膠壓延機	spreading calender
涂料	塗料	paint
涂料技术	塗裝技術	coating technology
途径	途徑	route
土地污染	土地污染	land pollution
土霉素	土黴素,地靈黴素	terramycin
土壤	土壤	soil
土壤承受压力	土壤承受壓力	soil bearing pressures
土壤处理	土壤處理	soil treatment
土壤肥力	土壤肥度	soil fertility
土壤腐蚀	土壤腐蝕	soil corrosion
土壤改良剂	土壤改良劑	soil conditioner
土壤酵母	土壤酵母	soil yeast
土壤力学	土壤力學	soil mechanics
土壤溶胶	土壤溶膠	soil sol
土壤酸度	土壤酸度	soil acidity
土壤微生物	土壤微生物	soil organism
土壤稳定剂	土壤安定劑	soil stabilizer
土壤污染	土壤污染	soil contamination, soil pollution
土壤消毒剂	土壤消毒劑	soil sterilant
土石膏	土[狀]石膏	gypsite
土水泥	土壤水泥	soil cement

大　陆　名	台　湾　名	英　文　名
土质颜料	土質顔料	earth pigment
吐酒石	吐酒石,酒石酸銻(Ⅲ)鉀鹽	emetic tartar, tartar emetic
湍动流化床	紊動流[體]化床	turbulent fluidized bed
湍流	紊流	turbulent flow
湍流边界层	紊流邊界層	turbulent boundary layer
湍流标度	紊流標度	turbulence scale
湍流促进器	紊流促進元件	turbulence promoter
湍流度	紊流度	degree of turbulence
湍流反应器	紊流反應器	turbulent flow reactor
湍流核心	紊流核心	turbulent core
湍流剪应力	紊流剪應力	turbulent shear stress
湍流扩散	紊流擴散	turbulent diffusion
湍流扩散系数	紊流擴散係數	turbulent diffusivity
湍流能谱	紊流能譜	turbulent-energy spectrum
湍流强度	紊流強度	intensity of turbulence
湍流松弛现象	紊流弛豫現象	turbulent relaxation phenomenon
湍流应力	紊流應力	turbulent stress
湍流运动	紊流運動	turbulent motion
团聚	團聚	agglomeration
团聚物	團聚物	agglomerate
团聚值	團聚值	agglomerating value
推迟时间	阻滯時間	retardation time
推斥势	推斥勢	repulsive potential
推动力	驅動力	driving force
推进剂	推進劑	propellant
推理控制	推算控制	inferential control
退化	退化,減併	degeneracy, degeneration, obsolescence
退化形式	退化形式	degeneration form
退火	退火	annealing
退火点	退火點	annealing point
退火炉	退火窯	lehr
褪色剂	脱色劑	stripping agent
褪色试验	剝離試驗	stripping test
托	托	torr
拖刮机	拖刮機	drag scraper
脱氨基	去胺[作用]	deamination
脱氨酶	去胺酶	deaminase

大 陆 名	台 湾 名	英 文 名
脱玻化	去玻化	devitrification
脱臭过程	[天然氣]脱臭程序	sweetening process
脱氮	脱氮[反應],去氮[作用]	denitrogenation
脱氮细菌	脱氮細菌	denitrifying bacteria
脱氮作用	脱氮[作用]	denitrification
脱丁烷塔	去丁烷塔	debutanizer
脱芳构化	去芳香化[作用]	dearomatization
脱辅基酶	脱輔基酶	apoenzyme
脱附等温线	脱附等溫線	desorption isotherm
脱附控制	脱附控制	desorption control
脱甲烷	去甲烷[作用]	demethanization
脱甲烷塔	去甲烷塔	demethanizer
脱胶	褪膠	boiling-off
脱蜡	脱蠟	dewaxing
脱硫	脱硫[作用],去硫[作用]	desulfurization, sulfur elimination
脱硫过程	[天然氣]脱臭程序	sweetening process
脱卤	脱鹵[反應],去鹵[反應]	dehalogenation
脱氯塔	去氯器	dechlorinator
脱模板	脱模板	stripper plate
脱模机	汽提塔	stripper
脱模剂	①脱模劑,離型劑 ②脱色劑	mold-release agent, parting agent, release agent, stripping agent
脱漆剂	除漆劑	paint remover
脱气	除氣	deaeration, degasification
脱气器	除氣器	degasifier
脱气塔	除氣器	deaerator
脱氢	脱氫[反應],去氫[作用]	dehydrogenation
脱氢催化剂	脱氫觸媒	dehydrogenation catalyst
脱氢环化作用	脱氫環化[作用]	dehydrocyclization
脱氢酶	脱氫酶	dehydrogenase
脱溶剂器	脱溶劑器	desolventizer
脱色	脱色,去色	decolorization
脱色剂	脱色劑,去色劑	decolorizer
脱水	脱水	dehydration, dewatering

大　陆　名	台　湾　名	英　文　名
脱水酒精	脱水酒精,無水酒精	dehydrated alcohol
脱水器	脱水器	dehydrator
脱水收缩	[膠體]脱水收縮	syneresis
脱水物	脱水物	dehydrate
脱羧	脱羧[反應],去羧[作用]	decarboxylation
脱碳	去碳	decarburization
脱烷基化	脱烷[作用]	dealkylation
脱硝	脱硝[反應],去硝[作用]	denitration
脱硝器	脱硝器	denitrator
脱硝塔	脱硝塔,去硝塔	denitration tower
脱硝作用	脱硝[作用]	denitrification
脱盐	脱鹽	desalination
脱氧	去氧[作用]	deoxygenation
脱乙烷塔	去乙烷塔	deethanizer
脱脂	脱脂	degrease
脱脂奶	脱脂乳	skim milk
妥尔油	松油	tall oil
妥尔油松香	松香	tall-oil rosin
妥尔油皂	松香油皂	tall-oil soap

W

大　陆　名	台　湾　名	英　文　名
瓦楞纸	瓦楞紙	corrugating paper
外表面	外表面	external surface
外表面积	外表面積	external area
外表面浓度	外表面濃度	external surface concentration
外部施胶	外部上膠	external sizing
外部坐标	外部坐標	external coordinates
外换热器	外熱交換器	external heat exchanger
外回流	[塔]外回流	outside reflux, external reflux
外回流比	[塔]外回流比	outside reflux ratio
外径	外徑	external diameter, outside diameter
外控制环路	外控制環路	external control loop
外扩散	外部擴散	external diffusion
外力	外力	external force

大　陆　名	台　湾　名	英　文　名
外流	外部流動	external flow
外能	外能	external energy
外扰	外部擾動	external disturbance
外热法	外熱法	external heating
外渗	外滲透[作用]	exosmosis
外推	外插法	extrapolation
外消旋化	[外]消旋反應	racemization
外消旋混合物	[外]消旋混合物	racemic mixture
外循环	外循環	external recycle
外循环反应器	外循環反應器	external recycle reactor
外压力	外壓[力]	external pressure
外延	磊晶	epitaxy
外延生长	磊晶生長	epitaxial growth
外在容积	外在容量	external capacity
弯管	彎管	bend
弯管机	彎管機	pipe bending machine
弯管相当管长	彎管相當管長	bend equivalent pipe length
弯曲试验	彎曲試驗	bending test
弯曲因子	扭曲因數	tortuosity factor
弯曲应力	彎曲應力	bending stress, flexural stress
弯头	彎頭,肘管	elbow
完全混合反应器	完全混合反應器	perfectly mixed reactor
完全混合消化	完全混合消化槽	complete-mixing digester
完全解偶	完全解偶	perfect decoupling
完全控制	理想控制	perfect control
顽[火]辉石	頑火輝石	enstatite
烷基苯磺酸盐	烷[基]苯磺酸鹽	alkylbenzene sulfonate
烷基化	烷[基]化	alkylation
烷基转移	轉烷化[作用]	transalkylation
碗形磨	碗[形]磨	bowl mill
万古霉素	萬古黴素	vancomycin
万向接头	萬向接頭	universal joint
王水	王水	aqua regia
网鞍填料	麥氏網鞍填料	McMahon packing
θ网环	θ網環,狄克森網環	Dixon ring
网孔	篩孔	mesh
网孔塔板	網孔塔板	perform tray
网络	網路	network

大　陆　名	台　湾　名	英　文　名
网筛	網篩	mesh screen
网状层	網狀層	stratum reticular
网状催化剂	網狀觸媒	gauze catalyst
往复板式萃取塔	往復板萃取塔	reciprocating plate extraction column
往复板式塔	往復板萃取塔	reciprocating plate column
往复泵	往復泵,循環泵	reciprocating pump, recirculating pump
往复加料器	往復進料器	reciprocating feeder
往复耙	往復耙	reciprocating rake
往复筛	往復篩	reciprocating screen
往复式	往復運動	reciprocating
往复式活塞压缩机	往復式活塞壓縮機	reciprocating piston compressor
往复式压缩机	往復[式]壓縮機	reciprocating compressor
往复式真空泵	往復式真空泵	reciprocating vacuum pump
往复循环反应器	往復反應器	reciprocating reactor
危险性评估	危險性評估	hazard evaluation
危险指数	危險指數	hazard index
威尔逊方程	威爾森方程	Wilson equation
微波	微波	microwave
微波波谱学	微波光譜學	microwave spectroscopy
微波干燥	微波乾燥	microwave drying
微波谱	微波譜	microwave spectrum
微差测温法	微差測溫法	differential thermometry
微差间隙	微差間隙	differential gap
微差接触设备	微差接觸設備,微分接觸設備	differential contact equipment
微差温度计	微差溫度計	differential thermometer
微滴	小滴	droplet
微电路	微電路	microcircuit
微分	微分	derivative
微分表面	微分表面,微差表面	differential surface
微分产率	微分產率	differential yield
微分沉降	微分沈降,微差沈降	differential settling
微分法	微分[法]	differentiation, differential technique, differential method
微分反应器	微分反應器,微差反應器	differential reactor
微分方程	微分方程[式]	differential equation
微分分析仪	微分分析儀,微差分析	differential analyzer

大　陆　名	台　湾　名	英　文　名
	儀	
微分控制	微分控制	derivative control, D control, differential control
微分冷凝	微分冷凝,微差冷凝	differential condensation
微分能量平衡	微分能量均衡	differential energy balance
微分膨胀计测定法	微分熱膨脹法	differential dilatometry
微分器	微分器	differentiator
微分溶解热	微分溶解熱	differential heat of solution
微分筛析	微分篩析,微差篩析	differential screen analysis
微分时间	微分時間	derivative time
微分时间常数	微分時間常數	derivative time constant
微分吸附	微分吸附,微差吸附	differential adsorption
微分吸收	微分吸收,微差吸收	differential absorption
微分稀释热	微分稀釋熱	differential heat of dilution
微分蒸馏	微分蒸餾,微差蒸餾	differential distillation
微分质量平衡	微分質量均衡	differential mass balance
微分重量分布	微差重量分佈	differential weight distribution
微粉磨	微磨機	micronizer
微观混合	微觀混合	micromixing
微观可逆性	微觀可逆性	microscopic reversibility
微观可逆性原理	微觀可逆性原理	principle of microscopic reversibility
微观流体	微觀流體	microfluid
微观平衡	微觀平衡	microscopic balance
微观态	微觀狀態	microscopic state
微观现象	微觀現象	microscale phenomenon
微积分	微積分	calculus
微胶囊	微膠囊	microcapsule
微结构	微結構	microstructure
微结构材料	微結構化材料	microstructured materials
微晶	微晶	crystallite, microcrystal
微晶结构	微晶結構	microcrystalline structure
微晶蜡	微晶蠟	microcrystalline wax
微孔	微孔	micropore
微孔材料	海綿狀材料	cellular materials
微孔分离器	微孔分離器	microporous separator
微孔过滤器	微孔過濾器	microporous filter
微孔扩散	孔[隙]擴散	pore diffusion
微孔扩散阻力	孔[隙]擴散阻力	pore diffusion resistance

大　陆　名	台　湾　名	英　文　名
微孔滤膜	微孔透膜	microporous membrane
微孔［透壁］分气法	細孔［透壁］氣體分離法	atmolysis
微粒	顆粒,粒狀物	particulate
微粒学	微粒學	micromeritics
微量分析	微量分析［法］	microanalysis
微量化学	微量化學	microchemistry
微滤	①微量過濾 ②微［孔］過濾,微［過］濾	microfiltration, microstraining
微滤器	濾微器	microstrainer
微米	微米	micron
微囊化	微膠囊化	microencapsulation
微球	微球	microsphere
微扰	微擾	perturbation
微扰变量	微擾變數	perturbation variable
微扰法	微擾法	perturbation method
微扰理论,摄动理论	微擾理論	perturbation theory
微扰硬链理论	微擾硬鏈理論	perturbed hard chain theory, PHC theory
微乳	微乳液	microemulsion
微商热重法	微分熱重量法	derivative thermogravimetry
微生物	微生物	microbe, microorganism
微生物动力学	微生物動力［學］	microbial dynamics, microbial kinetics
微生物发酵	微生物發酵［法］	microbial fermentation
微生物法	微生物法	microbial process
微生物反应	微生物反應	microbial reaction
微生物反应器	微生物反應器	microbial reactor
微生物腐蚀	微生物腐蝕	microbial corrosion
微生物过程	微生物法	microbiological process
微生物结垢	微生物積垢	microbial fouling
微生物膜	微生物膜	microbial film
微生物学	微生物學	microbiology
微型催化反应器	微型觸媒反應器	microcatalytic reactor
微型反应工程	微反應工程	microreaction engineering
微型反应器	微型反應器	microreactor
微型计算机	微電腦	microcomputer
微型仪器	微型儀器	miniature instrument
微载体	微載體	microcarrier
微正则配分函数	微正則配分函數	microcanonical partition function

大　陆　名	台　湾　名	英　文　名
微正则系综	微正則系綜	microcanonical ensemble
韦[伯]	韋伯	Weber
韦伯数	韋伯數	Weber number
韦格斯坦法	韋格斯坦法	Wegstein method
韦斯模数	維茲模數	Weisz modulus
桅杆清漆	桅桿清漆	spar varnish
唯象模型	現象學模式	phenomenological model
唯象系数	現象學係數	phenomenological coefficient
维纶	維尼綸	vinylon
维生素	維生素,維他命	vitamin
维修	維護,保養	maintenance
维修费	維護成本	maintenance cost
维修管理	維護管理	maintenance management
伟晶岩	偉晶岩	pegmatite
尾管	尾管	tail pipe
尾料	尾料	tailing
尾流,尾涡	尾流	wake
尾气	尾氣	end gas, tail gas
尾随涡	拖尾漩渦	trailing vortex
尾涡(=尾流)		
卫生纸	衛生紙	sanitary paper
卫星工厂	衛星工廠	satellite plant
未反应核模型	未反應核模型	unreacted core model
未控流	未控流	wild stream
位力方程	位力方程	virial equation
位力系数	位力系數	virial coefficient
位势	位能	potential
位头	勢差,位能差	potential head
位相关系	相關係	phase relation
位形配分函数,构型配 分函数	組態配分函數	configurational partition function
位形性质	組態性質	configurational property
位移	位移	displacement
位移计	位移[流量]計	displacement meter
位移流量计	位移流量計	displacement flowmeter
位移压力计	位移壓力計	displacement pressure gage
位置模型理论	位置模型理論	site-model theory
位阻	位阻,立體阻礙	steric hindrance

大　陆　名	台　湾　名	英　文　名
喂料装置	稱量進料機	weighing feeder
温标	溫標	temperature scale
温差	溫差	temperature difference
温差电堆	熱電堆	thermopile
温差驱动力	溫差驅動力	temperature difference driving force
温度	溫度	temperature
温度边界层	溫度邊界層	thermal boundary layer
温度变送器	溫度傳送器	temperature transmitter
温度程序	溫度排程	temperature schedule
温度滴定法	溫度滴定[分析]法	thermometric titrimetry
温度范围	溫度範圍	temperature range
温度分布	溫度分佈	temperature distribution
温度计	溫度計	temperature gauge, thermometer
温度计套管	溫度計套管	thermowell
温度记录器	溫度記錄器	temperature recorder
温度校正	溫度校正,溫度修正	temperature calibration, temperature correction
温度控制	溫度控制	temperature control
温度控制器	溫度控制器	temperature controller
温度廓线	溫度剖面圖	temperature profile
温度历程	溫度歷程	temperature progression
温度失控(=飞温)		
温度梯度	溫度梯度	temperature gradient
温度调节阀	熱調閥	thermo regulating valve
温度调节器	溫度調節器,調溫器	temperature regulator, thermoregulator
温度指示控制器	溫度指示控制器	temperature indicating controller
温度指示器	溫度指示器	temperature indicator
温度组成图	溫度-組成圖	temperature-composition diagram
温熵图	溫熵[關係]圖	temperature-entropy diagram
温湿图,T-H 图	溫濕圖	temperature-humidity chart
温室效应	溫室效應	greenhouse effect
文丘里	文氏管	Venturi
文丘里管	文氏管	Venturi tube
文丘里流量计	文氏流量計,文氏計	Venturi flowmeter, Venturi meter
文丘里喷嘴	文氏噴嘴	Venturi nozzle
文丘里洗涤器	文氏管洗滌器	Venturi scrubber
文献	文獻	literature
文献调查	文獻調查	literature survey

大　陆　名	台　湾　名	英　文　名
紊流	紊流	turbulent flow
稳定操作条件	穩定操作條件	stable operating condition
稳定成分	安定成分	stabilizing ingredient
稳定重整油	穩定重組油	stabilized reformate
稳定度	穩定度,安定度	degree of stability
稳定化槽	穩定化槽	stabilization tank
稳定化池	穩定化池,穩定化槽	stabilization basin, stabilization pond
稳定极限	穩定性極限	stability limit
稳定剂	穩定劑,安定劑,穩定器	stabilizing agent, stabiliser
稳定界限	穩定邊限	margin of stability
稳定近似	穩態近似法	steady state approximation
稳定流	穩流	steady flow
稳定汽油	穩定汽油	stabilized gasoline
稳定区域	穩定區[域]	stability region, stable region
稳定溶液	穩定溶液	stable solution
稳定乳液	穩定乳液	stable emulsion
稳定塔	穩定器	stabilizer
稳定条件	穩定條件	stability condition
稳定同位素	穩定同位素	stable isotope
稳定稳态	穩定穩態	stable steady state
稳定响应	穩定應答	stable response
稳定性	穩定性,穩定度,安定性	stability
稳定性分析	穩定性分析	stability analysis
稳定性判据	穩定準則	stability criterion
稳定性试验	安定性試驗	stability test
稳定[作用]	穩定化,安定化	stabilization
稳健性	韌性	robustness
稳泡剂	泡沫穩定劑	foam stabilizer
稳态,定常态	穩態,穩定狀態	steady-state, stable state
稳态多重性	穩態多重性	multiplicity of steady states
稳态分析	穩態分析	steady state analysis
稳态过程	穩態程序	steady state process
稳态近似	穩態近似法	stationary state approximation
稳态扩散速率	穩態擴散速率	steady-state diffusion rate
稳态流动	穩態流	steady state flow
稳态模型	穩態模式	steady-state model
稳态柔量	穩態柔量	steady-state compliance
稳态蠕变	穩態潛變	steady-state creep

大　陆　名	台　湾　名	英　文　名
稳态唯一性	穩態唯一性	uniqueness of steady state
稳态稳定性	穩態穩定性	stability of steady state
稳态误差	穩態誤差	steady state error
稳态响应	穩態應答	steady state response
稳态性能	穩態效能	steady-state performance
稳态运行	穩態操作	steady state operation
稳态增益	穩態增益	steady state gain
稳态周期操作	穩定週期操作	steady periodic operation
涡度	漩渦度	vorticity
涡壳泵	渦捲泵	volute pump
涡流	渦流,渦電流	eddy flow, eddy current
涡流长度	渦流長度	eddy length
涡流传递	渦流輸送	eddy transport
涡流扩散	渦流擴散	eddy diffusion
涡流扩散系数	渦流擴散係數	eddy diffusivity
涡流损耗	渦流損失	eddy current loss
涡流停留时间	渦流滯留時間	eddy residence time
涡流吸收	渦流吸收	eddy absorption
涡流消除器	漩渦碎機	vortex breaker
涡流卸掉	渦流分離	eddy shedding
涡流应力	渦流應力	eddy stress
涡流运动	渦流運動	eddy motion
涡流运动黏度	渦流動黏度	eddy kinematic viscosity
涡流转速计	渦流轉速計	eddy current tachometer
涡流阻力	渦流阻力	eddy resistance
涡轮	渦輪機	turbine
涡轮泵	渦輪泵	turbo-pump, turbine pump
涡轮干燥机	渦輪乾燥機	turbo-dryer
涡轮鼓风机	渦輪鼓風機	turbo-blower
涡轮机	渦輪機	turbo-machine
涡轮搅拌器	渦輪攪拌器	turbine agitator
涡轮离心泵	渦輪離心泵	turbine centrifugal pump
涡轮流量计	渦輪[式]流量計	turbine flowmeter
涡轮压缩机	渦輪壓縮機	turbo-compressor
涡轮叶轮	渦輪[式]葉輪	turbine impeller
涡黏性	渦流黏度	eddy viscosity
涡片	漩渦片	vortex sheet
涡线	漩渦線	vortex line

大　陆　名	台　湾　名	英　文　名
涡旋	漩渦	vortex
涡旋搅拌器	漩渦攪拌器	vortex agitator
涡旋势	漩渦勢	vortex potential
涡旋脱落	漩渦剝離	vortex shedding
涡旋尾迹	漩渦尾跡	vortex trail
涡旋运动	漩渦運動	vortex motion
蜗结	小渦輪,蛇輪	snail
蜗壳	渦捲	volute
沃尔展开式	沃爾展開式	Wohl expansion
卧式混合器	臥式混合器	horizontal mixer
卧式离心筛	臥式離心篩	horizontal centrifugal screen
卧式炉	臥式爐	horizontal furnace
卧式蒸煮器	臥式蒸煮器	horizontal digester
乌木蜡	烏木臘	ebony wax
污点	斑點	speck
污垢	[污]垢	scale
污垢系数	積垢係數,比例係數	fouling coefficient, fouling factor
污泥层	污泥層	sludge blanket
污泥处理	污泥處理	sludge disposal
污泥处理过程	污泥處理法	sludge disposal process
污泥焚烧炉	污泥焚燒爐,污泥焚化爐	sludge furnace, sludge incinerator
污泥过滤	污泥過濾	sludge filtration
污泥密度指数	污泥密度指數	sludge density index
污泥浓缩	污泥增黏	sludge thickening
污泥浓缩器	污泥濃縮器	sludge thickener
污泥蓬松现象	污泥蓬鬆[現象]	sludge bulking
污泥容积指数	污泥體積指數	sludge volume index
污泥消化	污泥消化	sludge digestion
污泥增稠	增黏污泥	thickening sludge
污染	污染	contamination, pollution
污染工程	污染工程	pollution engineering
污染减少	污染減量	pollution abatement
污染控制	污染控制,污染防治	pollution control
污染水	污水	polluted water
污染物	污染物	contaminant, pollutant
污染物标准指数	污染物標準指數	pollutant standards index
污染源	污染源	pollution source

大　陆　名	台　湾　名	英　文　名
污水池	機油箱,廢油池	sump
污水处理	污水處理	sewage disposal
污水管	污水管	sewer pipe
污油	污油	slop
无电涂	無電塗層,無電塗裝	electroless coating
无定形	無定形	amorphism
无定形聚合物	無定形聚合物	amorphous polymer
无定形碳	無定形碳	amorphous carbon
无动混合器	無動混合器	motionless mixer
无反应性	無反應性	nonreactivity
无反应性组分	無反應性成分	nonreactive component
无纺布	不織布	nonwoven fabrics
无缝管	無縫管	seamless pipe
无缝夹套锅	無縫[夾]套鍋	jacketed seamless kettle
无光油漆	無光油漆	matt paint
无光釉	無光釉	matt glaze
无规共聚物	隨機共聚物	random copolymer
无规降解	隨機降解	random degradation
无规扩散	隨機擴散	random diffusion
无规[立构]聚合物	雜排聚合物	atactic polymer
无滑动条件	無滑動[邊界]條件	no slip condition
无灰滤纸	無灰濾紙	ashless filter paper
无混合流动反应器	無混合流動反應器	unmixed flow reactor
无机染料	礦物染料	mineral dye
无机酸	礦酸	mineral acid
无菌操作	無菌操作	aseptic operation
无菌密封	無菌密封	aseptic seal
无孔碳	高密度碳	karbate
无量纲参数	無因次參數	dimensionless parameter
无量纲数	無因次數	dimensionless number
无量纲数群	無因次群	dimensionless group
无摩擦流动	無摩擦流動	frictionless flow
无铅汽油	無鉛汽油	unleaded gasoline
无热溶液	無熱溶液	athermal solution
无溶剂基准	無溶劑基準	solvent-free basis
无溶剂黏合剂	無溶劑黏合劑	solventless adhesive
无溶剂漆	無溶劑漆	solventless paint
无溶剂涂料	無溶劑塗料	solventless coating

大 陆 名	台 湾 名	英 文 名
无水酒精(=无水乙醇)	無水酒精,絕對酒精	absolute alcohol
无水乙醇,无水酒精	無水酒精,絕對酒精	absolute alcohol
无弹簧执行机构	無彈簧致動器	springless actuator
无梯度反应器	無梯度反應器	gradientless reactor
无梯度接触器	無梯度接觸器	gradientless contactor
无限回流	無限回流	infinite reflux
无限稀释	無限稀釋	infinite dilution
无相互作用系统	無相互作用系統	noninteracting system
无旋流	無旋轉流	irrotational flow
无旋运动	無旋轉運動	irrotational motion
无烟火药	無煙火藥	smokeless powder
无烟煤	無煙煤	anthracite coal
无烟燃料	無煙燃料	smokeless fuel
无烟推进剂	無煙推進劑	smokeless propellant
无盐聚电解质溶液	無鹽高分子電解質溶液	salt-free polyelectrolyte solution
无液流量计	無液流量計	aneroid flowmeter
无液气压计	無液氣壓計	aneroid barometer
无液压力计	無液壓力計	aneroid manometer
无约束优化	無約束最適化	unconstrained optimization
无载体催化剂	無擔載觸媒	unsupported catalyst
无阻尼系统	無阻尼系統	undamped system
无阻尼响应	無阻尼應答	undamped response
无阻尼自然频率	無阻尼自然頻率	undamped natural frequency
物理参数	物理參數	physical parameter
物理常数	物理常數	physical constant
物理处理	物理處理	physical treatment
物理处理过程	物理處理法	physical treatment process
物理传递步骤	物理輸送步驟	physical transport step
物理定律	物理定律	physical law
物理分离	物理分離	physical separation
物理过程	物理程序	physical process
物理化学	物理化學	physical chemistry
物理化学吸收	物理化學吸收	physicochemical absorption
物理模型	物理模式	physical model
物理气相沉积	物理氣相沈積	physical vapor deposition
物理示踪剂	物理示蹤劑	physical tracer
物理试验	物理試驗	physical test
物理吸附	物理吸附,物理吸收	physisorption, physical adsorption

大　陆　名	台　湾　名	英　文　名
物理吸收	物理吸收	physical absorption
物理吸收剂	物理吸收劑	physical absorbent
物理系统	物理系統	physical system
物理性质	物理性質	physical property
物理㶲	物理可用能	physical exergy
物理约束	物理限制	physical constraint
物量流程图	計量流程圖	quantitative flow diagram
物料	物料	material
物料平衡	物質均衡	material balance
物料平衡流程图	物質均衡流程圖	material balance flowsheet
物料与能量平衡	質能均衡	material and energy balance
物态参量	狀態參量	state property
物位传送器	液位傳送器	level transmitter
物位控制器	液位控制器	level controller
物质	物質	matter
物质导数	物質導數	material derivative
物质分离剂	物質分離劑	mass separating agent
物质函数	物質函數	material function
物质客观性原理	物質客觀性原理	principle of material objectivity
物种	物種,種	species
误差	誤差	error
误差比	誤差比	error ratio
误差常数	誤差常數	error constant
误差传播	誤差傳播	error propagation
误差传递	誤差傳播	propagation of error
误差积分	誤差積分	error integral
误差判据	誤差準則	error criterion
误差平方积分	誤差平方積分	integral of square error
误差平方控制	誤差平方控制	error-squared control
误差信号	誤差信號	error signal
误校正	誤校正,校正失誤	miscalibration
雾	[煙]霧	mist
雾化	霧化	atomization
雾化空气	霧化空氣	atomizing air
雾化喷嘴	霧化噴嘴,噴嘴	atomizing spray nozzle, spray nozzle
雾化器	霧化器,噴槍	atomizer, spray gun
雾化燃烧器	霧化燃燒器	atomizing burner
雾化油	霧化油	atomized oil

大　陆　名	台　湾　名	英　文　名
雾化蒸汽	霧化蒸汽	atomizing steam
雾粒	霧粒	mist particle
雾沫混合器	霧沫混合器	entrainment mixer
[雾沫]夹带	[霧沫]挾帶	entrainment
雾沫蒸发器	霧沫蒸發器	entrainment evaporator
雾状流	霧沫流動	mist flow

X

大　陆　名	台　湾　名	英　文　名
吸附	吸附[作用],吸著	adsorption, sorption
吸附表面	吸附表面	adsorption surface
吸附波	吸附波	adsorption wave
吸附部位	吸附[部]位	adsorption site
吸附等容线	吸附等容線	adsorption isostere
吸附等温线	吸附等溫線	adsorption isotherm, sorption isotherm
吸附等压线	吸附等壓線	adsorption isobar
吸附法	吸附法	adsorption method
吸附管	吸附管	adsorption tube
吸附过滤	吸附過濾	adsorption filtration
吸附剂	吸附劑,吸著劑	adsorbent, adsorptive agent, sorbent
吸附控制	吸附控制	adsorption control
吸附能力	吸附度	adsorbability
吸附平衡	吸附平衡	adsorption equilibrium
吸附热	吸附熱	heat of adsorption
吸附容量	吸附容量	adsorption capacity, adsorptive capacity
吸附色谱法	吸附層析法	adsorption chromatography
吸附设备	吸附塔	adsorber
吸附势	吸附勢	adsorption potential
吸附速率	吸附速率	adsorption rate
吸附图表	吸附圖表	adsorption chart
吸附-脱附	吸附-脫附	adsorption-desorption
吸附系数	吸附係數	adsorption coefficient
吸附性	吸附性	adsorptivity
吸附质	[被]吸附質	adsorbate
吸附滞后现象	吸附遲滯現象	sorption hysteresis
吸附中心	吸附中心	adsorption center
吸附柱	吸附[管]柱	adsorption column

大　陆　名	台　湾　名	英　文　名
吸光度	吸光度	absorbance
吸滤器	吸濾器	suction filter
吸墨纸	吸墨紙	blotting paper
吸气器	抽氣器	aspirator
吸气速率	吸氣速率	air suction rate
吸热反应	吸熱反應	endothermic reaction
吸热过程	吸熱程序	endothermic process
吸入速度	吸入速度	suction velocity
吸入压力	吸入壓力	suction pressure
吸入压头	吸取高差	suction head
吸湿	吸濕	sorption of water
吸湿性	吸濕性	hygroscopicity
吸收	吸收	absorption
吸收比	吸收比	absorption ratio
吸收不连续性	吸收不連續性	absorption discontinuity
吸收层	吸收床	absorbent bed
吸收场	吸收場	absorption field
吸收池	吸收槽	absorption cell
吸收等温线	吸收等溫線	absorption isotherm
吸收度曲线	吸收率曲線	absorbency curve
吸收反应	吸收反應	absorption reaction
吸收光谱	吸收光譜	absorption spectrum
吸收过程	吸收程序	absorption process
吸收过滤器	吸收過濾器	absorbent filter
吸收基质	吸收基質	absorption base
吸收剂	吸收劑	absorbent
吸收剂量	吸收劑量	absorbed dose
吸收校正	吸收修正	absorption correction
吸收截面	吸收截面	absorption cross-section
吸收力	吸收能力	absorption power
吸收率	吸收率	absorptivity
吸收[谱]带	吸收帶,吸光帶	absorption band
吸收曲线	吸收曲線	absorption curve
吸收热	吸收熱	absorption heat, heat of absorption
吸收色谱法	吸收層析[法]	absorption chromatography
吸收式过滤器	吸收式濾器	absorptive-type filter
吸收试验	吸收試驗	absorption test
吸收数	吸收數	absorption number

大　陆　名	台　湾　名	英　文　名
吸收速率	吸收速率	absorption rate
吸收塔	吸收塔	absorber，absorption tower
吸收系数	吸收係數	absorption coefficient
吸收系统	吸收系統	absorption system
吸收性软膏基	吸收性軟膏基	absorption ointment base
吸收性纤维素	吸收性纖維素	absorbent cellulose
吸收因子	吸收因子	absorption factor
吸收油	吸收油	absorption oil
吸收指示剂	吸收指示劑	absorption indicator
吸收指数	吸收指數	absorption index
吸收制冷	吸收［式］製冷	absorption refrigeration
吸收柱	吸收管	absorption column
吸收装置	吸收裝置	absorbing apparatus
吸水率	吸水率	water absorption
吸引管线	吸入管線	suction line
吸引力	吸引力	attractive force
吸引力区［域］	吸引區	region of attraction
吸引升液器	吸入升力	suction lift
吸引势	吸引勢，吸入勢	attractive potential，suction potential
吸着	吸著	sorption
吸着剂	吸著劑	sorbent
析出	沈澱	precipitation
析因设计	析因設計	factorial design
析因实验	析因實驗	factorial experiment
烯类聚合物	乙烯系聚合物	vinyl polymer
硒聚合物	硒聚合物	selenium polymer
稀糊化淀粉	稀糊化澱粉	thin-boiling starch
稀胶法	稀膠法	broth dilution
稀释	稀釋	dilution
稀释剂（＝减黏剂）		
稀释热	稀釋熱	heat of dilution
稀释效应	稀釋效應	dilution effect
稀疏矩阵	稀疏矩陣	sparse matrix
稀相	稀［釋］相	dilute phase
稀相流化床	稀相流［體］化床	dilute-phase fluidized bed
锡箔	錫箔	tinfoil
锡媒染剂	錫媒染劑	tin mordant
锡皂	錫皂	tin soap

大　陆　名	台　湾　名	英　文　名
熄灭	熄滅,熄火	extinction
熄灭温度	熄滅溫度	extinction temperature
席夫碱	希夫鹼	Schiff base
席夫碱聚合物	希夫鹼聚合物	Schiff base polymer
洗出液	析出液	eluate
洗涤剂	清潔劑	detergent
洗涤碱	洗滌鹼,碳酸鈉	washing soda
洗涤器	洗滌器	scrubber
洗涤速率	洗滌速率	washing rate
洗涤塔	洗滌塔,柱狀洗滌塔	column scrubber, scrubbing tower, washing tower
洗毛	洗滌	scouring
洗糖法	洗糖	affination
洗糖数	洗糖數	affination number
洗糖值	洗糖值	affination value
洗脱	洗提,洗析	elution
洗脱剂	流洗液,溶析液	eluant
洗脱色谱法	洗析層析[法]	elution chromatography
洗衣碱	洗衣鹼	laundry soda
洗衣皂	洗衣皂	laundry soap
系统	系統	system
系统边界	系統邊界	system boundary
系统参数	系統參數	system parameter
系统动力学	系統動態[學]	system dynamics
系统分析	系統分析	system analysis
系统工程	系統工程	system engineering
系统功能	系統函數	system function
系统函数	系統函數	system function
系统稳定性	系統穩定性	system stability
系统性能	系統性能	system performance
系统优化	系統最適化	system optimization
细胞材料	細胞物質	cellular materials
细胞分离	細胞分離	cell separation
细胞负载	細胞負載	cell loading
细胞密度	細胞密度	cell density
细胞培养	細胞培養	cell culture
细胞破碎	細胞破碎	cell disruption
细胞溶解	細胞溶解,細胞分解	cytolysis, cell lysis

大　陆　名	台　湾　名	英　文　名
细胞融合	細胞融合	cell fusion
细胞收集	細胞收集	cell harvesting
细胞碎片	細胞碎片	cell debris
细胞悬浮培养	細胞懸浮培養	cell suspension culture
细胞匀浆	細胞均質	cell homogenate
细胞周期	細胞週期	cell cycle
细长度	細長度	slenderness
细菌	細菌	bacteria
细菌床	細菌床	bacteria bed
细菌过滤	除菌過濾	bacteriologic filtration
细菌过滤器	濾菌器	bacteria filter
细菌含量	細菌含量	bacteria content
细菌数	菌數	bacteria count
细菌[性]发酵	細菌[性]發酵	bacterial fermentation
细菌氧化	細菌氧化[反應]	bacterial oxidation
细颗粒	細粒	fine particle
细孔方向分布	孔隙方向分佈	pore orientation distribution
细孔入口	孔隙入口	pore entrance
细孔形状因子	孔[隙]形狀因數	pore shape factor
细碎机	細級壓碎機	fine crusher
细致平衡	細部均衡	detailed balancing
霞石	霞石	nepheline
下传动式石磨机	底動石磨機	under-driven buhrstone mill
下限	下限	lower bound, lower limit
下相(=底相)		
下行式反应器	降流反應器	downflow reactor
下游	下游	downstream
下游处理	下游處理	downstream processing
下游过程	下游程序	downstream process
下游压力	下游壓力	downstream pressure
纤维	纖維	fibre
纤维编织	緞紋組織	satin weave
纤维长度	纖維長度	staple length
[纤维]长度分布图	絲毛長度圖	staple diagram
[纤维长度]分析器	選別器	sorter
[纤维]分级	揀選,分類	sorting
纤维素	纖維素	cellulose
纤维素黄酸钠	黃酸纖維素鈉鹽	sodium cellulose xanthate

大 陆 名	台 湾 名	英 文 名
纤维素胶体	纖維素膠體	cellulose colloid
纤维素酶	纖維素酶	cellulase
纤维素黏合剂	纖維素黏合劑	cellulose adhesive
纤维素喷漆	纖維素[噴]漆	cellulose lacquer
纤维素衍生物	纖維素衍生物	cellulose derivative
纤维增强塑料	纖維強化塑膠	fiber reinforced plastics
弦线	弦線	chord line
显函数	顯函數	explicit function
显热	顯熱	sensible heat
显式法	顯式法	explicit method
显微光度计	微光度計	microphotometer
显微术	顯微術	microscopy
显影剂	顯影劑	developer
现金比率	現金比率	cash ratio
现金回收期	現金回收期	cash recovery period
现金流通图	現金流量圖	cash-flow diagram
现值	現值	present value, present worth
现值法	現值法	present worth method
现值因子	現值因數	present worth factor
线汇座	線匯座	line sink
线路损失	線路損失	line loss
线上计算机控制	線上控制	on-line computer control
线上调谐	線上調諧	on-line tuning
线速度	線速度	linear velocity
线型低密度聚乙烯	線型低密度聚乙烯	linear low density polyethylene
线性	線性	linearity
线性凹槽	線性凹槽	linear notch
线性超前	線性超前	linear lead
线性动量原理	線性動量原理	linear momentum principle
线性阀	線性閥	linear valve
线性分析	線性分析	linear analysis
线性共沸液	線性共沸液	linear azeotrope
线性规划	線性規畫	linear programming, LP
线性化	線性化	linearization
线性化系统	線性化系統	linearized system
线性回归分析	線性回歸分析[法]	linear regression analysis
线性控制	線性控制	linear control
线性控制阀	線性控制閥	linear control valve

大　陆　名	台　湾　名	英　文　名
线性模型	線性模型	linear model
线性黏弹性	線性黏彈性	linear viscoelasticity
线性吸收系数	線性吸收係數	linear absorption coefficient
线性系统	線性系統	linear system
线源	線源	line source
限度控制	限度控制	limit control
限量成分	限量成分	limiting component
限制	限制	constraint
限制器	限制器	limiter
相对饱和度	相對飽和度	relative saturation
相对粗糙度	相對粗糙度	relative roughness
相对丰度	相對豐度	relative abundance
相对过饱和度	相對過飽和度	relative supersaturation
相对黑度	相對黑度	relative blackness
相对挥发度	相對揮發度	relative volatility
相对黏度	相對黏度	relative viscosity
相对偏差	相對偏差	relative deviation
相对湿度	相對濕度	relative humidity
相对速度	相對速度	relative velocity
相对速度因子	相對速度因數	relative velocity factor
相对速率常数	相對速率常數	relative rate constant
相对稳定性	相對穩定性	relative stability
相对误差	相對誤差	relative error
相对吸附能力	相對吸附度	relative adsorptivity
相对吸收率	相對吸收率	relative absorptivity
相对增益	相對增益	relative gain
相对蒸汽压	相對蒸氣壓	relative vapour pressure
相关函数	相關函數,關聯函數	correlation function
相关数据	相關數據,關聯數據	correlating data
相互扩散	相互擴散	interdiffusion
相互凝聚	相互凝聚[作用]	intercoagulation
相互作用	相互作用	interaction
相互作用度	相互作用度	degree of interaction
相互作用环路	相互作用環路	interacting loops
相互作用力	相互作用力	interaction force
相互作用能	相互作用能	interaction energy
相互作用势	相互作用勢	interaction potential
相互作用系数	交互作用係數	interaction coefficient

大　陆　名	台　湾　名	英　文　名
相互作用系统	相互作用系統	interacting system
相互作用因数	相互作用因數	interaction factor
相互作用指数	相互作用指數	interaction index
相邻矩阵	相鄰矩陣	adjacency matrix
相容性	相容性	compatibility
相似变换	相似變換	similarity transformation
相似定律	相似定律	similar law
相似理论	相似性理論	similarity theory
相似性	相似性	similarity
相似性解	相似解	similarity solution
相	相	phase
相变	相變化,相變換	phase change, phase transformation
相速度	相速度	phase velocity
相特性	相行為	phase behavior
相间传质	相間質傳	interphase mass transfer
相间反应	相間反應	interphase reaction
相间交换系数	相間交換係數	interphase exchange coefficient
相间扩散	相間擴散	interphase diffusion
相间温度	相間溫度	interphase temperature
相间相内扩散	相間相內擴散	interphase-intraphase diffusion
相间相内有效性	相間相內有效度因數	interphase-intraphase effectiveness
相角	相角	phase angle
相角轨迹	相角軌跡	phase angle locus
相空间	相[位]空間	phase space
相律	相律	phase rule
相内产率	相內產率	interphase yield
相内温度	相內溫度	interphase temperature
相内有效性	相內有效度因數	interphase effectiveness
相平衡	相平衡	phase equilibrium
相平面	相平面	phase plane
相平面标绘	相平面圖	phase plane plot
相平面分析	相平面分析	phase plane analysis
相平面图	相平面圖	phase plane portrait
相图	相圖	phase diagram
相位	相位	phase
相[位]差	相位差	phase difference
相位超前	相位超前	phase advance, phase lead
相位交叉	相位交叉	phase crossover

大　陆　名	台　湾　名	英　文　名
相位交叉点	相位交叉點	phase crossover point
相位交叉频率	相位交叉頻率	phase crossover frequency
相位裕量	相位裕量	phase margin
相位滞后	相位滯延	phase lag
相移	相位位移	phase shift
相转移	相[間]轉移	phase transfer
相转移催化	相[間]轉移催化[作用]	phase transfer catalysis
相转移催化反应	相[間]轉移催化反應	phase transfer catalyzed reaction
相转移催化剂	相[間]轉移觸媒	phase transfer catalyst
香草	香草	vanilla
香草醛	香草精	vanillin
香精回收	香精回收	essence recovery
香料	香料	perfume
香茅醇	香茅醇	citronellol
香茅醛	香茅醛	citronellal
香茅油	香茅油	citronella oil
香柠檬油	香柑油	bergamot oil
厢式干燥器	廂式乾燥器,盤[式]乾燥器	compartment dryer, tray drier
箱式压滤机	廂式壓濾機	chamber filter press
箱式研磨机	箱式研磨機	magazine grinder
详细工程设计	細部工程設計	detailed engineering design
响应面	應答面	response surface
响应曲线	應答曲線	response curve
响应时间	應答時間	response time
响应速度	應答速率	response speed
向列态	向列液晶態	nematic state
向列型液晶	向列[型]液晶	nematic liquid crystal
向热性	向熱性	thermotropism
向日葵油	葵花[子]油	sunflower oil
向心加速度	向心加速度	centripetal acceleration
向心力	向心力	centripetal force
项目	專案,計畫	project
项目工程	專案工程	project engineering
橡胶	橡膠	rubber
橡胶衬里管	橡膠襯管	rubber-lined pipe
橡胶胶乳	橡膠乳膠	rubber latex

大　陆　名	台　湾　名	英　文　名
橡胶磨	橡膠磨碾機	rubber mill
橡胶态液体	橡膠態液體	rubber-like liquid
橡胶状态	橡膠狀態	rubbery state
橡皮管	橡皮管	rubber tube
橡实油	橡實油	acorn oil
肖伯回弹性	肖伯回彈性	Schob resilience
肖伯弹力试验机	肖伯彈性計	Schob elastometer
肖氏弹性计	蕭氏彈性計	Shore elastometer
肖氏硬度	蕭氏硬度	Shore hardness
肖氏硬度计	蕭氏硬度計	Shore durometer
削皮机	削片機	skiving machine
消除反应	消去反應,脫去反應	elimination reaction
消除器	除污劑	eradicator
消毒	消毒	disinfection
消毒剂	消毒劑	disinfectant
消防水	消防水	hydrant water
消光	退光,去光	delustering
消光剂	退光劑,去光劑	delusterant
消光系数	消光係數	extinction coefficient
消化	消化	digestion
消化槽	消化槽	digester
消泡剂	消泡劑	defoaming agent
消泡剂(＝防沫剂)		
消泡器	消泡機	foam breaker
硝饼	硝餅	niter cake
硝饼炉	硝餅爐	niter cake furnace
硝锅	硝鍋	niter pot
硝化法	硝化法	nitration process
硝化甘油炸药	硝化甘油炸藥,代納邁炸藥	dynamite, nitroglycerine explosive
硝化菌	[氮]硝化菌	nitrifying bacteria
硝化器	硝化器	nitrator
硝化作用	硝化,[氮]硝化	nitration, nitrification
硝基火药	硝基炸藥	nitroexplosive
硝基染料	硝基染料	nitro dye
硝炉	硝爐	niter oven
硝棉	[強]硝化棉,火棉	gun cotton
硝石	硝石	saltpeter

大　陆　名	台　湾　名	英　文　名
硝酸	硝酸	aqua fortis
硝酸纤维素	硝酸纖維素,硝化纖維素	cellulose nitrate, nitrocellulose
硝酸纤维素火药	硝化纖維素火藥,無煙火藥	nitrocellulose powder
硝酸纤维素塑料	硝化纖維素塑膠,硝酸纖維素塑膠	cellulose nitrate plastic
小角光散射	小角光散射[法],小角度光散射	small angular light scattering
小角散射	小角[度]散射	small angle scattering, SAS
小角 X 射线散射	小角 X 射線散射[法]	small angle X-ray scattering, SAXS
小角 X 射线衍射	小角度 X 光繞射	small angle X-ray diffraction, SAXD
[小时]体积空[间]速[度]	體積空間時速	volumetric hourly space velocity
小型超级计算机	迷你超級電腦	minisupercomputer
小中取大效用判据	小中取大效用判據	maximin utility criterion
效率	效率	efficiency
楔	楔形,楔形體	wedge
协同效应	增效作用	synergistic effect
斜板	斜板	inclined plate
斜管压力计	斜管壓力計	inclined manometer
斜管蒸发器	斜管蒸發器	inclined tube evaporator
斜坡函数	斜坡函數	ramp function
斜坡扰动	斜坡擾動	ramp disturbance
斜坡响应	斜坡應答	ramp response
斜栅冷却器	斜柵冷卻器	inclined grate cooler
斜弯头	斜彎頭	miter elbow
谐波分析	諧波分析	harmonic analysis
谐波应答	諧波應答	harmonic response
谐和平均	諧和平均	harmonic mean
谐和循环	諧和循環	harmonic cycling
谐振子	諧波振盪器	harmonic oscillator
泄放器	洩放器	bleeder
泄放装置	洩放裝置	relief device
泄料池	噴漿池	blow pit
泄料箱	洩料槽,排放槽	blowdown tank
泄漏试验	漏洩試驗	leakage test
泄压阀	排氣閥,釋氣閥	air relief valve

大　陆　名	台　湾　名	英　文　名
泻流	瀉流,逸散	effusion
卸料器	卸料器	dumper
卸载	卸载	unloading
辛烷值	辛烷值	octane number
辛烷值增进剂	辛烷值增進劑	octane enhancer
锌白	鋅白	zinc white
锌钡白	鋅鋇白	lithopone
锌铬黄	鋅鉻黄,鋅黄	zinc chrome, zinc yellow
锌海绵	鋅海綿	zinc sponge
锌合金	鋅合金	zinc alloy
锌华	鋅華	zinc flower
锌绿	鋅綠	zinc green
锌冕玻璃	鋅冕玻璃	zinc-crown glass
锌青铜	鋅青銅	zinc bronze
锌铁尖晶石	鋅鐵礦	franklinite
锌蓄电池	鋅蓄電池	zinc storage battery
新霉素	新黴素	neomycin
新生霉素	新生黴素	novobiocin
新闻纸	新聞紙	news-printing paper
新鲜蒸汽	活蒸汽	live steam
信号传递滞后	信號傳送落後	signal transfer lag
信号流程图	信號流程圖	signal flow diagram, signal flow graph
信号烟	信號煙	signal smoke
信息	資訊	information
信息板	資訊板	information board
信息储存	資訊儲存	information storage
信息处理	資訊處理	information processing
信息聚合物	資訊聚合物	informational polymer
信息流图	資訊流程圖	information flow diagram
星形聚合物	星形聚合物	star polymer
行星式搅拌器	行星式攪拌器	planet agitator
T 形件,三通	T 形件,三通	T-piece
S 形塔板	單向流塔板	uniflux tray
形状数	高寬比	aspect number
形状双折射	形狀雙折射	shape birefringence
形状稳定性	形狀穩定性	shape stability
形状系数	形狀因數	shape factor
形状阻力	形狀阻力	form drag

大　陆　名	台　湾　名	英　文　名
U 型管	U 形管	U-tube
U 型管换热器	U 形管熱交換器	U-tube heat exchanger, hairpin tube heat exchanger
T 型管相当长度	T 形管相當長度	tee equivalent pipe length
U 型结晶器	U 形結晶器	U-type crystallizer
型坯	型坯	parison
V 型缺口	V 形凹槽	V notch
杏仁油	杏仁油	almond oil, appricot kernel oil
性能	性能	performance
性能试验	性能試驗	performance test
性能试验报告	性能試驗報告	performance test report
性能系数	性能係數	coefficient of performance, COP
性能指标	性能指數	performance index
性质变化	性質變化	property change
性质关联式	性質關係[式]	property relationship
休止角	靜止角	angle of repose
修补	修補	patching
修匀常数	修匀常數	smoothing constant
修匀信号	修匀信號	smoothed signal
锈	鏽	rust
溴化	溴化[作用]	bromination
虚拟变量	虛擬變量	pseudo variable
虚拟变数	啞變數	dummy variable
虚拟反应	虛擬反應	pseudo reaction
虚拟活性	虛擬活性	dummy activity
虚拟机理	虛擬機構	pseudomechanism
虚拟系数	虛擬參數	pseudo-parameter
需氧量	需氧量	oxygen demand
序贯抽样	順序取樣	sequential sampling
序贯分析	順序分析	sequential analysis
序列	序列	sequence
序列长度	序列長度	sequence length
序列长度分布	序列長度分佈	sequence-length distribution
序列共聚物	序列共聚物	sequential copolymer
序列互穿网络	時序[形成]互穿網結構	sequential interpenetrating network
序列聚合	序列聚合	sequential polymerization
絮凝	絮凝[作用]	flocculation

大　陆　名	台　湾　名	英　文　名
絮凝槽	絮凝槽	flocculation tank
絮凝测试	絮凝試驗	floc test
絮凝沉降	絮凝沈降	flocculent settling
絮凝反应	絮凝反應	flocculation reaction
絮凝机理	絮凝機構	flocculating mechanism
絮凝剂	絮凝劑	flocculant
絮凝粒子	絮凝粒子	flocculent particle, floc particle
絮凝器	絮凝器	flocculator
絮凝物	絮凝物,絮凝體	flocculate, floc
絮凝物尺寸	絮凝體大小	floc size
絮凝物生长	絮凝體成長	floc growth
絮凝悬浮液	絮凝懸浮液	flocculent suspension
蓄电池	蓄電池	storage battery, storage cell
蓄热器	蓄熱器,復熱器	heat regenerator, heat accumulator, recuperator
蓄热式换热器	蓄熱式熱交換器	regenerative heat exchanger
蓄水池	貯槽	reservoir
悬滴试验	懸滴試驗	pendant drop test
悬浮	懸浮	suspension
悬浮催化剂	懸浮觸媒	suspended catalyst
悬浮固体	懸浮固體	suspended solid
悬浮胶体	懸[浮]膠體	suspension colloid
悬浮聚合	懸浮聚合	suspension polymerization
悬浮黏合剂	懸浮黏合劑	suspension adhesive
悬浮区法	浮動帶域[純化]法	floating zone method
悬浮生长	懸浮生長	suspended growth
悬浮生长系统	懸浮生長系統	suspended-growth system
悬浮体	懸[浮]膠體	suspensoid
悬浮稳定剂	懸浮安定劑	suspension stabilizer
悬浮物负载	懸浮物負載	suspension load
悬浮系统	懸浮系統	suspension system
悬浮线	懸線	suspension wire
悬筐蒸发器	籃式蒸發器	basket-type evaporator
旋光计	旋光計,偏光計	polarimeter
旋光率	比旋光[度]	specific rotation
旋光能力	旋光能力	rotatory power
旋光色散	旋光色散,旋光分散	optical rotatory dispersion
旋光异构体	光學異構物	optical isomer

大　陆　名	台　湾　名	英　文　名
旋筐反应器	旋籃反應器	rotating-basket reactor
旋塞	旋塞,[活]栓	cock
旋塞阀	塞閥	plug valve
旋涂	旋轉塗布	spin coating
[旋]涡	渦流	eddy
旋涡空穴	漩渦空洞	vortex cavity
旋液分离器	旋液分離器	hydrocyclone
旋转萃取器	旋轉萃取器	rotating extractor
旋转阀	旋轉閥	rotary valve
旋转分配器	旋轉分配器	rotary distributor
旋转过滤机	旋濾機	rotary filter
旋转焊接	摩擦焊	spin welding
旋转结晶器	旋轉結晶器	rotary crystallizer
旋转进料器	旋轉飼機	rotary feeder
旋转冷却器	旋轉冷卻器	rotary cooler
旋转连续过滤机	連續旋濾機	rotary continuous filter
旋转流	旋轉流動	rotational flow
旋转黏度计	旋轉黏度計	rotary viscometer, rotation viscometer
旋转排代泵	旋轉排量泵	rotary displacement pump
旋转筛	旋轉篩	rotary screen, rotary sieve
旋转筛摇床	旋轉搖篩器	rotary sieve shaker
旋转式喷头	旋轉噴灑器	rotating sprinkler
旋转洗涤机	旋轉洗滌機	rotary washer
旋转效率	旋轉效率	rotary efficiency
旋转压滤机	旋轉壓濾機	rotary filter press
旋转压碎机	旋轉壓碎機	rotary crusher
旋转延伸	紡伸製程	spin-drawing
旋转摇床	旋轉搖動器	rotary shaker
旋转真空泵	旋轉真空泵	rotary vacuum pump
选矿	選礦	ore dressing
选区衍射	選區繞射	selected-area diffraction
选择定则	選擇法則	selection rule
选择加氢	選擇[性]氫化	selective hydrogenation
选择精馏	選擇[性]精餾	selective rectification
选择聚合	選擇[性]聚合	selective polymerization
选择裂化	選擇[性]裂解	selective cracking
选择器	選擇器	selector
选择吸收	選擇[性]吸收	selective absorption

大　陆　名	台　湾　名	英　文　名
选择吸收剂	選擇[性]吸收劑	selective absorbent
选择性	選擇性	selectivity
选择性比	選擇性比	selectivity ratio
选择性萃取	選擇[性]萃取	selective extraction
选择性发酵	選擇[性]發酵	selective fermentation
选择性分散曲线	選擇性分散曲線	selectivity dispersion curve
选择性浮选	選擇[性]浮選	selective flotation
选择性浸取	選擇[性]瀝取	selective leaching
选择性控制	選擇[性]控制	selective control
选择性控制器	選擇[性]控制器	selective controller
选择性溶剂	選擇性溶劑	selective solvent
选择性系数	選擇性係數	selectivity coefficient
选择性中毒	選擇[性]中毒	selective poisoning
旋风除尘器	旋風除塵器	dust cyclone
旋风分离	旋風分離	cyclone separation
旋风分离器	旋風分離器	cyclone separator, separation cyclone
旋风排气机	旋風排氣機	cyclone exhauster
旋风器	旋風器,旋流器	cyclone
旋风器效率	旋風器效率	cyclone efficiency
旋风洗涤器	旋風洗滌器	cyclone scrubber
旋风蒸发器	旋風蒸發器	cyclone evaporator
学习控制系统	學習控制系統	learning control system
血浆	血漿	plasma
血清	血清	serum
血清白蛋白黏合剂	血清白蛋白黏合劑	serum albumin adhesive
血清球蛋白	血清球蛋白	serum globulin
熏蒸剂	燻蒸劑	fumigant
熏制	燻製	smoking
驯化	馴化	acclimation
驯化污泥	馴化污泥	acclimation sludge
循环	循環	cycling, recycle, recycling
循环泵混合器	循環泵混合器	circulating pump mixer
循环比	循環比	recirculating ratio, recirculation ratio, recycle ratio
循环操作	循環操作,週期操作	cyclic operation
循环层	循環層	circulation layer
循环滴	循環[流動]滴	circulating drop
循环法	循環法	circulation method

大　陆　名	台　湾　名	英　文　名
循环反应器	循環反應器	recirculation reactor
循环管	循環管	circulation pipe, circulation tube
循环过程	循環程序	cyclic process, recycle process
循环合成气	循環合成氣	recycle synthesis gas
循环回流	循環回流	circulating reflux
循环料	回收[進]料	recycle feedstock
循环流动	循環流動	circulation flow, recirculation flow
循环流化床	循環流[體]化床	circulating fluidized bed
循环模式	循環模式	recirculation pattern
循环判据	循環準則	circulating criterion
循环式混合器	循環式混合器	circulation mixer
循环物流	循環流	recycle stream
循环系统	循環系統	recycle system
循环因子	循環因數	recirculation factor
循环油	循環油	recycle oil
蕈状蒸馏塔	蕈狀蒸餾塔	mushroom distilling column

Y

大　陆　名	台　湾　名	英　文　名
压差	微差壓,壓[力]差	differential pressure, pressure difference
压电晶体	壓電晶體	piezoelectric crystal
压焓图	壓力-焓圖	pressure-enthalpy diagram
压降	壓[力]降	pressure drop
压紧	壓緊	compaction
压块	壓塊	briquetting
压力变送器	壓力傳送器	pressure transmitter
压力补偿	壓力補償	pressure compensation
压力测量	壓力量測	pressure measurement
压力传感器	壓力傳感器	pressure transducer
压力范围	壓力範圍	pressure range
压力分布	壓力分佈	pressure distribution
压力计[表]	壓力計,[流體]壓力計	pressure gage, pressure meter, manometer
压力计压头	壓力計高差	manometric head
压力记录器	壓力記錄器	pressure recorder
压力校正	壓力校正,壓力修正	pressure calibration, pressure correction
压力控制	壓力控制	pressure control
压力控制器	壓力控制器	pressure controller

大　陆　名	台　湾　名	英　文　名
压力扩散	壓力擴散	pressure diffusion
压力能	壓[力]能	pressure energy
压力平衡阀	壓力均衡閥	pressure balanced valve
压力平衡器	均壓器	pressure equalizer
压力-容积功	壓力-體積功	pressure-volume work
压力容器	壓力容器,壓力槽	pressure vessel
压力式温度计	壓力式溫度計	pressure type thermometer
压力损失	壓力損失	pressure loss
压力梯度	壓力梯度	pressure gradient
压力-温度图	壓力-溫度圖	pressure-temperature diagram
压力系数	壓力係數	pressure coefficient
压力效率	壓力效率	pressure efficiency
压力效应	壓力效應	pressure effect
压力信号	壓力信號	pressure signal
压力-真空阀	壓力真空閥	pressure-vacuum valve
压力指示计	壓力指示器	pressure indicator
压力-组成图	壓力-組成圖	pressure-composition diagram
压滤饼	壓濾餅	filter press cake
压滤布	壓濾布	filter press cloth
压滤机	壓濾機	pressure filter
压敏胶	壓敏膠,感壓膠	pressure sensitive adhesive, PSA
压模	壓模	press mold
压模机	模壓機	molding press
压片	壓片,乾壓製錠	tabletting, slugging tablet making, shee-ting
压容图	壓力-體積圖	pressure-volume diagram
压实度	壓密度	degree of consolidation
压缩	壓縮	compression
压缩比	壓縮比	compression ratio
压缩波	壓縮波	compression wave
压缩功	壓縮功	compression work
压缩过程	壓縮程序	compression process
压缩机	壓縮機	compressor
压缩机规格	壓縮機規格	compressor specification
压缩空气	壓縮空氣	compressed air
压缩空气喷雾器	壓縮空氣噴霧器	compressed air sprayer
压缩力	壓縮力	compressive force
压缩率判据	壓縮度準則	compressibility criterion

大　陆　名	台　湾　名	英　文　名
压缩强度	抗壓強度	compressive strength
压缩区	壓縮區	compression zone
压缩热	壓縮熱	heat of compression
压缩渗透测量法	滲透率壓縮量測	compression-permeability technique
压缩系数	壓縮係數	compressibility coefficient
压缩压力	壓縮壓力	compressive pressure
压缩因子	壓縮因子	compressibility factor
压头	壓力高差	pressure head
压纹机(=压花机)		
压延机	壓延機,硏光機,軋光機	calender
压延孔板波纹填料	壓延孔板波紋填料	protruded corrugated sheet packing
压延效应	壓延效應	calender effect
压榨	壓榨	expression
压榨机	擠壓機	squeezer
压铸	壓鑄,模鑄	die casting
压铸成型	壓鑄成型	die casting molding
芽殖	出芽	budding
芽殖细胞	出芽細胞	budding cell
芽殖真菌	出芽眞菌	budding fungi
雅可比矩阵	賈可比矩陣	Jacobian matrix
亚单元	次單元	subunit
亚单元结构	次單元結構	subunit structure
亚基	次單元	subunit
亚基结构	次單元結構	subunit structure
亚甲蓝	亞甲藍	methylene blue
亚晶胞	亞晶胞	subcell
亚硫酸氢盐	亞硫酸氫鹽	hydrosulfite
亚硫酸氢盐漂白	亞硫酸氫鹽漂白	hydrosulfite bleaching
亚硫酸盐法	亞硫酸鹽法	sulfite process
亚硫酸盐浆	亞硫酸[鹽]紙漿	sulfite pulp
亚麻	亞麻	flax
亚麻布	亞麻布	linen
亚麻纤维	亞麻纖維	flax fiber
亚麻子油	亞麻仁油	flax-seed oil, linseed oil
亚声速喷嘴	次音速噴嘴	subsonic nozzle
亚微观的	亞微觀的	submicroscopic
亚微观断裂	次微米斷裂	submicrofracture
亚微观胶束	亞微觀微胞	submicroscopic micelle

大 陆 名	台 湾 名	英 文 名
亚微观结构	亞微觀結構	submicroscopic structure
亚微裂纹	次微米裂紋	submicrocrack
亚稳单相极限线	離相曲線	spinodal
亚稳定性	介穩定性	metastability
亚稳界限	介穩定界限	metastable limit
亚稳区	介穩定區[域]	metastable region
亚稳态	介穩定狀態	metastable state
亚硝胺红	亞硝胺紅	nitrosamine red
亚硝基染料	亞硝基染料	nitroso dye
压花	壓花,壓紋	embossing
压花机,压纹机	壓花機,壓紋機	embossing calender
烟	煙	fume
烟道	煙道	flue
烟道气	煙道氣	stack gas, flue gas
烟点	發煙點	smoke point
烟煤	煙煤	soft coal, bituminite
烟幕弹	煙幕彈	smoke bomb
烟片[橡胶]	煙片[橡膠]	smoked sheet
烟曲霉素	煙黴素	fumagillin
烟鞣制	煙鞣	smoke tanning
烟雾	煙霧	smog
烟雾发生器	發煙器	smoke generator
烟雾剂	發煙劑	smoke agent
烟雾烛[缸]	發煙罐	smoke candle
腌制	鹽漬	salting
延长线	延長線	extension wire
延迟	延遲,遲延	delay
延迟时间	延遲時間,遲延時間	delay time
延迟作用	延遲作用,遲延作用	delay action
延伸颈	延伸頸	extension neck
延伸流动	延伸流動	extensional flow
延伸率	延伸性	extensibility
延时曝气	延時曝氣	extended aeration
延展性材料	延展性材料	ductile materials
严格法	嚴格法	rigorous method
岩浆	岩漿	magma
研究	研究	research
研磨	研磨	grinding, grind

大　陆　名	台　湾　名	英　文　名
研磨辅料	研磨添加劑	grinding additive
研磨辊	碾磨輥	mill roller
研磨介质	研磨介質,磨媒	grinding medium
研磨能力	碾磨能力,研磨容量	milling capacity, mill capacity
研磨室	碾磨室	mill room
研磨效率	研磨效率	mill efficiency
盐	鹽	salt
盐捕集器	鹽接受器	salt catcher
盐皮	鹽[醃]皮	salted hide
盐桥	鹽橋	salt bridge
盐溶	鹽溶	salting in
盐水	鹽水,滷水	brine
盐酸	鹽酸	muriatic acid
盐析	鹽析	salting out
盐析剂	鹽析劑	salting out agent
盐洗	鹽洗	brine wash
盐效应	鹽效應	salt effect
盐釉	鹽釉	salt glaze
颜料	顏料	pigment
衍生物	衍生物	derivative
掩蔽能力	遮蓋[能]力	masking power
厌氧处理	厭氧處理	anaerobic treatment
厌氧发酵	厭氧發酵[法]	anaerobic fermentation
厌氧分解	厭氣分解	anaerobic decomposition, anaerobic break-down
厌氧腐蚀	厭氧腐蝕	anaerobic corrosion
厌氧菌	厭氧菌	anaerobic bacteria, anaerobe
厌氧培养	厭氧培養	anaerobic culture
厌氧污泥	厭氧污泥	anaerobic sludge
厌氧污泥消化	厭氧污泥消化	anaerobic sludge digestion
厌氧消化	厭氧消化	anaerobic digestion
厌氧消化槽	厭氧消化槽	anaerobic digester
厌氧氧化	厭氣氧化[反應]	anaerobic oxidation
验收抽样	驗收取樣	acceptance sampling
验收抽样计划	驗收取樣計畫	acceptance sampling plan
验收试验	驗收試驗	acceptance test
验收质量水准	合格品質水準	acceptable quality level
堰	堰	weir

大　陆　名	台　湾　名	英　文　名
堰高	堰高	weir height
扬程	揚程,高差,上升距離	head, lift
扬析常数	淘析常數	elutriation constant
羊毛蜡	羊毛蠟	wool wax
羊毛油	羊毛油	wool oil
羊毛脂	羊毛脂,[粗]羊毛脂	wool fat, wool grease, lanolin
羊毛脂醇	羊毛醇	wool alcohol
羊皮纸	羊皮紙	parchment paper
阳极	陽極	anode
阳极电位	陽極電位	anode potential
阳极反应	陽極反應	anodic reaction
阳极腐蚀	陽極腐蝕	anodic corrosion
阳极氧化	陽極氧化[反應]	anode oxidation
阳离子	陽離子,正離子	cation
阳离子淀粉	陽離子澱粉	cationic starch
阳离子交换	陽離子交換	cation exchange
阳离子交换剂	陽離子交換劑	cation exchanger
阳离子交换器	陽離子交換器	cation exchanger
阳离子交换树脂	陽離子交換樹脂	cation exchange resin
阳离子聚合	陽離子聚合[作用]	cationic polymerization
阳离子型表面活性剂	陽離子界面活性劑	cationic surfactant, cationic surface active agent
洋菜(＝琼脂)		
氧传递	氧傳送	oxygen transfer
氧分析仪	氧分析儀	oxygen analyzer
氧含量	含氧量	oxygen content
氧合作用	加氧[反應],充氧	oxygenation
氧化	氧化[作用]	oxidation
氧化苯乙烯聚合物	氧化苯乙烯聚合物	styrene oxide polymer
氧化铂催化剂	氧化鉑觸媒	platinum oxide catalyst
氧化催化剂	氧化觸媒	oxidation catalyst
氧化电位	氧化電位	oxidation potential
氧化淀粉	氧化澱粉	oxidized starch
氧化发光	氧化發光	oxyluminescence
氧化反应	氧化反應	oxidation reaction
氧化锆玻璃	鋯玻璃	zirconia glass
氧化锆耐火材料	氧化鋯耐火物	zirconium oxide refractory
氧化沟	氧化渠	oxidation ditch

大　陆　名	台　湾　名	英　文　名
氧化过程	氧化程序	oxidation process
氧化还原滴定	氧化還原滴定[法]	oxidation reduction titration
氧化还原电池	氧化還原電池	oxidation reduction cell
氧化还原电位	氧化還原電位	redox potential, oxidation reduction potential
氧化还原反应	氧化還原反應	redox reaction, oxidation reduction reaction
氧化还原酶	氧化還原酶	oxydo-reductase
氧化剂	氧化劑	oxidant
氧化降解	氧化降解	oxidative degradation
氧化酶	氧化酶	oxidase, oxydase
氧化期	氧化期	oxidation period
氧化热裂化	氧化熱裂解	oxidative pyrolysis
氧化塘	氧化池	oxidation pond
氧化油	氧化油	oxidized oil
氧量耗尽	氧耗盡	oxygen depletion
氧收率系数	氧產率係數	oxygen yield coefficient
样品系统	樣品系統	sample system
窑	窯	kiln
[窑]炉	爐	furnace
摇动筛	搖動篩	shaking screen
摇筛器	搖篩器	sieve shaker
摇饲机	搖飼機	cradle feeder
遥测	遙測[術]	telemetering
遥控	遙控	remote control
药典	藥典	pharmacopoeia
药动学	藥物動力學	pharmacokinetics
药剂学	藥劑學	pharmaceutics
药理学	藥理學,藥物學	pharmacology
药物化学	藥物化學	pharmaceutical chemistry
药物制剂	藥物製劑	pharmaceutical preparation
要素	要素	element
椰子油	椰子油	coconut oil
噎塞	噎塞	choke
冶金过程	冶金程序	metallurgical process
冶金焦	冶金焦	metallurgical coke
冶金学	冶金學	metallurgy
冶炼厂	冶煉廠,熔爐,熔煉廠	smelter, smeltery

大　陆　名	台　湾　名	英　文　名
冶炼过程	熔煉法	smelting process
冶炼炉	熔煉爐	smelter hearth
冶炼炉气	熔煉爐氣	smelter gas
叶红素	葉紅素	erythrophyll
叶蜡石	葉蠟石	pyrophyllite
叶滤机	葉濾器	leaf filter
叶绿素	葉綠素	chlorophyll
叶轮	葉輪,槳	impeller, paddle wheel
叶轮混合器	葉輪混合器	impeller mixer
叶轮搅拌器	葉輪攪拌器	impeller agitator
叶轮式流量计	葉輪流量計	impeller type flowmeter
叶轮直径	葉輪直徑	impeller diameter
叶轮轴	葉輪軸	impeller shaft
叶片	葉片	blade
叶片式鼓风机	葉輪式鼓風機	vane type blower
叶片式流动	葉輪流動	vane flow
叶片转子	葉片轉子	bladed rotor
曳力	［拖］曳力	drag force
曳力降低	［拖］曳力降減	drag reduction
页岩	頁岩	shale
页岩气	頁岩氣	shale gas
页岩油	頁岩油	shale oil
液氨	液氨	liquid ammonia
液氮	液［態］氮	liquid nitrogen
液滴瓦解	液滴瓦解	drop breakup
液泛	溢流	flooding
液封	液封	liquid seal
液固反应	液固反應	liquid-solid reaction
液固旋风器	液固旋風器	liquid-solid cyclone
液化	液化	liquefaction
液化点	液化點	liquefaction point
液化反应器	液化反應器	liquefaction reactor
液化过程	液化程序	liquefaction process
液化剂	液化器	liquefier
液化气体	液化氣［體］	liquefied gas
液化器	液化器	liquefier
液化燃料	液化燃料	liquefied fuel
液化热	液化熱	heat of liquefaction

大　陆　名	台　湾　名	英　文　名
液化石油气	液化石油氣	liquefied petroleum gas
液化天然气	液化天然氣	liquefied natural gas
液环压缩机	液環壓縮機	liquid ring compressor
液解	溶劑解	lyolysis
液晶	液晶	liquid crystal
液晶高分子	液晶聚合物	liquid crystal polymer
液晶显示	液晶顯示器	liquid crystal display
液晶转变	液晶轉變	liquid crystal transition
液面计	液位計	liquid level gage
液面控制	液位控制	liquid level control
液膜	液[態薄]膜	liquid membrane, liquid film
液膜接触器	液膜接觸器	liquid film contactor
液膜系数	液膜係數	liquid film coefficient
液膜阻力	液膜阻力	liquid film resistance
液气反应	液氣反應	liquid-gas reaction
液汽平衡	液汽平衡	liquid-vapor equilibrium
液时空速	液體空間時速	liquid hourly space velocity
液态反应	液態反應	liquid reaction
液态空气	液態空氣	liquid air
液体	液體	liquid
[液体]比重计	液體比重計	araeometer
液体萃取物	液體萃取物	liquid extract
液体浮选	液體浮選[法]	liquid flotation
液体界面控制	液體界面控制	liquid interface control
液体静力学	液體靜力學	hydrostatics
液体静压[强]	液體靜壓[力]	hydrostatic pressure
液体流化床	液體流[體]化床	liquid fluidized bed
液体密度计	比重計	hydrometer
液体平衡	液液平衡	liquid-liquid equilibrium
液体燃料	液體燃料	liquid fuel
液体溶液	液態溶液	liquid solution
液体石蜡	液態石蠟	liquid paraffin
液体推进剂	液體推進劑	liquid propellant
液体温度计	液體溫度計	liquid thermometer
液体压头	液體高差	liquid head
液位	液位	liquid level
液位计	液位計	level gauge
液位控制	液位控制	level control

大　陆　名	台　湾　名	英　文　名
液位指示器	液位指示器	liquid level indicator
液下泵	沈式泵	submerged pump
液相	液相	liquid phase
液相反应	液相反應	liquid phase reaction
液相反应器	液相反應器	liquid phase reactor
液相裂解	液相裂解	liquid phase cracking
液相色谱法	液相層析[法]	liquid chromatography
液相色谱图	液相層析圖	liquid chromatogram
液相色谱仪	液相層析儀	liquid chromatograph
液相线	液相線	liquidus
液压	液壓	hydraulic pressure
液压泵	水力泵	hydraulic pump
液压控制	液壓控制	hydraulic control
液压控制器	液壓控制器	hydraulic controller
液压离心机	液壓離心機	hydraulic centrifuge
液压流体	液壓流體	hydraulic fluid
液压筒	液壓筒	hydraulic cylinder
液压运送	液壓運送	hydraulic conveying
液压制动器	液壓致動器	hydraulic brake
液氧	液[態]氧	liquid oxygen
液液萃取	液液萃取	liquid-liquid extraction
液液反应	液液反應	liquid-liquid reaction
液液分离器	液液分離器	liquid-liquid separator
液柱静压头	流體靜力高差	hydrostatic head
液柱压力计	液柱壓力計	liquid column gage
一步法[酚醛]树脂	一步法[酚醛]樹脂	single-stage resin
一级标准	一級標準,原級標準	primary standard
一级处理	初級處理	primary treatment
一级动力学	一級動力學,一階動力學	first order kinetics
一级反应	一級反應,一階反應	first order reaction
一级过程	一階程序	first-order process
一级控制系统	一階控制系統	first-order control system
一级落后	一階落後	first-order lag
一级相变	一級相變,一階相變	first order phase transition
一级应答	一階應答	first-order response
一级正应力系数	一次法向應力係數	primary normal stress coefficient
一级转变温度	一階轉變溫度,一階轉	first-order transition temperature

大　陆　名	台　湾　名	英　文　名
	移溫度	
一阶系统	一階系統	first-order system
一维模型	一維模型	one-dimensional model
一致性	一致性	consistency
一致性试验	一致性試驗法	consistency test
伊利石	伊來石,白雲母石	illite
衣物洗涤剂	洗衣清潔劑	laundry detergent
仪表	儀表	instrumentation
仪表板	儀表板	panel board
仪表流程图	儀器流程圖	instrument flowsheet
仪表配置图	儀表配置圖	instrumentation diagram
仪器	儀器	instrument
仪器分析	儀器分析	instrumental analysis
仪器伺服机构	儀器伺服機制	instrument servomechanism
胰岛素	胰島素	insulin
移标式液位计	移標式液位計	displacement float level gage
移动床	移動床	moving bed
移动床法	移動床法	moving bed process
移动床反应器	移動床反應器	moving bed reactor
移动床气化器	移動床氣化器	moving bed gasifier
移动床吸附器	移動床吸附器	moving bed adsorber
移动催化剂床	移動觸媒床	moving catalyst bed
移动界面	移動界面,移動邊界	moving boundary
移动筛	移動篩	traveling screen
移动相	移動相	moving phase
乙丙橡胶	乙烯-丙烯橡膠	ethylene-propylene rubber
乙二醇	乙二醇	glycol
乙二磺酸钠聚合物	乙烯基磺酸鈉聚合物	sodium ethylene sulfonate polymer
乙基纤维素	乙基纖維素	ethyl cellulose
乙腈	乙腈	acetonitrile
乙炔黑	乙炔黑	acetylene black
乙酸丁酸纤维素	醋酸丁酸纖維素	cellulose acetate butyrate
乙酸人造丝,醋酸嫘縈	乙酸嫘縈,醋酸嫘縈	acetate rayon
乙酸纤维素,醋酸纤维素	乙酸纖維素,醋酸纖維素	acetyl cellulose
乙酸酯膜,醋酸酯膜	乙酸酯膜,醋酸酯膜	acetate film
乙烯基化合物	乙烯基化合物	vinyl compound
乙烯系树脂	乙烯系樹脂	vinyl resin

大　陆　名	台　湾　名	英　文　名
乙酰化	乙醯化	acetylation
乙酰值	乙醯值	acetyl value
艺术印刷纸	美術印刷紙,銅版紙	art-printing paper
异丙苯	異丙苯	cumene, isopropylbenzene
异常反应	異常反應	abnormal reaction
异构重整	異構重組	isoforming
异构化	異構化,異構作用	isomerization
异构体	異構物	isomer
异化[作用]	異化[作用]	dissimilation
异径管	漸縮管	reducer, pipe reducer
异径管件	漸縮管件	reducing fitting
异径三通管	漸縮三通管,漸縮 T 型管	reducing tee
异径弯头	漸縮彎頭	reducing elbow
异氰酸酯聚合物	異氰酸酯聚合物	isocyanate polymer
异氰酸酯黏合剂	異氰酸酯黏合劑	isocyanate adhesive
异戊[二烯]橡胶	異戊二烯橡膠,異平橡膠	isoprene rubber
异相成核	異相成核	heterogeneous nucleation
异型发酵菌	異型發酵菌	heterofermentative bacteria
异质群体	雜菌群,雜族群	heterogeneous population
抑制	抑制[作用]	inhibition
抑制剂	抑制劑,制止器	restrainer, inhibitor
易碎性	易碎性	fragility
疫苗	疫苗	vaccine
逸出气体分析法	逸出氣分析[法]	evolved gas analysis
逸出气体检测	逸出氣檢測[法]	evolved gas detection
逸度	逸壓	fugacity
逸度系数	逸壓係數	fugacity coefficient
溢料槽	溢料槽	spew groove
溢料缝	溢料	spew
溢料面	溢料面	spew area
溢料线	溢料線	spew line
溢流	溢流	overflow
溢流槽	接收槽	run-down tank
溢流管	溢流管,回流管	overflow pipe, return tube
溢流警报	溢流警報	overflow alarm
溢流速度	溢流速度	overflow velocity, overflow rate

大　陆　名	台　湾　名	英　文　名
溢流吸入管线	溢流吸入管線	flooded suction line
溢流堰	溢流堰	overflow weir
因变量	因變數	dependent variable
因次	因次	dimension
因次分析(=量纲分析)		
因次均一性原理	因次均一性原理	principle of dimensional homogeneity
因果性	因果性	causality
因子	因子	factor
F因子	F因子	F factor
j因子	j因數	j-factor
阴丹士林蓝	陰丹士林藍	indanthrene blue
阴丹士林染料	陰丹士林染料	indanthrene dye
阴极	陰極	cathode
阴极反应	陰極反應	cathodic reaction
阴离子	陰離子	anion
阴离子[催化]聚合	陰離子聚合[作用]	anionic polymerization
阴离子淀粉	陰離子澱粉	anionic starch
阴离子交换剂	陰離子交換劑	anion exchanger
阴离子交换树脂	陰離子交換樹脂	anion exchange resin
阴离子型表面活性剂	陰離子界面活性劑	anionic surface-active agent，anionic surfactant
音速	音速	acoustic velocity
银颈	銀頸	silver necking
银值	銀值	silver number
引发剂	引發劑,起始劑	initiator
引风	抽氣通風	induced draft
引风式凉水塔	抽氣通風冷卻塔	induced draft cooling tower
引力场	重力場	gravitational field
引燃剂	引燃劑	combustion initiator
引诱剂	誘[引]劑	attractant
饮用水	飲用水	potable water
隐蔽层	隱藏層	hidden layer
隐函数	隱函數	implicit function
隐枚举法	隱枚舉法	implicit enumeration method
隐色基	隱色鹼	leuco base
隐色体	隱色化合物	leuco compound
隐式法	隱式法	implicit method
荧光	螢光	fluorescence

大　陆　名	台　湾　名	英　文　名
荧光分光镜	螢光譜儀	fluorescence spectroscope
荧光光谱	螢光光譜	fluorescent spectrum
荧光黄	螢光黃	fluorescein
荧光晶体	螢光晶體	fluorescent crystal
荧光染料	螢光染料	fluorescent dye
荧光增白剂	螢光增白劑	fluorescent whitening agent
应变光学系数	應變光學係數	strain optical coefficient
应变能	應變能	strain energy
应变能函数	應變能函數	strain energy function
应变黏合剂	應變黏合劑	strain adhesive
应变软化	應變軟化	strain softening
应变双折射	應變雙折射	strain birefringence
应变椭球	應變橢圓體	strain ellipsoid
应变仪	應變[測量]計	strain gauge
应变硬化	應變硬化	strain hardening
应变张量	應變張量	strain tensor
应力	應力	stress
应力传递机理	應力轉移機制	stress-transfer mechanism
应力分布	應力分佈	stress distribution
应力分析	應力分析	stress analysis
应力腐蚀	應力腐蝕	stress corrosion
应力功率	應力功率	stress power
应力光学系数	應力光學係數	stress optical coefficient, SOC
应力集中	應力集中	stress concentration
应力集中系数	應力集中因數	stress concentration factor
应力结晶性	應力結晶性	stress crystallinity
应力开裂	應力開裂	stress cracking
应力强度因子	應力強度因子	stress intensity factor
应力软化	應力軟化	stress softening
应力石墨化	應力石墨化	stress graphitization
应力史	應力歷程	stress history
应力双折射	應力雙折射	stress birefringence
应力松弛	應力弛豫	stress relaxation
应力松弛计	應力弛豫計	stress relaxometer
应力松弛模量	應力弛豫模數	stress relaxation modulus
应力松弛曲线	應力弛豫曲線	stress relaxation curve
应力椭球体	應力橢球	stress ellipsoid
应力消除试验	應力釋放試驗	stress relief test

大　陆　名	台　湾　名	英　文　名
应力-形变曲线	應力-形變曲線	stress-deformation curve
应力银纹	應力細紋	stress crazing
应力-应变曲线	應力-應變曲線	stress-strain curve
应力-应变曲线关系式	應力應變曲線關係	stress-strain curve relation
应力-应变响应	應力-應變響應	stress-strain response
应力-应变行为	應力-應變行為	stress-strain behavior
应力诱导极化	應力誘發極化	stress-induced polarization
应力诱导结晶	應力誘發結晶	stress-induced crystallization
应力诱导取向	應力誘發定向	stress-induced orientation
应力诱导生长	應力誘發生長	stress-induced growth
应力张量	應力張量	stress tensor
应力致白	應力致白	stress whitening
应用	應用	application
应用化学	應用化學	applied chemistry
应用热力学	應用熱力學	applied thermodynamics
应用微生物学	應用微生物學	applied microbiology
萤石	螢石,氟石	fluorspar
映射	映射	mapping
硬度	硬度	degree of hardness, hardness
硬度标度	硬度標度	hardness scale
硬度计	硬度計	durometer, sclerometer
硬度试验	硬度試驗	hardness test
硬化	膠接[作用]	cementation
硬化	硬化	hardening
硬化剂	硬化劑,加強材,防撓 材,硬挺劑	hardener, stiffener, stiffening agent
硬化油	硬化油	hardened oil
硬件	硬體	hardware
硬蜡	硬蠟	hard wax
硬沥青	瀝青	pitch
硬泥法	硬泥法	stiff-mud process
硬球	硬球	hard sphere
硬球理论	硬球理論	hard sphere theory
硬石膏	熟石膏,硬膏劑	plaster
硬水	硬水	hard water
硬水铝石	水鋁石	diaspore
硬橡胶	硬橡膠	hard rubber, vulcanite
硬性链	剛性鏈	stiff chain

大 陆 名	台 湾 名	英 文 名
硬性洗涤剂	硬性清潔劑	hard detergent
硬质胶	硬橡膠	ebonite
硬质耐火黏土	燧石耐火黏土	flint fireclay
硬质黏土	燧石黏土	flint clay
㶲分析	有效能分析	exergy analysis
㶲衡算	有效能均衡	exergy balance
㶲损失	有效能損失	exergy loss
永久变形	永久變形	permanent deformation
永久扰动	永久擾動	permanent disturbance
永久损失	永久損失	permanent loss
永久硬度	永久硬度	permanent hardness
涌波	湧波	surge wave
涌料	沖洗	flushing
用剂	用劑	dosing
油	油	oil
油包水乳状液	油包水乳液	water in oil emulsion
油饼	油餅	oil cake
油浮选	油浮選[法]	oil flotation
油脚	油腳, 油渣	oil foot
油轮	油輪	oil tanker, tanker
油媒染剂	油媒染劑	oil mordant
油喷雾器	油霧化器	oil atomizer
油漆	油漆	paint
油气	油氣	oil gas
油鞣	油鞣	chamber leather
油润滑脂	棕櫚油潤滑脂	palm oil grease
油砂	油砂	oil sand
油水分离器	油水分離器	oil-water separator
油田	油田	oil field
油雾分离器	分霧器	mist separator
油性树脂	油性樹脂	oleo-resin
油页岩	油頁岩	oil shale
油浴	油浴	oil bath
油毡	油氈, 油布	linoleum
油脂	油脂	fats and oils
油脂涂料	油性漆	oil paint
油纸	油紙	oiled paper
游标尺	游標尺	Vernier scale

大 陆 名	台 湾 名	英 文 名
游标卡尺	游標卡尺	Vernier caliper
游离碱	自由鹼	free alkali
游离酶	游離酶	free enzyme
游离水分	自由水分	free moisture
游离水含量	自由水含量	free moisture content
游离酸	自由酸	free acid
有毒材料	有毒材料,毒物	toxic materials
有规结构流体	結構化流體	structured fluid
有规立构共聚	立構特定共聚	stereospecific copolymerization
有规立构聚合物	立體規則性聚合物	stereoregular polymer
有规立构橡胶	立構特定橡膠	stereospecific rubber
有害气体	有害氣體	noxious gas
有害性废弃物	有害性廢棄物	hazardous waste
有核结晶	成核結晶	nucleate crystallization
有机补助剂	有機補助劑	organic builder
有机超负载	有機物超負載	organic overload
有机肥料	有機肥料	organic fertilizer
有机负荷量	有機物負載	organic load
有机汞	有機汞	organomercury
有机固体	有機固體	organic solid
有机硅瓷漆	聚矽氧瓷漆	silicone enamel
有机硅聚合物	矽氧聚合物	silicone polymer
有机硅黏合剂	聚矽氧黏合劑	silicone adhesive
有机硅树脂	矽氧樹脂	silicone resin
有机硅塑料	矽氧塑膠	silicone plastic
有机硅脱模剂	矽氧脫膜劑	silicone release
有机硅弹性体	聚矽氧彈性體	silicone elastomer
有机金属化合物	有機金屬化合物	organometallic compound
有机黏合剂	有機黏結劑	organic binder
有机黏土	有機黏土	organoclay
有机溶胶	有機溶膠	organosol
有机酸	有機酸	organic acid
有机碳	有機碳	organic carbon
有机污泥	有機污泥	organic sludge
有理函数	有理函數	rational function
有理数	有理數	rational number
有鳞[片]纤维[黏胶]	有鱗[片]纖維[黏膠]	scale fiber
有限边界元素	有限邊界元素	finite boundary element

大　陆　名	台　湾　名	英　文　名
有限差分	有限差分	finite difference
有限差分法	有限差分法	method of finite difference
有限扰动	有限擾動	finite disturbance
有限时间稳定性	有限時間穩定性	finite-time stability
有限元法	有限元[素]法	finite element method
有向图	有向圖	digraph
有效表面	有效表面	effective surface
有效导热系数	有效導熱係數	effective thermal conductivity
有效功率	有效功率	effective power
有效挥发度	有效揮發度	effective volatility
有效颗粒直径	有效粒徑	effective particle diameter
有效孔半径	有效孔徑	effective pore radius
有效扩散系数	有效擴散係數	effective diffusivity
有效利息	有效利息	effective interest
有效粒度	有效晶粒大小	effective grain size
有效氯	有效氯[量]	available chlorine
有效密度	有效密度	effective density
有效能	有效能	availability, exergy
有效能分析	有效能分析	availability analysis
有效浓度	有效濃度	effective concentration
有效曝气	有效曝氣	effective aeration
有效系数	有效度因數	effectiveness factor
有效性	有效性,有效度	effectiveness
有效值	有效值	effective value
有载体催化剂	負載型觸媒	supported catalyst
有载体金属催化剂	負載型金屬觸媒	supported metal catalyst
有载体双金属催化剂	負載型雙金屬觸媒	supported bimetallic catalyst
有择溶解度	優先溶解度	preferential solubility
诱变剂	誘變劑	mutagen
诱导发力	誘發阻力	induced drag
诱导反应	誘發反應	induced reaction
诱导偶极	誘發偶極	induced dipole
诱导期	誘導期	induction period
釉	釉	glaze
釉烧窑	釉窯	glost kiln
淤浆法	漿料法	slurry process
淤浆聚合	漿料聚合[反應]	slurry polymerization
淤泥法	污泥法	sludge process

大　陆　名	台　湾　名	英　文　名
淤渣生成促进剂	污泥促進劑	sludge promoter
淤渣再循环	污泥循環	sludge recycle
鱼肝油	鱈魚肝油	cod-liver oil
鱼藤酮	魚藤酮	rotenone
玉米淀粉	玉米澱粉	corn starch
玉米糖	玉米糖	corn sugar
玉米糖浆	玉米糖漿	corn syrup
玉米油	玉米油	corn oil
浴	浴	bath
预焙	預焙	prebaking
预测	預測	prediction
预测控制	預測控制	predictive control
预处理	預處理,初步處理	pretreatment, preliminary treatment
预防	預防	prevention
预防处理	預防處理	preventative treatment
预分馏	預分餾	prefractionation
预混模制物	預混成型	premix molding
预冷器	預冷器	precooler
预滤	預濾	prefiltration
预热	預熱	preheating
预热炉	預熱爐	preheat furnace
预热期	預熱期	preheating period
预热器	預熱器	preheater
预热区	預熱區	preheating zone
预热蒸发器	預熱蒸發器	preheating evaporator
预闪蒸塔	預閃蒸塔	preflash tower
预设计成本估算	預設計成本估算	predesign cost estimate
预先控制	預先控制	anticipatory control
预应力外壳	預應力外殼	stressed shell
预载荷	預負載	preload
预蒸发器	預蒸發器	preevaporator
阈能	閾能,低限能	threshold energy
阈频[率]	閾頻,低限頻率	threshold frequency
阈限值	閾值,低限值	threshold limit value
阈[值]	閾值,低限	threshold
裕量	容許量	allowance
元件	元件	element
元素	元素	element

大　陆　名	台　湾　名	英　文　名
元素周期表	週期表	periodic table of elements
原电池	電池	galvanic cell
原废水	原廢水	raw wastewater
原理	原理	principle
原料	原料,[儲]存料	raw material, stock, feedstock
原料成本	原料成本	raw material cost
原料储罐	原料儲存	raw material storage
原料检查报表	原料檢查報表	raw material check list
原料气	原料氣	raw gas
原生动物	原生動物	protozoa
原生质	原生質	protoplasm
原生质体	原生質體	protoplast
原水	原水	raw water
原污泥	原污泥	raw sludge
原污水	原污水	raw sewage
原型	原型,雛型	prototype
原型试验	原型實驗,雛型實驗	prototype experiment
原液	原液	primary liquid
原油	原油	crude oil
原油蒸馏	原油蒸餾	crude distillation
原油资源	原油資源	crude resources
原汁	原汁	absolute juice
原子间距	原子間距	atomic spacing
原子量	原子量	atomic weight
原子量表	原子量表	atomic weight table
原子热容	原子熱容[量]	atomic heat capacity
原子相互作用	原子相互作用	atomic interaction
原子序数	原子序	atomic number
原子质量	原子質量	atomic mass
圆孔口	圓孔口	rounded orifice
圆盘	圓盤	disk
圆盘干燥器	圓盤乾燥機	disk dryer
圆盘给料机	盤飼機	disk feeder
圆盘浇口	圓盤澆口	disk gate
圆盘流动式冷凝器	圓盤流動冷凝器	disc flow condenser
圆盘磨	圓盤磨	disc attrition mill
圆盘塔	圓盤塔	disc column
圆桶喷雾器	圓桶噴霧器	barrel sprayer

大　陆　名	台　湾　名	英　文　名
圆筒干燥器	圓筒乾燥器	cylinder drier
圆筒筛	礦石篩, 轉筒篩	trommel
圆形标度指示计	圓形標 [度指] 示計	circular scale indicator
圆形度	[真]圓度	circularity
圆形图	圓形圖	circular chart
圆周速度	圓周速度, 周邊速率	circumferential velocity, peripheral speed, peripheral velocity
圆锥破碎机	錐碎機	cone crusher
约化质量	約化質量	reduced mass
约束[条件]	限制	constraint
月桂油	月桂油	bay oil, laurel oil
月桂籽油	月桂子油	bayberry oil
云母	雲母	mica
云母铀矿	瀝青鈾礦	uranite
匀化	均質化	homogenization
匀化器	均質機	homogenizer
匀浆	匀漿, 均質[物]	homogenate
运动	運動	motion
运动方程	運動方程式	equation of motion
运动黏度	動黏度	kinematic viscosity
运动黏度计	動黏度計	kinematic viscometer
运动相似	運動相似性	kinematic similarity
运动性质	運動性質	kinematic property
运动学	運動學	kinematics
运输	運輸, 輸送	transportation
运输性质	輸送性質	transport property
运输滞后	輸送滯延	transportation lag
运行特性	性能特徵	performance characteristics
运转	運轉, 操作	run
蕴压	蘊壓[力]	intrinsic pressure

Z

大　陆　名	台　湾　名	英　文　名
杂醇油	雜醇油	fusel oil
杂多酸催化剂	異聚酸觸媒	heteropolyacid catalyst
杂酚油	雜酚油	creosote
杂交瘤	雜交瘤	hybrid tumor

大　陆　名	台　湾　名	英　文　名
杂菌污染	雜菌污染	microbial contamination
杂散光	雜散光	stray light
杂质	雜質	impurity
甾醇,固醇	固醇類	sterol
载气	載氣	carrier gas
载热体	熱載體	heat carrier
载体	載體,載具	carrier, vehicle
载体分馏	載體蒸餾	carrier distillation
再沉积	再沈積	redeposition
再充氧作用	再充氧[作用]	reoxygenation
再锻烧	再燃燒[反應]	recalcination
再锻烧过程	再煅燒程序	recalcining process
再沸冷凝器	再沸冷凝器	reboiler condenser
再沸器,重沸器	再沸器	reboiler
再沸器动力学	再沸器動力學	reboiler dynamics
再归一化	再歸一化,重整	renormalization
再结晶	再結晶	recrystallization
再聚合	再聚合[反應]	repolymerization
再聚集	再聚集[作用]	reaggregation
再曝气	再曝氣	reaeration
再热器	再熱器	reheater
再生	再生	regeneration
再生材料	再生材料	reclaimed materials
再生胶	再生橡膠	reclaimed rubber
再生炉	再生爐,蓄熱爐	regeneration furnace
再生器	再生器	regenerator
再生热	再生熱	regenerative heat
再生系统	再生系統	regenerative system
再生纤维	再生纖維	regenerated fiber
再生纤维素	再生纖維素	regenerated cellulose
再生橡胶	再生橡膠	regenerated rubber
再生循环	再生循環	regenerative cycle
再现度	再現度	degree of reproducibility
再现性	再現性	reproducibility
再现因素	繁殖因數	reproduction factor
再悬浮	再懸浮	resuspension
再循环	循環	recirculation
再蒸馏收率	重餾產率	rerun yield

大　陆　名	台　湾　名	英　文　名
再蒸馏水	再蒸餾水	double distilled water
再蒸馏塔	重餾塔	rerun column, rerunning tower
再装满,回填	再裝滿	refill
在线	線上	on-line
在线适应	線上調適	on-line adaptation
藻胶	藻素	algin
藻类	藻類	algae
藻酸纤维	藻酸纖維	alginate fiber, algin fiber
皂糊	皂糊	soap paste
皂化	皂化[作用]	saponification
皂化剂	皂化劑	saponifying agent
皂化聚合物	皂化聚合物	saponified polymer
皂化纤维素乙酸酯	皂化乙酸纖維素	saponified cellulose acetate
皂化乙酸人造丝	皂化乙酸嫘縈	saponified acetate rayon
皂化值	皂化值	saponification number
皂胶束	皂微胞	soap micelle
皂脚	皂腳	niger
皂脚油	皂腳油	niger oil
皂片	皂片	soap flake
皂液萃取	皂萃取	soap extraction
造粒	造粒	granulation, pelletizing
造粒机	造粒機	pelletizer
造粒塔	製粒塔	prilling tower
造粒转窑	粒化轉窯	nodulizing kiln
噪声	噪音	noise
噪声温度计	噪音溫度計	noise thermometer
噪声污染	噪音污染	noise pollution
择形催化剂	形狀選擇觸媒	shape-selective catalyst
增稠	增濃,增黏	thickening
增稠过程	增濃程序,增黏程序	thickening process
增稠剂	增稠劑	thickening agent
增稠能力	增稠能力	thickening capacity
增稠器	增濃器	thickener
增感染料	敏化染料	sensitizing dye
增量剂	增效劑	extender
增强材料	增强材料	reinforcing materials
增强塑料	强化塑膠	reinforced plastic
增强填充料	增强填料	reinforcing filler

大　陆　名	台　湾　名	英　文　名
增强因子	增進因數	enhancement factor
增溶反应	助溶反應	solubilizing reaction
增溶剂	助溶劑	solubilizer, solubilizing agent
增溶作用	溶解化[作用]	solubilization
增升装置	高升裝置	high lift device
增湿	增濕	humidification, humidifying
增湿过程	增濕程序	humidification process
增湿器	增濕器	humidifier
增湿图	增濕圖	humidifying chart
增塑剂	塑化劑,可塑劑	plasticizer
增塑溶胶	塑料溶膠	plastisol
增塑作用	塑化	plasticization
增碳燃气	增碳燃氣	carburetted gas
增碳水煤气	增碳水煤氣	carburetted water gas
增效混合剂	增效混合物	synergistic mixture
增效机理	增效機構	synergistic mechanism
增效剂	增效劑	synergist
增效添加剂	增效添加劑	synergistic additive
增效性稳定剂	增效穩定劑	synergistic stabilizer
增效性阻燃剂	增效阻燃劑	synergistic flame retardant
增效作用	增效作用	synergism
增压泵	增壓壓縮機	booster compressor
增益	增益	gain
增益补偿	增益補償	gain compensation
增益矩阵	增益矩陣	gain matrix
增益相图	增益相位圖	gain-phase plot
增益裕量	增益裕量	gain margin
增益越界	增益交越	gain crossover
增益越界点	增益交越點	gain crossover point
增益越界频率	增益交越頻率	gain crossover frequency
增殖收率系数(=生长 　收率系数)		
增资	增資	incremental investment
甑,干馏釜	甑,蒸餾罐	retort
甑馏法	甑[蒸]餾法	retort process
渣棉	渣棉	slag wool
渣油	殘渣,殘留物	residue
闸阀	閘閥	gate valve

大　陆　名	台　湾　名	英　文　名
闸函数	闸函數	barrier function
闸喉喷嘴	闸喉噴嘴	choked nozzle
闸门	闸,澆口	gate
炸药	炸藥	explosive
展幅机	擴幅機	stenter
张力	張力	tension
张力计	張力計	tensiometer
张力试验机	張力試驗機	tension testing machine
张量	張量	tensor
樟脑油	樟腦油	camphor oil
涨落	漲落	fluctuation
涨落速度	波動速度	fluctuating velocity
胀塑性	膨脹性	dilatancy
胀塑性流体	脹塑性流體,剪力增黏流體	dilatant fluid
账面价值	賬面價值	book value
沼气	沼氣,生質氣	marsh gas, biogas
赵[广绪]-西得方法	趙[廣緒]-西得方法	Chao-Seader method
照相增感剂	感光敏化劑	photographic sensitizer
照相纸	感光紙	photographic paper
折旧	折舊	depreciation
折旧备用金	折舊準備金	depreciation reserve
折旧费	折舊成本	depreciation cost
折流分离	擋板分離	baffle separation
折流分离器	折流分離器	deflection separator
折流蒸发器	擋板蒸發器	baffled evaporator
折射计	折射計	refractometer
折射率	折射率	refractive index
折射温度系数	折射溫度係數	refraction temperature coefficient
折现收益率	折現收益率	discounted cash flow rate of return
折现因子	折現因數	discount factor
折现盈亏平衡点	折現損益均衡點	discount breakeven point
赭石	赭石	ocher
褶	褶,疊	plait
褶点	褶點	plait point
蔗糖	蔗糖	cane sugar, saccharose, sucrose
蔗渣	蔗渣	bagasse
蔗[渣]板	蔗[渣]板	bagasse board

大　陆　名	台　湾　名	英　文　名
蔗渣蒸煮罐	蔗渣蒸煮器	bagasse digester
针点浇口	針點澆口	pin-point gate
针孔数	針孔數	needle number
针入度	穿透度	degree of penetration
针入度测定计	針入計	penetrometer
针入度指数	針入[度]指數	penetration index
针形阀	針閥	needle valve
珍珠岩	珠岩	perlite
真沸点	真沸點	true boiling point
真菌	真菌	fungi
真空	真空	vacuum
真空泵	真空泵	vacuum pump
真空成型	真空成型	vacuum forming, vacuum molding
真空阀门	真空閥	vacuum valve
真空放电	真空放電	vacuum discharge
真空分馏器	真空分餾器	vacuum fractionator
真空封接	真空密封	vacuum seal
真空浮选	真空浮選	vacuum flotation
真空干燥	真空乾燥	vacuum drying
真空干燥器	真空乾燥器,真空乾燥裝置	vacuum desiccator, vacuum drying apparatus
真空干燥箱	真空乾燥烘箱	vacuum drying oven
真空高差	真空高差	vacuum head
真空鼓式过滤器	真空濾桶	vacuum drum filter
真空管	真空管	vacuum tube
真空锅	真空罐	vacuum pan
真空过程	真空程序	vacuum process
真空过滤机	真空過濾機	vacuum filter
真空烘箱	真空烘箱	vacuum oven
真空计	真空計	vacuometer, vacuum gauge
真空焦油	真空焦油	vacuum tar
真空结晶	真空結晶	vacuum crystallization
真空结晶器	真空結晶器	vacuum crystallizer
真空解除	破真空	vacuum relief
真空浓缩	真空濃縮	vacuum concentration
真空盘架干燥器	真空盤式乾燥器,真空箱乾燥器	vacuum tray drier, vacuum shelf drier
真空润滑油	真空潤滑脂	vacuum grease

大　陆　名	台　湾　名	英　文　名
真空闪蒸	真空驟汽化	vacuum flash vaporization
真空脱水器	真空脱水器	vacuum dehydrator
真空压	真空壓力	vacuum pressure
真空压力计	真空[壓力]計	vacuum manometer
真空压制	真空壓製	vacuum pressing
真空压制机	真空壓製機	vacuum press
真空油	真空油	vacuum oil
真空甑	真空甑	vacuum retort
真空蒸发	真空蒸發	vacuum evaporation
真空蒸发器	真空蒸發器	vacuum evaporator
真空蒸馏	真空蒸餾	vacuum distillation
真空蒸馏釜	真空蒸餾器	vacuum still
真空蒸馏塔	真空蒸餾塔	vacuum column
真密度	真密度	true density
真实气体	真實氣體	real gas
真实组成	真實組成	real composition
振荡	振盪	oscillation
振荡反应	振盪反應	oscillating reaction
振荡频率	振盪頻率	oscillation frequency
振荡元件	振盪元件	oscillatory element
振动给料器	振動進料機	vibratory feeder
振动流化床	振動流[體]化床	vibrated fluidized bed
振动能	振動能	vibrational energy
振动能级	振動能階	vibrational energy level
振动配分函数	振動分配函數	vibration partition function
振动器	振動器	vibrator
振动筛	振盪篩	oscillating screen, vibrating screen
振动输送机	搖運機	shaking conveyor
振幅	振幅	amplitude
振幅比	振幅比	amplitude ration
振铃极点	振鈴極點	ringing pole
振实密度	敲緊密度	tap density
镇静钢	全静碳鋼,脱氧鋼	killed steel
震动器	搖動器	shaker
震凝性流体	搖變增黏流體	rheopectic fluid
蒸氨塔	氨蒸餾塔	ammonia still
蒸发	蒸發	evaporation
蒸发管	蒸發管	evaporating pipe

大 陆 名	台 湾 名	英 文 名
蒸发结晶	蒸發結晶	evaporative crystallization
蒸发结晶器	蒸發結晶器	evaporative crystallizer
蒸发冷冻	蒸發冷凍	evaporative refrigeration
蒸发冷凝器	蒸發冷凝器	evaporative condenser
蒸发冷却	蒸發冷卻	evaporation cooling, evaporative cooling
蒸发率	蒸發速率	evaporation rate, rate of evaporation
蒸发器	蒸發器	evaporator
蒸发器污垢	蒸發器積垢	evaporator scale
蒸发热	蒸發熱	heat of evaporation
蒸发室	蒸發室	evaporator room
蒸发速度测定器	蒸發速率測定器	atometer
蒸发损失	蒸發損失	evaporation loss
蒸发塔	蒸發塔	evaporating column
蒸发系数	蒸發係數	evaporation coefficient
蒸发效率	蒸發效率	evaporation efficiency
蒸馏	蒸餾	distillation
蒸馏釜	蒸餾釜,蒸餾器	still, distillation still
蒸馏阱	蒸餾阱	distillation trap
蒸馏冷凝器	蒸餾冷凝器	distilling condenser
蒸馏瓶	蒸餾瓶	distilling flask
蒸馏器	蒸餾器,蒸餾裝置	distiller, distilling apparatus, distillating apparatus
蒸馏区	蒸餾區	distillation zone
蒸馏试验	蒸餾試驗	distillation test
蒸馏水	蒸餾水	distilled water
蒸馏塔板	蒸餾塔板	distillation tray
蒸馏值	蒸餾值	distillation value
蒸馏柱	蒸餾柱	distillation column
蒸气	蒸氣,汽	vapor
蒸气扩散泵	蒸氣擴散泵	vapor diffusion pump
蒸气旁路	蒸氣旁路	vapor bypass
蒸气压	蒸氣壓	vapor pressure
蒸气压缩	蒸氣壓縮	vapor compression
蒸汽	蒸汽	steam
蒸汽吹洗	蒸汽吹驅	steam purge
蒸汽动力发电厂	蒸汽發電廠	steam power plant
蒸汽阀	蒸汽閥	steam valve
蒸汽管	蒸汽管	steam tube

大　陆　名	台　湾　名	英　文　名
蒸汽锅炉	蒸汽鍋爐	steam boiler
蒸汽机	蒸汽機	steam engine
蒸汽夹套	蒸汽夾套	steam jacket
蒸汽经济	蒸汽經濟	steam economy
蒸汽冷凝器	蒸汽冷凝器	steam condenser
蒸汽冷凝水	蒸汽冷凝液	steam condensate
蒸汽裂解	蒸汽裂解	steam cracking
蒸汽硫化	蒸汽硫化	steam cure
蒸汽灭菌	蒸汽滅菌	steam sterilization
蒸汽喷射泵	蒸汽噴射泵,蒸汽噴射器	steam jet ejector, steam jet pump
蒸汽喷射器	蒸汽噴射器	steam ejector
蒸汽品质	蒸汽乾度	steam quality
蒸汽驱动泵	蒸汽驅動泵	steam-driven pump
蒸汽省热器	蒸汽省熱器	steam economizer
蒸汽室	蒸汽室	steam chamber
蒸汽释放阀	蒸汽釋放閥	steam relief valve
蒸汽熟化	蒸汽熟化	steam cure
蒸汽水分离器	[蒸]汽水分離器	steam separator
蒸汽涡轮	蒸汽渦輪	steam turbine
蒸汽消毒器	蒸汽滅菌器	steam sterilizer
蒸汽旋管	蒸汽旋管,蒸汽盤管	steam coil
蒸汽浴	蒸汽浴	steam bath
蒸汽再生法	蒸汽再生[法]	steam regeneration
蒸汽转化	蒸汽重組	steam reforming
蒸汽总管	蒸汽集管[箱]	steam header
蒸腾	蒸散[作用]	transpiration
蒸煮	蒸煮	cook
蒸煮槽	蒸煮槽	cooking tank
蒸煮过程	蒸煮程序	cooking process
蒸煮酸	蒸煮酸	cooking acid
蒸煮液	蒸煮液	cooking liquor
整合	整合	integration
整流器	整流器	rectifier
整数规划	整數規劃	integer programming
整体浓度	體濃度	concentration bulk
整体强度	整體強度	bulk intensity
整体通气器	整體通氣器	bulk aerator

大　陆　名	台　湾　名	英　文　名
整体指数	整體指數	bulk index
整体重量	總體重量	bulk weight
整装催化剂	整裝觸媒,蜂巢狀觸媒	monolithic catalyst
整装填料	結構填充物	structured packing
正比计数器	比例計數器	proportional counter
正常操作	正常操作	normal operation
正常反应动力学	正常反應動力學	normal reaction kinetics
正常渠深	正常渠深	normal channel depth
正常线性化	常態線性化	normal linearization
正常状态	常態	normal condition
正齿轮泵	正齒輪泵	spur gear pump
正反馈	正回饋	positive feedback
[正负电子]对湮没	成對毀滅	pair annihilation
正负共沸物	正負共沸液	positive-negative azeotrope
正割模量	正割模數	secant modulus
正共沸混合物	正共沸液	positive azeotrope
正规溶液	正規溶液	regular solution
正规溶液理论	正規溶液理論	regular solution theory
正激波	正震波	normal shock wave
正交配置	正交配置[法]	orthogonal collocation
正交性	正交性	orthogonality
正偏差	正偏離	positive deviation
正式报告	正式報告	formal report
正态分布	常態分佈	normal distribution
正态分布函数	常態分佈函數	normal distribution function
正位移	正排量	positive displacement
正位移计	正排量[流量]計	positive displacement meter
正弦波	正弦波	sine wave, sinusoidal wave
正弦波发生器	正弦波發生器	sine wave generator
正弦分析	正弦分析	sinusoidal analysis
正弦扰动	正弦擾動	sinusoidal disturbance
正弦式变形	正弦形變	sinusoidal deformation
正弦响应	正弦應答	sine response, sinusoidal response
正弦信号	正弦信號	sinusoidal signal
正向反应	正反應	forward reaction
正则变换	正則變換	canonical transformation
正则方程	正則方程式	canonical equation
正则配分函数	正則分配函數	canonical partition function

大　陆　名	台　湾　名	英　文　名
正则系综	正則系綜	canonical ensemble
正则形式	正則形式	canonical form
正则坐标	正則坐標	canonical coordinate
支撑液膜	固定式液膜	immobilized liquid membrane
支承板	支撑板	support plate
支持膜	支持膜	supporting film
支付	支付	payout
支管	支管	side tube
支链	支鏈	branched chain
支配方程	統御方程式	governing equation
支座	拱臺	abutment
芝麻油	芝麻油	sesame oil
知识工程	知識工程	knowledge engineering
知识库	知識庫	knowledge base
织物过滤器	織物過濾器	fabric filter
脂肪	脂肪	fat
脂肪醇	脂[肪]醇	fatty alcohol
脂肪酶	脂肪酶	lipase
脂肪酸	脂[肪]酸	fatty acid
脂肪族烃	脂族烃	aliphatic hydrocarbon
直边孔口	方[緣]孔[口]	square edge orifice
直方图	直方圖	histogram
直浇道	澆口	sprue
直角坐标	直角坐標	rectangular coordinate
直接成本	直接成本	direct cost
直接法	直接法	direct method
直接还原	直接還原[法]	direct reduction
直接火焰炉	明火加熱爐	direct-fired furnace
直接加热型蒸发器	直接加熱型蒸發器	evaporator with direct heating
直接浇口	直接澆口,直接模口	direct gate
直接控制	直接控制	direct control
直接煤液化	直接煤液化[法]	direct coal liquefaction
直接氢化	直接氫化[反應]	direct hydrogenation
直接染料	直接染料	direct dye
直接热解	直接熱解	direct pyrolysis
直接生产成本	直接生產成本	direct production cost
直接数字控制	直接數位控制	direct digital control
直接搜索法	直接搜索法	direct search method

大　陆　名	台　湾　名	英　文　名
直接液化	直接液化[法]	direct liquefaction
直接作用控制器	直接作用控制器	direct-acting controller
直径	直徑	diameter
直链聚合物	直鏈聚合物	straight-chain polymer
直列管排	直列管排	in-line tube arrangement
直馏[法]	直餾	straight-run distillation
直馏油	直餾	straight run
直闪石	直閃石	anthophyllite
直线折旧	直線折舊[法]	straight-line depreciation
K 值	汽液平衡比值, K 值	K value
pH 值	pH 值, 酸度值	pH value
植物激素	植物激素	plant hormone
植物生长激素	植物生長激素	plant growth hormone
植物生长调节剂	植物生長調節劑	plant growth regulator
植物油	植物油	vegetable oil
植物甾醇	植固醇	phytosterol
止逆阀, 单向阀	止回閥, 單向閥	check valve
纸电泳	紙電泳[法]	paper electrophoresis
纸分配色谱	紙分配層析[法]	paper partition chromatography
纸浆	紙漿	pulp
纸浆[粕]液	紙漿[粕]液	slush pulp
纸色谱法	紙層析[法]	paper chromatography
指前因子	指數前因數, 頻率因數	preexponential factor
指示计	指示器	indicator
指示剂	指示劑	indicator
指数分布	指數分佈	exponential distribution
指数函数	指數函數	exponential function
指数律	指數律	exponential law
指数滞后	指數滯延	exponential lag
酯	酯	ester
酯化[作用]	酯化[作用]	esterification
酯交换	轉酯化[作用]	transesterification
酯值	酯值	ester number, ester value
制革	鞣製	tanning
制管机	製管機	tubing machine
制浆	製漿	pulping
制浆法	製漿法	pulping process
制冷循环	冷凍循環	refrigeration cycle

大　陆　名	台　湾　名	英　文　名
制霉菌素	奈黴素	nystatin
制药工业	製藥工業	pharmaceutical industry
制造	製造	manufacture, fabrication
制造成本	製造成本	manufacturing cost
制造技术	製造技術	manufacturing technology
质量	質量	mass
质量传递(＝传质)		
质量分率	質量分率	mass fraction
质量规格	品質規格	quality specification
质量衡算	質量均衡	mass balance
质量控制	品質管制	quality control
质量扩散系数	質量擴散係數	mass diffusivity
质量流	質量流	mass flow
质量流量计	質量流量計	mass flowmeter
质量流率	質量流率	mass flow rate
质量流速	質量流速	mass velocity
质量摩尔浓度	重量莫耳濃度	molal concentration
质量浓度	質量濃度	concentration mass, mass concentration
质量平衡	質量均衡	mass balance
质量平均速度	質量平均速度	mass average velocity
质量迁移	質量輸送	mass transport
质量去除曲线	質量去除曲線	mass removal curve
质量生成曲线	質量生成曲線	mass generation curve
质量守恒	質量守恆	mass conservation
质量守恒原理	質量守恆原理	principle of conservation of mass
质量通量	質量通量	mass flux
质量吸收系数	質量吸收係數	mass absorption coefficient
质[量中]心	質[量中]心	center of mass
质量作用定律	質量作用定律	law of mass action
质量作用动力学	整體作用動力學	mass action kinetics
质谱法	質譜法	mass spectrometry
质谱图	質譜	mass spectrum
质谱仪	質譜儀	mass spectrometer
质心	形心,矩心	centroid
质子溶剂	質子性溶劑	protic solvent
滞后	遲滯,滯延	hysteresis, lag
滞后补偿	滯延補償	lag compensation
滞后期	滯延期	lag phase

大　陆　名	台　湾　名	英　文　名
滞留槽	貯留槽	holding tank
滞留池	貯留池	holding pond
滞留量	貯留量,滯留量	hold-up
滞留时间	貯留時間	hold-up time
滞止膜	[停]滯膜	stagnant film
滞止相	[停]滯相	stagnant phase
置换	置換	displacement
置换	置換	replacement
置换反应	置換反應	displacement reaction
置信水平	信賴水準	confidence level
置信限	信賴界限	confidence limit
置信域	信賴區域	confidence region
中国化工学会	中國化工學會	Chemical Industry and Engineering Society of China, CIESC
中和	中和[作用]	neutralization
中和滴定	中和滴定[法]	neutralization titration
中和点	中和點	neutralization point
中和反应	中和反應	neutralization reaction
中和剂	中和劑	neutralizing agent
中和耐火材料	中性耐火材	neutral refractory
中和区	中和區	neutral zone
中和热	中和熱	heat of neutralization
中和值	中和值	neutralization number, neutralization value
中级筛选	中級篩選	medium screening, medium sizing
中级压碎	中級壓碎	medium crushing
中级研磨	中級研磨	medium grinding
中间产物	中間產物	intermediate product
中间产物储存	中間產物儲存	intermediate storage
中间澄清器	中級澄清器,中級沈降槽	intermediate clarifier
中间冷却器	中間冷卻器,側流冷卻器,中間冷凝器	intercooler, side cooler, intercondenser
中间试验装置	先導工廠,試驗工廠	pilot plant
中间体	中間體	intermediate
中间相	介相	mesophase
中间压碎机	中級壓碎機	intermediate crusher
中空纤维组件	中空纖維組件	hollow-fiber module
中空铸型法	中空鑄型[法]	slush molding

大　陆　名	台　湾　名	英　文　名
中弯头	中彎頭	medium sweep elbow
中位值	中位數, 中值	median
中温菌	嗜中溫菌	mesophile, mesophilic bacteria
中性玻璃	中性玻璃	neutral glass
中性肥料	中性肥料	neutral fertilizer
中性肥皂	中性肥皂	neutral soap
中性耐火砖	中性耐火磚	neutral firebrick
中性施胶	中性膠料	neutral size
中性亚硫酸盐半化学法	中性亞硫酸[鹽]半化學法	neutral sulfite semichemical process
中性亚硫酸盐法	中性亞硫酸[鹽]法	neutral sulfite process
中值选择器	中值選擇器	median selector
中质油	中質油	middle oil
中毒	中毒	poisoning
终点	終點	end point
终点控制	終點控制	end-point control
终端	終端	terminal
终端速度	終端速度	terminal velocity
终端下降速度	最終下降速度	terminal falling velocity
终沸点	終沸點	final boiling point
终值原理	終值定理	final-value theorem
终止反应	終止反應	termination reaction
终止剂	終止劑	termination agent
钟乳石	鐘乳石	stalactite
钟形流量计	鐘式流量計	bell flowmeter
种群动态	種群動態	population dynamics
种群密度函数	種群密度函數	population density function
种群平衡	種群均衡, 粒數衡算	population balance
种子聚合	核種聚合	seeding polymerization
仲氦	仲氦	parahelium
仲氢	仲氫	parahydrogen
重化学品	重化學品	heavy chemical
重晶石	重晶石	barite
重均沸点	重量平均沸點	weight average boiling point
重均分子量	重量平均分子量	weight average molecular weight
重力	重力	gravitation, gravitational force, gravity
重力常数	重力常數	gravitational constant
重力沉降	重力沈降	gravitation settling, gravity settling

大　陆　名	台　湾　名	英　文　名
重力沉降分离器	重力沈降分離器	gravity settler separator
重力沉降器	重力沈降器	gravity settler
重力电池	重力電池	gravity cell
重力分离	重力分離	gravitational separation, gravity separation
重力分离器	重力分離器	gravity separator
重力分凝	重力離析	gravity segregation
重力过滤	重力過濾	gravity filtration
重力过滤器	重力[過]濾器	gravitational filter
重力换算因数	重力換算因數	gravitational conversion factor
重力计	比重計	gravity meter
重力流动	重力流動	gravity flow
重力式过滤器	重力過濾器	gravity filter
重力循环	重力循環	gravity circulation
重力压头	重力高差	gravity head
重力增稠	重力增稠	gravity thickening
重力增稠器	重力增稠器	gravity thickener
重量分率	重量分率	weight fraction
重量分析[法]	重量分析[法]	gravimetric analysis
重量摩尔浓度	重量莫耳濃度	molality
重量平均直径	重量平均直徑	weight mean diameter
重量损失	失重	weight loss
重馏分	重餾分	heavy ends
重氢	重氫	heavy hydrogen
重燃料油	重燃料油	heavy fuel oil
重水	重水	heavy water
重苏打灰	重鹼灰	dense soda ash
重烃	重烴	heavy hydrocarbon
重直馏油	重直餾油	heavy straight run
重质油	重油	heavy oil
周期	週期	period
周期操作	週期操作	periodic operation
周期反应	週期反應	cyclic reaction
周期负荷	週期負載	periodic load
周期解	週期解	periodic solution
周期流	週期流[動]	periodic flow
周期流反应器	週期流[動]反應器	periodic flow reactor
周期凝聚	週期凝聚[作用]	periodic coagulation
周期时间	週期	cycle length, cycle time

大　陆　名	台　湾　名	英　文　名
周期式反应器	週期式反應器	cyclic reactor
周期现象	週期現象	periodic phenomenon
周期性	週期性	periodicity
周期性应力	週期性應力	cyclic stress
周期振荡	週期振盪	periodic oscillation
周转率	周轉率	turnover ratio
轴	軸	shaft
轴衬	軸襯	bush
轴承	軸承	bearing
轴承衬套	軸承襯套	bearing bushing
轴承滚珠	軸承球	bearing ball
轴承圈	軸承圈	bearing collar
轴承式混合器	軸承式混合器	bearing mixer
轴承托座	軸承架	bearing bracket
轴承箱	軸承箱	bearing box, bearing housing
轴对称流	軸向對稱流動	axial symmetric flow
轴功	轉軸功	shaft work
轴角	軸角	axial angle
轴流泵	軸流泵	axial flow pump
轴流反应器	軸流反應器	axial-flow reactor
轴流风机	軸[向]流風扇	axial-flow fan
轴流式压缩机	軸[向]流壓縮機	axial-flow compressor
轴向对称	軸向對稱	axial symmetry
轴向分散	軸向分散	axial dispersion
轴向分散系数	軸向分散係數	axial dispersion coefficient
轴向混合	軸向混合	axial mixing
轴向流	軸[向]流	axial flow
轴向速度	軸向速度	axial velocity
轴制动马力	軸制動馬力	shaft brake horsepower
肘形管	肘形管	elbow pipe
骤混	驟混	flash mixing
骤混器	驟混器	flash mixer
骤冷	驟冷	flash chilling
骤氯化	驟氯化[反應]	flash chlorination
猪油	豬油	lard oil
猪脂	豬油	lard
竹节丝	粗節紗	slub yarn
竹桃霉素	安黴素,奧連黴素	oleandomycin

大　陆　名	台　湾　名	英　文　名
逐步反应	逐步反應	stepwise reaction
逐步共聚合	逐步共聚[作用]	step copolymerization
逐步加成聚合	逐步加成聚合[反應]	stepwise addition polymerization
逐步加成聚合物	逐步加成聚合物	step addition polymer
逐步聚合	逐步聚合[反應]	step reaction polymerization, stepwise polymerization
逐步冷冻法	漸進冷凍	progressive freezing
逐次反应	逐次反應	successive reaction
逐次聚合	逐次聚合	successive polymerization
逐次开闭操作	逐次開關操作	sequencing on-off operation
逐次撕裂强度	逐次撕裂強力	successive tear strength
逐级反应	逐次反應,階段反應	staircase reaction
逐级计算	分階計算	stage calculation
逐级接触	分階接觸	stage contacting
逐级接触器	分階接觸器	stage contactor
主变量	主要變數	primary variable
主产品	主產物	primary product
主导极点	主要極點	dominant pole
主发酵	主發酵	principal fermentation
主反应	主要反應,初步反應	primary reaction
主干	針桿,莖	stem
主根	主根	dominant root
主环路	主環路	primary loop
主加料机	主飼機,主進料器	primary feeder
主令控制器	主控制器	master controller
主要本征值	主宰固有值,優勢固有值	dominant eigenvalue
主要尺寸	主因次	primary dimension
主要冲击	主要衝擊	primary impact
主要设计变量	主要設計變數	primary design variable
主要元素	主要元素	primary element
主应力	主應力	principal stress
主增塑剂	主塑化劑	primary plasticizer
主轴	主軸	principal axis
助促进剂	助促進劑	secondary accelerator
助催化剂	輔觸媒,促進劑	co-catalyst, promoter
助抗氧剂	助抗氧劑	secondary antioxidant
助滤剂	助濾劑	filter aid

大　陆　名	台　湾　名	英　文　名
助磨剂	助磨劑	grinding aid
助凝剂	助凝劑	coagulant aid
助色团	助色團,助色基	auxochrome
助悬剂	懸浮劑	suspending agent
注残料顶杆	殘料頂桿	sprue ejector
注浆成型过程	注漿成型法	slip casting process
注射成型	射出成型	injection molding
注射尺寸	射出尺寸	shot size
注射量	射出容量,射出[重]量	shot capacity, shot weight
注射器	注射器	injector, syringe
注射速率	射出速率	shot rate
注射体积	射出體積	shot volume
注射体积控制器	射出體積控制器	shot volume controller
注射周期	射出週期	shot cycle
驻点	停滯點,駐定點	stagnation point, stationary point
驻点温度	停滯溫度	stagnation temperature
驻点性质	停滯性質	stagnation property
驻点压力	停滯壓力	stagnation pressure
柱晶白霉素	白黴素	leucomycin
柱面坐标	圓柱坐標	cylindrical coordinates
柱容量	管柱容量	column capacity
柱塞	柱塞	plunger
柱塞泵	柱塞泵	plunger pump
柱色谱法	管柱層析術	column chromatography
柱式蒸馏器	柱式蒸餾器	column still
柱特性因素	柱特性因數	column characterization factor
铸钢	鑄鋼	cast steel
铸件	鑄件	casting
[铸件等]气孔	氣孔	blow hole
铸模	鑄模	casting mold
铸塑聚合	鑄塑聚合[作用]	cast polymerization
铸塑树脂	鑄模樹脂	cast resin
铸铁	鑄鐵	cast iron
专家系统	專家系統	expert system
专利	專利	patent
专利权使用费	權利金	royalty
专用化学品	特用化學品	speciality chemical
砖	磚	brick

大 陆 名	台 湾 名	英 文 名
转变点	轉變點	transition point
转变范围	轉變範圍	transition range
转变热	［相］轉變熱	heat of transition
转变温度	轉變溫度,轉移溫度	transition temperature
转带浸取器	轉帶萃取器	belt extractor
转化	轉化	conversion
转化参数	轉化參數	conversion parameter
转化点	反轉點	inversion point
转化率	轉化率	conversion
转化率	轉化速率	conversion rate
转化率-温度图	轉化率-溫度圖	conversion-temperature chart
转化酶	轉化酶	invertase
转化器	轉化器	converter
转化热	轉化熱	heat of conversion
转化深度	轉化程度	conversion level
转化时间	轉化時間	conversion time
转化数	［觸］媒轉化數	turnover number
转化糖	轉化糖	invert sugar
转化温度	轉化溫度	conversion temperature
转化原理	轉化原理	conversion principle
Z 转换	Z 變換	Z transformation
转换涂层	轉化塗層	conversion coating
转角接头	轉角接頭	corner tap
转接线	插線	patch cord
转位	差排	dislocation
转向阀	轉向閥	diversion valve
转矩	扭矩	torque
转矩管	扭矩管	torque tube
转矩流变仪	扭矩流變儀	torque rheometer
转动惯量	轉動慣量	moment of inertia
转动能	旋轉能	rotational energy
转动能级	旋轉能階	rotational energy level
转动配分函数	轉動配分函數	rotational partition function
转鼓过滤机	旋桶濾機	rotary drum filter
转鼓混合机	轉鼓混合機	tumbler mixer
转盘	旋轉盤	rotating disc
转盘过滤机	旋盤濾機	rotary disc filter
转盘流量计	轉盤流量計	rotary disc meter

大　陆　名	台　湾　名	英　文　名
转盘塔	轉盤塔	rotating disc contactor
转速表	轉速計, 轉數計	tachometer
转筒筛	轉筒篩	bolting reel, screening trommel
转筒吸收器	轉桶吸收器	rotating-drum absorber
转筒洗涤	轉桶洗滌	drum washing
转筒真空过滤机	真空旋桶濾機	rotary vacuum drum filter
转子	轉子	rotor
转子流量计	浮子流量計	rotameter
转子叶片	轉子葉片	rotor blade
装配	安裝	mounting
装配线	裝配線	assembly line
装填效应	裝填效應	packing effect
装卸设备	裝卸設備	handling facility
装置建立成本	配置成本	setup cost
状态	狀態	state
状态变量	狀態變數	state variable
状态方程	狀態方程式	equation of state
PR[状态]方程	彭[定宇]-羅[賓遜]狀態方程	Peng-Robinson equation
状态函数	狀態函數	state function
追赶法	追蹤法	chasing method
锥板式黏度计	錐板黏度計	cone-and-plate viscometer
锥顶罐	錐頂槽	cone roof tank
锥管螺纹	攻牙器	pipe tap
锥形给料器	錐形進料器	cone feeder
锥形管	錐形管	tapered pipe
锥形流动	錐形流動	cone flow
锥形瓶	錐形瓶	Erlenmeyer flask
锥形雾化器	錐形霧化器	cone atomizer
准化学近似	準化學近似	quasi-chemical approximation
准化学溶液模型	準化學溶液模型	quasi-chemical solution model
准静态	準靜態	quasi-stationary state
准静态过程	準靜態過程	quasi-static process
准静态热重量分析法	似靜態熱重量法	quasi-static thermogravimetry
准确度	準確度	accuracy
准则	準則	criterion
灼烧[灭菌法]	燃燒	combustion
浊度	濁度	turbidity

大　陆　名	台　湾　名	英　文　名
浊度计	濁度計	turbidimeter, turbodometer
着火危险	火災	fire hazard
着色法	著色法	staining method
着色试验	著色試驗	staining test
资本	資本	capital
资本比率	資本比率	capital ratio
资本回收因数	資本回收因數	capital recovery factor
资本所得	資本所得	capital gain
资本投资成本因数	[資本]投資成本因數	capital investment cost factor
资产负债表	資產負債表, 平衡表	balance sheet
资产会计	資產會計	asset accounting
资产型式	資產型式	asset type
资金的时间价值	貨幣時間價值	time value of money
资源成本	資源成本	resources cost
籽棉	籽棉	seed cotton
子程序	子程式	subprogram, subroutine
子细胞	子細胞	daughter cell
子元素	子元素	daughter element
紫胶	蟲膠, 蟲漆	shellac, lac
紫胶蜡	蟲膠臘	shellac wax
紫胶清漆	蟲膠清漆	shellac varnish
紫霉素	紫黴素	viomycin
紫苏子油	紫蘇油	perilla oil
紫外灯	紫外線燈	ultraviolet lamp
紫外光	紫外光	ultraviolet light
紫外光谱仪	紫外線光譜儀	UV spectrophotometer
紫外线	紫外線	ultraviolet ray
紫外线比色计	紫外線比色計	ultraviolet colorimeter
紫外线发生器	紫外線產生器	ultraviolet generator
紫外[线]分光光度计	紫外線分光光度計	ultraviolet spectrophotometer
紫外线熟化	紫外線熟化	ultraviolet curing
紫外线吸收	紫外線吸收	ultraviolet absorption
紫外线吸收剂	紫外線吸收劑	ultraviolet absorber
紫外线荧光	紫外線螢光	ultraviolet fluorescence
T 字模	T 字模	T-die
自变量	自變數	independent variable
自持续	自持續	self sustaining
自持续反应	自持續反應	self-sustained reaction

大　陆　名	台　湾　名	英　文　名
自持振荡	持續振盪	sustained oscillation
自催化	自催化［作用］	autocatalysis
自动过程控制器	自動程序控制器	automatic process controller
自动过滤器	自動過濾器	automatic filter
自动恒温器	自動恆溫器	automatic thermostat
自动记录器	自動記錄器	auto recorder
自动加料器	自動進料器	automatic feeder
自动加热性	自熱性	self-heating property
自动进料	自動進料	automatic feed
自［动］聚合	自聚合［作用］	auto-polymerization
自动控制	自動控制	automatic control
自动控制器	自動控制器	automatic controller
自动控制系统	自動控制系統	automatic control system
自动皮带秤	自動稱重計	weightometer
自动-手控开关	自動手動開關	auto-manual switch
自动调节器	自動調節器	automatic regulator
自动选择器	自動選擇器	auto-selector
自动选择器控制	自動選擇器控制	auto-selector control
自发成核［作用］	突發成核［作用］	sporadic nucleation
自发反应	自發反應	spontaneous reaction
自发过程	自發程序	spontaneous process
自发极化	自發極化	spontaneous polarization
自发结晶	自發結晶	spontaneous crystallization
自发乳化	自發乳化	spontaneous emulsification
自发形核	自發成核	spontaneous nucleation
自发终止	自發終止［作用］	spontaneous termination
自分解	自分解	autolysis
自固化	自硬化	self-cure, self-curing
自固化环氧树脂	自硬化環氧樹脂	self-cure epoxy resin
自固化黏合剂	自固化黏合劑	self-curing adhesive
自加速作用	自加速［作用］	autoacceleration
自交联丙烯酸树脂	自交聯丙烯酸樹脂	self-crosslinking acrylic resin
自交联丙烯酸酯橡胶	自交聯丙烯酸酯橡膠	self-crosslinking acrylate rubber
自胶合	自膠合［作用］	auto-agglutination
自校正控制器	自調諧控制器	self-tuning controller
自校正调节器	自調諧調節器	self-tuning regulator
自净化	自淨化［作用］	self purification
自聚合	自聚合［作用］	self-polymerization

大　陆　名	台　湾　名	英　文　名
自扩散	自擴散	self-diffusion
自来水	自來水	tap water
自老化	自老化	self ageing
自硫化	自硫化[作用]	auto-vulcanization, self-vulcanizing
自硫化	自硬化	self-curing
自磨机	自磨機	autogenous mill
自黏合剂	自黏著劑	self-adhesive
自黏合纤维	自黏合纖維	self-bonded fiber
自黏胶带	自黏膠帶	self-adhesive tape
自凝聚	自凝聚[作用]	auto-coagulation
自平衡电桥	自均衡電橋	self-balancing bridge
自平衡电位计	自均衡電位計	self-balancing potentiometer
自起动泵	自起動泵	self-priming pump
自然对流	自然對流	natural convection
自然老化	自然老化	natural aging
自然冷却	自然冷卻	natural cooling
自然凝结	自發凝聚[作用]	spontaneous coagulation
自然频率	自然頻率	natural frequency
自然通风	自然通風	natural draft, natural ventilation
自然通风冷却塔	自然通風冷卻塔	natural draft cooling tower
自然循环蒸发器	自然循環蒸發器	natural circulation evaporator
自然蒸发	自然蒸發	natural evaporation, spontaneous evaporation
自然周期	自然週期	natural period
自然资源	天然資源	natural resources
自燃	自燃	self-ignition, spontaneous combustion
自燃温度	自燃溫度,[自]燃點	self-ignition temperature, autoignition temperature
自热反应	自熱反應	autothermal reaction
自溶脂肪酶	自[分]解脂酶	autolytic lipase
自身低聚化	自低聚合[反應]	self-oligomerization
自适应控制	自適應控制,適應控制	self-adaptive control, adaptive control
自适应去耦器	適應解偶器	adaptive decoupler
自调整	自調節	self regulation
自稳定	自穩定	auto-stabilization
自吸搅拌器,空心叶轮搅拌器	空心自吸攪拌器	hollow agitator
自熄	自熄	self-extinguishing

大　陆　名	台　湾　名	英　文　名
自熄性	自熄性	self extinguish ability
自旋回波	自旋回波	spin echo
自旋回波法	自旋回波法	spin echo method
自旋晶格弛豫	自旋晶格弛豫	spin-lattice relaxation
自旋晶格弛豫时间	自旋晶格弛豫时间	spin-lattice relaxation time
自旋耦合	自旋耦合	spin coupling
自旋耦合常数	自旋耦合常數	spin coupling constant
自旋去耦	自旋去耦合	spin decoupling
自旋-自旋弛豫	自旋間弛豫	spin-spin relaxation
自旋-自旋弛豫时间	自旋間弛豫時間	spin-spin relaxation time
自旋-自旋耦合	自旋-自旋耦合	spin-spin coupling
自氧化	自氧化[作用]	autoxidation
自引发	自引發	self-initiation
自由表面	自由表面	free surface
自由部位浓度	自由部位濃度	free site concentration
自由沉降	自由沈降	free settling
自由沉降速度	自由沈降速度	free falling velocity
自由程	自由徑	free path
自由度	自由度	degree of freedom
自由对流	自由對流	free convection
自由焓(=吉布斯自由能)		
自由基	自由基	free radical
自由基捕获剂	自由基捕獲劑	radical scavenger
自由基反应	自由基反應	free radical reaction
自由基机理	自由基機構	free radical mechanism
自由基聚合	自由基聚合[反應]	free radical polymerization
自由流	自由流	free stream
自由流动	自由流動	free flow
自由流动压头	自由流動高差	free flow head
自由能	自由能	free energy
自由能组成图	自由能-組成圖	free energy-composition diagram
自由水	自由水	free water
自由涡旋	自由漩渦	free vortex
自由压碎	自由壓碎	free crushing
自诱导反应	自誘發反應	auto-induced reaction
自增长	自增長	self-propagation
自增强聚合物	自強化聚合物	self-reinforcing polymer

大　陆　名	台　湾　名	英　文　名
自治系统	自律系統	autonomous system
自终止	自終止[作用]	self-termination
综合墙	複合壁	compositive wall
棕榈蜡	棕櫚蠟	palm wax
棕榈仁油	棕櫚仁油	palm kernel oil, palm nut oil
棕榈油	棕櫚油	palm oil
棕土(颜料)	富錳棕土	umber
总产量分率	總產量分率	overall fractional yield
总传递单元	總傳送單元	overall transfer unit
总传递函数	總傳遞函數,總轉移函數	overall transfer function
总传热系数	總熱傳係數	overall heat transfer coefficient
总传质单元数	總傳送單元數	number of overall transfer units
总传质系数	總質傳係數	overall mass transfer coefficient
总发射系数	總發射係數	total emissivity
总付参数	種群參數	population parameter
总固体	總固體[量]	total solid
总过程	總程序	overall process
总回收	總回收[率]	overall recovery
总能	總能	total energy
总能量平衡	總能量均衡	overall energy balance
总热平衡	總熱量均衡	overall heat balance
总热值	總熱值	gross heating value
总熵平衡	總熵均衡	overall entropy balance
总速率	總速率	overall rate
总体反应	總體反應	global reaction
总体反应速率	總反應速率	global reaction rate
总体流动	總體流動	bulk flow
总体速率	總速率	global rate
总体稳定性	總體穩定性	global stability
总体最优[值]	總體最佳[值]	global optimum
总吸收系数	總吸收係數	overall absorption coefficient
总系数	總係數	overall coefficient
总效率	總效率	overall efficiency
总压法	總壓法	total pressure method
总压力	總壓力	total pressure
总压头	總高差	total head
总硬度	總硬度	total hardness

大　陆　名	台　湾　名	英　文　名
总有机碳	總有機碳	total organic carbon
总质量平衡	總質量均衡	overall mass balance
总资本投资额	總資本投資額	total capital investment
总组成	總組成	overall composition
纵向翅片	縱向鰭片	longitudinal fin
纵向分散系数	縱向分散係數	longitudinal dispersion coefficient
纵向应力	縱向應力	longitudinal stress
纵摇	添加酵母[釀酒]	pitching
阻迟运动	阻滯運動	retarded motion
阻抗	阻抗	impedance
阻力	阻力	resistance
阻力系数	阻力係數	drag coefficient, resistance coefficient
阻尼	阻尼	damping
阻尼比	阻尼比	damping ratio
阻尼常数	阻尼常數	damping constant
阻尼力	阻尼力,減振力	damping force
阻尼器	阻尼器,減振器	damper
阻尼系数	阻尼係數	damping coefficient
阻尼液	阻尼流體	damping fluid
阻尼振荡	阻尼振盪	damped oscillation
阻尼振荡器	阻尼振盪器	damped oscillator
阻燃剂	阻燃劑,耐燃劑	fire retardant, flame resisting agent
阻塞压碎	阻塞壓碎	choke crushing
阻滞沉降	阻滯沈降	impeded settling
阻滞剂	阻滯劑	retarder
阻滞因子	阻滯因數	retardation factor
组成	組成	composition
组成规格	組成規格	composition specification
组成轨道	組成軌跡	composition trajectory
组成控制	組成控制	composition control
组分	組分,成分	constituent, component
组分分涂黏合剂	雙組分黏合劑	separated application adhesive
组分指定图	成分指定圖	component assignment diagram
组合规则	組合規則	combining rule
组合项	組合項	combinatorial term
组态扩散	組態擴散	configurational diffusion
组织	組織	tissue, organization
最大比生长速率	最大比生長速率	maximum specific growth rate

大 陆 名	台 湾 名	英 文 名
最大概率数	最可能數	most probable number
最大混合度	最大混合度	maximum mixedness
最大似然原理	最大概似	maximum likelihood principle
最低沸点共沸物	最低沸點共沸液	minimum boiling azeotrope
最低凝固点	最低凝固點	minimum freezing point
最低容许收益率	最低容許收益率	minimum acceptable rate of return, MARR
最陡上坡点	最陡上坡點	point of steepest ascent
最陡下坡点	最陡下坡點	point of steepest descent
最概然分布	最可能分佈	most probable distribution
最高沸点共沸物	最高沸點共沸液	maximum boiling azeotrope
最高流速	尖峰流量	peak flow rate
最高允许温度	最高容許溫度	maximum allowable temperature
最高允许压力	最高容許壓力	maximum allowable pressure
最佳操作条件	最適操作條件,最適操作狀況	optimal operation condition
最佳工况	最適條件,最適狀況	optimal condition
最佳回流	最適回流	optimal reflux, optimum reflux
最佳回流比	最適回流比	optimum reflux ratio
最佳搜寻法	最適搜尋法	optimal searching method
最佳调谐	最適調諧	optimal tuning
最佳途径	最適途徑	optimal path, optimum path
最适策略	最適策略	optimal policy, optimum policy
最适温度	最適溫度	optimal temperature, optimum temperature
最速上升法	最陡上升法	steepest ascent method
最速下降法	最陡下降法	steepest descent method
最小二乘法	最小平方法	least squares method
最小二乘分析	最小平方[分析]法	least squares analysis
最小回流	最小回流	minimum reflux
最小回流比	最小回流比	minimum reflux ratio
最小流化速度	最低流體化速度	minimum fluidization velocity, minimum fluidizing velocity
最小流化态	最低流體化態	minimum fluidization
最小偏差	最小偏差	minimum deviation
最小塔板数	最少板數	minimum tray number
最小值原理	極小原理	minimum principle
最优化	最適化	optimization
最优控制	最適控制	optimal control, optimum control

大　陆　名	台　湾　名	英　文　名
最优切换	最適切換	optimum switching
最优设定点	最適設定點	optimal set point, optimum set point
最优设计	最適設計	optimal design, optimum design
最优系统	最適系統	optimal system, optimum system
最优性	最適性	optimality
最优值	最適值	optimal value, optimum value
最优转化率	最適轉化率	optimal conversion, optimum conversion
最优组成	最適組成	optimal composition
最终产品	最終產物	end product
最终澄清器	最終澄清器	final clarifier
最终处理	最終處理	ultimate disposal
最终控制元件	最終控制元件	final control element
最终条件	最終條件	final condition, terminal condition
最终周期	最終週期	ultimate period
最终周期解	最終週期解	ultimate periodic solution
最终周期应答	最終週期應答	ultimate periodic response
最终状态	[最]終[狀]態	final state
左旋糖	左旋醣	levo-rotatory sugar
坐标系	坐標系	coordinate system

副 篇

A

英 文 名	大 陆 名	台 湾 名
abatement	减量	①减量 ②降低
ablation	烧蚀	剥蚀
abnormal reaction	异常反应	異常反應
abrasion	磨损	磨损
abrasion index	磨耗指数	磨耗指數
abrasion resistance	耐磨性	①耐磨性 ②磨耗阻力
abrasion test	磨耗试验	磨耗試驗
abrasive	磨料	[研]磨料
abrasive cleaner	擦洗剂	擦潔劑
abrasiveness	抗磨性	抗磨性
abrasive paper	砂纸	砂紙
Absinthium	苦艾	苦艾
absolute alcohol	无水乙醇,无水酒精	無水酒精,絕對酒精
absolute boiling point	绝对沸点	絕對沸點
absolute color value	绝对色值	絕對色值
absolute counting	绝对计数	絕對計數
absolute deviation	绝对偏差	絕對偏差
absolute dry fiber	绝对干纤维	絕對乾纖維
absolute entropy	绝对熵	絕對熵
absolute humidity	绝对湿度	絕對濕度
absolute juice	原汁	原汁
absolute moisture content	绝对湿含量	絕對含水量
absolute pressure	绝对压力	絕對壓力
absolute pressure gauge	绝对压力计	絕對壓力計
absolute reaction rate	绝对反应速率	絕對反應速率
absolute specificity	绝对专一性	絕對專一性
absolute stability	绝对稳定性	絕對穩定性
absolute temperature	绝对温度	絕對溫度

英　文　名	大　陆　名	台　湾　名
absolute thermodynamic temperature scale	绝对热力学温标	絕對熱力溫標
absolute turbidity	绝对浊度	絕對濁度
absolute value	绝对值	絕對值
absolute viscosity	绝对黏度	絕對黏度
absolute zero	绝对零度	絕對零度
absorbance	吸光度	吸光度
absorbed dose	吸收剂量	吸收劑量
absorbency curve	吸收度曲线	吸收率曲線
absorbent	吸收剂	吸收劑
absorbent bed	吸收层	吸收床
absorbent cellulose	吸收性纤维素	吸收性纖維素
absorbent filter	吸收过滤器	吸收過濾器
absorber	吸收塔	吸收塔
absorbing apparatus	吸收装置	吸收裝置
absorption	吸收	吸收
absorption band	吸收[谱]带	吸收帶,吸光帶
absorption base	吸收基质	吸收基質
absorption cell	吸收池	吸收槽
absorption chromatography	吸收色谱法	吸收層析[法]
absorption coefficient	吸收系数	吸收係數
absorption column	吸收柱	吸收管
absorption correction	吸收校正	吸收修正
absorption countercurrent	逆流吸收	逆流吸收
absorption cross-section	吸收截面	吸收截面
absorption curve	吸收曲线	吸收曲線
absorption discontinuity	吸收不连续性	吸收不連續性
absorption factor	吸收因子	吸收因子
absorption field	吸收场	吸收場
absorption heat	吸收热	吸收熱
absorption index	吸收指数	吸收指數
absorption indicator	吸收指示剂	吸收指示劑
absorption isotherm	吸收等温线	吸收等溫線
absorption number	吸收数	吸收數
absorption oil	吸收油	吸收油
absorption ointment base	吸收性软膏基	吸收性軟膏基
absorption power	吸收力	吸收能力
absorption process	吸收过程	吸收程序
absorption rate	吸收速率	吸收速率

英　文　名	大　陆　名	台　湾　名
absorption ratio	吸收比	吸收比
absorption reaction	吸收反应	吸收反應
absorption refrigeration	吸收制冷	吸收[式]製冷
absorption spectrum	吸收光谱	吸收光譜
absorption system	吸收系统	吸收系統
absorption test	吸收试验	吸收試驗
absorption tower	吸收塔	吸收塔
absorptive-type filter	吸收式过滤器	吸收式濾器
absorptivity	吸收率	吸收率
ABS resin(=acrylonitrile-butadiene-sty-rene resin)	ABS 树脂	丙烯腈–丁二烯–苯乙烯樹脂
abundance	丰度	豐度
abundance curve	丰度曲线	豐度曲線
abutment	支座	拱臺
acaricide	杀螨剂	殺蟎劑
accelerated aging	加速老化	加速老化
accelerated cement	快干水泥,速凝水泥	快乾水泥,速凝水泥
accelerated gum	速成胶质	速成膠
acceleration error coefficient	加速度误差系数	加速誤差係數
accelerator	①加速剂 ②加速器	①加速劑 ②加速器
accelerator of vulcanization	硫化加速剂	硫化加速劑
accelerometer	加速度计	加速計
acceptable quality level	验收质量水准	合格品質水準
acceptable return	可接受的回报	可接受[的]報酬
acceptance sampling	验收抽样	驗收取樣
acceptance sampling plan	验收抽样计划	驗收取樣計畫
acceptance test	验收试验	驗收試驗
acceptor	受体	受體
acclimation	驯化	馴化
acclimation sludge	驯化污泥	馴化污泥
accumulation rate	累积速率	累積速率
accumulative crystallization	聚集结晶	聚集結晶
accuracy	准确度	準確度
acentric factor	偏心因子	偏心因子
acetal	缩醛	縮醛
acetal resin	缩醛树脂	縮醛樹脂
acetate film	乙酸酯膜,醋酸酯膜	乙酸酯膜,醋酸酯膜
acetate rayon	乙酸人造丝,醋酸嫘縈	乙酸嫘縈,醋酸嫘縈

英 文 名	大 陆 名	台 湾 名
acetic acid bacteria	醋酸菌	醋酸菌
acetonitrile	乙腈	乙腈
acetylation	乙酰化	乙醯化
acetyl cellulose	乙酸纤维素,醋酸纤维素	乙酸纖維素,醋酸纖維素
acetylene black	乙炔黑	乙炔黑
acetyl value	乙酰值	乙醯值
achromycin	四环素	四環素,鉑黴素
acid acceptor	酸受体	酸受體
acid catalysis	酸催化	酸催化
acid chrome dye	酸性铬媒染料	酸性鉻媒染料
acid color	酸性染料	酸性染料
acid component	酸性成分	酸性成分
acid cooler	酸冷却器	酸冷卻器
acid dye	酸性染料	酸性染料
acid feed box	供酸罐	飼酸箱
acid fertilizer	酸性肥料	酸性肥料
acid firebrick	酸性耐火砖	酸性耐火磚
acid-forming bacteria	产酸菌	產酸[細]菌
acid fume	酸烟	酸煙
acid gas	酸性气体	酸性氣體
acidic catalysis	酸性催化	酸性催化[作用]
acidic catalyst	酸性催化剂	酸性觸媒
acidification	酸化	酸化
acidifier	酸化剂	酸化劑
acidity	①酸度 ②酸性	①酸度 ②酸性
acidity index	酸度指数	酸度指數
acidity test	酸度试验	酸度試驗
acid measuring tank	量酸槽	量酸槽
acid mist	酸雾	酸霧
acid number	酸值	酸值
acid process	酸法	酸法
acid-proof brick	耐酸砖	耐酸磚
acid-proof cement	耐酸水泥	耐酸水泥
acid-proof enamel	耐酸搪瓷	耐酸搪瓷
acid-proof paper	耐酸纸	耐酸紙
acid-proof slab	耐酸板	耐酸板
acid pump	酸泵	酸泵

英　文　名	大　陆　名	台　湾　名
acid rain	酸雨	酸雨
acid refractory	酸性耐火材料	酸性耐火材
acid refractory brick	酸性耐火砖	酸性耐火磚
acid resistance	耐酸性	耐酸性
acid resistant	耐酸物	耐酸
acid sludge	酸性污泥	酸性污泥
acid tank	储酸罐	酸液儲槽
acid tank car	酸槽车,酸罐车	酸槽車
acidulant	酸化剂	酸化劑
acidulating tank	酸化槽	加酸槽
acidulation	酸化	加酸
acidulator	酸化器	加酸器
acorn oil	橡实油	橡實油
acousimeter	测声计	測聲計
acoustic control	声控	聲控
acoustic flow	声速流动	聲速流動
acoustics	声学	聲學
acoustic velocity	音速	音速
acrylic ester	丙烯酸酯	丙烯酸酯
acrylic ester rubber	丙烯酸酯橡胶	丙烯酸酯橡膠
acrylic resin	丙烯酸[酯]树脂	丙烯酸樹脂,壓克力樹脂
acrylonitrile	丙烯腈	丙烯腈
acrylonitrile-butadiene-styrene resin(ABS resin)	丙烯腈–丁二烯–苯乙烯树脂	丙烯腈–丁二烯–苯乙烯樹脂
actinometer	光强测定仪	光量計
actinomycetin	白放线菌素	放線菌素
actinomycin	放线菌素	放線菌黴素
activated adsorption	活性吸附	活化吸附
activated alumina	活性矾土	活性礬土
activated aluminum oxide	活性氧化铝	活性氧化鋁
activated biological filtration	活性生物过滤	活性生物過濾
activated carbon	活性炭	活性碳
activated carbon column	活性炭塔	活性碳塔
activated char	活性焦	活化[焦]炭
activated chemisorption	活化化学吸附	活化化學吸附
activated complex	活化络合物	活化錯合物,活化複合體

英 文 名	大 陆 名	台 湾 名
activated molecule	活化分子	活化分子
activated silica	活化硅石	活性矽石
activated sludge	活性污泥	活性污泥
activated sludge plant	活性污泥处理场	活性污泥處理場
activated sludge process	活性污泥法	活性污泥法
activation	活化[作用]	活化[作用]
activation analysis	活化分析	活化分析
activation cross section	活化截面	活化截面
activation energy	活化能	活化能
activation volume	活化体积	活化體積
activator	活化剂	活化劑
active agent	活性剂	活性劑
active alkali	活性碱	活性鹼
active alumina	活性矾土	活性礬土
active carbon	活性炭	活性碳
active catalyst	活性催化剂	活性觸媒
active catalytic center	活性催化中心	觸媒活性中心
active center	活性中心	活性中心
active-center complex	活性中心络合物	活性中心錯合物
active clay	活性黏土	活性黏土
active complex	活性络合物	活性錯合物
active dry yeast	活性干酵母	活性乾酵母
active oxygen	活性氧	活性氧
active point	活性点	活性點
active site	活性[部]位	活性[部]位
active-site theory	活性部位理论	活性[部]位理論
active solid	活性固体	活性固體
active sulfur	活性硫	活性硫
active surface area	活性表面积	活性表面積
activity	①活性 ②活度	①活性 ②活度
activity coefficient	活度系数	活性係數
activity correction factor	活性校正因子	活性校正因子
activity cracking	裂解活性	裂解活性
activity decay	活性衰变	活性衰變,活性衰退
activity distribution	活性分布	活性分佈
activity factor	活性因子	活性因子,活性因數
activity index	活性指数	活性指數
actual lift	实际上升	實際上升

英　文　名	大　陆　名	台　湾　名
actual plate	实际塔板	實際塔板
actual plate number	实际塔板数	實際[塔]板數
actuating signal	驱动信号	致動信號
actuator	驱动器	致動器
adaptive control	自适应控制	適應控制
adaptive decoupler	自适应去耦器	適應解偶器
adapt tuning	适应调谐	適應調諧
added value	附加值	附加價值
addition polymerization	加成聚合	加成聚合[作用]
addition reaction	加成反应	加成反應
additive	添加剂	添加劑
additive decolorization	加成脱色	加成脱色
additive law	加成定律	加成律
additive property	加成性	加成性
adduct	加合物	加成物
adductive crystallization	加合结晶	締合結晶
adhesion	黏附	黏著
adhesional wetting	黏附润湿	黏著潤濕
adhesion energy	黏附能	黏著能
adhesion force	黏附力	黏著力
adhesion strength	黏附强度	黏著強度
adhesion test	黏附试验	黏著試驗
adhesion work	黏附功	黏附功
adhesive	黏合剂	黏著劑
adhesive agent	黏合剂	黏著劑
adhesive attraction	黏合吸引	黏著吸引
adhesive formulation	黏合剂配方	黏著劑配方
adhesive strength	黏附强度	黏著強度
adhesive tape	胶带	膠帶
adhesive tension	黏附张力	黏著張力
adhesive test	黏附试验	黏著試驗
adiabatic apparatus	绝热仪器	絕熱裝置
adiabatic calorimeter	绝热式量热计	絕熱卡計
adiabatic chemical reaction	绝热化学反应	絕熱化學反應
adiabatic column	绝热塔	絕熱塔
adiabatic compression	绝热压缩	絕熱壓縮
adiabatic condition	绝热条件[状态]	絕熱狀況
adiabatic contraction	绝热收缩	絕熱收縮

英　文　名	大　陆　名	台　湾　名
adiabatic conversion	绝热转化	絕熱轉化
adiabatic cooling	绝热冷却	絕熱冷卻
adiabatic cooling curve	绝热冷却曲线	絕熱冷卻曲線
adiabatic cooling line	绝热冷却线	絕熱冷卻線
adiabatic drying	绝热干燥	絕熱乾燥
adiabatic evaporation	绝热蒸发	絕熱蒸發
adiabatic expansion	绝热膨胀	絕熱膨脹
adiabatic factor	绝热因子	絕熱因數
adiabatic flow	绝热流	絕熱流
adiabatic humidification	绝热增湿	絕熱增濕
adiabatic operating line	绝热操作线	絕熱操作線
adiabatic operation	绝热操作	絕熱操作
adiabatic plot	绝热图	絕熱圖
adiabatic process	绝热过程	絕熱程序
adiabatic reaction	绝热反应	絕熱反應
adiabatic reactor	绝热反应器	絕熱反應器
adiabatic rectification	绝热精馏	絕熱精餾
adiabatic saturation	绝热饱和	絕熱飽和
adiabatic saturation temperature	绝热饱和温度	絕熱飽和溫度
adiabatic system	绝热系统	絕熱系統
adiabatic temperature rise	绝热温升	絕熱溫升
adiabatic throttling device	绝热节流装置	絕熱節流裝置
adiabatic throttling process	绝热节流过程	絕熱節流程序
adiabatic twin calorimeter	双层绝热量热器	絕熱雙卡計
adiabatic wall	绝热壁	絕熱壁
adjacency matrix	相邻矩阵	相鄰矩陣
adjustable bearing	可调轴承	可調軸承
adjustable orifice	可调锐孔	可調孔口
adjustable parameter	可调参数	可調參數
adjustable weight	可调节权重	可調節權重
adjustment	调整	調整
administrative cost	管理成本	行政成本
admiralty test	干湿老化试验	乾濕老化試驗
admissible roughness	容许粗糙度	容許粗糙度
adsorbability	吸附能力	吸附度
adsorbate	吸附质	[被]吸附質
adsorbent	吸附剂	吸附劑
adsorber	吸附设备	吸附塔

英　文　名	大　陆　名	台　湾　名
adsorption	吸附	吸附[作用]
adsorption capacity	吸附容量	吸附容量
adsorption center	吸附中心	吸附中心
adsorption chart	吸附图表	吸附圖表
adsorption chromatography	吸附色谱法	吸附層析法
adsorption coefficient	吸附系数	吸附係數
adsorption column	吸附柱	吸附[管]柱
adsorption control	吸附控制	吸附控制
adsorption-desorption	吸附–脱附	吸附–脱附
adsorption equilibrium	吸附平衡	吸附平衡
adsorption filtration	吸附过滤	吸附過濾
adsorption isobar	吸附等压线	吸附等壓線
adsorption isostere	吸附等容线	吸附等容線
adsorption isotherm	吸附等温线	吸附等溫線
adsorption method	吸附法	吸附法
adsorption potential	吸附势	吸附勢
adsorption rate	吸附速率	吸附速率
adsorption site	吸附部位	吸附[部]位
adsorption surface	吸附表面	吸附表面
adsorption tube	吸附管	吸附管
adsorption wave	吸附波	吸附波
adsorptive agent	吸附剂	吸附劑
adsorptive capacity	吸附容量	吸附容量
adsorptivity	吸附性	吸附性
advanced treatment	高级处理	高級處理
advanced wastewater treatment	高级废水处理	高級廢水處理
adverse pressure gradient	逆压梯度	逆壓力梯度
aerated flow	掺气流	掺氣流,通氣流,曝氣流
aerated lagoon	曝气池	曝氣污水塘,通氣污水塘
aerated stabilization pond	曝气稳定化池	曝氣穩定化池,通氣穩定化池
aerated system	曝气系统	曝氣系統,通氣系統
aerated water	曝气水	氣溶水,汽水
aeration	①充气 ②曝气	①通氣 ②曝氣
aeration candle	通气管	通氣管
aeration device	充气装置	通氣裝置
aeration number	充气数	通氣數

英　文　名	大　陆　名	台　湾　名
aeration rate	充气率	通氣速率
aeration tank	充气槽	通氣槽
aeration wheel	充气轮	曝氣輪
aerator	①充气器 ②曝气机	①通氣機 ②曝氣機
aerial condenser	空气冷凝器	空氣冷凝器
aerial pollution	空气污染	空氣污染
aerobic bacteria	好氧菌,好气菌	好氧菌,嗜氧菌
aerobic biological oxidation	好氧生物氧化	好氣生物氧化[法]
aerobic culture	好氧培养	好氧培養,嗜氧培養
aerobic decomposition	好氧分解	需氧分解
aerobic digestion	好氧消化	需氧消化
aerobic fermentation	好氧发酵	需氧發酵[法]
aerobic oxidation	好氧氧化	需氧氧化[反應]
aerobic sludge digestion	好氧污泥消化	需氧污泥消化
aerobic treatment	好氧处理	好氧處理
aerodynamic diameter	空气动力直径	空氣動力直徑
aerodynamics	空气动力学	空氣動力學
aerometer	气体比重计	量氣計
aerosol	气溶胶	氣溶膠,霧劑
aerosol bomb	喷雾弹	噴霧彈
aerosol propellant	喷雾推进剂	噴霧[推進]劑,推噴劑
aerosol valve	喷雾阀	噴霧閥
affination	①[离心]精炼法 ②洗糖法	①糖精煉法 ②洗糖
affination number	洗糖数	洗糖數
affination value	洗糖值	洗糖值
affinity	亲和势	親和力
affinity adsorption	亲和吸附	親和吸附
affinity chromatography	亲和色谱[法]	親和[力]層析術
affinity interaction	亲和作用	親和[相互]作用
affinity labeling	亲和标记	親和標記法
affinity membrane	亲和膜	親和[薄]膜
affinity precipitation	亲和沉淀	親和[性]沈澱
affinity ultrafiltration	亲和超滤	親和超[過]濾
after burner	后燃装置	後燃裝置
after burning	后期燃烧	後燃燒
after contraction	后收缩	後收縮
after cooler	后置冷却器	後冷卻器

英　文　名	大　陆　名	台　湾　名
after fermentation	后发酵	後發酵
after filter	后过滤器	後[過]濾器
after filtration	后过滤	後過濾
after hardening	后硬化	後硬化
after product	后产物	後產物
after purification	后净化	後純化
after shrinkage	后收缩	後收縮
after treatment	后处理	後處理
after vulcanization	后硫化	後硫化
agar	琼脂,洋菜	瓊脂,洋菜
agar diffusion	琼脂扩散	瓊脂擴散
agar diffusion technology	琼脂扩散技术	瓊脂擴散技術
agar electrophoresis	琼脂电泳	瓊脂電泳
agar filtration	琼脂过滤	瓊脂過濾
agar medium	琼脂培养基	瓊脂培基
agarose	琼脂糖	瓊脂糖
agarose gel	琼脂糖凝胶	瓊脂糖凝膠
age distribution	年龄分布	年齡分佈
age distribution function	年龄分布函数	年齡分佈函數
age hardening	时效硬化	時效硬化
agglomerate	团聚物	團聚物
agglomerating value	团聚值	團聚值
agglomeration	团聚	團聚
aggregate	聚集体	聚集體
aggregation	聚集	聚集[作用]
aggregation number	聚集数	聚集數
aggregative fluidization	聚式流态化	聚集流體化
aging	老化	老化
aging equipment	老化装置	老化裝置
aging machine	熟化器	熟化器
aging oven	老化箱	熟化烘箱
aging test	老化试验	老化試驗
aging test machine	老化试验机	老化試驗機
agitated crystallizer	搅拌结晶器	攪拌結晶器
agitated drier	搅拌干燥器	攪拌乾燥器
agitated reactor	搅拌反应器	攪拌反應器
agitated vessel	搅拌釜	攪拌槽
agitating	搅拌	攪拌

英　文　名	大　陆　名	台　湾　名
agitating tank	搅拌槽	攪拌槽
agitator	搅拌器	攪拌器
agitator drier	搅拌干燥器	攪拌乾燥器
agricultural waste	农业废物	農業廢料
AI(= artificial intelligence)	人工智能	人工智慧
air agitation	充气搅拌	空氣攪拌,氣動攪拌
air agitator	充气搅拌器	空氣攪拌器,氣動攪拌器
air bag	气囊	氣囊
air binding	空气黏合	氣結
air blasting	鼓风	鼓風
air blower	鼓风机	鼓風機
air bubble system	气泡系统	氣泡[發生]系統
air cap	气罩	氣罩
air compressor	空气压缩机	空氣壓縮機
air conditioner	空调器	空調[機]
air conditioning unit	空调机组	空調機組
air conveyor	气力输送装置	氣動運送機
air-cooled heat exchanger	空冷换热器	氣冷式熱交換器
air cooled jacket	气冷夹套	氣冷[夾]套
air cooler	空气冷却器	空氣冷卻器
air cooling	空气冷却	空氣冷卻
air diffuser	空气扩散器	空氣擴散器
air discharge orifice	排气锐孔	排氣銳孔
air displacement pump	排气泵	排氣泵
air displacement system	排气系统	排氣系統
air distillation	常压蒸馏	常壓蒸餾
air dryer	空气干燥器	空氣乾燥機
air drying	风干	風乾
air-drying loss	风干失重	風乾損失
air drying tower	空气干燥塔	空氣乾燥塔
air ejector	空气喷射器	空氣噴射器
air elutriation	空气扬析	空氣揚析,空氣淘析
air-entraining agent	空气雾沫剂	空氣霧沫劑
air equilibrium distillation	常压平衡蒸馏	常壓平衡蒸餾
air evaporation	空气蒸发	空氣蒸發[法]
air expander	空气膨胀机	空氣膨脹機
air film	空气膜	空氣膜

英 文 名	大 陆 名	台 湾 名
air filter	空气过滤器	空氣過濾器
air filtration	空气过滤	空氣過濾
air flow	空气流	空氣流動
air flow meter	空气流量计	空氣流量計
air-fuel ratio	空气–燃料比	空氣–燃料比
air furnace	空气炉	空氣爐
air gage	气压计	氣壓計
air gap	气隙	空氣隙
air gas	风煤气	風煤氣
air heater	空气加热器	空氣加熱器
air hose	空气软管	空氣軟管
air-lift	气升	氣升
air-lift agitation	气升搅拌	氣升攪拌
air-lift agitator	气升搅拌器	氣升攪拌器
air-lift pump	气升泵	氣升泵
air lock	气闸	氣閘
air meter	气量计	氣量計
air mixing depth	空气混合深度	空氣混合深度
air nozzle	空气喷嘴	空氣噴嘴
air-operated valve	气动阀	氣動閥
air permeability	透气性	透氣性
air pollutant	空气污染物	空氣污染物
air pollution	空气污染	空氣污染
air pollution abatement	空气污染减量	空氣污染減量
air pollution control	空气污染控制	空氣污染控制
air preheater	空气预热器	空氣預熱器
air pressure	空气压力	空氣壓力
air pressure drop	空气压力降	空氣壓力降
air pressure dynamics	空气动力学	氣壓動態[學]
air pressure gage	气压计	氣壓計
air pump	气泵	氣泵
air purge	空气吹扫	空氣吹驅
air quality	大气质量	空氣品質
air quality index	大气质量指数	空氣品質指數
air regulator	空压调节器	空氣調節器
air release valve	排气阀	排氣閥,釋氣閥
air relief valve	泄压阀	排氣閥,釋氣閥
air ring	空气冷却环	空氣冷卻環

英 文 名	大 陆 名	台 湾 名
air saturator	空气饱和器	空氣飽和器
air scrubber	空气洗涤器	空氣洗滌器
air separation	空气分离	空氣分離
air separation tank	空气分离罐	空氣分離槽
air separation tower	空气分离塔	空氣分離塔
air separator	空气分离器	空氣分離器
air shrinkage	风干收缩	風乾收縮
air slide	气动滑送	氣動滑送
air sparger	空气鼓泡器	空氣鼓泡器,空氣噴佈器
air sterilizer	空气灭菌器	空氣滅菌器
air stream	气流	氣流
air suction filter	空气吸滤器	空氣吸濾器
air suction rate	吸气速率	吸氣速率
air supply	空气供应	空氣供應
air-swept mill	气吹磨	風掃磨
air washing	空气洗涤	空氣洗滌
alarm	报警器	警報器
albite	钠长石	鈉長石
alcoholase	醇酶	醇酶,酒精酶
alcoholysis	醇解	醇解
aldol	羟醛	醛醇
algae	藻类	藻類
algin	藻胶	藻素
alginate fiber	藻酸纤维	藻酸纖維
algin fiber	藻酸纤维	藻酸纖維
algorithm	算法	演算[法]
algorithmic synthesis technique	算法合成技术	算法合成技術
alignment chart	列线图	列線圖
aliphatic hydrocarbon	脂肪族烃	脂族烴
alizarine dyestuff	茜素染料	茜素染料
alizarine lake	茜素色淀	茜素色澱
alizarine oil	太古油	茜草油
alkali	碱	鹼
alkali cellulose	碱纤维素	鹼纖維素
alkali lignin	碱木质素	鹼[溶]木質素
alkalimeter	碱量计	鹼計
alkaline flooding	碱液泛	鹼液氾流[法]

英　文　名	大　陆　名	台　湾　名
alkaline glaze	碱釉	鹼性釉
alkaline pulping	碱法制浆	鹼式製漿
alkalinity	碱度	鹼度
alkali process	碱法	鹼法
alkali resistance	抗碱性	抗鹼性,耐鹼性
alkali silicate	碱性硅酸盐	鹼矽酸鹽
alkallimetry	碱测定法	鹼定量法
alkaloid	生物碱	生物鹼
alkyd resin	醇酸树脂	醇酸樹脂
alkylation	烷基化	烷[基]化
alkylbenzene sulfonate	烷基苯磺酸盐	烷[基]苯磺酸鹽
allowance	①容差 ②裕量	容許量
alloy	合金	合金
alloy steel	合金钢	合金鋼
almond oil	杏仁油	杏仁油
aloxite	铝砂	鋁砂
alternate operating column	交替操作塔	交替操作塔
alternating copolymerization	交替共聚合	交替共聚合[反應]
alternating system	交替系统	交替系統
alternation	交替	交替[現象]
alternative copolymer	交替共聚物	交替共聚物
alternative hypothesis	备择假设	備擇假設,對立假設
alternative investment	交替投资	交替投資
alternative operation	交替操作	交替操作
alum	明矾	明礬
alumina	矾土	礬土
aluminum foil	铝箔	鋁箔
aluminum mordant	铝媒染剂	鋁媒染劑
aluminum pipe	铝管	鋁管
aluminum soap	铝皂	鋁皂
alum tanning	矾鞣	明礬鞣製
alundum	刚铝石,刚玉	剛鋁石,剛玉
alundum filter	刚铝石过滤器,刚玉过滤器	剛鋁石過濾器,剛玉過濾器
Amagat law	阿马加定律	愛瑪哥定律
amalgam	汞齐	汞齊
amber	琥珀	琥珀
ambient pressure	环境压力	環境壓力

英 文 名	大 陆 名	台 湾 名
ambient temperature	环境温度	環境溫度
amination	胺化[作用]	胺化[作用]
amino acid	氨基酸	胺基酸
amino resin	氨基树脂	胺基樹脂
ammonia extraction process	氨萃取过程	氨萃取法，氨萃取程序
ammonia liquor	氨水	氨液
ammonia reactor	氨反应器	氨反應器
ammonia saturator	氨饱和器	氨飽和器
ammonia-soda process	氨碱法	氨鹼法
ammonia still	蒸氨塔	氨蒸餾塔
ammonia synthesis	氨合成	氨合成
ammoniated brine	氨盐水	氨化鹽水
ammoniated fertilizer	氨化肥料	氨化肥料
ammoniated superphosphate	含铵过磷酸钙	氨化過磷酸鈣
ammoniation	氨化[作用]	氨化[作用]
ammonia washer	氨洗涤器	洗氨器
ammonia water	氨水	氨水
ammonium alum	铵矾	銨礬
ammonium soap	铵皂	銨皂
ammonolysis	氨解[作用]	氨解
amorphism	无定形	無定形
amorphous carbon	无定形碳	無定形碳
amorphous polymer	无定形聚合物	無定形聚合物
amortization	分期偿还	分期償還
amosite	铁石棉	鐵石綿
amperometric titration	电流滴定[法]	電流滴定
amphotericin	两性霉素 B	雙性[殺]黴素 B
ampicillin	氨苄青霉素	胺苄青黴素
amplification	放大	放大
amplifier	放大器	放大器
amplitude	振幅	振幅
amplitude ration	振幅比	振幅比
amylase	淀粉酶	澱粉酶
amylolysis	淀粉水解	澱粉水解
anaerobe	厌氧菌	厭氧菌
anaerobic bacteria	厌氧菌	厭氧菌
anaerobic breakdown	厌氧分解	厭氧分解
anaerobic corrosion	厌氧腐蚀	厭氧腐蝕

英　文　名	大　陆　名	台　湾　名
anaerobic culture	厌氧培养	厭氧培養
anaerobic decomposition	厌氧分解	厭氣分解
anaerobic digester	厌氧消化槽	厭氧消化槽
anaerobic digestion	厌氧消化	厭氧消化
anaerobic fermentation	厌氧发酵	厭氧發酵[法]
anaerobic oxidation	厌氧氧化	厭氣氧化[反應]
anaerobic sludge	厌氧污泥	厭氧污泥
anaerobic sludge digestion	厌氧污泥消化	厭氧污泥消化
anaerobic treatment	厌氧处理	厭氧處理
analog computation	模拟计算	類比計算[法]
analog computer	模拟计算机	類比計算機
analog simulation	模拟仿真	類比模擬
analogy	类比	類比
analogy system	类比系统	類比系統
analogy theory	类比理论	類比理論
analysis mode	分析模式	分析模式
analysis of reaction network	反应网络分析	反應網路分析
analytical balance	分析天平	分析天平
analytical distillation	分析蒸馏	分析蒸餾
analytical method	解析法	解析法
analytical solution of group contribution method	基团[贡献]解析法	基團[貢獻]解析法
analyzer	分析器	分析器
anchorage-dependent cell	贴壁细胞	貼壁細胞,固面依附細胞
anchor agitator	锚式搅拌器	錨式攪拌器
andalusite	红柱石	紅柱石
anemometer	风速计	風速計
anemoscope	风向仪	風向計
aneroid barometer	无液气压计	無液氣壓計
aneroid flowmeter	无液流量计	無液流量計
aneroid manometer	无液压力计	無液壓力計
anesthetic	麻醉剂	麻醉劑
angle factor	角系数	角度因數
angle of arrival	到达角	到達角
angle of attack	冲角	攻角,衝角
angle of deflection	偏离角	偏離角,偏向角
angle of departure	出发角	發射角

英　文　名	大　陆　名	台　湾　名
angle of fall	落角	落角
angle of inclination	倾角	傾[斜]角
angle of internal friction	内磨擦角	内摩擦角
angle of nip	夹角	挾角
angle of release	释放角	釋放角
angle of repose	休止角	静止角
angle of slide	滑动角	滑動角
angle of spatula	刮铲角	刮勺角
angle of wall friction	壁摩擦角	壁摩擦角
angle valve	角阀	角閥
angular deformation	角[向]变形	角[度]形變
angular distortion	角畸变	角畸變
angular frequency	角频率	角頻率
angular impulse	角冲量	角衝量
angular momentum	角动量	角動量
angular scattering pattern	角散射	角散射
angular velocity	角速度	角速度
aniline black	苯胺黑	苯胺黑
aniline blue	苯胺蓝	苯胺藍
aniline dye	苯胺染料	苯胺染料
aniline green	苯胺绿	苯胺綠
aniline orange	苯胺橙	苯胺橙
aniline point	苯胺点	苯胺點
aniline red	苯胺红	苯胺紅
aniline violet	苯胺紫	苯胺紫
animal glue	动物胶	動物膠
animal hide	动物皮	動物皮
animal manure	动物厩肥	動物廐肥
animal oil	动物油	動物油
anion	阴离子	陰離子
anion exchanger	阴离子交换剂	陰離子交換劑
anion exchange resin	阴离子交换树脂	陰離子交換樹脂
anionic polymerization	阴离子[催化]聚合	陰離子聚合[作用]
anionic starch	阴离子淀粉	陰離子澱粉
anionic surface-active agent	阴离子型表面活性剂	陰離子界面活性劑
anionic surfactant	阴离子型表面活性剂	陰離子界面活性劑
anisotropic absorption	各向异性吸收	各向異性吸收,異向性 吸收

英　文　名	大　陆　名	台　湾　名
anisotropic membrane	各向异性膜	各向異性膜,異向性膜
anisotropy	各向异性	各向異性,異向性
ANN(=artificial neural network)	[人工]神经网络	類神經網路
annealing	退火	退火
annealing point	退火点	退火點
annuity table	年金表	年金表
annular flow	环状流	環狀流
annular reactor	环状反应器	環狀反應器
annular space	环隙	環隙
anode	阳极	陽極
anode oxidation	阳极氧化	陽極氧化[反應]
anode potential	阳极电位	陽極電位
anodic corrosion	阳极腐蚀	陽極腐蝕
anodic reaction	阳极反应	陽極反應
anomalous viscosity	反常黏度	反常黏度
anthophyllite	直闪石	直閃石
anthracene oil	蒽油	蒽油
anthracite coal	无烟煤	無煙煤
anthraquinone dye	蒽醌染料	蒽醌染料
anthraquinone vat dye	蒽醌还原染料	蒽醌缸染料
antibiotic	抗生素	抗生素
anti-blocking agent	防结块剂	防結塊劑
antibody	抗体	抗體
anticipatory control	预先控制	預先控制
anticoagulant agent	抗凝剂	抗凝劑
anticoagulant effect	抗凝效应	抗凝效應
anti-competitive inhibition	反竞争性抑制	反競爭抑制[作用]
anticorrosion	防腐蚀	防[腐]蝕
anticorrosive agent	防蚀剂	防蝕劑
anticorrosive paint	防蚀漆	防蝕漆
antifoam agent	防沫剂,消泡剂	防沫劑,消泡劑
antifreeze	防冻	防凍劑
antifreeze mixture	抗冻混合物	防凍混合物
antifreezer	防冻器	防凍器
antifreezing agent	防冻剂	防凍劑
antigen	抗原	抗原
antiknock	抗爆[震]	抗[爆]震
antiknock compound	抗爆[震]化合物	抗[爆]震化合物

英　文　名	大　陆　名	台　湾　名
antiknock value	抗爆[震]值	抗[爆]震值
antimicrobial action	抗微生物作用	抗微生物作用
antimicrobial activity	抗微生物活性	抗微生物活性
antimicrobial drug	抗微生物药	抗微生物藥
antimonial soap	锑皂	銻皂
antimony electrode	锑电极	銻電極
antimony white	锑白	銻白
antimony yellow	锑黄	銻黃
antioxidant	抗氧[化]剂	抗氧化劑
antiozonant	抗臭氧剂	抗臭氧[化]劑
antirust grease	防锈脂	防鏽脂
antisoftener	防软[化]剂	防軟[化]劑
antistaining agent	防污染剂	防污劑
antistatic agent	抗静电剂	去靜電劑
anti-surge control	防喘振控制	抗激變控制
anti-windup	抗卷绕	抗繞緊
Antoine equation	安托万方程	安東尼方程式
apatite	磷灰石	磷灰石
apoenzyme	脱辅基酶	脫輔基酶
apparent activation energy	表观活化能	視活化能
apparent activity	表观活度	視活性
apparent composition	表观组成	視組成
apparent density	表观密度	視密度
apparent diffusivity	表观扩散系数	視擴散係數
apparent gravity	表观重力	視重力
apparent mass	表观质量	視質量
apparent order	表观级数	視階
apparent porosity	表观孔隙率	視孔隙度
apparent purity	表观纯度	視純度
apparent shear rate	表观剪切率	視剪[切]率
apparent shear stress	表观剪切应力	視剪[切]應力
apparent solubility	表观溶解度	視溶解度
apparent specific gravity	表观比重	視比重
apparent specific heat	表观比热	視比熱
apparent stress	表观应力	視應力
apparent viscosity	表观黏度	視黏度
apparent volume	表观体积	視體積
apparent weight	表观重量	視重量

英　文　名	大　陆　名	台　湾　名
apple seed oil	苹果籽油	蘋果籽油
application	应用	應用
applied chemistry	应用化学	應用化學
applied microbiology	应用微生物学	應用微生物學
applied thermodynamics	应用热力学	應用熱力學
appricot kernel oil	杏仁油	杏仁油
approximate method	近似法	近似法
approximate solution	近似解	近似解
approximation	近似	近似
approximation formula	近似式	近似式
apron carrier	裙式输送机	裙式輸送機
apron conveyor	裙式传送机	裙式運送機
apron feeder	裙式给料机	裙飼機
aprotic solvent	非质子溶剂	非質子性溶劑
aqua ammonia	氨水	氨水
aqua fortis	硝酸	硝酸
aqua regia	王水	王水
aqueous ammonia	氨水	氨水
aqueous extract	水萃取物	水相萃取物
aqueous solution	水溶液	水溶液
aqueous two-phase extraction	双水相萃取	雙水相萃取,兩水相萃取
aqueous two-phase system	双水相系统,双水相体系	雙水相系統,兩水相系統
aquolization process	加水裂化过程	加水裂化過程
araeometer	[液体]比重计	液體比重計
arc furnace	电弧炉	電弧爐
arch brick	拱砖	拱磚
arc resistance	耐电弧性	耐電弧性
arctic sperm oil	北极鲸蜡油	北極鯨蠟油,抹香鯨油
area-averaged equation	面积平均方程式	面積平均方程式
area flowmeter	面积流量计	面積流量計
area meter	面积计	面積計
area type flowmeter	面积流量计	面積流量計
arithmetic mean	算术平均	算術平均
arithmetic mean temperature difference	算术平均温差	算術平均溫差
arithmetic mean value	算术平均值	算術平均值
armature	电枢	電樞

英 文 名	大 陆 名	台 湾 名
aromatics	芳[香]烃	芳[香]烃,芳香族[化合物]
aromatization	芳构化	芳化[作用]
Arrhenius equation	阿伦尼乌斯方程	阿瑞尼斯方程式
arrow diagram	箭头[流程]图	箭頭[流程]圖
arsenical insecticide	砷杀虫剂	砷殺蟲劑
arsenical rodenticide	砷杀鼠剂	砷殺鼠劑
artificial aging	人工老化	人工老化
artificial drier	人工干燥器	人工乾燥器
artificial fertilizer	人工肥料	人造肥料
artificial fiber	人造纤维	人造纖維
artificial fuel	人造燃料	人造燃料
artificial intelligence(AI)	人工智能	人工智慧
artificial kidney	人工肾脏	人造腎臟
artificial leather	人造革	人造皮
artificial neural network(ANN)	[人工]神经网络	人工類神經網路
artificial organ	人工器官	人造器官
artificial sweetener	人工甜味剂	人工甜味料,人工甘味劑
art-printing paper	艺术印刷纸	美術印刷紙,銅版紙
asbestos	石棉	石棉,石綿
asbestos filter	石棉过滤器	石綿過濾器,石棉過濾器
ascending chromatography	上行色谱法	上升層析[法]
ascending-descending chromatography	升降色谱法	升降層析[法]
ascending kiln	上行窑	上行窯,階梯窯
ascending pipe	上行管	上升管
aseptic operation	无菌操作	無菌操作
aseptic seal	无菌密封	無菌密封
ash	灰分	灰分
ash analysis	灰分分析	灰分分析
ash diffusion control	灰分扩散控制	灰分擴散控制
ashless filter paper	无灰滤纸	無灰濾紙
ash-seed oil	槐子油	槐子油
ash zone	灰区	灰區
aspect number	①高宽比 ②形状数	高寬比
aspect ratio	①长宽比 ②高径比	長寬比
aspergillus	曲菌	麴菌

英　文　名	大　陆　名	台　湾　名
asphalt	沥青	瀝青,柏油
asphalt paint	沥青漆	瀝青漆,柏油漆
asphaltum	［石油］沥青	地瀝青
aspirator	吸气器	抽氣器
assembly line	装配线	裝配線
assembly of interacting particles	非独立粒子系集	交互作用粒子系集
assembly of localize particles	定域粒子系集	定域粒子系集
asset accounting	资产会计	資產會計
asset type	资产型式	資產型式
assimilation	同化	同化
associated solution model	缔合溶液模型	締合溶液模型
asymmetrical flow	不对称流动	不對稱流動
asymmetric membrane	非对称膜	非對稱膜
asymptote	渐近线	漸近線
asymptotic plot	渐近线图	漸近線圖
asymptotic solution	渐近解	漸近解
asymptotic stability	渐近稳定性	漸近穩定性
atactic polymer	无规［立构］聚合物	雜排聚合物
athermal solution	无热溶液	無熱溶液
atmolysis	微孔［透壁］分气法	細孔［透壁］氣體分離法
atmosphere	①大气 ②大气压	①大氣 ②大氣壓(壓力單位)
atmospheric condenser	空气冷凝器	空氣冷凝器
atmospheric cooling tower	空气冷却塔	空氣冷卻塔
atmospheric diffusion	大气扩散	大氣擴散
atmospheric drier	常压干燥器	常壓乾燥器
atmospheric flash tower	常压闪蒸塔	常壓驟餾塔
atmospheric fractionator	常压分馏塔	常壓分餾塔,常壓分餾器
atmospheric mashing	常压芽浆糖化	常壓製醪
atmospheric operation	常压操作	常壓操作
atmospheric pressure	大气压	大氣壓力,常壓
atmospheric steam	常压蒸汽	常壓蒸汽
atometer	蒸发速度测定器	蒸發速率測定器
atomic heat capacity	原子热容	原子熱容［量］
atomic interaction	原子相互作用	原子相互作用
atomic mass	原子质量	原子質量

英　文　名	大　陆　名	台　湾　名
atomic number	原子序数	原子序
atomic spacing	原子间距	原子間距
atomic weight	原子量	原子量
atomic weight table	原子量表	原子量表
atomization	雾化	霧化
atomized oil	雾化油	霧化油
atomizer	雾化器	霧化器
atomizing air	雾化空气	霧化空氣
atomizing burner	雾化燃烧器	霧化燃燒器
atomizing spray nozzle	雾化喷嘴	霧化噴嘴
atomizing steam	雾化蒸汽	霧化蒸汽
attenuation	衰减	衰減[作用]
attenuation effect	衰减效应	衰減效應
attenuation factor	衰减因子	衰減因數
attenuator	衰减器	衰減器
attractant	引诱剂	誘[引]劑
attractive force	吸引力	吸引力
attractive potential	吸引势	吸引勢
attrition	磨损	磨耗
attrition mill	碾磨机	磨碎機
attrition of catalyst	催化剂磨损	觸媒磨耗
attrition resistance	抗磨损	抗磨損
attrition resistant	抗磨剂	抗磨劑
auger machine	螺旋挤出机	鑽探機
aureomycin	金霉素	金黴素
autoacceleration	自加速作用	自加速[作用]
auto-agglutination	自胶合	自膠合[作用]
autocatalysis	自催化	自催化[作用]
autoclave	高压釜	高壓釜
autoclave expansion test	高压釜膨胀试验	高壓膨脹試驗
auto-coagulation	自凝聚	自凝聚[作用]
autogenous mill	自磨机	自磨機
autoignition temperature	自燃温度	[自]燃點
auto-induced reaction	自诱导反应	自誘發反應
autolysis	自分解	自分解
autolytic lipase	自溶脂肪酶	自[分]解脂酶
auto-manual switch	自动–手控开关	自動手動開關
automatic control	自动控制	自動控制

英　文　名	大　陆　名	台　湾　名
automatic controller	自动控制器	自動控制器
automatic control system	自动控制系统	自動控制系統
automatic feed	自动进料	自動進料
automatic feeder	自动加料器	自動進料器
automatic filter	自动过滤器	自動過濾器
automatic process controller	自动过程控制器	自動程序控制器
automatic regulator	自动调节器	自動調節器
automatic thermostat	自动恒温器	自動恆溫器
autonomous system	自治系统	自律系統
auto-polymerization	自[动]聚合	自聚合[作用]
auto recorder	自动记录器	自動記錄器
auto-selector	自动选择器	自動選擇器
auto-selector control	自动选择器控制	自動選擇器控制
auto-stabilization	自稳定	自穩定
autothermal reaction	自热反应	自熱反應
auto-vulcanization	自硫化	自硫化[作用]
autoxidation	自氧化	自氧化[作用]
auxiliary absorber	辅助吸收器	輔助吸收器
auxiliary electrode	辅助电极	輔助電極
auxochrome	助色团	助色團,助色基
availability	有效能	有效能
availability analysis	有效能分析	有效能分析
available chlorine	有效氯	有效氯[量]
average boiling point	平均沸点	平均沸點
average deviation	平均偏差	平均偏差
average error	平均误差	平均誤差
average molecular weight	平均分子量	平均分子量
average residence time	平均停留时间	平均滯留時間
average specific filtration resistance	平均过滤比阻	平均過濾比阻
average temperature difference	平均温差	平均溫差
average velocity	平均速度	平均速度
averaging level control	均匀液位控制	平均[化]液位控制
aviation fuel	航空燃料	航空燃料
aviation gasoline	航空汽油	航空汽油
axial angle	轴角	軸角
axial dispersion	轴向分散	軸向分散
axial dispersion coefficient	轴向分散系数	軸向分散係數
axial flow	轴向流	軸[向]流

英 文 名	大 陆 名	台 湾 名
axial-flow compressor	轴流式压缩机	軸[向]流壓縮機
axial-flow fan	轴流风机	軸[向]流風扇
axial flow pump	轴流泵	軸流泵
axial-flow reactor	轴流反应器	軸流反應器
axial mixing	轴向混合	軸向混合
axial symmetric flow	轴对称流	軸向對稱流動
axial symmetry	轴向对称	軸向對稱
axial velocity	轴向速度	軸向速度
axis of symmetry	对称轴	對稱軸
azeotrope	共沸物	共沸液
azeotrope tower	共沸塔	共沸塔
azeotropic composition	共沸组成	共沸組成
azeotropic copolymer	恒组分共聚物	恆組成共聚物
azeotropic copolymerization	恒组分共聚	恆組成共聚[作用]
azeotropic distillation	共沸蒸馏	共沸蒸餾
azeotropic drying	共沸干燥	共沸乾燥
azeotropic line	共沸线	共沸線
azeotropic mixture	共沸混合物	共沸混合物
azeotropic parameter	共沸参数	共沸參數
azeotropic phenomenon	共沸现象	共沸現象
azeotropic point	共沸点	共沸點
azeotropic process	共沸[蒸馏]过程	共沸程序
azeotropic property	共沸性质	共沸性質
azeotropic range	共沸范围	共沸範圍
azeotropic state	共沸状态	共沸狀態
azeotropic system	共沸系统	共沸系統
azeotropic temperature	共沸温度	共沸溫度
azeotropy	共沸[性]	共沸[現象]
azo component	偶氮成分	偶氮成分
azo dye	偶氮染料	偶氮染料
azoic dye	偶氮染料	不溶性偶氮染料
azoic dyeing	偶氮染色	不溶性偶氮染色
azotobacteria	固氮菌	固氮[細]菌
azotometer	定氮仪	氮量計
azurite	蓝铜矿	藍銅礦

B

英 文 名	大 陆 名	台 湾 名
bacillus	杆菌	桿菌
Bacillus subtilis	枯草杆菌	枯草桿菌
bacitracin	杆菌肽	枯草桿菌肽
back diffusion	反向扩散	逆擴散
backfiller	回填机	回填機
back fire	回火	回火,逆火
backflow	回流	逆流
backflushing	反洗	反洗,逆洗
backlash	齿隙	背隙
backmix reactor	返混反应器	逆混反應器
back pressure	背压	背壓
back pressure-valve	背压阀	背壓閥
back-propagation algorithm	反向传播算法	反向傳播演算法
backward feed	逆向进料	逆向進料
backward reaction	逆向反应	逆[向]反應
backwash rate	反洗速率	逆洗速率
bacteria	细菌	細菌
bacteria bed	细菌床	細菌床
bacteria content	细菌含量	細菌含量
bacteria count	细菌数	菌數
bacteria filter	细菌过滤器	濾菌器
bacterial fermentation	细菌[性]发酵	細菌[性]發酵
bacterial oxidation	细菌氧化	細菌氧化[反應]
bacteriologic filtration	细菌过滤	除菌過濾
baffle	挡板	擋板
baffle chamber	挡板室	擋板室
baffle column	挡板塔	擋板塔
baffle column mixer	挡板塔混合器	擋板塔混合器
baffled evaporator	折流蒸发器	擋板蒸發器
baffle mark	挡板痕	擋板痕
baffle mixer	挡板混合器	擋板式混合器
baffle nozzle system	挡板喷嘴系统	擋板噴嘴系統
baffle pan	挡板塔盘	擋板盤

英 文 名	大 陆 名	台 湾 名
baffle plate	防冲板	擋板
baffle plate column mixer	挡板混合塔	擋板混合塔
baffle plate tower	挡板塔	擋板塔
baffle separation	折流分离	擋板分離
bagasse	蔗渣	蔗渣
bagasse board	蔗[渣]板	蔗[渣]板
bagasse digester	蔗渣蒸煮罐	蔗渣蒸煮器
bag filter	袋滤器	袋[狀]濾器
bag filter separator	袋式过滤分离器	袋式過濾分離器
bag molding	袋式成型	袋式成型
bag type dust collector	袋式集尘器	袋式集塵器
bakelite	酚醛树脂	酚醛樹脂,電木
baking furnace	烘炉	烘爐
baking oven	烘箱	烘箱
baking soda	焙用碱	焙鹼
baking varnish	烤清漆	烤清漆
balanced bellow valve	均衡伸缩阀	均衡伸縮閥
balanced growth	均衡生长	均衡生長
balanced value	均衡值	均衡值
balance equation	平衡方程式	平衡方程式
balance fertilizer	均衡肥料	均衡肥料
balance sheet	资产负债表	資產負債表,平衡表
balancing motor	平衡电动机	平衡馬達
balancing ring	平衡环	平衡環
ball bearing	滚珠轴承	球軸承
ball clay	球土	球土
balling	成球	成球,球狀化
ball mill	球磨机	球磨機
ball mill reactor	球磨反应器	球磨反應器
ball mill refiner	球磨精制机	球磨精製機
Banbury mixer	[班伯里]密[闭式混]炼机	密閉式混煉機
band conveyor	皮带运输机	帶運機
banded matrix	带状矩阵	帶式矩陣
bandwidth	带宽	帶寬
bang-bang control	继电控制	開關式控制
bank-note paper	钞票纸	鈔票紙
bare pipe	裸管	裸管

英　文　名	大　陆　名	台　湾　名
bare surface	裸面	裸面
bare thermocouple	裸热电偶	裸熱電偶
bare tube	裸管	裸管
barite	重晶石	重晶石
barium alum	钡矾	鋇礬
barking drum	桶式去皮机	桶式去皮機
barometer	气压计	氣壓計
barometer tube	气压管	氣壓管
barometric condenser	大气冷凝器	氣壓冷凝器
barometric discharge pipe	大气排放管	氣壓排洩管
barometric effect	气压效应	氣壓效應
barometric equation	气压方程式	氣壓方程式
barometric height	气压高度	氣壓高度
barometric leg	大气腿	氣壓眞空柱
barometric leg pump	气压柱泵	氣壓泵
barometric pipe	大气腿	氣壓管
barometric pressure	大气压	大氣壓力
barometrograph	气压记录器	氣壓記錄器
barrel mill	桶磨机	桶磨機
barrel sprayer	圆桶喷雾器	圓桶噴霧器
barrier function	闸函数	閘函數
base	①碱 ②基质	①鹼 ②鹼基
base catalysis	碱催化	鹼催化
base-level override	基准位超越	基準位超越
base-level resonance	基准位共振	基準位共振
basic catalyst	碱性催化剂	鹼[性]觸媒
basic dye	碱性染料	鹼性染料
basic firebrick	碱性耐火砖	鹼性耐火磚
basic solid	碱性固体	鹼性固體
basket centrifuge	篮式离心机	籃式離心機
basket elevator	篮式升降机	籃式升降機
basket extractor	篮式提取器	籃式萃取器
basket filter	篮式过滤器	籃式[過]濾器
basket reactor	吊篮式反应器	籃式反應器
basket-type evaporator	悬筐蒸发器	籃式蒸發器
batch agitator	间歇搅拌器	批式攪拌器
batch centrifuge	间歇离心机	批式離心機
batch charger	间歇加料机	批式加料機

英 文 名	大 陆 名	台 湾 名
batch coke still	间歇焦化蒸馏器	批式焦化蒸餾器
batch control	间歇控制	批式控制
batch crystallization	间歇结晶	批式結晶
batch crystallizer	间歇结晶器	批式結晶器
batch culture	分批培养	批式培養[菌]
batch distillation	间歇蒸馏	批式蒸餾
batch drier	间歇干燥器	批式乾燥器
batch extraction	间歇萃取	批式萃取
batch extractor	间歇萃取器	批式萃取器
batch feeder	间歇进料器	批式進料器
batch filter	间歇过滤器	批式過濾機
batch filtration	间歇过滤	批式過濾
batch fractionation	间歇分馏	批式分餾
batch mixer	间歇混合器	批式混合器
batch number	批号	批號
batch operation	间歇操作	批式操作
batch process	间歇过程	批式程序
batch processing	间歇加工	批式加工[法]
batch production	间歇生产	批式生產
batch purification	间歇净化	批式純化
batch reactor	间歇反应器	批式反應器
batch rectification	间歇精馏	批式精餾
batch settling	间歇沉降	批式沈降
batch size	批量大小	批式產能
batch still	间歇蒸馏器	批式蒸餾器
batch still distillation	间歇釜式蒸馏	批式蒸餾
batch system	间歇系统	批式系統
batch vaporization	间歇汽化	批式汽化
bath	浴	①浴 ②槽
battery of continuous stirred tank reactors (battery of CSTRs)	连续搅拌反应器组	連續攪拌反應器組
battery of CSTRs（=battery of continuous stirred tank reactors）	连续搅拌反应器组	連續攪拌反應器組
bauxite	铝矾土	鋁礬土
bauxitic clay	铝质黏土	鋁氧黏土
bayberry oil	月桂籽油	月桂子油
bay oil	月桂油	月桂油
bearing	轴承	軸承

英　文　名	大　陆　名	台　湾　名
bearing ball	轴承滚珠	軸承球
bearing box	轴承箱	軸承箱
bearing bracket	轴承托座	軸承架
bearing bushing	轴承衬套	軸承襯套
bearing collar	轴承圈	軸承圈
bearing housing	轴承箱	軸承箱
bearing mixer	轴承式混合器	軸承式混合器
bed-collapsing technique	床层塌落技术	床層塌落技術
bed grid support	床栅支架	床栅支架
bed support	床层支架	床支架
bees wax	蜂蜡	蜂蠟
beet sugar	甜菜糖	甜菜糖
bell and hopper	进料器	進料器
bell and spigot joint	承插接头	插承接頭
bell flowmeter	钟形流量计	鐘式流量計
bell manometer	浮钟压力计	鐘式壓力計
bellows	风箱	風箱
bellows manometer	波纹管式压力计	伸縮式壓力計
bellows type flowmeter	波纹管式流量计	伸縮式流量計
belt conveyor	皮带输送机	帶[式]運[送]機
belt dryer	带式干燥器	帶式乾燥機
belt extractor	转带浸取器	轉帶萃取器
belt filter	带式过滤机	帶[式過]濾機
bench scale reactor	台架反应器	實驗室規模反應器
bench scale test	台架试验	實驗室規模試驗
bend	弯管	彎管
bend equivalent pipe length	弯管相当管长	彎管相當管長
bending stress	弯曲应力	彎曲應力
bending test	弯曲试验	彎曲試驗
Benedict-Webb-Rubin equation（BWR 　equation）	BWR 方程	本[尼迪克特]韋[伯] 　魯[賓]方程
Benson solubility coefficient	本森溶解度系数	本森溶解係數
benzofuran	苯并呋喃	苯并呋喃
benzoin	安息香	安息香
benzoyl peroxide	过氧化苯甲酰	過氧化苯甲醯[基]
bergamot oil	香柠檬油	香柑油
Berl saddle	弧鞍填料	弧鞍[形]填料
Bernoulli equation	伯努利方程	白努利方程[式]

英　文　名	大　陆　名	台　湾　名
beryl	绿柱石	綠柱石,綠寶石
BET equation(=Brunauer-Emmett-Teller equation)	BET 方程	BET 方程[式]
bias	偏压	偏壓
bias in proportional	比例控制偏离	比例控制偏離
bicontinuous structure	双连续结构	雙連續結構
bidispersion	双分散	雙分散
bifurcation	分叉	分歧
bifurcation analysis	分叉分析	分歧分析
bifurcation point	歧点	歧點
bilateral slit	双向狭缝	雙向狹縫
bimetal	双金属片	雙金屬
bimetallic catalyst	双金属催化剂	雙金屬催化劑,雙金屬觸媒
bimetallic cluster	双金属簇	雙金屬團
bimetallic strip	双金属片	雙金屬片
bimetallic thermometer	双金属温度计	雙金屬溫度計
bimetallic thermoregulator	双金属调温计	雙金屬調溫器
bimodal distribution	双峰分布	雙峰分佈
bimodal pore-size distribution	双峰孔径分布	雙峰孔徑分佈
bimolecular reaction	双分子反应	雙分子反應,二分子反應
bin activator	料仓松动器	料倉鬆動器
binary azeotrope	二元共沸液	二元共沸物
binary catalyst	二元催化剂	二元觸媒
binary cycle	双循环	雙循環,二元循環
binary distillation	二元蒸馏	二元蒸餾
binary mixture	二元混和物	雙成分混合物
binary system	二元系[统]	雙成分系統
binder	黏合剂	黏合劑,黏結劑
binder power	结合能力	結合能力
binding agent	黏合剂	黏合劑,黏結劑
binding energy	结合能	結合能
binding force	结合力	結合力
bin discharger	料仓卸料器	料倉卸料器
Bingham fluid	宾厄姆流体	賓漢流體
binodal solubility curve	双结点溶度曲线	雙結點溶解度曲線
bioassay	生物测定	生物檢定

英　文　名	大　陆　名	台　湾　名
bioavailability	生物可利用率	生體可用率
biocatalysis	生物催化	生物催化[作用]
biocatalyst	生物催化剂	生物觸媒
biocatalytic reaction	生物催化反应	生物催化反應
biocell	生物电池	生物電池
biochemical combustion	生物燃烧	生化燃燒
biochemical engineering	生化工程	生化工程
biochemical oxygen demand	生化需氧量	生化需氧量
biochemical reaction	生化反应	生化反應
biochemical reaction engineering	生化反应工程	生化反應工程
biochemical reactor	生化反应器	生化反應器
biochemical separation	生化分离	生化分離
biochemical thermodynamics	生化热力学	生化熱力學
biochemistry	生物化学	生物化學
bioconversion	生物转化	生物轉化
biodegradability	生物降解性	生物降解性
biodegradable detergent	生物降解洗涤剂	[可]生物降解清潔劑
biodegradable plastic	生物降解塑料	[可]生物降解塑膠
biodegradation	生物降解	生物降解
biodeterioration	生物变质	生物變質
bio-disc process	生物转盘法	生物轉盤法
bioenergetics	生物能学	生物能量學
bioengineering	生物工程	生物工程
biofluid mechanics	生物流体力学	生物流體力學
biofunctional reagent	生物功能试剂	生物功能試劑
biogas	沼气	生質氣
bioleaching	生物浸取	生物瀝取
biological agent	生物制剂	生物製劑
biological corrosion	生物腐蚀	生物腐蝕
biological dispersion	生物分散	生物分散
biological film	生物膜	生物膜
biological filter	生物过滤器	生物[過]濾器
biological flocculation	生物絮凝[作用]	生物絮凝[作用]
biological materials	生物材料	生物材料
biological membrane	生物膜	生物[薄]膜
biological oxidation	生物氧化	生物氧化[反應]
biological polymer	生物基聚合物	生物聚合物
biological process	生物过程	生物程序

英　文　名	大　陆　名	台　湾　名
biological reactor	生物反应器	生物反應器
biological solid	生物固体	生物固體
biological time lag	生物延时	生物時延
biological tower	生物塔	生物塔
biological treatment	生物处理	生物處理
biological treatment process	生物处理过程	生物處理程序
bioluminescence	生物发光	生物發光
biomacromolecule	生物大分子	生物巨分子
biomass	生物质	生質
biomass conversion	生物质转化	生質轉化
biomass conversion process	生物质转化过程	生質轉化程序
biomass energy	生物质能	生質能
biomass fuel	生物质燃料	生質燃料
biomass gasification	生物质气化	生質氣化
biomass pyrolysis	生物质热解	生質熱裂解
biomaterials	生物材料	生物材料
biomedical device	生物医药器件	生醫裝置
biomedical engineering	生物医学工程	生醫工程
biomedical polymer	生物医用聚合物	生醫聚合物
biomedical technology	生物医学技术	生醫技術
biomedicine	生物医学	生物醫學
bio-osmosis	生物渗透	生物滲透
biopolymer	生物聚合物	生物聚合物
bioprocess	生物过程	生物程序
bioreactor	生物反应器	生物反應器
biosensor	生物传感器	生物感測器
bioseparation	生物分离	生物分離
biospecificity binding	生物专一性连结	生物專一性連結
biosynthesis	生物合成	生物合成
biosynthetic reaction	生物合成反应	生物合成反應
biota	生物群	生物相
biotechnology	生物技术	生物技術
Biot number	毕奥数	畢奧數
biotoxicity	生物毒性	生物毒性
biotoxicity test	生物毒性试验	生物毒性試驗
biotransformation	生物转化	生物轉化
bipartite graph	偶图	偶圖
bipolar electrode	双极性电极	雙極性電極

英　文　名	大　陆　名	台　湾　名
bipolar moment	双极矩	雙極矩
birefraction	双折射	雙折射
birefringence	双折射	雙折射
birefringent fluid	双折射流体	雙折射流體
birotation	双旋光	雙旋光[作用]
birth rate	生成速率	生成[速]率
bistable device	双稳定装置	雙穩定裝置
bituminite	烟煤	煙煤
biuret	缩二脲	縮二脲
black ash	黑灰	黑灰
black body	黑体	黑體
black body coefficient	黑体系数	黑體係數
black body radiation	黑体辐射	黑體輻射
black box	黑箱	黑箱
black box method	黑箱法	黑箱法
black-box model	黑箱模型	黑箱模型
black liquor	黑液	黑液
blade	叶片	葉片
blade agitator	板片搅拌器	葉片攪拌器
bladed rotor	叶片转子	葉片轉子
blast	①鼓风 ②爆炸	①鼓風 ②爆炸
blast blower	鼓风机	鼓風機
blast burner	喷灯	噴燈
blast furnace	①高炉 ②鼓风炉	①高爐 ②鼓風爐
blast furnace coke	高炉焦	高爐焦
blast furnace gas	高炉煤气	高爐氣,鼓風爐氣
blast furnace slag	高炉渣	高爐渣,鼓風爐渣
blasticidin-S	杀稻瘟菌素 S	殺稻瘟菌素 S
blasting compound	爆炸化合物	爆炸化合物
bleaching	漂白	漂白
bleaching action	漂白作用	漂白作用
bleaching agent	漂白剂	漂白劑
bleaching liquor	漂白液	漂白液
bleaching powder	漂白粉	漂白粉
bleach powder	漂白粉	漂白粉
bleeder	泄放器	洩放器
blend	①掺合物 ②共混物	掺合物
blended gasoline	调和汽油	掺配汽油

英 文 名	大 陆 名	台 湾 名
blended paint	调和漆	調合漆
blender	掺和机	摻合機
blending	①掺合 ②共混	①摻合 ②摻配
blending agent	掺和剂	摻合劑
blending mixer	掺和式混合器	摻合式混合器
blending process	掺和过程	摻配程序
blending system	掺和系统	摻配系統
blending tank	掺和槽	摻配槽
block copolymer	嵌段共聚物	嵌段共聚物,團聯共聚物
block diagonal matrix	分块对角矩阵	分塊對角方陣
block diagram	框图	方塊圖,塊解圖
block flow diagram	程序框图	程序方塊圖
block polymerization	嵌段聚合	嵌段聚合[反應],團聯聚合[反應]
block tridiagonal matrix	方块三对角矩阵	分塊三對角方陣
blotting paper	吸墨纸	吸墨紙
blow cock	排气栓	排氣栓
blowdown	排料	洩料,排放
blowdown tank	泄料箱	洩料槽,排放槽
blow hole	[铸件等]气孔	氣孔
blowing agent	发泡剂	發泡劑
blow mold	吹模	吹模
blow molding	吹塑成型	吹氣成型法
blown glass	吹制玻璃	吹製玻璃
blown oil	吹制油	吹製油,吹煉油
blow pipe	吹管	吹管
blow pit	泄料池	噴漿池
blue gas	水煤气	水煤氣
blue vitriol	蓝矾	藍石,膽礬
Bodenstein number	博登斯坦数	博登施泰數
body-centered lattice	体心晶格	體心晶格
boehmite	水铝石	水鋁石
boiled oil	熟油	熟[煉]油
boiler	锅炉	鍋爐
boil feed water	锅炉给水	鍋爐給水
boiling	沸腾	沸騰
boiling-off	脱胶	褪膠

英　文　名	大　陆　名	台　湾　名
boiling point	沸点	沸點
boiling point apparatus	沸点测定器	沸點測定器,沸點裝置
boiling point-composition diagram	沸点-组成图	沸點-組成圖
boiling point curve	沸点曲线	沸點曲線
boiling point depression	沸点降低	沸點下降
boiling point diagram	沸点图	沸點圖
boiling point elevation	沸点升高	沸點上升
boiling point index	沸点指数	沸點指數
boiling point lowering	沸点降低	沸點下降
boiling point rise	沸点升高	沸點上升
boiling range	沸腾范围	沸點範圍
boiling starch	沸腾淀粉	糊化澱粉
bolometer	辐射热计	輻射熱測定計
bolt	①螺栓 ②筛	①螺栓 ②篩
bolting cloth	筛布	篩布
bolting mill	筛选机	篩選機
bolting reel	转筒筛	轉筒篩
bolting silk	筛用丝	篩用絲
Boltzmann distribution	玻耳兹曼分布	波茲曼分佈
bomb aging	弹内[加氧]陈化	彈内[加氧]老化
bomb calorimeter	弹式量热器	彈[式]卡計
bomb reactor	弹式反应器	彈[式]反應器
bond	键	鍵
bond angle	键角	鍵角
bond energy	键能	鍵能
bonding force	键结力	鍵結力
bonding strength	黏合强度	黏合強度,鍵結強度
bone black	骨炭	骨炭,骨黑
bone char	骨炭	骨炭
book value	账面价值	帳面價值
booster	升压器	增壓器,助力器
booster compressor	增压泵	增壓壓縮機
booster ejector	升压喷射器	增壓噴射器
booster pump	升压泵	增壓泵
boost gage	升压压力表	增壓壓力計
boracite	方硼石	方硼石
Borad ring	[博拉德]双层网环	[博拉德]雙層網環
borate	硼酸盐	硼酸鹽

英 文 名	大 陆 名	台 湾 名
borax	硼砂	硼砂
borax glass	硼玻璃	硼玻璃
bornite	斑铜矿	斑銅礦
boron carbide	碳化硼	碳化硼
borosilicate glass	硼硅玻璃	硼矽酸玻璃
Bose-Einstein distribution	玻色-爱因斯坦分布	玻[色]-愛[因斯坦]分布
bottle neck	瓶颈	瓶頸
bottom discharge	底部卸料	底部卸料
bottom-level control	底部液位控制	塔底液位控制
bottom phase	底相,下相	底相
bottom product	底部产物	底部產物
bottom steam	底部蒸汽	底部蒸汽
bottom valve	底阀	底部閥
bottom yeast	沉底酵母	沈底酵母
boundary condition	边界条件	邊界條件
boundary layer	边界层	邊界層
boundary layer concentration	边界层浓度	邊界層濃度
boundary layer equation	边界层方程	邊界層方程[式]
bound moisture	结合水分	結合水分
Bourdon gauge	弹簧管压力计	彈簧管壓力計,波頓壓力計
bowl mill	碗形磨	碗[形]磨
branch and bound method	分支定界法	分支定界法
branched chain	支链	支鏈
breakage	破裂	破裂
breakthrough capacity	穿透容量	貫流容量
breakthrough curve	穿透曲线	貫流曲線
breakthrough point	穿透点	貫流點
breakthrough time	穿透时间	貫流時間
breakup of emulsion	破乳	破乳化
breathing tank	呼吸式油罐	通氣槽
brewery	酿造厂	釀酒廠
brewer yeast	啤酒酵母	釀酒酵母
brewing	酿造	釀造
brick	砖	磚
bridge	电桥	電橋
brightness	亮度	亮度

英 文 名	大 陆 名	台 湾 名
brine	盐水	鹽水,滷水
brine wash	盐洗	鹽洗
briquetting	压块	壓塊
brittle materials	脆性物料	脆性材料
brittleness	脆性	脆性
brittle point	脆点	脆化點
bromination	溴化	溴化[作用]
bronze	青铜	青銅
broth dilution	稀胶法	稀膠法
Brownian diffusion	布朗扩散	布朗擴散
Broyden method	布罗伊登法	布羅伊登法
Brunauer-Emmett-Teller equation（BET equation）	BET 方程	BET 方程[式]
bubble	气泡	氣泡
bubble aeration	气泡曝气	氣泡通氣
bubble cap	泡罩	泡罩
bubble cap absorption column	泡罩吸收塔	泡罩吸收塔
bubble cap column	泡罩塔	泡罩塔
bubble cap contactor	泡罩接触器	泡罩接觸器
bubble cap plate	泡罩板	泡罩板
bubble cap plate distillation tower	泡罩板蒸馏塔	泡罩板蒸餾塔
bubble cap tower	泡罩塔	泡罩塔
bubble cap tray	泡罩板	泡罩板
bubble cloud	气泡云	氣泡雲
bubble coalescence	气泡聚并	氣泡聚結
bubble column	鼓泡塔	氣泡塔
bubble counter	气泡计数器	氣泡計數器
bubble film	[气]泡膜	氣泡膜
bubble flow	气泡流	氣泡流
bubble forming agent	气泡剂	發泡劑
bubble fractionation	气泡分馏	氣泡分餾[法]
bubble gage	气泡指示计	氣泡指示計
bubble plate	泡罩板	泡罩板
bubble plate column	泡罩塔	泡罩塔
bubble point	泡点	泡點
bubble point pressure	泡点压力	泡點壓力
bubble point temperature	泡点温度	泡點溫度
bubble reactor	气泡反应器	氣泡反應器

英　文　名	大　陆　名	台　湾　名
bubble tower overhead	泡罩塔顶馏分	泡罩塔頂餾分
bubble tray	鼓泡塔盘	泡罩板
bubble type viscometer	气泡型黏度计	泡式黏度計
bubbling	鼓泡	鼓泡作用
bubbling absorber	鼓泡式吸收器	鼓泡吸收器
bubbling bed	沸腾床	鼓泡床
bubbling cap	泡罩	泡罩
bubbling fluidization	鼓泡流态化	鼓泡流體化
bubbling fluidized bed	鼓泡流化床	鼓泡流[體]化床
bucket conveyor	斗式运输机	斗式運送機
bucket elevator	斗式提升机	斗式升降機
bucket-elevator extractor	提斗浸取器	斗式萃取器
budding	芽殖	出芽
budding cell	芽殖细胞	出芽細胞
budding fungi	芽殖真菌	出芽眞菌
buffer	缓冲器	緩衝劑
buffer action	缓冲作用	緩衝作用
buffer capacity	缓冲容量	緩衝容量
buffer index	缓冲指数	緩衝指數
buffer layer	缓冲层	緩衝層
buffer region	缓冲区	緩衝區[域]
buffer solution	缓冲溶液	緩衝[溶]液
buffer tank	缓冲罐	緩衝槽
buffer zone	缓冲区	緩衝區
builder	填充剂,辅料	助強劑
bulk aerator	整体通气器	整體通氣器
bulk analysis	全分析	總體分析
bulk concentration	体内浓度	總體濃度
bulk density	堆密度	[總]體密度
bulk diffusion	体扩散	體擴散
bulk flow	总体流动	總體流動
bulk index	整体指数	整體指數
bulking filler	填充剂,辅料	增容填料
bulk intensity	整体强度	整體強度
bulk modulus	本体模量	總體模數
bulk phase	体相	[總]體相
bulk property	本体性质	總體性質
bulk storage	散装储存	散裝儲存

英　文　名	大　陆　名	台　湾　名
bulk temperature	本体温度	總體溫度
bulk viscosity	体[积]黏性	總體黏度
bulk weight	整体重量	總體重量
buoyancy	浮力	浮力
buoyant effect	浮力效应	浮力效應
buoyant force	浮力	浮力
burner	燃烧器	燃燒器
burning point	燃点	燃燒點,火點
burning rate	燃烧率	燃燒速率
burning shrinkage	燃烧收缩	燃燒收縮
burnish	抛光	研光
burnt ocher	煅赭石	煅赭石
burnt sienna	煅富铁黄土	煅富鐵黃土
burst	爆裂	爆裂
bursting pressure	爆裂压力	爆裂壓力
bursting strength	爆裂强度	爆裂強度
bush	轴衬	軸襯
butterfly valve	蝶[形]阀	蝶型閥
butt joint	对接[头]	對接[頭]
butyl cellulose	丁基纤维素	丁基纖維素
butyl rubber	丁基橡胶	丁基橡膠
BWR equation(=Benedict-Webb-Rubin equation)	BWR 方程	本[尼迪克特]韋[伯]魯[賓]方程
bypass	旁路	旁路
bypass control	旁通控制	旁路控制
bypass filter	旁通过滤器	旁通過濾器
bypassing effect	旁通效应	旁通效應
bypass pipe	旁通管	旁通管
bypass ratio	旁通比	旁通比
bypass valve	旁通阀	旁通閥
by-product	副产物	副產物

C

英　文　名	大　陆　名	台　湾　名
CAD(=computer-aided design)	计算机辅助设计	電腦輔助設計
cadmium cell	镉电池	鎘電池
cadmium soap	镉皂	鎘皂

英 文 名	大 陆 名	台 湾 名
cadmium yellow	镉黄	鎘黃
CAE(=computer-aided engineering)	计算机辅助工程	電腦輔助工程
cage effect	笼效应	籠[蔽]效應
cage hydraulic press	笼式水压机	籠式水壓機
CAI(=computer-aided instruction)	计算机辅助教学	電腦輔助教學
caking	结块	結塊
caking power	结块能力	結塊能力
caking property	结块性	結塊性
calandria	排管	排管
calandria pan	排管式蒸发罐	排管式蒸發罐
calandria type evaporator	排管型蒸发器	排管型蒸發器
calcimeter	二氧化碳测定计	二氧化碳測定計
calcination	煅烧	煅燒
calcinator	煅烧炉	煅燒爐
calcined clay	煅烧黏土	煅燒黏土
calcined coke	煅烧焦	煅燒焦
calcined plaster	煅石膏	煅石膏
calciner	煅烧炉	煅燒爐
calcining zone	煅烧区	煅燒區
calcite	方解石	方解石
calcium carbide	碳化钙,电石	碳化鈣,電石
calcium carbonate	碳酸钙	碳酸鈣
calcium lactate	乳酸钙	乳酸鈣
calcium soap	钙皂	鈣皂
calculation	计算	計算
calculator	计算器	計算器
calculus	微积分	微積分
calculus of variations	变分学	變分學,變分法
calender	压延机	壓延機,研光機,軋光機
calender effect	压延效应	壓延效應
calibration	校准	校準
calibration chart	校准图	校準圖
calibration curve	校正曲线	校準曲線
calomel cell	甘汞电池	甘汞電池
calomel electrode	甘汞电极	甘汞電極
calorie	卡[路里]	卡
calorimeter	量热计	熱量計,卡計
calorimetry	量热学	熱量測定法,量熱法

英　文　名	大　陆　名	台　湾　名
CAM（＝computer-aided manufacturing）	计算机辅助制造	電腦輔助製造
camphor oil	樟脑油	樟腦油
cane sugar	蔗糖	蔗糖
cannery waste water	罐头厂废水	罐頭廠廢水
canonical coordinate	正则坐标	正則坐標
canonical ensemble	正则系综	正則系綜
canonical equation	正则方程	正則方程式
canonical form	正则形式	正則形式
canonical matrix	典范矩阵	正則矩陣
canonical partition function	正则配分函数	正則分配函數
canonical transformation	正则变换	正則變換
capacitance	电容	電容
capacitance element	电容元件	電容元件
capacitance liquid level gage	电容液面计	電容[式]液面計
capacitance moisture meter	电容湿度计	電容[式]濕度計
capacitance pressure transducer	电容压力转换器	電容[式]壓力轉換器
capacitance probe	电容探针	電容探針
capacitor	电容器	電容器
capacitor impulse type tachometer	电容脉动式转速计	電容器脈衝[式]轉速計
CAPD（＝computer-aided process design）	计算机辅助过程设计	電腦輔助程序設計
capillarimeter	毛细测液器	毛細計
capillary	毛细管	毛細管
capillary absorption	毛细管吸收	毛細吸收
capillary ascent	毛细管上升	毛細上升
capillary attraction	毛细管吸引	毛細吸引,毛細引力
capillary condensation	毛细管冷凝	毛細冷凝[作用]
capillary elevation	毛细管上升	毛細上升
capillary flask	毛细瓶	毛細瓶
capillary force	毛细管力	毛細[管]力
capillary module	毛细管膜组件	毛細管模組
capillary number	毛细管准数	毛細數
capillary penetration	毛细穿透	毛細穿透
capillary pressure	毛细压力	毛細壓力
capillary tension	毛细张力	毛細張力
capillary tube	毛细管	毛細管
capillary tube method	毛细管法	毛細管法
capillary-type bulb	毛细管型球	毛細管式球

英 文 名	大 陆 名	台 湾 名
capillary viscometer	毛细管黏度计	毛細管黏度計
capital	资本	資本
capital gain	资本所得	資本所得
capital investment cost factor	资本投资成本因数	[资本]投资成本因数
capital ratio	资本比率	資本比率
capital recovery factor	资本回收因数	資本回收因數
caprolactam	己内酰胺	己內醯胺
capsule	胶囊	膠囊
caramel	焦糖	焦糖
carbazole	咔唑	咔唑
carbide brick	碳化硅砖	碳化矽磚
carbide silicon	碳化硅	碳化矽
carbitol	卡必醇	卡必醇
carbohydrate	碳水化合物	碳水化合物,醣
carbonaceous fuel	含碳燃料	含碳燃料
carbon adsorption	碳吸附	碳吸附
carbonation process	碳酸化过程	碳酸法
carbon balance	碳平衡	碳平衡
carbon black	炭黑	碳黑
carbon brick	炭砖	碳磚
carbon brush	炭刷	碳刷
carbon cycle	碳循环	碳循環
carbon electrode	炭电极	碳電極
carbon fiber	碳纤维	碳纖維
carbon fiber reinforced plastic	碳纤维强化塑料	碳纖維強化塑膠
carbon filter	[活性]炭过滤器	碳濾器
carbon filtration	炭滤	碳濾[法]
carbonium dyes	碳阳离子染料	碳陽離子染料
carbon paper	复写纸	複寫紙
carbon refractory	碳质耐火材料	碳質耐火物
carbon steel	碳钢	碳鋼
carbonylation	羰基化	羰基化[作用]
carboxylase	羧基酶	羧基酶
carboxylate	羧酸盐	羧酸鹽
carboxylation	羧基化	羧化[作用]
carboxymethyl cellulose	羧甲基纤维素	羧甲纖維素
carburetted gas	增碳燃气	增碳燃氣
carburetted water gas	增碳水煤气	增碳水煤氣

英　文　名	大　陆　名	台　湾　名
carnallite	光卤石	光鹵石
carnauba wax	巴西棕榈蜡	棕櫚蠟,卡拿巴蠟
carnotite	钒钾铀矿	釩酸鉀鈾礦
carrier	载体	載體
carrier distillation	载体分馏	載體蒸餾
carrier gas	载气	載氣
carryover	带出	帶出
cascade	串级	串級
cascade combination	串级组合	串級組合
cascade compensation	串级补偿	串級補償
cascade concentration plate	串级浓缩板	串級濃縮板
cascade concentrator	串级浓缩器	串級濃縮器
cascade control	串级控制	串級控制
cascade cooler	串级冷却器	串級冷卻器
cascade cycle	串级循环	串級循環
cascade evaporator	串级蒸发器	串級蒸發器
cascade reactor	串级反应器	串級反應器
cascade refrigeration	串级冷冻	串級冷凍
cascade system	串级系统	串級系統
cascade tower	串级塔	串級塔
cascade-type plate	串级分馏板	串型分餾板
case hardening	表面硬化	表面硬化
casein	酪蛋白	酪蛋白
casein adhesive	酪蛋白黏合剂	酪蛋白黏合劑,酪蛋白 黏著劑
cash-flow diagram	现金流通图	現金流量圖
cashmilon	开斯米纶	開司米綸
cash ratio	现金比率	現金比率
cash recovery period	现金回收期	現金回收期
casing	①护罩 ②机壳	護罩
casing head gas	井口气	井口天然氣
casing head gasoline	天然汽油	[井口]天然汽油
cassava starch	木薯淀粉	樹薯粉
cassia oil	肉桂皮油	桂皮油
casting	①浇铸 ②铸件	①鑄造 ②鑄件
casting mold	铸模	鑄模
cast iron	铸铁	鑄鐵
castor oil	蓖麻子油	蓖麻油

英　文　名	大　陆　名	台　湾　名
cast polymerization	铸塑聚合	鑄塑聚合[作用]
cast resin	铸塑树脂	鑄模樹脂
cast steel	铸钢	鑄鋼
catabolism	分解代谢	分解代謝[作用]
catalysis	催化	催化[作用]
catalyst	催化剂	觸媒
catalyst activation	催化剂活化	觸媒活化
catalyst activity	催化剂活性	觸媒活性
catalyst attrition	催化剂磨损	觸媒磨耗
catalyst bed	催化剂床	觸媒床
catalyst binder	催化剂黏结剂	觸媒黏結劑
catalyst burn-off still	催化剂烧尽釜	觸媒燒盡釜
catalyst carrier	催化剂载体	觸媒載體
catalyst circulation	催化剂循环	觸媒循環
catalyst coking	催化剂结焦	觸媒結焦
catalyst contamination index	催化剂污染指数	觸媒污染指數
catalyst deactivation	催化剂失活	觸媒失活
catalyst decay	催化剂衰退	觸媒衰退
catalyst diluent	催化剂稀释剂	觸媒稀釋劑
catalyst effectiveness	催化剂有效性	觸媒有效性
catalyst extender	催化剂补充剂	觸媒增效劑
catalyst formulation	催化剂配方	觸媒配方
catalyst fouling	催化剂污染	觸媒積垢
catalyst gauze	催化剂网	觸媒網
catalyst gauze reactor	催化剂网反应器	觸媒網反應器
catalyst geometry	催化剂几何形状	觸媒[幾何]形狀
catalyst impregnation	催化剂浸渍	觸媒含浸
catalyst inhibitor	催化剂抑制剂	觸媒抑制劑
catalyst iron-alumina	铁铝氧催化剂	鐵鋁氧觸媒
catalyst iron-ammonia	铁氨催化剂	鐵氨觸媒
catalyst life	催化剂寿命	觸媒壽命
catalyst particle	催化剂粒子	觸媒粒子
catalyst pellet	催化剂颗粒	觸媒粒
catalyst poisoning	催化剂中毒	觸媒中毒
catalyst pore structure	催化剂细孔结构	觸媒細孔結構
catalyst porosity	催化剂孔隙率	觸媒孔隙率,觸媒孔隙度
catalyst reactivation	催化剂再生	觸媒再活化

英　文　名	大　陆　名	台　湾　名
catalyst receiver	催化剂接收器	觸媒接受器
catalyst regeneration	催化剂再生	觸媒再生
catalyst stabilizer	催化剂稳定剂	觸媒安定劑
catalyst support	催化剂载体	觸媒擔體
catalytic action	催化作用	催化作用
catalytic activity	催化活性	催化活性
catalytic agent	催化剂	催化劑
catalytic area	催化面积	催化面積
catalytic carrier	催化载体	催化載體
catalytic chemisorption	催化化学吸附	催化化學吸附
catalytic conversion	催化转化	催化轉化
catalytic converter	催化转化器	催化轉化器
catalytic cracking	催化裂化	催化裂解
catalytic cracking process	催化裂化过程	催化裂解程序
catalytic cycle	催化循环	催化循環
catalytic decomposition	催化分解	催化分解
catalytic kinetics	催化动力学	催化動力學
catalytic metal	催化金属	催化性金屬
catalytic oxidation	催化氧化	催化氧化[反應]
catalytic polymerization	催化聚合	催化聚合
catalytic process	催化过程	催化程序
catalytic promoter	催化促进剂	催化助劑
catalytic reaction	催化反应	催化反應
catalytic reactor	催化反应器	催化反應器
catalytic reforming	催化重整	催化重組,觸媒重組
catalytic reforming process	催化重整过程	催化重組法,觸媒重組法
catalytic site	催化部位	催化部位
catalytic specificity	催化专一性	催化特異性
catalytic surface	催化表面	催化表面
catalytic system	催化系统	催化系統
catastrophe theory	突变论	劇變理論
catastrophic model	突变模型	劇變模型
cathode	阴极	陰極
cathodic reaction	阴极反应	陰極反應
cation	阳离子	陽離子,正離子
cation exchange	阳离子交换	陽離子交換
cation exchanger	①阳离子交换剂 ②阳	①陽離子交換劑 ②陽

英 文 名	大 陆 名	台 湾 名
	离子交换器	離子交換器
cation exchange resin	阳离子交换树脂	陽離子交換樹脂
cationic polymerization	阳离子聚合	陽離子聚合[作用]
cationic starch	阳离子淀粉	陽離子澱粉
cationic surface active agent	阳离子型表面活性剂	陽離子界面活性劑
cationic surfactant	阳离子型表面活性剂	陽離子界面活性劑
causality	因果性	因果性
causticization	苛化	苛性[作用]
caustic soda	氢氧化钠,烧碱	燒鹼,苛性鈉
cavitation	汽蚀	成腔現象
cavitation corrosion	空蚀	空腔腐蝕
cavity radiation	空穴辐射	空腔輻射
cell culture	细胞培养	細胞培養
cell cycle	细胞周期	細胞週期
cell debris	细胞碎片	細胞碎片
cell density	细胞密度	細胞密度
cell disruption	细胞破碎	細胞破碎
cell fusion	细胞融合	細胞融合
cell harvesting	细胞收集	細胞收集
cell homogenate	细胞匀浆	細胞均質
cell house	电槽室	電槽室
cell loading	细胞负载	細胞負載
cell lysis	细胞溶解	細胞分解
cell model	胞腔模型	細胞模型
cellophane	赛璐玢	賽璐玢,玻璃紙
cell separation	细胞分离	細胞分離
cell suspension culture	细胞悬浮培养	細胞懸浮培養
cellular materials	①微孔材料 ②细胞材料	①海綿狀材料 ②細胞物質
cellulase	纤维素酶	纖維素酶
celluloid	赛璐珞	賽璐珞
cellulose	纤维素	纖維素
cellulose acetate	醋酸纤维素	醋酸纖維素
cellulose acetate butyrate	乙酸丁酸纤维素	醋酸丁酸纖維素
cellulose adhesive	纤维素黏合剂	纖維素黏合劑
cellulose colloid	纤维素胶体	纖維素膠體
cellulose derivative	纤维素衍生物	纖維素衍生物
cellulose lacquer	纤维素喷漆	纖維素[噴]漆

英　文　名	大　陆　名	台　湾　名
cellulose nitrate	硝酸纤维素	硝酸纖維素
cellulose nitrate plastic	硝酸纤维素塑料	硝化纖維素塑膠,硝酸纖維素塑膠
cellulose propionate	丙酸纤维素	丙酸纖維素
cellulose xanthate	黄原酸纤维素	黃原酸纖維素
cement	水泥	水泥
cementation	①胶结 ②硬化	膠接[作用]
cementation index	[水泥]硬化指数	[水泥]硬化指數
cementation process	胶结过程	膠接程序
cement pump	水泥输送泵	水泥泵
center of asymptote	渐近线中心	漸近線中心
center of buoyancy	浮[力中]心	浮力中心
center of mass	质[量中]心	質[量中]心
centipoise（cP）	厘泊	厘泊
centrifugal blower	离心式鼓风机	離心鼓風機
centrifugal clarification	离心净化	離心淨化
centrifugal collector	离心收集器	離心收集器
centrifugal compressor	离心压缩机	離心壓縮機
centrifugal cyclone	离心旋风器	離心旋風器
centrifugal dehydrator	离心脱水器	離心脫水機
centrifugal dryer	离心干燥器	離心乾燥機
centrifugal extractor	离心萃取器	離心萃取機
centrifugal filter	离心过滤机	離心過濾機
centrifugal force	离心力	離心力
centrifugal head	离心扬程	離心揚程
centrifugalization	离心[分离]	離心化
centrifugal liquid separator	离心式液体分离器	離心式液體分離器
centrifugal mill	离心碾磨机	離心磨機
centrifugal molding	离心成型	離心成型
centrifugal propeller	离心式螺旋桨	離心式螺旋槳
centrifugal pump	离心泵	離心泵
centrifugal purification	离心纯化	離心純化
centrifugal screen	离心筛	離心篩
centrifugal scrubber	离心涤气机	離心洗氣器
centrifugal separator	离心分离机	離心分離器
centrifugal settling	离心沉降	離心沈降
centrifugal solid separator	离心式固体分离器	離心式固體分離器
centrifugal spray	离心喷雾	離心噴霧

英　文　名	大　陆　名	台　湾　名
centrifugal spray tower	离心喷淋塔	離心噴霧塔
centrifugation	离心[分离]	離心[分離]
centrifuge	离心机	離心機
centrifuge	离心机	離心機
centrifuge basket	离心机盘	離心機籃
centrifuge blade	离心机叶片	離心機葉片
centrifuge head	离心机转头	離心機轉頭
centrifuge tube	离心[机]管	離心管
centripetal acceleration	向心加速度	向心加速度
centripetal force	向心力	向心力
centroid	质心	形心,矩心
cephalin	脑磷脂	腦磷脂
ceramic	陶瓷制品	陶瓷製品
ceramic glaze	陶瓷釉	陶瓷釉
ceramic refractory	陶瓷耐火材料	陶瓷耐火物
ceramics	陶瓷	陶瓷
ceramic tile	瓷砖	瓷磚
ceramic tube	陶瓷管	陶瓷管
ceramic ware	陶瓷器皿	陶瓷器
cermet	金属陶瓷	陶瓷金屬,陶金
cermet resistor	金属陶瓷电阻器	陶瓷金屬電阻器
cetane index	十六烷指数	十六烷指數
cetane number	十六烷值	十六烷值
chain conveyor	链式输送机	鏈式運送機,鏈運機
chain depolymerization	链解聚[作用]	鏈解聚合[反應]
chain initiation	链引发	鏈引發
chain polymerization	链聚合	鏈聚[合][作用]
chain propagation	链增长	鏈傳播[反應]
chain reaction	链反应	鏈反應
chain termination	链终止	鏈終止[作用]
chain termination reaction	链终止反应	鏈終止反應
chain transfer	链转移	鏈轉移[作用],鏈傳遞[作用]
chalcolite	铜铀云母,辉铜矿	銅鈾雲母
chalcopyrite	黄铜矿	黃銅礦
chalk	白垩	白堊
chamber filter press	箱式压滤机	廂式壓濾機
chamber leather	①油鞣 ②麂皮	①油鞣 ②麂皮

英　文　名	大　陆　名	台　湾　名
chamber process	铅室法	鉛室法
chamotte	①熟料 ②耐火黏土	①燒粉 ②燒黏土
change-can mixer	换罐混合器	換罐式混合器
changeover cost	更迭成本	更迭成本
changeover time	更迭时间	更迭時間
channel	①槽 ②通道	①槽 ②通道
channel black	槽法炭黑	槽黑
channel flow	渠道流	渠流
channel type flowmeter	槽式流量计	槽式流量計
Chao-Seader method	赵[广绪]-西得方法	趙[廣緒]-西得方法
char	炭	炭
characteristic constant	特性常数	特性常數
characteristic curve	特性曲线	特性曲線
characteristic equation	特性方程	特性方程式,特徵方程式
characteristic factor	特性因数	特性因子
characteristic function	特性函数	特性函數
characteristic length	特征长度	特性長度
characteristic reaction time	特性反应时间	特性反應時間
characteristic root	特性根	特性根
characteristic time	特征时间	特性時間
characteristic value	特征值	特徵值
characterization	特性化	特性分析,表徵分析
characterization factor	特性因素	特性化因數
charcoal	木炭	木炭
charge conveyor	加料运输机	加料運送機
charged membrane	荷电膜	荷電膜
charge elevator	加料升降机	加料升降機
charge hopper	加料斗	加料斗
charger	①加料机 ②充电器	①加料機 ②充電器
chasing method	追赶法	追蹤法
check valve	止逆阀,单向阀	止回閥,單向閥
chelating agent	螯合剂	螯合劑,鉗合劑
chemical	化学品	化學品
chemical absorbent	化学吸收剂	化學吸收劑
chemical absorption	化学吸收	化學吸收
Chemical Abstracts	化学文摘	化學摘要
chemical adsorption	化学吸附	化學吸附

英　文　名	大　陆　名	台　湾　名
chemical bond	化学键	化學鍵
chemical bonding	化学成键	化學鍵結
chemical change	化学变化	化學變化
chemical clarifier	化学澄清器	化學澄清器
chemical coagulation	化学凝聚	化學凝聚[作用]
chemical conditioning	化学调理	化學調理[法]
chemical diffusivity	化学扩散系数	化學擴散係數
chemical dynamics	化学动态学	化學動態學
chemical energy	化学能	化學能
chemical engineer	化学工程师	化學工程師
chemical engineering	化学工程	化學工程
chemical engineering analysis	化工分析	化工分析
chemical engineering construction cost index	化学工程建造成本指数	化學工程營建成本指數
chemical engineering kinetics	化工动力学	化工動力學
chemical engineering process analysis	化工过程分析	化工程序分析
chemical engineering science	化学工程学	化工科學
chemical engineering thermodynamics	化工热力学	化工熱力學
chemical equation	化学方程式	化學方程式,化學反應式
chemical equilibrium	化学平衡	化學平衡
chemical feedstock	化学原料	化學原料
chemical flooding	化学液泛	化學氾流[法]
Chemical Industry and Engineering Society of China(CIESC)	中国化工学会	中國化工學會
chemical kinetics	化学动力学	化學動力學
chemical mechanical polishing	化学机械抛光	化學機械研磨
chemical operation	化学操作	化學操作
chemical oscillation	化学振荡	化學振盪
chemical oscillator	化学振荡器	化學振盪器
chemical oxygen demand（COD）	化学需氧量	化學需氧量
chemical parameter	化学参数	化學參數
chemical physics	化学物理[学]	化學物理學
chemical plant	化[学]工厂	化學工廠
chemical polarization	化学极化	化學極化
chemical potential	化学势	化學勢
chemical precipitation	化学沉淀	化學沈澱[法]
chemical process	化学过程	化學程序

英　文　名	大　陆　名	台　湾　名
chemical process industry	化学[过程]工业	化學程序工業
chemical promoter	化学促进剂	化學促進劑
chemical property	化学性质	化學性質
chemical pulping process	化学制浆法	化學製漿法
chemical reaction	化学反应	化學反應
chemical reaction engineering	化学反应工程	化學反應工程
chemical reaction equilibrium	化学反应平衡	化學反應平衡
chemical reaction stability	化学反应稳定性	化學反應穩定性
chemical reactor	化学反应器	化學反應器
chemical reactor stability	化学反应器稳定性	化學反應器穩定性
chemical seal	化学密封	化學密封
chemical separation	化学分离	化學分離
chemical smoke	化学烟雾	化學煙霧
chemical species	化学物种	化學物種
chemical stability	化学稳定性	化學穩定性,化學安定性
chemical technology	化工技术	化工技術
chemical transport	化学输送	化學輸送
chemical treatment	化学处理	化學處理
chemical treatment process	化学处理法	化學處理法
chemical vapor	化学蒸汽	化學蒸氣
chemical vapor deposition	化学气相沉积	化學氣相沈積
chemiluminescence	化学发光	化學發光
chemisorption	化学吸附	化學吸附
chemistry of hydrogen energy	氢能化学	氫能化學
chemostat	恒化器	恆化器
china	瓷器	瓷器
china clay	瓷土	瓷土
chinaware	瓷器	瓷器
chip drier	木片干燥器	木片乾燥器
chiral separation	手性分离	手性分離,掌性分離
chloral	三氯乙醛	[三]氯[乙]醛
chloramine	氯胺	氯胺
chloranil	四氯苯醌	[四]氯[苯]醌
chloride contact chamber	氯化物接触室	氯接觸槽
chlorinated hydrocarbon	氯化烃	氯[化]烴
chlorinated paraffin	氯化石蜡	氯[化]蠟
chlorinated phenol	氯化酚	氯化酚

英　文　名	大　陆　名	台　湾　名
chlorinated quinone	氯化苯醌	氯化[苯]醌
chlorinated rubber	氯化橡胶	氯化橡膠
chlorination	氯化	氯化[反應]
chloromycetin	氯霉素	氯黴素
chlorophyll	叶绿素	葉綠素
chloroprene rubber	氯丁橡胶	氯丁二烯橡膠,氯平橡膠
chlorosulfonated polyethylene	氯磺化聚乙烯	氯磺化聚乙烯
chlorosulfonation	氯磺酰化	氯磺化
chlorosulfonic polyethylene	氯磺化聚乙烯	氯磺化聚乙烯
choke	噎塞	噎塞
choke crushing	阻塞压碎	阻塞壓碎
choked nozzle	闸喉喷嘴	閘喉噴嘴
choke flow	扼流	扼流
cholesteric state	胆甾状态	膽甾態,膽固醇態
cholesterol	胆固醇	膽固醇,膽甾醇
chord line	弦线	弦線
chromatoelectrophoresis	色谱电泳	層析電泳
chromatofocusing	色谱聚焦	層析聚焦
chromatogram	色谱图	層析圖
chromatograph	色谱仪	層析儀,色譜儀
chromatographic column	色谱柱	層析管柱
chromatographic column pulse reactor	色谱柱式脉冲反应堆	層析柱式脈動反應器
chromatographic separation	色谱分离	層析分離
chromatography	色谱[法]	層析術
chrome alum	铬矾	鉻礬
chrome brick	铬砖	鉻磚
chrome dye	铬染料	鉻媒染料
chrome green	铬绿	鉻綠
chrome leather	铬革	鉻革,鞣皮
chrome-magnesite brick	铬镁砖	鉻鎂磚
chrome mordant dyeing	铬媒染色	鉻媒染法
chrome orange	铬橙	鉻橙
chrome pigment	铬颜料	鉻顏料
chrome plating	镀铬	鍍鉻
chrome red	铬红	鉻紅
chrome refractory	铬质耐火材料	鉻質耐火物
chrome steel	铬钢	鉻鋼

英　文　名	大　陆　名	台　湾　名
chrome tannage	铬鞣	鉻鞣法
chrome tanning	铬鞣制	鉻鞣[製]
chrome yellow	铬黄	鉻黃
chromogen	色原	色素原
chromometer	比色计	比色計
chromophore	发色团	發色團
chrysoberyl	金绿宝石	金綠[寶]石
chute	滑槽	滑槽
CIESC (= Chemical Industry and Enginee-ring Society of China)	中国化工学会	中國化工學會
cinnamon oil	肉桂油	肉桂油,桂皮油
circular chart	圆形图	圓形圖
circularity	圆形度	[真]圓度
circular scale indicator	圆形标度指示计	圓形標[度指]示計
circulating criterion	循环判据	循環準則
circulating drop	循环滴	循環[流動]滴
circulating fluidized bed	循环流化床	循環流[體]化床
circulating pump mixer	循环泵混合器	循環泵混合器
circulating reactor	环流反应器	循環反應器
circulating reflux	循环回流	循環回流
circulation flow	循环流动	循環流動
circulation layer	循环层	循環層
circulation method	循环法	循環法
circulation mixer	循环式混合器	循環式混合器
circulation pipe	循环管	循環管
circulation tube	循环管	循環管
circumferential velocity	圆周速度	圓周速度
cis-butadiene rubber	顺丁橡胶	順丁二烯橡膠
cis-form	顺式	順式
cis-isoprene rubber	顺式异戊二烯橡胶	順異戊二烯橡膠,順異平橡膠
citronellal	香茅醛	香茅醛
citronella oil	香茅油	香茅油
citronellol	香茅醇	香茅醇
Clapeyron-Clausius equation	克拉珀龙–克劳修斯方程	克[拉伯隆]–克[勞修斯]方程式
clarification	澄清	澄清[作用]
clarification capacity	澄清能力	澄清能力

英 文 名	大 陆 名	台 湾 名
clarification filtration	澄清过滤	澄清過濾
clarification process	澄清过程	澄清程序
clarification zone	澄清区	澄清區
clarifier	澄清器	澄清器
clarifying capacity	澄清能力	澄清能力
clarifying filter	澄清过滤器	澄清過濾器
classification	分级	分類
classifier	分级器	分類器
clathrate	包合物	晶籠化合物
clathration	包合作用	晶籠［作用］
Clausius inequality	克劳修斯不等式	克勞修斯不等式
clay	黏土	黏土
cleaning process	清洗过程	清洗程序
clearance	间隙	間隙
clearance meter	间隙测量计	間隙計
clearance volume	间隙体积	間隙體積
climbing-film evaporator	升膜蒸发器	升膜蒸發器
clinging cavity	附着空洞	附著空洞
clone	克隆	克隆
close cut distillate	精密分馏馏出物	精密餾出物
close cut fraction	精密分馏馏分	精密餾分
close cut separation	精密分馏分离	精密分離
closed boundary	闭式边界	閉邊界
closed circuit	闭路	閉路
closed circuit crushing	闭路压碎	閉路壓碎,循環壓碎
closed circuit grinding	闭路研磨	閉路研磨,循環研磨
closed conduct	封闭导管	封閉導管
closed impeller	封闭式叶轮	閉式葉輪
closed loop	闭环	閉環
closed loop control	闭环控制	閉環控制,閉路控制
closed loop frequency response	闭环频率应答	閉環頻率應答
closed loop stability	闭环稳定性	閉環穩定性
closed loop system	闭环系统	閉環系統,閉路系統
closed loop transfer function	闭环传递函数	閉環轉移函數
closed reactor	密闭反应器	密閉反應器
closed system	封闭系统	封閉系統
closed vessel	闭式容器	密閉容器
cloth filter	布滤机	布濾機

英　文　名	大　陆　名	台　湾　名
cluster of particle	颗粒簇	粒子團簇
coagulant	凝聚剂	凝聚劑
coagulant aid	助凝剂	助凝劑
coagulating agent	凝聚剂	凝聚劑
coagulation tank	凝聚槽	凝聚槽
coal bed	煤床	煤床
coal conversion	煤转化	煤轉化
coal-derived gas	煤衍生气	煤衍生氣
coal-derived liquid	煤衍生液[体]	煤衍生液[體]
coal devolatilization	煤的脱挥发作用	煤去揮發物作用
coal distillation	煤蒸馏	煤蒸餾
coalescence	聚并	凝聚,聚結
coalescence model	聚并模式	聚結模式
coalescer	聚并器	凝聚器,聚結器
coal field	煤田	煤田
coal friability	煤脆性	煤脆性
coal gas	煤气	煤氣
coal gasification	煤气化	煤氣化[法]
coal gasification process	煤气化过程	煤氣化程序
coal gasifier	煤气化炉	煤氣化器
coal liquefaction	煤液化	煤液化[法]
coal liquefaction process	煤液化过程	煤液化程序
coal porosity	煤孔隙度	煤孔隙度
coal processing technology	煤加工技术	煤處理技術
coal pyrolysis	煤热解	煤熱解
coal tar	煤焦油	煤焦油,煤溚
coal-water slurry	水煤浆	煤水漿
coarse crusher	粗级压碎机	粗級壓碎機
coarse particle	粗颗粒	粗粒子
coarse screen	粗筛网	粗篩
coarse suspension	粗悬浮液	粗懸浮液
coastal pollution	海岸污染	海岸污染
coating technology	涂料技术	塗裝技術
co-catalyst	助催化剂	輔觸媒
cock	旋塞	旋塞,[活]栓
coconut oil	椰子油	椰子油
cocurrent flow	并流	同向流
cocurrent operation	并流操作	同向流操作

英 文 名	大 陆 名	台 湾 名
cocurrent process	并流过程	同向流程序
COD（=chemical oxygen demand）	化学需氧量	化學需氧量
cod-liver oil	鱼肝油	鱈魚肝油
coefficient condensation	冷凝系数	冷凝係數
coefficient contraction	收缩系数	收縮係數
coefficient of performance（COP）	性能系数	性能係數
coefficient of variation	变动系数	變異係數
coenzyme	辅酶	輔酶
coexistence equation	共存方程	共存方程式
coextraction	共萃取	共萃取
cofactor	辅因子	輔因子
cohesion	内聚力	內聚力,內聚性
cohesion work	内聚功	內聚功
cohesive density	内聚能密度	內聚能密度
coil condenser	蛇管冷凝器,盘管冷凝器	蛇管冷凝器,盤管冷凝器
coiled pipe	蛇管,盘管	蛇管,盤管
coiled radiator	蛇管散热器,盘管散热器	蛇管散熱器,盤管散熱器
coil evaporator	蛇管蒸发器,盘管蒸发器	蛇管蒸發器,盤管蒸發器
coil expansion pipe	盘形膨胀管	盤形膨脹管
co-ion	共离子	共離子
coke	焦炭	焦炭
coke drum	结焦槽	結焦槽
coke knocker	除焦机	除焦機
coke number	焦值	焦值
coke oven	[炼]焦炉	煉焦爐
coke oven gas	焦炉煤气	焦爐氣
coke oven tar	焦炉油	焦爐瀝
coke porosity	焦炭孔隙度	焦炭孔隙度
coker	焦化器	煉焦器
coke still	焦化蒸馏器	焦化蒸餾器
coking	结焦	結焦,焦化,煉焦
coking coal	[炼]焦煤	煉焦煤
coking process	炼焦过程	焦化程序,煉焦法
coking reaction	焦化反应	焦化反應
cold-flow model experiment	冷模试验	冷模試驗

英　文　名	大　陆　名	台　湾　名
cold junction	冷接点	冷接點
cold reflux	冷回流	冷回流
cold rubber	冷制橡胶	冷製橡膠
cold-soda process	冷碱法	冷鹼法
cold vulcanization	冷硫化	冷硫化［法］
colistin	黏菌素	黏菌素
collagen	胶原	膠原蛋白
collecting ring	收集环	收集環
collection efficiency	捕集效率	收集效率
collector	收集器	收集器
collector electrode	集尘极	收集電極
colligative property of solution	溶液的依数性	依數性
colloid	胶体	膠體
colloidal clay	胶态白土	膠質黏土
colloidal dispersion	胶态分散	膠態分散
colloidal mill	胶体磨	膠體磨機
colloidal particle	胶体粒子	膠體粒子
colloidal phenomenon	胶体现象	膠體現象
colloidal platinum	胶体铂	膠態鉑
colloidal precipitation	胶体沉淀	膠體沈澱
colloidal solution	胶体溶液	膠體溶液
colloidal state	胶体状态	膠態
colloidal suspension	胶体悬浮液	膠體懸浮液
colloid chemistry	胶体化学	膠體化學
colloidization	胶化	膠體化
colloid solution	胶体溶胶	膠體溶液
colloid stabilization	胶体稳定化	膠體穩定化［法］
color comparator	比色器	比色器
color densitometer	比色密度计	比色密度計
colorimeter	比色计	比色計
column capacity	柱容量	管柱容量
column characterization factor	柱特性因素	柱特性因數
column chromatography	柱色谱法	管柱層析術
column internals	塔内件	塔内件
column packing	塔填料	塔填充物
column scrubber	洗涤塔	柱狀洗滌塔
column still	柱式蒸馏器	柱式蒸餾器
combinatorial term	组合项	組合項

英　文　名	大　陆　名	台　湾　名
combining rule	组合规则	組合規則
combustible liquid code	可燃液体规范	可燃液體規範
combustion	①燃烧 ②灼烧[灭菌法]	燃燒
combustion chamber	燃烧室	燃燒室
combustion characteristics	燃烧特性	燃燒特性
combustion curve	燃烧曲线	燃燒曲線
combustion efficiency	燃烧效率	燃燒效率
combustion engine	内燃机	燃燒機
combustion front	燃烧面	燃燒鋒面
combustion initiator	引燃剂	引燃劑
combustion mechanism	燃烧机理	燃燒機構
combustion process	燃烧过程	燃燒程序
combustion rate	燃烧速率	燃燒速率
combustion reaction	燃烧反应	燃燒反應
combustion reactor	燃烧反应器	燃燒反應器
combustion zone	燃烧区	燃燒區
commensalism	共栖	共生,共棲
commercial development	商业发展	商業發展
comminuting machine	粉碎机	粉碎機
comminution	粉碎	粉碎
comminutor	粉碎机	粉碎機
commodity chemical	大宗化学品	大宗化學品
commodity polymer	大宗聚合物	大宗聚合物
common stock	普通原料	普通原料
common utility	公共设施	公用設施
compact heat exchanger	紧凑型换热器	緊緻熱交換器
compaction	压紧	壓緊
compactness	紧密度	緊密度
comparator	比较器	比較器
compartment dryer	厢式干燥器	廂式乾燥器
compartmented tank reactor	多室槽式反应器	分倉槽式反應器
compatibility	相容性	相容性
compensated thermometer system	补偿温度计系统	補償溫度計系統
compensation	补偿	補償
compensation effect	补偿效应	補償效應
compensation servo-mechanism	伺服机构补偿	伺服機構補償
compensator	补偿器	補償器

英　文　名	大　陆　名	台　湾　名
competitive adsorption	竞争吸附	競爭吸附
competitive inhibition	竞争性抑制	競爭性抑制[作用]
competitiveness	竞争性	競爭性
competitive reaction	竞争反应	競爭反應
complementary feedback	互补反馈	互補回饋
complete mixing	全混	完全混合
complete-mixing digester	完全混合消化	完全混合消化槽
complete mixing flow	全混流	完全混合流
complex balancing	络合物平衡	錯合物均衡
complex conjugate root	复数共轭根	複數共軛根
complex method	复合形法	複合形法
complexometric titration	络合滴定	錯合[法]滴定
complex pipe system	复合管道系统	複雜管路系統
complex plane	复平面	複數平面
complex potential	复位势	複勢
complex process	复合过程	複雜程序
complex reaction	复杂反应	複雜反應,錯合反應
complex reaction network	复杂反应网络	複雜反應網路
complex variable	复变量	複變數
complex viscosity	复数黏度	複數黏度
component	组分	成分
component assignment diagram	组分指定图	成分指定圖
composite-averaged equation	复合平均方程式	複合平均方程式
composite liquid	复合液体	複合液體
composite materials	复合材料	複合材料
composite membrane	复合膜	複合薄膜
composite sample	复合样品	複合樣品
composite transfer coefficient	复合传递系数	複合傳送係數
composite wall	复合壁	複合壁
composition	组成	組成
composition control	组成控制	組成控制
composition specification	组成规格	組成規格
composition trajectory	组成轨道	組成軌跡
compositive wall	综合墙	複合壁
compost	堆肥	堆肥
compound	化合物	化合物
compound ball mill	复合球磨机	複合球磨機
compound distillating apparatus	复式蒸馏器	複式蒸餾器

英 文 名	大 陆 名	台 湾 名
compound interest	复利率	複利[率]
compound interest factor	复利因子	複利因數
compound rectifying column	复式精馏塔	複式精餾塔
compressed air	压缩空气	壓縮空氣
compressed air sprayer	压缩空气喷雾器	壓縮空氣噴霧器
compressibility	可压缩性	壓縮性
compressibility coefficient	压缩系数	壓縮係數
compressibility criterion	压缩率判据	壓縮度準則
compressibility factor	压缩因子	壓縮因子
compressible filter cake	可压缩滤饼	可壓縮濾餅
compressible flow	可压缩流动	[可]壓縮流動
compressible fluid	可压缩流体	可壓縮流體
compression	压缩	壓縮
compression-permeability technique	压缩渗透测量法	滲透率壓縮量測
compression process	压缩过程	壓縮程序
compression ratio	压缩比	壓縮比
compression wave	压缩波	壓縮波
compression work	压缩功	壓縮功
compression zone	压缩区	壓縮區
compressive force	压缩力	壓縮力
compressive pressure	压缩压力	壓縮壓力
compressive strength	压缩强度	抗壓強度
compressor	压缩机	壓縮機
compressor specification	压缩机规格	壓縮機規格
computation	计算	計算
computer aided design(CAD)	计算机辅助设计	電腦輔助設計
computer aided engineering(CAE)	计算机辅助工程	電腦輔助工程
computer aided instruction(CAI)	计算机辅助教学	電腦輔助教學
computer aided manufacturing(CAM)	计算机辅助制造	電腦輔助製造
computer control	计算机控制	計算機控制
computer program	计算机程序	電腦程式
computer simulation	计算机模拟	電腦模擬
concentrate	浓缩液	濃縮物
concentration	浓度	濃度,濃縮
concentration boundary layer	浓度边界层	濃度邊界層
concentration bulk	整体浓度	體濃度
concentration cell	浓差电池	濃[度]差電池
concentration difference	浓度差	濃度差

英　文　名	大　陆　名	台　湾　名
concentration diffusion	浓度扩散	濃度擴散
concentration gradient	浓度梯度	濃度梯度
concentration mass	质量浓度	質量濃度
concentration polarization	浓差极化	濃度極化
concentration profile	浓度[分布]剖面[图]	濃度輪廓
concentration-rate curve	浓度速率曲线	濃度速率曲線
concentrator	浓缩器	濃縮器
concentric cylinder viscometer	同心柱黏度计	同軸圓柱黏度計
concentric dial	同心刻度盘	同心儀
concentric float	同心浮球	同心浮子
concentric orifice plate	同心孔口板	同心孔口板
concentric reducer	同心变径管	同心漸縮管
concentric ring	同心环	同心環
concentric-tube column	同心管柱	同心管柱
concrete	混凝土	混凝土
concrete pipe	水泥管	水泥管
concurrent reaction	并发反应	併發反應
condensate	冷凝液	冷凝液
condensation coefficient	缩合系数	縮合係數
condensation point	冷凝点	冷凝點
condensation polymerization	缩聚[反应]	縮[合]聚合[作用]
condensation reaction	缩合反应	縮合反應
condensation tube	冷凝管	冷凝管
condensed phase	凝结相	凝相
condenser	①冷凝器 ②电容器	①冷凝器 ②電容器
condenser paper	电容器纸	電容器紙
condensing chamber	冷凝室	冷凝室
condensing temperature	冷凝温度	冷凝溫度
condensing tube	冷凝管	冷凝管
conditional stability	条件稳定性	條件穩定[性]
conditioning agent	调节剂	調節劑
conductance	电导	電導
conductibility	传导性	傳導性
conduction	传导	傳導
conductive polymer	导电聚合物	導電聚合物
conductivity cell	电导池	電導[測定]槽
conductometer	电导计	電導計,熱導計
conductometric titration	电导滴定	電導滴定[法]

英　文　名	大　陆　名	台　湾　名
conductor catalyst	导体催化剂	導體觸媒
conduit	导线管	導管
cone-and-plate viscometer	锥板式黏度计	錐板黏度計
cone atomizer	锥形雾化器	錐形霧化器
cone crusher	圆锥破碎机	錐碎機
cone feeder	锥形给料器	錐形進料器
cone flow	锥形流动	錐形流動
cone roof tank	锥顶罐	錐頂槽
confidence level	置信水平	信賴水準
confidence limit	置信限	信賴界限
confidence region	置信域	信賴區域
configurational diffusion	组态扩散	組態擴散
configurational partition function	位形配分函数,构型配分函数	組態配分函數
configurational property	位形性质	組態性質
conformal mapping	共形映射	保角映射
conformation	构象	構形
congealing point	冻凝点	凍凝點
congealing temperature	冻凝温度	凍凝溫度
congelation	冻凝作用	凍凝[作用]
congelation point	冻凝点	凍凝點
conjugate depth	共轭深度	共軛深度
conjugate pair	共轭对	共軛對
conjugate phase	共轭相	共軛相
conjugate solution	共轭溶液	共軛溶液
consecutive reaction	连串反应	逐次反應
conservation equation	守恒方程	守恆方程式
conservation law	守恒律	守恆律
conservation law for angular momentum	角动量守恒定律	角動量守恆定律
consistency	①一致性 ②稠度	①一致性 ②稠度
consistency test	①一致性试验 ②稠度试验	①一致性試驗法 ②稠度試驗[法]
consolute temperature	临界共溶温度	[臨界]共溶溫度
constant	常数	常數
constant boiling point	恒沸点	共沸點
constant composition solution	恒组成溶液	恆組成溶液
constant flow rate filtration	恒流率过滤	恆流速過濾
constant head tank	定位槽	定高差槽

英　文　名	大　陆　名	台　湾　名
constant molar overflow	恒摩尔溢流	恆莫耳溢流
constant phase angle curve	等相角曲线	等相角曲線
constant pressure	恒压	恆壓
constant pressure filtration	恒压过滤	恆壓過濾
constant rate drying period	恒速干燥[阶]段	恆速乾燥期
constant shear viscometer	恒剪切力黏度计	恆剪力黏度計
constant speed pump	恒速泵	恆速泵
constant temperature bath	恒温浴	恆溫浴,恆溫槽
constant temperature oven	恒温烘箱	恆溫烘箱
constituent	组分	组分,成分
constitutive equation	本构方程	物性方程式
constraint	①限制 ②约束[条件]	限制
constriction	收缩	收縮
contact angle	接触角	接觸角
contact catalysis	接触催化	接觸催化[作用]
contact chamber	接触室	接觸室
contact condenser	接触冷凝器	接觸冷凝器
contact filtration	接触过滤	接觸過濾
contacting pattern	接触方式	接觸方式
contacting phase	接触相	接觸相
contact nucleation	接触成核	接觸成核[作用]
contactor	接触器	接觸器
contactor efficiency	接触器效率	接觸器效率
contact resistance	接触电阻	接觸電阻,接觸阻力
contact stabilization process	接触稳定过程	接觸穩定程序
container	容器	容器
contaminant	污染物	污染物
contaminated soil	受污染土壤	受污土壤
contamination	污染	污染
content	含量	含量
contingency factor	临时费用因数	應急費用因數
continuity equation	连续性方程	連續方程式
continuity law	连续性定律	連續定律
continuous belt drum filter	连续皮带鼓式过滤器	連續輸送鼓式過濾器
continuous carbon filter	连续式炭滤器	連續式碳濾器
continuous carbon filtration	连续炭滤法	連續碳濾[法]
continuous cash flow	连续现金流动	連續現金流
continuous centrifuge	连续离心机	連續離心機

英 文 名	大 陆 名	台 湾 名
continuous crystallization	连续结晶	連續結晶
continuous crystallizer	连续结晶器	連續結晶器
continuous culture	连续培养	連續式培養
continuous-cycling method	连续循环法	連續環圈法
continuous decantation	连续倾析	連續傾析[法]
continuous digester	连续蒸煮器	連續消化槽,連續蒸煮器
continuous distillation	连续蒸馏	連續蒸餾
continuous drier	连续干燥器	連續乾燥器
continuous extraction	连续萃取	連續萃取
continuous extractor	连续萃取器	連續萃取器
continuous filament	连续细丝	連續細絲
continuous filter	连续过滤器	連續過濾器
continuous filtration	连续过滤	連續過濾
continuous flow reactor	连续流动反应器	連續流動反應器
continuous flow stirred tank reactor	连续流动式搅拌槽反应器	連續攪拌槽反應器
continuous flow system	连续流动系统	連續流動系統
continuous fractionation	连续分馏	連續分餾
continuous grinder	连续研磨机	連續研磨機
continuous kiln	连续窑	連續窯
continuous operation	连续操作	連續操作
continuous phase	连续相	連續相
continuous polymerization	连续聚合	連續聚合[法]
continuous process	连续过程	連續程序
continuous processing	连续加工	連續加工[法]
continuous reactor	连续式反应器	連續式反應器
continuous stirred reactor	连续搅拌反应器	連續攪拌反應器
continuous stirred tank reactor（CSTR）	连续搅拌槽反应器	連續攪拌槽反應器
continuous stripping still	连续汽提器	連續汽提器
continuous tunnel drier	连续隧道干燥器	連續隧道式乾燥器
continuous tunnel kiln	连续隧道窑	連續隧道窯
continuum	连续介质	連續介質
contour plot	等高线图	等高線圖
contraction coefficient	收缩系数	收縮係數
contraction equivalent pipe length	当量收缩管长	縮管相當管長
contraction loss	收缩损失	收縮損失
contraction stability	收缩稳定性	收縮穩定性

英　文　名	大　陆　名	台　湾　名
control	控制	控制
control action	控制作用	控制作用
control algorithm	控制算法	控制算則
control chart	控制图	控制圖
control configuration	控制构型	控制組態
control function	控制函数	控制函數
control interval	控制区间	控制區間
controllability	可控性	可控性
control law	控制律	控制律
controlled field	受控区域	受控區域
controlled release	控[制]释[放]	控制釋放
controlled variable	控制变量	受控變數
controller adaptation	控制器匹配	控制器調適
controller calibration	控制器校正	控制器校正
controller gain	控制器增益	控制器增益
controller mechanism	控制器机理	控制器機構
controller parameter	控制器参数	控制器參數
controller ringing	控制器振铃	控制器振鈴
controller setting	控制器设定	控制器設定
controller tuning	调节器参数整定	控制器調諧
control limit	控制界限	控制界限
control loop	控制回路	控制回路,控制環路
control panel	控制仪表板	控制儀表板
control point	控制点	控制點
control-reactor	反应器控制	反應器控制
control scheme	控制线路	控制方案
control strategy	控制策略	控制策略
control surface	控制表面	控制表面
control system	控制系统	控制系統
control valve	控制阀	控制閥
control variable	控制变量	控制變數
control volume	控制体积	控制體積
convected coordinates	对流坐标	對流坐標
convection	对流	對流
convection heat transfer	对流传热	對流熱傳
convection mass transfer	对流传质	對流質傳
convection section	对流段	對流段
convective acceleration	对流加速度	對流加速

英 文 名	大 陆 名	台 湾 名
convective derivative	对流导数	對流導數
convective flux	对流通量	對流通量
convective heat transfer	对流传热	對流熱傳
convective heat transfer coefficient	对流传热系数	熱對流係數
convergence acceleration	收敛加速	收斂加速
convergence criterion	收敛判据	收斂準則
convergent tube	渐缩管	漸縮管
converging-diverging nozzle	缩扩喷嘴	縮擴噴嘴
conversion	①转化 ②转化率	①轉化 ②轉化率
conversion coating	转换涂层	轉化塗層
conversion constant	换算常数	換算常數
conversion curve	换算曲线	轉化曲線
conversion equation	换算方程式	換算方程式
conversion factor	换算因子	換算因數
conversion level	转化深度	轉化程度
conversion parameter	转化参数	轉化參數
conversion principle	转化原理	轉化原理
conversion rate	转化率	轉化速率
conversion table	换算表	換算表
conversion temperature	转化温度	轉化溫度
conversion-temperature chart	转化率–温度图	轉化率–溫度圖
conversion time	转化时间	轉化時間
converter	①转化器 ②合成塔	轉化器
conveyer	传送机	運送機
conveying	传送	運送
conveyor belt	传送带	輸送帶
convolution	卷积	捲積,摺積,疊積
cook	蒸煮	蒸煮
cooking acid	蒸煮酸	蒸煮酸
cooking liquor	蒸煮液	蒸煮液
cooking process	蒸煮过程	蒸煮程序
cooking tank	蒸煮槽	蒸煮槽
coolant	冷却剂	冷卻劑,冷媒
cooler	冷却器	冷卻器
cooling agent	冷却剂	冷卻劑,冷媒
cooling area	冷却面积	冷卻面積
cooling coil	冷却盘管	冷卻旋管
cooling curve	冷却曲线	冷卻曲線

英　文　名	大　陆　名	台　湾　名
cooling fin	散热片	散熱片
cooling medium	冷却介质	冷卻介質,冷媒
cooling rate	冷却速率	冷卻速率
cooling surface	冷却面	冷卻[表]面
cooling system	冷却系统	冷卻系統
cooling time	冷却时间	冷卻時間
cooling tower	冷却塔	冷卻塔
cooling water	冷却水	冷卻水
cooling zone	冷却段	冷卻區
cool liming	冷浸灰法	冷汁加灰法
cooperative adsorption	合作吸附	合作吸附
coordinates covalence	配位共价	配位共價
coordinate system	坐标系	坐標系
coordination catalyst	络合催化剂	配位觸媒
coordination compound	配位化合物	配位化合物
coordination polymerization	配位聚合	配位聚合[反應]
coordination reaction	配位反应	配位反應
COP(= coefficient of performance)	冷冻系数	性能係數
copolymer	共聚物	共聚物
copolymerization	共聚合	共聚合[反應]
coproduct	联产品	聯產物
coreactant	共反应剂	共反應物
corner flow	拐角流	繞角流
corner tap	转角接头	轉角接頭
corn oil	玉米油	玉米油
corn starch	玉米淀粉	玉米澱粉
corn sugar	玉米糖	玉米糖
corn syrup	玉米糖浆	玉米糖漿
corrective action	校正作用	校正作用
correlating data	相关数据	相關數據,關聯數據
correlation	关联	關聯,相關[性]
correlation function	相关函数	相關函數,關聯函數
corresponding state	对应态	對應[狀]態
corrodent	腐蚀剂	腐蝕質
corrosion	腐蚀	腐蝕
corrosion allowance	腐蚀裕量	允蝕度,腐蝕容許量
corrosion fatigue	腐蚀疲劳	腐蝕疲勞
corrosion inhibitor	缓蚀剂	抑蝕劑,腐蝕抑制劑

英 文 名	大 陆 名	台 湾 名
corrosion prevention	防腐蚀	防蚀
corrosion rate	腐蚀速率	腐蚀速率
corrosion test	腐蚀试验	腐蚀試驗
corrosive action	腐蚀作用	腐蚀作用
corrugated sheet	波纹板	浪板
corrugated tube	波纹管	波形管
corrugating paper	瓦楞纸	瓦楞紙
cortisone	可的松	可體松,皮質酮
cost	成本	成本
cost accounting	成本核算	成本會計
cost benefit analysis	成本效益分析	成本效益分析
cost curve	成本曲线	成本曲線
cost-effective management	成本效益管理	成本效益管理
cost effectiveness	成本效益	成本效益
cost engineer	成本工程师	成本工程師
cost estimate	成本估计	成本估计[值]
cost estimation	成本估计	成本估计
cost factor	成本因子	成本因數
cost index	成本指数	成本指數
cotton paper	棉纸	棉紙
cotton-seed oil	棉籽油	棉籽油
coulometric titration	库仑滴定	電量滴定[法]
Coulter counter	库尔特颗粒计数仪	庫爾特計數器
coumarone	苯并呋喃	苯并呋喃,薰草質,香豆質
coumarone-indene resin	苯并呋喃-茚树脂	苯并呋喃-茚樹脂
counter	计数器	計數器
countercurrent absorption	逆流吸收	對向流吸收
countercurrent column process	逆流塔过程	對向流塔程序
countercurrent decantation	逆流倾析	對向流傾析[法]
countercurrent extraction	逆流萃取	對向流萃取
countercurrent flow	逆流	對向流[動]
countercurrent flow reactor	逆流反应器	對向流反應器
countercurrent jet condenser	逆流喷射式冷凝器	對向流噴凝器
countercurrent network	逆流网络	對向流網路
countercurrent operation	逆流操作	對向流操作
countercurrent process	逆流过程	對向流程序
countercurrent washing	逆流洗涤	對向流洗滌

英 文 名	大 陆 名	台 湾 名
counter diffusion	逆扩散	逆[向]擴散
counter electromotive force	反电动势	反電動勢
counterflow	逆流	對向流[動]
counter range	计数器范围	計數器範圍
counter tube	计数管	計數[器]管
coupled equation	耦合方程式	偶合方程式
coupled system	耦合系统	偶合系統
coupling	①偶合 ②耦合 ③联管节	①偶合 ②接頭
coupling joint	接头	接頭
cP（＝centipoise）	厘泊	厘泊
cracked gas	裂解气,裂化气	裂解氣
cracked gasoline	裂解汽油	裂解汽油
cracking	裂解,裂化	裂解,裂煉
cracking activity	裂化活性	裂解活性
cracking catalyst	裂解催化剂,裂化催化剂	裂解觸媒
cracking chamber	裂解室	裂解室
cracking distillation	裂解蒸馏	裂解蒸餾
cracking fractionator	裂解分馏塔	裂解分餾塔
cracking furnace	裂解炉	裂解爐
cracking reaction	裂解反应	裂解反應
cracking severity	裂解度	裂解度
cradle feeder	摇饲机	搖飼機
creep	蠕变	蠕變,潛變
creeping flow	蠕流	蠕流
creeping motion	蠕变运动	蠕[流運]動
creep test	蠕变试验	潛變試驗
creosote	杂酚油	雜酚油
cresol	甲酚	甲酚
crevice corrosion	缝隙腐蚀	裂隙腐蝕
cricondenbar	临界凝析压力	臨界凝結壓
cricondentherm	临界凝析温度	臨界凝結溫度
cristobalite	方石英	白矽石
criterion	①准则 ②判据	準則
criterion for chemical reaction equilibrium	化学反应平衡判据	化學反應平衡準則
critical back pressure ratio	临界反压比	臨界反壓比,臨界背壓比

英　文　名	大　陆　名	台　湾　名
critical compressibility factor	临界压缩因子	臨界壓縮因子
critical compressibility ratio	临界压缩比	臨界壓縮比
critical concentration	临界浓度	臨界濃度
critical constant	临界常数	臨界常數
critical damping	临界阻尼	臨界阻尼
critical density	临界密度	臨界密度
critical depth	临界深度	臨界深度
critical diameter	临界直径	臨界直徑
critical flow	临界流	臨界流［動］
critical flow capacity	临界流动量	臨界流動量
critical frequency	临界频率	臨界頻率
critical gain	临界增益	臨界增益
critical humidity	临界湿度	臨界濕度
critical isotherm	临界等温线	臨界等溫線
critical mass	临界质量	臨界質量
critical micelle concentration	临界胶束浓度	臨界微胞濃度
critical mixing temperature	临界混合温度	臨界混合溫度
critical moisture content	临界湿含量	臨界水分
critical molar volume	临界摩尔体积	臨界莫耳體積
critical nucleus size	临界晶核尺寸	臨界晶核尺寸
critical packing size	临界填料尺寸	臨界填料尺寸
critical path	临界途径	臨界途徑
critical phenomenon	临界现象	臨界現象
critical point	临界点	臨界點
critical potential	临界电位	臨界勢
critical pressure	临界压力	臨界壓［力］
critical pressure ratio	临界压力比	臨界壓［力］比
critical pressure relief device	临界泄压器件	臨界洩壓裝置
critical property	临界性质	臨界性質
critical size	临界尺寸	臨界尺寸
critical slope	临界底坡	臨界斜率
critical solubility	临界溶解度	臨界溶解度
critical solution point	临界溶解点	臨界溶解點
critical solution temperature	临界共溶温度	臨界互溶溫度
critical sonic property	临界音速性质	臨界音速性質
critical specific volume	临界比容	臨界比容
critical speed	临界转速	臨界速率
critical state	临界状态	臨界條件

英　文　名	大　陆　名	台　湾　名
critical temperature	临界温度	臨界溫度
critical temperature difference	临界温差	臨界溫差
critical throat velocity	临界喉道流速	臨界喉道流速
critical time	临界时间	臨界時間
critical value	临界值	臨界值
critical velocity	临界速度	臨界速度
critical volume	临界体积	臨界體積,臨界容積
crosscurrent	错流	[交]錯流
crosscurrent extraction	错流萃取	[交]錯流萃取
crosscurrent operation	错流操作	[交]錯流操作
crossflow	错流	[交]錯流
crossflow adsorber	错流吸附器	[交]錯流吸附器
crossflow operation	错流操作	[交]錯流操作
cross linkage	交联	交聯
cross-linked polymer	交联聚合物	交聯聚合物
cross-linking	交联	交聯
cross-linking agent	交联剂	交聯劑
cross-linking density	交联密度	交聯密度
crossover	交迭	交越
crossover frequency	交迭频率	交越頻率
crossover gain	交迭增益	交越增益
cross section	截面	截面
cross sectional area	截面积	截面積
crude distillation	原油蒸馏	原油蒸餾
crude oil	原油	原油
crude resources	原油资源	原油資源
crush	破碎	壓碎
crusher	破碎机	壓碎機
cryogenic industry	低温工业	低溫工業
cryogenic process	深度冷冻	[超]低溫程序
cryolite	冰晶石	冰晶石
crystal	晶体	晶體
crystal face	晶面	晶面
crystal form	晶形	晶形
crystal growth	晶体生长	晶體成長
crystal habit	晶体习性	晶體習性,晶癖
crystal lattice	晶格	晶格
crystalline polymer	结晶聚合物	晶質聚合物

英 文 名	大 陆 名	台 湾 名
crystalline silica	晶硅石	晶矽石
crystalline structure	晶形结构	晶體結構
crystalline transition	晶形转变	晶形轉變,晶相轉變
crystallite	微晶	微晶
crystallization	结晶	結晶
crystallizer	结晶器	結晶器
crystallizing point	结晶温度	結晶點
crystallography	结晶学	結晶學
crystal nucleation	晶体成核	晶體成核
crystal seed	晶种	晶種
crystal shape	晶形	晶形
crystal size	晶体大小	晶體大小,晶體尺寸
crystal size distribution	晶体大小分布	晶體粒徑分佈
crystal system	晶系	晶系
crystal water	结晶水	結晶水
CSTR（＝continuous stirred tank reactor）	连续搅拌槽反应釜	連續流動攪拌反應器
CSTR design	连续搅拌槽反应釜设计	連續流動攪拌反應器設計
cubical expansion	体膨胀	體膨脹
cubical expansion coefficient	立方膨胀系数	體膨脹係數
cubic average boiling point	立方平均沸点	立方平均沸點
cubic expansion	立方膨胀	體膨脹
cullet	碎玻璃	玻璃屑
cultivation	培养	培養
cultivator	培养器	培養器
culture	培养	培養,培養菌
cumene	异丙苯	異丙苯
cumene hydroperoxide	过氧化氢异丙苯	異丙苯氫過氧化物
cumulative distribution	累积分布	累積分佈
cumulative probability	累积概率	累積機率
cumulative screen analysis	累积筛析	累積篩析
cumulative weight distribution	累积重量分布	累積重量分佈
cumulative weight percent oversize	累积重量百分率过粗部分	累積過粗重量百分率
cumulative weight percent undersize	累积重量百分率过细部分	累積過細重量百分率
cumulative weight undersize fraction	累积过细重量分率	累積過細重量分率
cuprammonium rayon	铜铵人造纤维	銅銨嫘縈

英 文 名	大 陆 名	台 湾 名
cuprite	赤铜矿	赤銅礦
cure meter	硫化仪	硬化計
curing test	熟化试验	硬化試驗
curing time	固化时间	固化時間
current asset	流动资产	流動資產
current meter	流速计	流速計,電流計
current ratio	流动比	流動比[率]
curvature	曲率	曲率
curve	曲线	曲線
curvilinear coordinates	曲线坐标	曲線坐標
cushion plate	缓冲板	緩衝板
cut diameter	切面直径	切面直徑
cut-off frequency	截止频率	截止頻率
cutting oil	切削油	切削油
CVD (=chemical vapor deposition)	化学气相沉积	化學氣相沈積
cycle length	周期时间	週期
cycle time	周期时间	週期
cyclic operation	循环操作	循環操作,週期操作
cyclic process	循环过程	循環程序
cyclic reaction	周期反应	週期反應
cyclic reactor	周期式反应器	週期式反應器
cyclic stress	周期性应力	週期性應力
cycling	循环	循環
cyclization	环化	環化
cyclization reaction	环化反应	環化反應
cyclize rubber	环化橡胶	環化橡膠
cycloheximide	放线酮	放線菌酮
cycloidal blower	摆线鼓风机	擺旋鼓風機
cyclone	旋风器	旋風器,旋流器
cyclone efficiency	旋风器效率	旋風器效率
cyclone evaporator	旋风蒸发器	旋風蒸發器
cyclone exhauster	旋风排气机	旋風排氣機
cyclone scrubber	旋风洗涤器	旋風洗滌器
cyclone separation	旋风分离	旋風分離
cyclone separator	旋风分离器	旋風分離器
cycloserine	环丝氨酸	環絲胺酸
cylinder drier	圆筒干燥器	圓筒乾燥器
cylindrical coordinates	柱面坐标	圓柱坐標

英 文 名	大 陆 名	台 湾 名
cytolysis	细胞溶解	細胞溶解

D

英 文 名	大 陆 名	台 湾 名
damped oscillation	阻尼振荡	阻尼振盪
damped oscillator	阻尼振荡器	阻尼振盪器
damper	阻尼器	阻尼器,減振器
damping	阻尼	阻尼
damping coefficient	阻尼系数	阻尼係數
damping constant	阻尼常数	阻尼常數
damping fluid	阻尼液	阻尼流體
damping force	阻尼力	阻尼力,減振力
damping ratio	阻尼比	阻尼比
data	数据	數據
data analysis	数据分析	數據分析
data fitting	数据拟合	數據擬合
data inaccuracy	数据不准性	數據不準性
data log	数据日志	數據日誌
data processing	数据处理	數據處理,資料處理
data reconciliation	数据校正	數據校正
data screening	数据筛选	數據篩選,資料篩選
datum	①基准 ②数据	①基準 ②數據(單)
datum temperature	基准温度	基準溫度
daughter cell	子细胞	子細胞
daughter element	子元素	子元素
dB (=decibel)	分贝	分貝
D control (=derivative control)	微分控制	微分控制
DDT (=dichloro-diphenyl-trichloroethane)	滴滴涕	DDT 殺蟲劑
deactivation	失活	失活,去活化
deactivation energy	减活化能	去活化能
deactivator	失活剂	去活化劑
dead band	死带	死帶
dead beat response	不摆应答	不擺應答
dead space	静滞区	靜滯區
dead time	停滞时间	遲延時間
dead water region	死水区	死水區[域]
dead weight gage	静重仪	靜重計

英　文　名	大　陆　名	台　湾　名
dead weight piston gage	静重活塞压力计	靜重活塞壓力計
dead zone	死区	死區
deaeration	脱气	除氣
deaerator	脱气塔	除氣器
deaggregation	解聚集	解聚集[作用]
dealkylation	脱烷基化	脫烷[作用]
deaminase	脱氨酶	去胺酶
deamination	脱氨基	去胺[作用]
dearomatization	脱芳构化	去芳香化[作用]
deashed coal	去灰煤	去灰煤
deashing	去灰分	去灰分
deashing process	去灰分过程	去灰分程序
death phase	死亡期	死亡期
death rate	死亡率	死亡速率
debug	检错	除錯
debutanizer	脱丁烷塔	去丁烷塔
decantation	倾析	傾析[法]
decanter	倾析器	傾析器
decarboxylation	脱羧	脫羧[反應],去羧[作用]
decarburization	脱碳	去碳
decay	衰变	衰變,衰退
decay constant	衰变常数	衰變常數
decay ratio	衰变比	衰變比,衰退比
decentralized control	分散控制	分散控制
dechlorinator	脱氯塔	去氯器
decibel	分贝	分貝
decision making	决策	決策
decision making under risk	风险型决策	風險性決策制定
decision making under uncertainty	不确定型决策	不確定性決策制定
decision tree	决策树	決策樹
decision variable	决策变量	決策變數
decline phase	衰亡期	衰減期
declining balance method	递减均衡法	遞減均衡法
decoking	除焦	除焦,去焦
decoking time	除焦时间	除焦時間,去焦時間
decolorization	脱色	脫色,去色
decolorizer	脱色剂	脫色劑,去色劑

英　文　名	大　陆　名	台　湾　名
decomposer	①分解器　②分解剂	①分解器　②分解劑
decomposition	分解	分解
decomposition efficiency	分解效率	分解效率
decomposition potential	分解电势	分解勢,分解電位
decomposition pressure	分解压力	分解壓[力]
decomposition voltage	分解电压	分解電壓
decompressor	减压器	減壓器
decontaminant	去污剂	去污劑
decontamination factor	去污指数	去污指數
decoupled loops	解耦环路	解偶環路,去偶環路
decoupled system	解耦系统	解偶系統,去偶系統
decoupler	解耦器	解偶器,去偶器
deep-well injection	深井注入[法]	深井注入[法]
deethanizer	脱乙烷塔	去乙烷塔
default value	缺省值	內定值,既定值
defecated juice	澄清汁	澄清汁
defecation	澄清	澄清
defecator	澄清器	澄清器
defect crystal	缺陷晶体	缺陷晶體
defect solid	缺陷固体	缺陷固體
deflagration	爆燃	爆燃
deflagrator	爆燃器	爆燃器
deflection potentiometer	偏离电位计	偏轉電位計
deflection separator	折流分离器	折流分離器
deflocculant	抗絮凝剂	解絮凝劑
deflocculation	解絮凝,反絮凝	解絮凝[作用]
deflocculation agent	抗絮凝剂	解絮凝劑
defluidization	反流态化	未流體化
defoaming agent	消泡剂	消泡劑
deformation	变形	變形
deformation coordinates	变形坐标	變形坐標
deformation eutectic	变形低共熔物	變形共熔物
deformation point	变形点	變形點
deformation range	变形范围	變形範圍
deformation rate	变形速率	變形速率
deformation work	变形功	變形功
degasification	脱气	除氣
degasifier	脱气器	除氣器

英　文　名	大　陆　名	台　湾　名
degeneracy	退化	退化,减併
degeneration	退化	退化
degeneration form	退化形式	退化形式
degradable plastic	可降解塑料	可降解塑膠
degradation	降解	降解,裂解
degrease	脱脂	脱脂
degree of consolidation	压实度	壓密度
degree of cross-linking	交联度	交聯度
degree of crystallization	结晶度	結晶度
degree of dispersion	分散度	分散度
degree of expansion	膨胀度	膨脹度
degree of freedom	自由度	自由度
degree of hardness	硬度	硬度
degree of interaction	相互作用度	相互作用度
degree of intermixing	互混度	互混度
degree of ionization	电离度	游離度
degree of mixing	混合度	混合度
degree of penetration	针入度	穿透度
degree of polymerization	聚合度	聚合度
degree of refining	精炼度	精煉度
degree of reproducibility	再现度	再現度
degree of reversibility	可逆度	可逆度
degree of saturation	饱和度	飽和度
degree of segregation	离析度	離析度
degree of stability	稳定度	穩定度,安定度
degree of supersaturation	过饱和度	過飽和度
degree of swelling	溶胀度	膨潤度
degree of turbulence	湍流度	紊流度
dehalogenation	脱卤	脱卤[反應],去卤[反應]
dehumidification	减湿	除濕
dehumidifier	除湿器	除濕器
dehydrate	脱水物	脱水物
dehydrated alcohol	脱水酒精	脱水酒精,無水酒精
dehydration	脱水	脱水
dehydrator	脱水器	脱水器
dehydrocyclization	脱氢环化作用	脱氫環化[作用]
dehydrogenase	脱氢酶	脱氫酶

英　文　名	大　陆　名	台　湾　名
dehydrogenation	脱氢	脱氢[反應],去氢[作用]
dehydrogenation catalyst	脱氢催化剂	脱氢觸媒
deionization	去离子作用	去離子[作用]
deionized water	去离子水	去離子水
delaminated clay	层离黏土	剝層黏土
delamination	分层	剝層[作用]
delay	延迟	延遲,遲延
delay action	延迟作用	延遲作用,遲延作用
delay time	延迟时间	延遲時間,遲延時間
delignification	去木素作用	去木質素[作用]
deliquescence	潮解	潮解
delusterant	消光剂	退光劑,去光劑
delustering	消光	退光,去光
demethanization	脱甲烷	去甲烷[作用]
demethanizer	脱甲烷塔	去甲烷塔
demineralization	去矿化	去礦[物]質
demineralized water	软化水	去礦[物]質水
demineralizer	软化器	去礦[物]質劑
demineral water	软化水	去礦[物]質水
demister	除沫器	除霧器
demonstration unit	示范装置	示範單元
demulsification	去乳化	去乳化[作用]
denaturant	变性剂	變性劑
denaturation	变性	變性
denatured alcohol	变性酒精	變性酒精
denitration	脱硝	脱硝[反應],去硝[作用]
denitration tower	脱硝塔	脱硝塔,去硝塔
denitrator	脱硝器	脱硝器
denitrification	①脱硝作用 ②脱氮作用	①脱硝[作用] ②脱氮[作用]
denitrifying bacteria	脱氮细菌	脱氮細菌
denitrogenation	脱氮	脱氮[反應],去氮[作用]
dense medium separation	密媒分离	重媒分離
dense-phase	密相	稠相
dense-phase flow	密相流动	重相流動

英 文 名	大 陆 名	台 湾 名
dense-phase fluidized bed	密相流化床	重相流[體]化床
dense soda ash	重苏打灰	重鹼灰
densimeter	密度计	密度計
densitometer	感光密度计	感光密度計
density	密度	密度
density gradient centrifugation	密度梯度离心	密度梯度離心
density index	密度指数	密度指數
deodorant	除臭剂	除臭劑
deodorization	除臭	除臭
deodorizer	除臭器	除臭器
deoxidation	去氧化	去氧化[作用]
deoxygenation	脱氧	去氧[作用]
departure function	偏离函数	偏離函數
dependent variable	因变量	因變數
depletion	耗尽	耗盡
depletion allowance	耗尽容许度	耗盡容許度
depolymerization	解聚	解聚合[作用]
depreciation	折旧	折舊
depreciation cost	折旧费	折舊成本
depreciation reserve	折旧备用金	折舊準備金
depropagation reaction	负增长反应	去傳遞反應
derivative	①衍生物 ②导数 ③微分	①衍生物 ②導數 ③微分
derivative control (D control)	微分控制	微分控制
derivative thermogravimetry	微商热重法	微分熱重量法
derivative time	微分时间	微分時間
derivative time constant	微分时间常数	微分時間常數
derived unit	导出单位	導出單位
desalination	脱盐	脱鹽
desaturation	去饱和作用	去飽和作用
describing function	描述函数	描述函數
desiccant	干燥剂	乾燥劑
desiccation apparatus	干燥器皿	乾燥器
desiccator	干燥器	乾燥器
design	设计	設計
design chart	设计图	設計圖
design criterion	设计准则	設計準則
design data	设计数据	設計數據

英 文 名	大 陆 名	台 湾 名
design engineer	设计工程师	設計工程師
design equation	设计方程式	設計方程式
design mode	设计模型	設計模式
design package	成套设计文件	設計套件
design parameter	设计参数	設計參數
design project	设计项目	設計專案
design standard	设计标准	設計標準
design tool	设计工具	設計工具
design variable	设计变量	設計變數
desired product	目标产物	目標產物
desired value	目标值	目標值
desolvation	去溶剂化	去溶劑合[作用],去媒合[作用]
desolventizer	脱溶剂器	脫溶劑器
desorption	解吸	脫附
desorption control	脱附控制	脫附控制
desorption factor	解吸因子	脫附因子
desorption isotherm	脱附等温线	脫附等溫線
destabilization	去稳定作用	去穩定作用
destructive distillation	分解蒸馏	分解蒸餾
desublimation	凝华作用	反升華[作用]
desulfurization	脱硫	脫硫[作用],去硫[作用]
desuperheat	去过热	去過熱
desuperheater	去过热器	去過熱器
detailed balancing	细致平衡	細部均衡
detailed engineering design	详细工程设计	細部工程設計
detector	检测器	偵檢器
detention period	停滞期间	停滯期間
detention time	停滞时间	停滯時間
detergency	去垢力	清潔力
detergent	洗涤剂	清潔劑
determinant	行列式	行列式
determination	测定	測定
deterministic model	确定性模型	確定性模型
detonation	起爆	起爆
detonation time	爆炸时间	爆炸時間
detoxification	解毒	解毒[作用]

英　文　名	大　陆　名	台　湾　名
detrition	磨耗	磨耗
developer	显影剂	顯影劑
development project	开发项目	發展專案
deviation	偏差	偏差
deviation form	偏差形式	偏差形式
deviation variable	偏差变数	偏差變數
deviatoric stress	偏应力	偏向應力,軸差應力
devitrification	脱玻化	去玻化
devulcanization	去硫化	去硫化
dewatering	脱水	脱水
dewaxing	脱蜡	脱蠟
dew point	露点	露點
dew point meter	露点计	露點計
dew point pressure	露点压力	露點壓力
dew point recorder	露点记录器	露點記錄器
dew point temperature	露点温度	露點溫度
dextran	葡聚糖	聚葡[萄]糖
dextrin	糊精	糊精
dextrose	葡萄糖	葡萄糖,右旋糖
dial indicator	刻度盘指示器	針盤指示器
dialysis	透析	透析[作用]
dialysis cell	透析槽	透析槽,滲析槽
dialysis culture	透析培养	透析培養
dialyzate	透析液	透析物,滲析物
dialyzator	透析器	透析器
dialyzer	透析器	透析器,滲析器
diameter	直径	直徑
diaphragm	隔膜片	隔膜,膜片
diaphragm box	隔膜盒	隔膜盒
diaphragm box level gauge	隔膜盒液位计	隔膜盒液位計
diaphragm cell	隔膜电解槽	隔膜[電解]槽
diaphragm chamber	隔膜室	隔膜室
diaphragm compressor	隔膜压缩机	隔膜壓縮機
diaphragm control valve	隔膜控制阀	隔膜控制閥
diaphragm operated valve	隔膜操作阀	隔膜操作閥
diaphragm pressure gage	膜式压力计	隔膜壓力計
diaphragm pump	隔膜泵	隔膜泵
diaphragm screen	隔膜筛	隔膜篩

英　文　名	大　陆　名	台　湾　名
diaphragm-spring actuator	膜簧式驱动器	膜簧式致動器
diaphragm type pressure transducer	隔膜式压力转换器	隔膜壓力轉換器
diaphragm type strain gage	隔膜式应变计	隔膜應變計
diaphragm valve	隔膜阀	隔膜閥
diaspore	硬水铝石	水鋁石
diastase	淀粉酶制剂	[澱粉]糖化酶
diathermal wall	绝热壁	絕熱壁
diatomaceous earth	硅藻土	矽藻土
diatomaceous-earth filter	硅藻土过滤器	矽藻土過濾器
diatomite	硅藻土	矽藻土
diazo compound	重氮化合物	重氮化合物
diazo dye	重氮染料	重氮染料
diazotization	重氮化	重氮化[反應]
dichloro-diphenyl-trichloroethane(DDT)	滴滴涕	DDT 殺蟲劑
dichotomous search	对分搜索	二分法搜尋
die	模[具]	模
die casting	压铸	壓鑄,模鑄
die casting molding	压铸成型	壓鑄成型
dielectric	介电体	介電質
dielectric absorption	介电吸收	介電吸收
dielectric constant	介电常数	介電常數
dielectric heating	介电加热	介電加熱
diesel engine	柴油机	柴油引擎,柴油機
diesel fuel	柴油	柴油[燃料]
diesel knock	柴油爆震	柴油震爆
die swell	模口膨胀	模頭膨脹
difference equation	差分方程	差分方程[式]
differential absorption	微分吸收	微分吸收,微差吸收
differential adsorption	微分吸附	微分吸附,微差吸附
differential analyzer	微分分析仪	微分分析儀,微差分析儀
differential centrifugation	差速离心分离	差速離心分離
differential condensation	微分冷凝	微分冷凝,微差冷凝
differential contact equipment	微差接触设备	微差接觸設備,微分接觸設備
differential control	微分控制	微分控制
differential dilatometry	示差热膨胀测量术	微差膨脹測定法
differential dilatometry	微分膨胀计测定法	微分熱膨脹法

英　文　名	大　陆　名	台　湾　名
differential distillation	微分蒸馏	微分蒸餾,微差蒸餾
differential energy balance	微分能量平衡	微分能量均衡
differential equation	微分方程	微分方程[式]
differential gap	微差间隙	微差間隙
differential head	差异压头	微分高差
differential heat of dilution	微分稀释热	微分稀釋熱
differential heat of solution	微分溶解热	微分溶解熱
differential manometer	差示压力计	微差壓力計
differential mass balance	微分质量平衡	微分質量均衡
differential method	微分法	微分法
differential pressure	压差	微差壓
differential pressure cell	差压池	微差壓計
differential pressure limit	差压极限	微差壓極限
differential pressure transducer	差压传感器	微差壓轉換器
differential reactor	微分反应器	微分反應器,微差反應器
differential scanning calorimetry	示差扫描量热法	微差掃描熱量法
differential screen analysis	微分筛析	微分篩析,微差篩析
differential settling	微分沉降	微分沈降,微差沈降
differential surface	微分表面	微分表面,微差表面
differential technique	微分法	微分法
differential thermal analysis	差热分析[法]	微差熱分析[法]
differential thermogravimetric analysis	差示热重分析法	微差熱重分析[法]
differential thermomechanical analysis （DTMA）	差示热机械分析法	微差熱機械分析[法]
differential thermometer	微差温度计	微差溫度計
differential thermometry	微差测温法	微差測溫法
differential U-tube	差示 U 形管	微差 U[形]管
differential valve	差动阀	[微]差壓閥
differential weight distribution	微分重量分布	微差重量分佈
differential yield	微分产率	微分產率
differentiation	微分法	微分[法]
differentiator	微分器	微分器
diffraction	绕射	繞射
diffuse layer	扩散层	擴散層
diffuser	扩散器	擴散器
diffuser casing	扩散器护罩	擴散器護罩
diffusion	扩散	擴散

英 文 名	大 陆 名	台 湾 名
diffusion barrier	扩散膜	擴散障壁
diffusion catalysis	扩散催化	擴散催化[作用]
diffusion coefficient	扩散系数	擴散係數
diffusion constant	扩散常数	擴散常數
diffusion control	扩散控制	擴散控制
diffusion-controlled reaction	扩散控制反应	擴散控制反應
diffusion current	扩散电流	擴散電流
diffusion equation	扩散方程	擴散方程[式]
diffusion film	扩散薄膜	擴散薄膜
diffusion flux	扩散通量	擴散通量
diffusion mass transfer	扩散传质	擴散質傳
diffusion potential	扩散电位	擴散勢
diffusion pressure	扩散压力	擴散壓
diffusion process	扩散过程	擴散程序
diffusion pump	扩散泵	擴散泵
diffusion rate	扩散速率	擴散速率
diffusion resistance	扩散阻力	擴散阻力
diffusion ring	扩散环	擴散環
diffusion-thermo effect	扩散热效应	擴散熱效應
diffusion time	扩散时间	擴散時間
diffusion vacuum pump	扩散真空泵	擴散真空泵
diffusivity	扩散系数	擴散係數
digester	消化槽	消化槽
digestion	消化	消化
digital computer	数字计算机	數位電腦,數位計算機
digital control	数字控制	數位控制
digital recorder	数字记录器	數位記錄器
diglyceride	甘油二酯	雙甘油酯
digraph	有向图	有向圖
dilatancy	胀塑性	膨脹性
dilatant fluid	胀塑性流体	脹塑性流體,剪力增黏流體
dilatation	膨胀	膨脹
dilatational stress	膨胀应力	膨脹應力
dilation	膨胀	膨脹
dilatometer	膨胀计	膨脹計
dilatometry	膨胀测定法	膨脹測量法
dilute phase	稀相	稀[釋]相

英　文　名	大　陆　名	台　湾　名
dilute-phase fluidized bed	稀相流化床	稀相流[體]化床
dilution	稀释	稀釋
dilution effect	稀释效应	稀釋效應
dimension	①因次 ②量纲	①因次 ②尺寸
dimensional analysis	量纲分析,因次分析	因次分析
dimensional consistency	量纲一致性	因次一致性
dimensional homogeneity	量纲均匀性	因次均匀性
dimensional invariance	量纲不变性	因次不變性
dimensional stability	尺寸稳定性	尺寸穩定性
dimensionless group	无量纲数群	無因次群
dimensionless number	无量纲数	無因次數
dimensionless parameter	无量纲参数	無因次參數
dimerization	二聚	二聚合
diode	二极管	二極體
diode function generator	二极函数发生器	二極函數產生器
dip bleaching	浸漂	浸漂
dip dyeing	浸染	浸染
dipeptidase	二肽酶	二肽酶
dip glazing	浸渍釉	浸釉
dipole	偶极	偶極
dipole moment	偶极矩	偶極矩
dipping	浸渍	浸漬
dipping tank	浸渍槽	浸漬槽
dip pipe level gage	浸管液位计	浸管液位計
dip tinning	浸渍镀锡	浸漬鍍錫
dip tube	倾斜管	傾斜管
direct-acting controller	直接作用控制器	直接作用控制器
direct coal liquefaction	直接煤液化	直接煤液化[法]
direct control	直接控制	直接控制
direct cost	直接成本	直接成本
direct digital control	直接数字控制	直接數位控制
direct dye	直接染料	直接染料
direct-fired furnace	直接火焰炉	明火加熱爐
direct gate	直接浇口	直接浇口,直接模口
direct hydrogenation	直接氢化	直接氫化[反應]
direct liquefaction	直接液化	直接液化[法]
direct method	直接法	直接法
direct production cost	直接生产成本	直接生產成本

英　文　名	大　陆　名	台　湾　名
direct pyrolysis	直接热解	直接熱解
direct reduction	直接还原	直接還原[法]
direct search method	直接搜索法	直接搜索法
disaccharide	双糖	雙醣
disc attrition mill	圆盘磨	圓盤磨
disc centrifuge	盘式离心机	盤式離心機
disc column	圆盘塔	圓盤塔
disc crusher	盘式压碎机	盤碎機
disc evaporator	盘式蒸发器	盤式蒸發器
disc filter	盘滤机	盤濾機
disc flow condenser	圆盘流动式冷凝器	圓盤流動冷凝器
discharge coefficient	①流量系数 ②孔流系数	排放係數
discharge head	排出压头	排放高差
discharge header	排放集管	排放集管[箱]
discharge loss	排放损失	排放損失
discharge pressure	排放压力	排放壓力
discharge time	排放时间	排放時間
discharge valve	排出阀	排放閥
disc meter	盘式流量计	盤式流量計
discontinuous process	不连续过程	不連續程序
discontinuous processing	不连续加工	不連續加工[法]
discount breakeven point	折现盈亏平衡点	折現損益均衡點
discounted cash flow rate of return	折现收益率	折現收益率
discount factor	折现因子	折現因數
discrete settling	不连续沉降	個別沈降
disinfectant	消毒剂	消毒劑
disinfection	消毒	消毒
disintegration mechanism	衰变机理	崩解機制
disintegrator	破碎机	崩解機
disk	圆盘	圓盤
disk column absorber	盘柱吸收器	盤柱吸收器
disk dryer	圆盘干燥器	圓盤乾燥機
disk feeder	圆盘给料机	盤飼機
disk filter	盘式过滤机	盤濾機
disk gate	圆盘浇口	圓盤澆口
dislocation	转位	差排
dispersant	分散剂	分散劑
dispersed-air flotation	曝气浮选[法]	曝氣浮選[法]

英　文　名	大　陆　名	台　湾　名
dispersed dye	分散性染料	分散染料
dispersed flow	分散流,弥散流	分散流
dispersed medium	分散介质	分散介質,分散媒
dispersed phase	分散相	分散相
dispersed slug flow	分散塞流	分散塞流
dispersed system	分散系统	分散系統
disperse phase	分散相	分散相
disperser	①分散剂 ②分散器	①分散劑 ②分散器
dispersibility	分散性	分散性
dispersing agent	分散剂	分散劑
dispersion	分散	分散[作用]
dispersion coefficient	分散系数,弥散系数	分散係數
dispersion curve	色散曲线	分散曲線
dispersion force	色散力	分散力
dispersion mill	分散磨	分散磨機
dispersion model	弥散模型	分散模型
dispersion reagent	分散剂	分散劑
dispersive power	分散本领	分散能力
displacement	①位移 ②置换	①位移 ②置換
displacement float level gage	移标式液位计	移標式液位計
displacement flowmeter	位移流量计	位移流量計
displacement meter	位移计	位移[流量]計
displacement pressure gage	位移压力计	位移壓力計
displacement reaction	置换反应	置換反應
dissimilation	异化[作用]	異化[作用]
dissipated energy	耗散能	散逸能
dissipation	耗散	散逸
dissipation function	耗散函数	散逸函數
dissociation temperature	解离温度	解離溫度
dissociative adsorption	解离吸附	解離吸附
dissolution	溶解	溶解
dissolved air	溶解空气	溶解空氣
dissolved-air flotation	溶气浮选[法]	溶氣浮選[法]
dissolved oxygen	溶解氧	溶氧
dissolved oxygen analyzer	溶解氧分析仪	溶氧分析儀
dissolved oxygen probe	溶解氧探头	溶氧探針
dissolved solid	溶解固体	溶解固體
dissolver	溶解器	溶解器

英　文　名	大　陆　名	台　湾　名
distance factor	距离因数	距離因數
distance velocity lag	距速滞后	距速滯延
distillate	馏出液	餾出液
distillating apparatus	蒸馏器	蒸餾裝置
distillation	蒸馏	蒸餾
distillation column	蒸馏柱	蒸餾柱
distillation range	馏程	蒸餾範圍
distillation still	蒸馏釜	蒸餾器
distillation test	蒸馏试验	蒸餾試驗
distillation trap	蒸馏阱	蒸餾阱
distillation tray	蒸馏塔板	蒸餾塔板
distillation value	蒸馏值	蒸餾值
distillation zone	蒸馏区	蒸餾區
distilled water	蒸馏水	蒸餾水
distiller	蒸馏器	蒸餾器
distilling apparatus	蒸馏器	蒸餾裝置
distilling condenser	蒸馏冷凝器	蒸餾冷凝器
distilling flask	蒸馏瓶	蒸餾瓶
distributed component	分布成分	分佈成分
distributed control system	集散控制系统	分散控制系統
distributed lag	分布滞后	分佈滯延
distributed-parameter model	分布参数模型	分佈參數模式
distributed-parameter system	分布参数系统	分散參數系統
distributed property	分布性质	分佈性質
distributed system	分布式系统	分散系統
distribution coefficient	分配系数	分配係數
distribution cost	分销成本	分銷成本
distribution curve	分布曲线	分佈曲線,分配曲線
distribution function	分布函数	分佈函數,分配函數
distribution law	分配定律	分配[定]律
distribution plate	分布板	分佈板
distribution ratio	分配比	分配率
distributive law	分配律	分配律
distributor	分布器,分配器	分佈器,分配器
distributor arm	分布器臂	分配器臂
disturbance	扰动	擾動
diversion valve	转向阀	轉向閥
dividing surface	分界表面,界面相	分界面

英　文　名	大　陆　名	台　湾　名
Dixon ring	θ网环	θ網環,狄克森網環
dolomite	白云石	白雲石
dolomite cement	白云石水泥	白雲石水泥
dolomite earthenware	白云石陶器	白雲石陶器
dolomitic lime	白云石石灰	白雲石石灰
dolomitic limestone	白云石质石灰石	白雲石石灰石
dominant eigenvalue	主要本征值	主宰固有值,優勢固有值
dominant pole	主导极点	主要極點
dominant root	主根	主根
donor	给体	［給］予體
dopant	掺杂剂	掺雜劑
dope	掺入	掺雜
doping	掺杂	掺雜
dosage	剂量	劑量
dose	剂量	劑量
dosing	用剂	用劑
double acting pump	双动泵	雙動泵
double arm kneading mixer	双臂捏合机	雙臂捏合機
double column	双重塔	雙重塔
double cone blender	双锥掺合机	雙錐掺合機
double cone classifier	双锥分级机	雙錐分級機
double declining balance method	双倍定率递减折旧法	雙倍遞減均衡法
double decomposition reaction	复分解反应	複分解
double distilled water	再蒸馏水	再蒸餾水
double effect evaporation	双效蒸发	雙效蒸發
double effect evaporator	双效蒸发器	雙效蒸發器
double entry bookkeeping	复式簿记	複式簿記
double filter	双式过滤器	雙式濾器
double-glazed insulating pane	双层隔热玻璃板	雙層隔熱玻璃板
double helical ribbon mixer	双螺带混合机	雙螺帶混合機
double motion agitator	双动搅拌器	雙動攪拌器
double paddle mixer	双桨混合器	雙漿混合機
double pipe cooler	套管冷却器	［雙］套管冷卻器
double-pipe heat exchanger	套管热交换器	［雙］套管熱交換器
double pipe reactor	套管反应器	［雙］套管反應器
double ported valve	双口阀	雙口閥
double promoted catalyst	双重促进催化剂	雙重促進觸媒

英 文 名	大 陆 名	台 湾 名
double roll crusher	双辊轧碎机	雙輥軋碎機
double seat valve	双座阀	雙座閥
double sizing	双重上胶	雙重上膠
double slidewire bridge	双滑线电桥	雙滑線電橋
double solvent extraction	双溶剂萃取	雙溶劑萃取
double suction impeller	双吸叶轮机	雙吸葉輪機
double superphosphate	富过磷酸钙	重過磷酸鹽
doubling calender	重合压延机	重合壓延機
doubling time	倍增时间	倍增時間
dough mixer	和面机	和麵機
downcomer	降液管	降流管
downcomer backup	降液管液柱高度	降液管液柱高度
downdraft kiln	倒风窑	倒焰窯
downflow	降液	降流
downflow reactor	下行式反应器	降流反應器
down-pipe	降液管	降流管
downstream	下游	下游
downstream pressure	下游压力	下游壓力
downstream process	下游过程	下游程序
downstream processing	下游处理	下游處理
draft fan	通风扇	通風扇
draft gage	①通风计 ②差式风压计	①通風計 ②差式風壓計
draft tube	导流筒	導流管
draft tube baffle	导流筒挡板	導流管擋板
draft-tube-baffled crystallizer	导流筒挡板结晶器	導流筒擋板結晶器
draft tube mixer	导流筒混合器	導流管混合器
drag coefficient	阻力系数	阻力係數
drag effect	减阻	拖曳效應
drag force	曳力	[拖]曳力
drag reduction	曳力降低	[拖]曳力降減
drag scraper	拖刮机	拖刮機
drag torque type viscometer	扭矩式黏度计	扭矩式黏度計
drain	排放口	排放口
draining table	排水台	排水檯
drain line	排水管道	排洩管線
drain pipe	排泄管	排洩管
drain pump	排泄泵	排洩泵
drain trap	排泄阱	排洩阱

英　文　名	大　陆　名	台　湾　名
drain valve	放泄阀	排洩閥
draw bar	拉杆	拉桿
draw gang	切工组	切工組
drawing apparatus	绘图仪	製圖儀,繪圖儀
drawing paper	绘图纸	繪圖紙
drawoff cock	排出栓	排出拴
drier bin	干燥箱	乾燥箱
drift	漂移	漂移
drip	滴注器	滴口
drip condenser	水淋冷凝器	滴水冷凝器
drip dry	滴干法	滴乾
driver	驱动器	驅動器
driver horsepower	驱动功率	驅動功率
driving force	推动力	驅動力
driving function	驱动函数	驅動函數
drop	①滴 ②差降	①滴 ②差降
drop breakup	液滴瓦解	液滴瓦解
droplet	微滴	小滴
drop phase	滴相	滴相
dropwise condensation	滴状冷凝	滴式冷凝
drum	筒	桶
drum barker	桶式去皮机	桶式去皮機
drum boiler	桶式锅炉	桶式鍋爐
drum drier	鼓式干燥器	鼓式乾燥器,桶式乾燥器
drum filter	鼓型过滤机	桶式過濾器
drum separator	鼓式分离器	桶式分離器
drum type counter	鼓式计数器	鼓式計數器
drum type recorder	鼓式记录器	鼓式記錄器
drum type scale indicator	鼓式标度指示计	鼓式標[度指]示計
drum washing	转筒洗涤	轉桶洗滌
dry-and-wet-bulb thermometer	干湿球温度计	乾濕球溫度計
dry-bulb temperature	干球温度	乾球溫度
dry-bulb thermometer	干球温度计	乾球溫度計
dry cell	干电池	乾電池
dry combustion	干式燃烧	乾式燃燒
dry distillation	干馏	乾餾
dryer	①干燥器 ②干燥剂	①乾燥機 ②乾燥劑

英 文 名	大 陆 名	台 湾 名
dry heat sterilization	干热灭菌	乾熱滅菌
drying	干燥	乾燥
drying agent	干燥剂	乾燥劑
drying cabinet	干燥橱	乾燥櫥
drying drum	干燥转鼓	乾燥桶
drying filter	干滤器	乾式過濾器
drying hack	干燥架	乾燥架
drying house	干燥房	乾燥房
drying kiln	干燥窑	烘窯
drying oil	干性油	乾性油
drying oven	干燥烘箱	乾燥烘箱
drying rate	干燥速率	乾燥速率
drying-rate curve	干燥速率曲线	乾燥速率曲線
drying room	干燥室	乾燥室
drying shed	干燥棚	乾燥棚
drying shrinkage	干燥收缩	乾燥收縮
drying stove	干燥炉	乾燥爐
drying tower	干燥塔	乾燥塔
drying tray	干燥盘	乾燥盤
drying tunnel	干燥烘道	乾燥烘道
drying zone	干燥区	乾燥區
dry oxidation	干式氧化	乾式氧化法
dry process	干燥过程	乾法
dry spinning	干纺	乾紡[絲]
dry steam	干蒸汽	乾蒸汽
dry surface	干燥面	乾燥面
dual-flow tray	穿流塔板	雙流塔板
dual-function catalyst	双功能催化剂	雙功能觸媒
duality	对偶[性]	對偶[性], 二元
dual-medium filter	双介质过滤器	雙介質過濾器
dual-mode control	双式控制	雙[模]式控制
dual-mode system	双模式系统	雙[模]式系統
dual recirculation	双重循环	雙重循環
dual-site mechanism	双部位机理	雙[部]位機構
duct	导管	導管
ductile materials	延展性材料	延展性材料
dummy activity	虚拟活性	虛擬活性
dummy plate	隔板	隔板

英　文　名	大　陆　名	台　湾　名
dummy variable	虚拟变数	啞變數
dumped packing	散装填料	堆積填充
dumper	卸料器	卸料器
duplex control	双用控制	雙工控制
duplex reciprocating pump	双缸往复泵	雙缸往復泵
duriron	耐酸硅铁	高矽鐵
durometer	硬度计	硬度計
dust	粉尘	粉塵
dust collection	集尘	集塵
dust collector	集尘器	集塵器
dust content	含尘量	含塵量
dust cyclone	旋风除尘器	旋風除塵器
dust cyclone separator	粉尘旋风分离器	粉塵旋風分離器
dust equipment	粉尘装置	粉塵裝置
dust-laden gas	含尘气体	含塵氣體
dust particle	粉尘粒子	粉塵粒子
dust particle size	粉尘粒子大小	粉塵粒子大小
dust separator	粉尘分离器	粉塵分離器
duty factor	负载因数	負載因子
dye	染料	染料
dyeability	可染性	可染性
dye affinity chromatography	染料亲和色谱法	染料親和層析法
dyeing	染色	染色[法]
dyeing machine	染色机	染色機
dye uptake method	染料摄入法	染料攝入法
dynamic analysis	动态分析	動態分析
dynamic behavior	动态行为	動態行為
dynamic compensation	动态补偿	動態補償
dynamic decoupler	动态去偶器	動態解偶器
dynamic element	动态元素	動態元素
dynamic equation	动力学方程	動態方程式
dynamic equilibrium	动态平衡	動態平衡
dynamic error	动态误差	動態誤差
dynamic error coefficient	动态误差系数	動態誤差係數
dynamic lag	动态滞后	動態滯延
dynamic model	动态模型	動態模型
dynamic modulus	动态模量	動態模數
dynamic optimization	动态最优化	動態最適化

英　文　名	大　陆　名	台　湾　名
dynamic pressure	动压[力]	動壓[力]
dynamic process	动态过程	動態程序
dynamic programming	动态规划	動態規劃
dynamic reflectance spectroscopy	动态反射光谱法	動態反射光譜法,動態反射光譜學
dynamic response	动态响应	動態應答
dynamic rigidity	动态刚性	動態剛性
dynamics	动力学	動力學
dynamic similarity	动力学相似性	動態相似性
dynamic simulation	动态模拟	動態模擬
dynamic system	动态系统	動態系統
dynamic test	动态试验	動態試驗
dynamic thermal mechanical method	动态热机械法	動態熱機械法
dynamic thermogravimetry	动态热重量分析法	動態熱重量法
dynamic viscosity	动力黏度	[動力]黏度
dynamite	硝化甘油炸药	硝化甘油炸藥,代納邁炸藥

E

英　文　名	大　陆　名	台　湾　名
earning rate	收益率	收益率
earthenware	粗陶器	陶器
earthenware clay	陶土	陶土
earth pigment	土质颜料	土質顏料
earth wax	地蜡	地蠟
ebonite	硬质胶	硬橡膠
ebony wax	乌木蜡	烏木臘
ebullated bed	沸腾床	沸騰床
ebullated bed reactor	沸腾床反应器	沸騰床反應器
ebullition	沸腾	沸騰
eccentricity	偏心率	偏心率
eccentric orifice plate	偏心孔口板	偏心孔口板
eccentric reducer	偏心渐缩管	偏心漸縮管
eccentric ring	偏心环	偏心環
eccentric scale	偏心秤	偏心刻度
economic evaluation	经济评价	經濟評估
economic feasibility	经济可行性	經濟可行性

英　文　名	大　陆　名	台　湾　名
economic impact	经济冲击	經濟衝擊
economic life	经济寿命	經濟壽命
economic pressure drop	经济压力降	經濟壓力降
economic reflux ratio	经济回流比	經濟回流比
economizer	节能器	節能器
ecosystem	生态系统	生態系統
eddy	[旋]涡	渦流
eddy absorption	涡流吸收	渦流吸收
eddy current	涡流	渦電流
eddy current loss	涡流损耗	渦流損失
eddy current tachometer	涡流转速计	渦流轉速計
eddy diffusion	涡流扩散	渦流擴散
eddy diffusivity	涡流扩散系数	渦流擴散係數
eddy flow	涡流	渦流
eddy kinematic viscosity	涡流运动黏度	渦流動黏度
eddy length	涡流长度	渦流長度
eddy motion	涡流运动	渦流運動
eddy residence time	涡流停留时间	渦流滯留時間
eddy resistance	涡流阻力	渦流阻力
eddy shedding	涡流卸掉	渦流分離
eddy stress	涡流应力	渦流應力
eddy transport	涡流传递	渦流輸送
eddy viscosity	涡黏性	渦流黏度
edge effect	边缘效应	邊緣效應
edge runner	轮碾机	輪輾機
edible fat	食用脂肪	食用脂肪
edible tallow	食用牛脂	食用牛脂
educt	浸提物	浸提物
eductor	喷射器	噴射器
effective aeration	有效曝气	有效曝氣
effective concentration	有效浓度	有效濃度
effective density	有效密度	有效密度
effective diffusivity	有效扩散系数	有效擴散係數
effective grain size	有效粒度	有效晶粒大小
effective interest	有效利息	有效利息
effectiveness	有效性	有效性,有效度
effectiveness factor	有效系数	有效度因數
effective packing	紧束效应	有效填充

英 文 名	大 陆 名	台 湾 名
effective particle diameter	有效颗粒直径	有效粒径
effective pore radius	有效孔半径	有效孔径
effective power	有效功率	有效功率
effective surface	有效表面	有效表面
effective thermal conductivity	有效导热系数	有效導熱係數
effective value	有效值	有效值
effective volatility	有效挥发度	有效揮發度
efficiency	效率	效率
efflorescence	风化	風化[作用]
effluence	流出	流出
effluent	流出物	流出物
effluent pipe	流出管	放流管,出料管
effluent standard	排放标准	排放標準
efflux	射流	射流,流出
efflux coefficient	射流系数	射流係數
efflux time	射流时间	射流時間
efflux velocity	射流速度	射流速度
efflux viscometer	流出式黏度计	流出黏度計
effusion	泻流	瀉流,逸散
eigenvalue	特征值	固有值
ejector	喷射器	噴射器,射出器
ejector arrangement	喷射器排列	噴射器排列,射出器排列
ejector capacity	喷射器容量	噴射器容量,射出器容量
elastic capsule	弹性胶囊	彈性膠囊
elastic energy	弹性能	彈性能
elastic fiber	弹性纤维	彈性纖維
elasticity	弹性	彈性
elasticity number	弹性数	彈性數
elasticity test	弹性试验	彈性試驗
elastic liquid	弹性液体	彈性液體
elastic modulus	弹性模量	彈性模數
elastic sulphur	弹性硫	彈性硫
elastomer	弹性体	彈性物,彈性體
elastomer blend	弹性体共混物	彈性體摻合物
elastoplastic	弹性塑料	彈性塑膠
elastopolymer	弹性聚合物	彈性聚合物

英　文　名	大　陆　名	台　湾　名
elbow	弯头	彎頭,肘管
elbow nozzle	弯头喷嘴	彎頭噴嘴
elbow pipe	肘形管	肘形管
electric actuator	电动致动器	電致動器
electric analogy	电类比	電類比
electric conductivity	电导率	電導係數
electric conductivity coefficient	电导系数	電導係數
electric control	电动控制	電[動]控制
electric controller	电控制器	電控制器
electric dust precipitator	静电除尘器	電集塵器
electric energy	电能	電能
electric field	电场	電場
electric-impedance heated reactor	电热反应器	電[阻抗]熱反應器
electric installation	电气设施	電力裝置
electric pipe precipitator	电气管道除尘器	電管集塵器,電管沈積器
electric plate precipitator	电极板除尘器	電板集塵器,電板沈積器
electric porcelain	电瓷	絕緣瓷
electric resistance pyrometer	电阻高温计	電阻高溫計
electric signal	电动信号	電[動]信號
electric transmission	电动传送	電[力]傳送
electric transmitter	电动传送器	電[力]傳送器
electric type flowmeter	电子流量计	電子流量計
electroaffinity	电亲和性	電親和力
electroanalysis	电解分析	電[解]分析
electrobalance	电动天平	電動天平
electrocapillarity	电毛细管现象	電毛細現象
electrocasting	①电涂装 ②电涂层	電鑄
electrocast mullite	电铸莫来石	電鑄富鋁紅柱石
electrochemical equivalent	电化当量	電化當量
electrochemical industry	电化学工业	電化[學]工業
electrochemical potential	电化学势	電化[學]電勢,電化[學]電位
electrochemical reaction	电化学反应	電化[學]反應
electrochemical reaction engineering	电化学反应工程	電化[學]反應工程
electrochemical reactor	电化学反应器	電化學反應器
electrochemistry	电化学	電化學

英　文　名	大　陆　名	台　湾　名
electrochlorination	电氯化反应	電氯化[反應]
electrochromatography	电色谱	電層析法
electrochromic display	电致变色显示	電致變色顯示器
electrocleaning	电清洗	電解清洗
electrocoating	电涂	電塗裝
electroconductive polymer	导电高分子	導電聚合物
electrocratic dispersion	电稳分散作用	電穩分散[作用],電穩分散液
electrode	电极	電極
electrodeposit	电沉积物	電沈積物
electrodeposition	电沉积	電沈積[法]
electrodialysis	电渗析	電透析[作用]
electrodialysis process	电渗析过程	電透析程序
electrodispersion	电分散	電分散
electroflotation	电浮选法	電浮選[法]
electrofluorination	电解氟化	電氟化[反應]
electroforming	电铸	電鑄
electrofusion	电融合	電融合
electrohydraulic actuator	电动液压引动器	電動液壓致動器
electrohydrodimerization	电解加氢二聚合	電解加氫二聚合[反應]
electrokinetic phenomenon	电动现象	電動力現象
electrokinetic potential	动电电位	動電勢,動電位
electrokinetics	动电学	電動力學
electroless coating	无电涂	無電塗層,無電塗裝
electroless plating	化学镀	無電[電]鍍
electroluminescence	电致发光	電致發光
electrolysis	电解	電解
electrolysis bath	电解槽	電解槽
electrolyte	电解质	電解質
electrolytic cell	电解池	電解槽
electrolytic conductance meter	导电计	電導計
electrolytic industry	电解工业	電解工業
electrolytic oxidation	电解氧化	電解氧化
electrolytic process	电解过程	電解程序
electrolytic separation	电解分离	電解分離
electrolyzer	电解槽	電解槽
electromagnetic flowmeter	电磁流量计	電磁流量計

英　文　名	大　陆　名	台　湾　名
electromagnetic radiation	电磁辐射	電磁輻射
electromagnetic separation	电磁分离[法]	電磁分離
electromagnetic separator	电磁磁选机	電磁分離器
electromagnetic spectrum	电磁谱	電磁[波]譜
electrometallurgy	电冶金	電冶金學
electromigration	电迁徙	電遷移[法]
electromotive force（EMF）	电动势	電動勢
electromotive force series	动电势序	電動勢序
electronic amplifier	电子放大器	電子放大器
electronic analog computer	电子模拟计算机	類比電子計算機
electronic computer	电子计算机	電子計算機
electronic control	电子控制	電子控制
electronic flowmeter	电子流量计	電子流量計
electronic materials	电子材料	電子材料
electronic partition function	电子配分函数	電子分配函數
electronic signal	电子信号	電子訊號
electronic simulator	电子模拟器	電子模擬器
electro-osmosis	电渗现象	電滲透
electrophoresis	电泳	電泳
electroplating	电镀	電鍍
electroplating engineering	电镀工程	電鍍工程
electropneumatic actuator	电动-气动致动器	電動氣動[式]致動器
electrosmosis	电渗	電滲[透]
electrostatic attraction	静电吸引	靜電吸引
electrostatic dust collector	静电除尘器	靜電集塵器
electrostatic effect	静电效应	靜電效應
electrostatic precipitation	静电除尘	靜電沈積[法]
electrostatic precipitator	静电沉降器,电除尘器	靜電集塵器,靜電沈積器
electrostatic separation	静电分离	靜電分離
electrostatic separator	静电分离器	靜電分離器
electrothermal process	电热过程	電熱法
electroviscous effect	电黏效应	電黏性效應
element	①元素 ②元件 ③要素	①元素 ②元件 ③要素
elementary reaction	基元反应	基本反應
elimination reaction	消除反应	消去反應,脫去反應
elongational flow	拉伸流动	拉伸流動
elongational viscosity	拉伸黏度	拉伸黏度

英　文　名	大　陆　名	台　湾　名
elongation rate	拉伸速率	拉伸速率
eluant	洗脱剂	流洗液,溶析液
eluate	洗出液	析出液
elution	洗脱	洗提,洗析
elution chromatography	洗脱色谱法	洗析層析[法]
elutriation	①淘析 ②淘洗	淘析[作用]
elutriation constant	扬析常数	淘析常數
elutriator	淘析器	淘析器
emanation thermal analysis	放射热分析	射氣熱分析[法]
embossing	压花	壓花,壓紋
embossing calender	压花机,压纹机	壓花機,壓紋機
embryo	胚胎	胚胎
emerald	①绿宝石 ②翡翠	①純綠寶石 ②祖母綠
emerald green	翡翠绿	翡翠綠
emergency operation	紧急操作	緊急操作
emery	金刚砂	剛砂
emery cloth	砂布	砂布
emery wheel	砂轮	砂輪
emetic tartar	吐酒石	吐酒石,酒石酸銻(III)鉀鹽
EMF(=electromotive force)	电动势	電動勢
emission	排放	排放,發射
emission factor	排放系数	排放係數,發射係數
emission inventory	排放物清单	排放物清單
emission spectrometer	发射光谱仪	發射分光計
emission spectrum	发射光谱	發射光譜
emission standard	排放标准	排放標準
emissive power	发射能力	發射能力
emissivity	发射率	發射率
emitter	发射体	發射體
empirical equation	经验方程式	經驗方程式,實驗方程式
empirical formula	经验式	實驗式
empirical method	经验法	經驗法
empirical model	经验模型	經驗模式
empirical rule	经验法则	經驗法則
emulsifiability	可乳化性	可乳化性
emulsification	乳化[作用]	乳化[作用]

英　文　名	大　陆　名	台　湾　名
emulsifier	①乳化器 ②乳化剂	①乳化器 ②乳化劑
emulsifying agent	乳化剂	乳化劑
emulsifying colloid	乳化胶体	乳化膠體
emulsifying ointment	乳化软膏	乳化軟膏
emulsion	乳［浊］液	乳液,乳膠
emulsion liquid membrane	乳化液膜	乳液薄膜
emulsion paint	乳胶漆	乳化漆
emulsion phase	乳相	乳［化］相
emulsion polymerization	乳液聚合	乳化聚合［作用］
emulsion stability	乳化稳定性	乳液穩定性
emulsoid	乳胶	乳化膠體,類乳化液
enamel drier	瓷漆干燥室	瓷漆乾燥劑
enamel glaze	珐琅釉	琺瑯釉
enamel paint	瓷漆	瓷漆
enargite	硫砷铜矿	硫砷銅礦
encapsulation	胶囊化	膠囊封裝
enclosed impeller	闭工叶轮	封閉式葉輪
end effect	端点效应	端［點］效應
end gas	尾气	尾氣
endoenzyme	内酶	内酶
endothermic process	吸热过程	吸熱程序
endothermic reaction	吸热反应	吸熱反應
end point	终点	終點
end-point control	终点控制	終點控制
end product	最终产品	最終產物
energetic constitutive equation	能量本构方程	能量本質方程式
energetic equation of state	能量状态方程	能量狀態方程式
energetics	能量学	能量學
energized molecule	①受激分子 ②高能化分子	①受激分子 ②高能化分子
energy	能量	能量
energy balance	能量平衡	能量均衡
energy balance equation	能量平衡方程式	能量均衡方程式
energy barrier	能垒	能量障壁
energy conservation	能量守恒	能量守恆
energy consumption	能量损耗	能量消耗［量］
energy content	能量含量	能［量］含量
energy conversion	能量转换	能量轉換

英　文　名	大　陆　名	台　湾　名
energy conversion engineering	能量转换工程	能量轉換工程
energy converter	能量转化器	能量轉化器
energy density	能量密度	能量密度
energy dissipation	能量耗散	能量耗散
energy distribution	能量分布	能量分佈
energy efficiency	能量效率	能量效率
energy engineering	能源工程	能源工程
energy equation	能量方程	能量方程式
energy equivalent	能量当量	能[量]當量
energy exchanger	能量交换器	能量交換器
energy flux	能流	能[量]通量
energy generation curve	能量生成曲线	能量生成曲線
energy grade line	能量级线	能量級線
energy integration	能量集成	能量集成
energy level	能级	能階
energy loss	能量损失	能量損失
energy loss curve	能量损失曲线	能量損失曲線
energy producing reaction	生能反应	生能反應
energy science	能源科学	能源科學,能量科學
energy separating agent	能量分离剂	能量分離劑
energy sink	能汇	能量匯座
energy sources	能源	能源
energy spectrum	能谱	能譜
energy state	能态	能態
energy storage	能量储存	能量儲存
energy technology	能源技术	能源技術
energy transfer	能量传递	能量傳送,能量傳遞
energy transformation	能量转换	能量轉換
engineering	工程	工程
engineering analysis	工程分析	工程分析
engineering design	工程设计	工程設計
engineering design cost	工程设计成本	工程設計成本
engineering experiment	工程实验	工程實驗
engineering materials	工程材料	工程材料
engineering plastic	工程塑料	工程塑膠
enhanced oil recovery	提高石油采收率	提高石油採收[法]
enhancement factor	增强因子	增進因數
enlargement loss	扩大损失	擴大損失

英　文　名	大　陆　名	台　湾　名
enriching section	提浓段	增濃段
enrichment	富集	增濃,加強
enrichment culture	富集培养	增濃培養
enrichment factor	富集因子	增濃因數
enstatite	顽[火]辉石	頑火輝石
entanglement	缠结	纏結
enthalpy	焓	焓
enthalpy-composition chart	焓–组成图	焓–組成圖
enthalpy-composition diagram	焓–组成图	焓–組成圖
enthalpy-concentration diagram	焓浓图	焓–濃度圖
enthalpy-entropy diagram	焓–熵图	焓–熵圖
enthalpy-humidity chart	焓–湿图	焓–濕度圖
enthalpy of activation	活化焓	活化焓
enthalpy of formation	生成焓	生成焓
enthalpy of reaction	反应焓	反應焓
entrained bed reactor	夹带床反应器	挾帶床反應器
entrainer	夹带剂	挾帶劑
entrainer pump	夹带剂泵	挾帶劑泵
entrainment	[雾沫]夹带	[霧沫]挾帶
entrainment evaporator	雾沫蒸发器	霧沫蒸發器
entrainment mixer	雾沫混合器	霧沫混合器
entrainment trap	夹带物分离器	[霧沫]挾帶阱
entrainment velocity	夹带速度	[霧沫]挾帶速度
entrance effect	进口效应	進口效應
entrance region	进口区	進口區[域]
entrapment	包埋	包埋
entropy	熵	熵
entropy balance	熵衡算	熵均衡
entropy change	熵变化	熵變化
entropy equation	熵方程式	熵方程式
entropy flow	熵流	熵流
entropy flux	①熵通量 ②熵流	熵通量
entropy generation	熵产生	熵生成
entropy of activation	活化熵	活化熵
entropy production	熵产生	熵產生
entry length	进口长度	進口長度
enumeration method	枚举法	列舉法
environment	环境	環境

英 文 名	大 陆 名	台 湾 名
environmental chemistry	环境化学	環境化學
environmental contamination	环境污染	環境污染
environmental cost	环境成本	環境成本
environmental engineering	环境工程	環境工程
environmental impact	环境冲击	環境衝擊
environmental pollution	环境污染	環境污染
environmental protection	环境保护	環境保護
environmental quality	环境质量	環境品質
environmental science	环境科学	環境科學
environmental state	环境态	環境態
environmental technology	环境技术	環境技術
enzymatic analysis	酶法分析	酶分析法
enzymatic electrocatalysis	酶促电催化	酶電催化
enzymatic hydrolysis	酶法水解	酶水解
enzymatic method	酶法	酶試驗法
enzymatic reaction	酶促反应	酶反應
enzyme	酶	酶
enzyme activity	酶活力	酶活性
enzyme catalysis	酶催化	酶催化[作用]
enzyme catalyzed reaction	酶催化反应	酶催化反應
enzyme deactivation	酶失活	酶失活
enzyme electrode	酶电极	酶電極
enzyme fermentation	酶发酵	酶發酵[法]
enzyme immunoassay	酶免疫分析法	酶免疫分析法
enzyme-linked immunosorbent assay	酶联免疫吸附测定	酶聯免疫吸附測定
enzyme membrane	酶膜	酶膜
enzyme product complex	酶产物复合物	酶產物複合物
enzyme reaction kinetics	酶反应动力学	酶動力學
enzyme selectivity	酶选择性	酶選擇性
enzyme specificity	酶专一性	酶特異性
enzyme stabilization	酶稳定	酶穩定化
enzyme-substrate complex	酶-底物复合物	酶-受質複合物
enzyme-substrate reaction	酶-底物反应	酶-受質反應
eosin dye	曙红染料	曙紅染料
eosin lake	曙红色淀	曙紅色澱
epichlorohydrin	环氧氯丙烷	環氧氯丙烷
epitaxial growth	外延生长	磊晶生長
epitaxy	外延	磊晶

英　文　名	大　陆　名	台　湾　名
epoxidation	环氧化	環氧化[作用]
epoxy resin	环氧树脂	環氧樹脂
equality constraint	等式约束	等式約束
equalization tank	平衡槽	均化槽
equalizing valve	平衡阀	平衡閥
equal-settling particle	等沉降粒子	等沈降粒子
equation	方程	方程式
equation of continuity	连续性方程	連續方程式
equation of motion	运动方程	運動方程式
equation of state	状态方程	狀態方程式
equation-oriented approach	联立方程法	聯立方程法
equation-solving approach	联立方程法	聯立方程法
equilateral triangular diagram	等边三角图	等邊三角圖
equilibration	平衡化	平衡化
equilibration rate	平衡速率	達平衡速率
equilibrium	平衡	平衡
equilibrium approach	平衡法	平衡法
equilibrium approximation	平衡近似	平衡近似[法]
equilibrium composition	平衡组成	平衡組成
equilibrium concentration	平衡浓度	平衡濃度
equilibrium condensation	平衡冷凝	平衡冷凝
equilibrium constant	平衡常数	平衡常數
equilibrium conversion	平衡转化率	平衡轉化率
equilibrium criterion	平衡判据	平衡準則
equilibrium curve	平衡曲线	平衡曲線
equilibrium diagram	平衡图	平衡圖
equilibrium distillation	平衡蒸馏	平衡蒸餾
equilibrium distribution	平衡分布	平衡分佈
equilibrium flash vaporization	平衡闪蒸	平衡驟汽化
equilibrium flash vaporization curve	平衡闪蒸曲线	平衡驟汽化曲線
equilibrium freezing curve	平衡冷凝曲线	平衡冷凍曲線
equilibrium line	平衡线	平衡線
equilibrium mixture	平衡混合物	平衡混合物
equilibrium moisture	平衡水分	平衡水分
equilibrium moisture content	平衡湿含量	平衡含水量
equilibrium plot	平衡图	平衡圖
equilibrium point	平衡点	平衡點
equilibrium separation process	平衡分离过程	平衡分離程序

英　文　名	大　陆　名	台　湾　名
equilibrium stage	平衡级	平衡階
equilibrium still	平衡釜	平衡釜,平衡蒸餾器
equilibrium system	平衡系统	平衡系統
equilibrium vaporization	平衡汽化	平衡汽化
equilibrium yield	平衡收率	平衡產率
equimolar counter diffusion	等摩尔逆向扩散	等莫耳逆向擴散
equipment cost	设备成本	設備成本
equipment flowsheet	设备流程图	設備流程圖
equipment reliability	设备可靠性	設備可靠性
equipment safety	设备安全	設備安全
equipotential line	等势线	等勢線,等位線
equivalent	①当量 ②等效	①當量 ②等效
equivalent conductance	①当量电导 ②等效电导	當量電導度
equivalent diameter	①当量直径 ②等效径	等效直徑
equivalent free-falling diameter	等效自由沉降直径	等效自由沈降直徑
equivalent isothermal temperature	等温温度相当值	相當溫度
equivalent length	当量长度	等效長度
equivalent non-circular duct diameter	非圆管当量直径	相當管徑
equivalent particle diameter	当量粒径	等效粒徑
equivalent pipe length	等值管长度	等效管長
equivalent point	等当点	當量點
eradicator	消除器	除污劑
erepsin	肠蛋白酶	腸蛋白酶
Erlenmeyer flask	锥形瓶	錐形瓶
erosion	磨蚀	侵蝕
error	误差	誤差
error constant	误差常数	誤差常數
error criterion	误差判据	誤差準則
error integral	误差积分	誤差積分
error propagation	误差传播	誤差傳播
error ratio	误差比	誤差比
error signal	误差信号	誤差信號
error-squared control	误差平方控制	誤差平方控制
eruption	喷发	噴發
erythromycin	红霉素	紅黴素
erythrophyll	叶红素	葉紅素
essence recovery	香精回收	香精回收
essential oil	精油	精油

英 文 名	大 陆 名	台 湾 名
ester	酯	酯
ester gum	甘油松香酯	松香硬酯,酯膠
esterification	酯化[作用]	酯化[作用]
ester number	酯值	酯值
ester value	酯值	酯值
estimated variance	估计方差	估計變異數
estimator	估计器	估計器
etchant	蚀刻剂	蝕刻劑
etching	①刻蚀 ②浸蚀	蝕刻
ether	醚	醚
etherification	醚化	醚化[作用]
ether number	醚数	醚值
ether value	醚值	醚值
ethyl cellulose	乙基纤维素	乙基纖維素
ethylene-propylene rubber	乙丙橡胶	乙烯–丙烯橡膠
eutectic alloy	共熔合金	共熔合金
eutectic equilibrium	共熔平衡	共熔平衡
eutectic mixture	共熔物	共熔混合物
eutectic point	共熔点	共熔點
eutectics	共熔物	共熔物
eutectic state	共熔状态	共熔狀態
eutectic system	共熔系统	共熔系統
eutectic temperature	共熔温度	共熔溫度
evaluation	评估	評估
evaporating column	蒸发塔	蒸發塔
evaporating pipe	蒸发管	蒸發管
evaporation	蒸发	蒸發
evaporation coefficient	蒸发系数	蒸發係數
evaporation cooling	蒸发冷却	蒸發冷卻
evaporation efficiency	蒸发效率	蒸發效率
evaporation loss	蒸发损失	蒸發損失
evaporation rate	蒸发率	蒸發速率
evaporative condenser	蒸发冷凝器	蒸發冷凝器
evaporative cooling	蒸发冷却	蒸發冷卻
evaporative crystallization	蒸发结晶	蒸發結晶
evaporative crystallizer	蒸发结晶器	蒸發結晶器
evaporative refrigeration	蒸发冷冻	蒸發冷凍
evaporator	蒸发器	蒸發器

英　文　名	大　陆　名	台　湾　名
evaporator room	蒸发室	蒸發室
evaporator scale	蒸发器污垢	蒸發器積垢
evaporator with direct heating	直接加热型蒸发器	直接加熱型蒸發器
evaporator with horizontal tubes	水平列管蒸发器	水平列管蒸發器
evolutionary method	调优法	調優法
evolutionary operation	调优操作	調優操作
evolved gas analysis	逸出气体分析法	逸出氣分析[法]
evolved gas detection	逸出气体检测	逸出氣檢測[法]
exact solution	精确解	精確解
excess air	过量空气	過量空氣
excess chemical potential	超额化学势	過剩化學勢
excess concentration technique	超额浓度法	過量濃度法
excess enthalpy	超额焓	超額焓,過剩焓
excess entropy	超额熵	超額熵,過剩熵
excess function	超额函数	過剩函數
excess Gibbs free energy	超额吉布斯自由能	過剩吉布斯自由能
excess property	超额性质	過剩性質
excess volume	超额体积	超額體積,過剩體積
excess water	过量水	過量水
exchange adsorption	交换吸附	交換吸附
exchange energy	交换能	互換能
exchanger	①交换器 ②交换剂	①交換器 ②交換劑
excited state	激发态	激態
exergy	有效能	有效能
exergy analysis	㶲分析	有效能分析
exergy balance	㶲衡算	有效能均衡
exergy loss	损失	有效能損失
exhalation valve	排气阀	排氣閥
exhauster	排气机	排氣機
exhaust pipe	排气管	排氣管
exhaust steam	排汽	排放蒸汽
exit gas	出口气体	出口氣體
exit tube	排气管	排氣管
exosmosis	外渗	外滲透[作用]
exothermicity	放热性	放熱度
exothermic process	放热过程	放熱程序
exothermic reaction	放热反应	放熱反應
expanded bed	膨胀床	膨脹床

英 文 名	大 陆 名	台 湾 名
expanded bed contactor	膨胀床接触器	膨脹床接觸器
expander	膨胀器	膨脹器,膨脹機
expansion	膨胀	膨脹,擴大
expansion coefficient	膨胀系数	膨脹係數
expansion factor	膨胀因子	膨脹因子
expansion joint	膨胀节	脹縮接頭
expansion loop	膨胀环	伸縮圈
expansion loss	膨胀损失	膨脹損失
expansion work	膨胀功	膨脹功
expected value criterion	期望值判据	期望值準則
experimental data	实验数据	實驗數據
experimental design	实验设计	實驗設計
experimental error	实验误差	實驗誤差
experimental reactor	实验用反应器	實驗反應器
expert system	专家系统	專家系統
explicit function	显函数	顯函數
explicit method	显式法	顯式法
explosion	爆炸	爆炸
explosion hazard	爆炸危险	爆炸傷害
explosion index	爆炸指数	爆炸指數
explosion limit	爆炸极限	爆炸界限
explosion pressure	爆炸压[力]	爆炸壓[力]
explosion proof	防爆	防爆
explosive	炸药	炸藥
explosive polymerization	爆聚[合]	爆聚合
exponential distribution	指数分布	指數分佈
exponential function	指数函数	指數函數
exponential lag	指数滞后	指數滯延
exponential law	指数律	指數律
exposed center	曝露中心	曝露中心
exposure limit	曝露界限	曝露界限
exposure test	曝露试验	曝露試驗
exposure time	曝露时间	曝露時間
expression	压榨	壓榨
extended aeration	延时曝气	延時曝氣
extended surface tube	扩展式表面管	延伸表面式管
extender	增量剂	增效劑
extensibility	①延伸率 ②可扩展性	延伸性

英　文　名	大　陆　名	台　湾　名
extension	伸长	延伸
extensional flow	延伸流动	延伸流動
extensional viscosity	拉伸黏度	延伸黏度
extension neck	延伸颈	延伸頸
extension wire	延长线	延長線
extensive property	广度性质	廣度性質,外延性質
extent of decomposition	分解程度	分解程度
extent of reaction	反应进度	反應程度
external area	外表面积	外表面積
external capacity	外在容积	外在容量
external control loop	外控制环路	外控制環路
external coordinates	外部坐标	外部坐標
external diameter	外径	外徑
external diffusion	外扩散	外部擴散
external disturbance	外扰	外部擾動
external energy	外能	外能
external flow	外流	外部流動
external force	外力	外力
external heat exchanger	外换热器	外熱交換器
external heating	外热法	外熱法
external pressure	外压力	外壓[力]
external recycle	外循环	外循環
external recycle reactor	外循环反应器	外循環反應器
external reflux	外回流	外回流
external sizing	外部施胶	外部上膠
external surface	外表面	外表面
external surface concentration	外表面浓度	外表面濃度
extinction	熄灭	熄滅,熄火
extinction coefficient	消光系数	消光係數
extinction temperature	熄灭温度	熄滅溫度
extracellular enzyme	胞外酶	胞外酶
extract	萃取物	萃取物
extraction	萃取	萃取
extraction battery	萃取器组	萃取器組
extraction column	萃取塔	萃取塔
extraction factor	萃取因子	萃取因子
extraction process	萃取过程	萃取程序
extraction rate	萃取率	萃取速率

英 文 名	大 陆 名	台 湾 名
extraction tower	萃取塔	萃取塔
extractive crystallization	萃取结晶	萃取結晶
extractive crystallizer	萃取结晶器	萃取結晶器
extractive distillation	萃取蒸馏	萃取蒸餾
extractive ultrafiltration	萃取超滤	萃取性超過濾
extractor	①萃取器 ②抽提塔	萃取器
extrapolation	外推	外插法
extruder	①挤出机 ②挤塑机	擠壓機
extrusion	挤出	擠壓
exudation	渗出	滲出

F

英 文 名	大 陆 名	台 湾 名
fabrication	制造	製造
fabric filter	织物过滤器	織物過濾器
factor	因子	因子
factorial design	析因设计	析因設計
factorial experiment	析因实验	析因實驗
facultative aerobic bacteria	兼性需氧细菌	兼性需氣菌
facultative bacteria	兼性细菌	兼性菌
facultative lagoon	兼性污水塘	兼性污水塘
facultative respiration	兼性呼吸	兼性呼吸
fading memory materials	记忆衰退材料	記憶衰退材料
failure diagnosis	故障诊断	故障診斷
failure mode and effect analysis	故障形式和影响分析	失效模式及效應分析
falling ball method	落球法	落球法
falling ball viscometer	落球黏度计	落球黏度計
falling film	降膜	降膜,落膜
falling-film evaporator	降膜蒸发器	降膜[式]蒸發器
falling film type absorber	降膜吸收器	降膜式吸收器
falling needle viscometer	落针黏度计	落針黏度計
falling rate drying period	降速干燥[阶]段	減速乾燥期
falling rate period	降速期	減速期
falling sphere viscometer	落球黏度计	落球黏度計
fan	排风机	風扇
fan blower	扇形排风机	扇式鼓風機
fast fluidization	快速流态化	快速流體化

英　文　名	大　陆　名	台　湾　名
fast fluidized bed	快速流化床	快速流[體]化床
fast reaction	快速反应	快[速]反應
fat	脂肪	脂肪
fatigue strength	疲劳强度	疲勞強度
fatigue test	疲劳试验	疲勞試驗
fats and oils	油脂	油脂
fatty acid	脂肪酸	脂[肪]酸
fatty alcohol	脂肪醇	脂[肪]醇
fault diagnosis	故障诊断	故障診斷
feasibility	可行性	可行性
feasibility analysis	可行性分析	可行性分析
feasibility study	可行性研究	可行性研究
feasibility survey	可行性调查	可行性調查
feasible path method	可行路径法	可行路徑法
feasible region	可行域	可行區域
fed-batch culture	分批补料式培养	饋料批式培養
feed	进料	進料,饋入
feedback	反馈	回饋
feedback bellow	反馈伸缩囊	回饋伸縮囊
feedback compensation	反馈补偿	回饋補償
feedback control	反馈控制	回饋控制
feedback controller	反馈控制器	回饋控制器
feedback control system	反馈控制系统	回饋控制系統
feedback element	反馈元件	回饋元件
feedback loop	反馈回路	回饋環路
feedback mechanism	反馈机理	回饋機構
feedback signal	反馈信号	回饋信號
feedback system	反馈系统	回饋系統
feed belt	进料带	進料帶
feed composition	进料组成	進料組成
feed conveyor	进料传送机	進料輸送機
feed disk	盘式进料器	進料盤
feed drum	筒式进料机	進料鼓
feeder	进料器	進料器,飼機
feedforward	前馈	前饋
feedforward control	前馈控制	前饋控制
feedforward system	前馈系统	前饋系統
feed hole	进料孔	進料孔

英　文　名	大　陆　名	台　湾　名
feed pipe	进料管	進料管
feed quality	进料质量	進料品質
feed rate	加料速度	進料速率
feed ratio	进料比	進料比
feed regulator	进料调节器	進料調節器
feed roller	进料滚柱	進料輥
feedstock	原料	原料
feed stream	进料流	進料流
feed system	进料系统	進料系統
feed tank	加料槽	進料槽
feed temperature	进料温度	進料溫度
feed tray	进料板	進料板
feed tray location	进料板位置	進料板位置
feed valve	进料阀	進料閥
feed water	给水	給水
feed water treatment	给水处理	給水處理
feed well	给水槽	給水井,進料井
feldspar	长石	長石
feldspathic glaze	长石釉	長石釉
Fenske equation	芬斯克方程	范氏方程式
Fenske packing	芬斯克填料	芬斯基填料
fermentation	发酵	發酵[作用]
fermentation apparatus	发酵器械	發酵裝置
fermentation cellar	发酵窖	發酵窖
fermentation chamber	发酵室	發酵室
fermentation efficiency	发酵效率	發酵效率
fermentation mechanism	发酵机理	發酵機構
fermentation medium	发酵培养基	發酵基,發酵介質
fermentation pathway	发酵途径	發酵途徑
fermentation pattern	发酵模式	發酵方式
fermentation process	发酵过程	發酵程序
fermentation tank	发酵槽	發酵槽
fermentation technology	发酵技术	發酵技術
fermentation tube	发酵管	發酵管
fermentation tube test	发酵管试验	發酵管試驗
fermentation vat	发酵槽	發酵缸
fermenter	发酵罐	發酵槽
fermentor	发酵罐	發酵槽

英　文　名	大　陆　名	台　湾　名
Fermi-Dirac distribution	费米–狄拉克分布	費米–狄拉克分佈
ferro-alloy	铁合金	鐵合金
ferromanganese	铁锰合金	錳鐵合金,錳鐵[齊]
ferrophosphorus	磷铁	磷鐵齊
ferro-silico-manganese	硅锰铁合金	矽錳鐵合金
ferrosilicon alloy	硅铁合金	矽鐵合金,矽鐵[齊]
fertilization	施肥	施肥
fertilizer	肥料	肥料
F factor	F 因子	F 因子
fiber reinforced plastics	纤维增强塑料	纖維強化塑膠
Fibonacci search method	斐波那契搜索法	費布納西搜尋法
fibre	纤维	纖維
Fick law	菲克定律	費克定律
filament	单丝	單絲,絲狀纖維
filamentous bacteria	丝状细菌	絲狀細菌
fill	填充	裝填
filler	填料	填料
film boiling	膜状沸腾	膜沸騰
film coefficient	膜系数	膜係數
film diffusion	膜扩散	膜擴散
film diffusion control	膜扩散控制	膜擴散控制
film heat transfer coefficient	膜传热系数	膜熱傳係數
film phase	膜相	膜相
film resistance	膜阻力	膜阻力
film scrubber	膜式洗涤器	膜式洗滌器
film strength	膜强度	膜強度
film temperature	膜温度	膜溫度
film test	膜试验	膜試驗
film theory	膜理论	膜理論
film type condensation	膜式冷凝	膜冷凝
film type evaporator	膜式蒸发器	膜蒸發器
filmwise condensation	膜状冷凝	膜式冷凝
film yeast	膜酵母	膜酵母
filter aid	助滤剂	助濾劑
filter area	过滤面积	過濾面積
filter bag	滤袋	濾袋
filter box	滤箱	濾箱
filter cake	滤饼	濾餅

英 文 名	大 陆 名	台 湾 名
filter cake conveyor	滤饼运输机	濾餅運送機
filter capacity	过滤器容量	過濾器容量
filter clogging	过滤器堵塞	過濾器阻塞
filter cloth	滤布	濾布
filter depth	滤池深度	濾池深度
filter drum	滤桶	濾桶
filter efficiency	过滤效率	過濾效率
filter flowsheet	过滤流程	過濾流程圖
filter frame	滤框	濾框
filter liquor	滤液	濾液
filter paper	滤纸	濾紙
filter performance	过滤性能	過濾性能
filter plate	滤板	濾板
filter press cake	压滤饼	壓濾餅
filter press cloth	压滤布	壓濾布
filter sand	滤砂	濾砂
filtration	过滤	過濾
filtration area	过滤面积	過濾面積
filtration cycle	过滤周期	過濾週期
filtration medium	过滤介质	過濾介質
filtration membrane	滤膜	濾膜
filtration rate	过滤速率	過濾速率
filtration sterilization	过滤灭菌	過濾滅菌
fin	翅片	①鰭片 ②散熱片
final boiling point	终沸点	終沸點
final clarifier	最终澄清器	最終澄清器
final condition	最终条件	最終條件
final control element	最终控制元件	最終控制元件
final state	最终状态	[最]終[狀]態
final-value theorem	终值原理	終值定理
financing cost	财务会计成本	資金成本
fine crusher	细碎机	細級壓碎機
fine dispersion	精细分散	精細分散
fine particle	细颗粒	細粒
fine screen	精筛	精篩
finite boundary element	有限边界元素	有限邊界元素
finite difference	有限差分	有限差分
finite disturbance	有限扰动	有限擾動

英 文 名	大 陆 名	台 湾 名
finite element method	有限元法	有限元[素]法
finite-time stability	有限时间稳定性	有限時間穩定性
finned pipe	翅片管	鰭片管
finned surface	翅片表面	鰭片表面
finned tube	翅片管	鰭片管
finned-tube heat exchanger	翅管式热交换器	鰭片管熱交換器
fire and explosion index	火灾及爆炸指数	火災及爆炸指數
firebrick	耐火砖	耐火磚
fireclay	耐火黏土	耐火黏土
fireclay brick	耐火砖	耐火[黏土]磚
fired furnace	明火加热炉	明火爐
fired heater	明火加热器	明火加熱器
fired-tubular reactor	管式反应器	明火管式反應器
fire extinguisher	灭火器	滅火器
fire hazard	着火危险	火災
fire protection	防火	防火
fire retardant	阻燃剂	阻燃劑
fire-retarding paint	防火漆	防火漆
first law of thermodynamics	热力学第一定律	熱力學第一定律
first-order control system	一级控制系统	一階控制系統
first order kinetics	一级动力学	一級動力學,一階動力學
first-order lag	一级落后	一階落後
first order phase transition	一级相变	一級相變,一階相變
first-order process	一级过程	一階程序
first order reaction	一级反应	一級反應,一階反應
first-order response	一级应答	一階應答
first-order system	一阶系统	一階系統
first-order transition temperature	一级转变温度	一階轉變溫度,一階轉移溫度
fission	裂变	分裂
fissionability	可裂变性	可[核]分裂性
fissionable materials	可裂变物质	可[核]分裂物質
fission energy	裂变能	[核]分裂能
fission product	裂变产物	[核]分裂產物
fissure	裂缝	裂縫
fitting equivalent length	配件当量长度	配件相當長度
fitting equivalent pipe length	配件当量管长	配件相當管長

英　文　名	大　陆　名	台　湾　名
fitting flowsheet	配件流程图	［管件］接頭流程圖
fitting resistance	配件阻力	［管件］接頭阻力
fixation	固定	固定
fixation bath	固定浴	固定浴
fixed bed	固定床	固定床
fixed-bed process	固定床过程	固定床法
fixed-bed reactor	固定床反应器	固定床反應器
fixed capital	固定资本	固定資本
fixed capital investment	固定资本投资额	固定資本投資額
fixed catalyst bed	固定催化剂床	固定觸媒床
fixed charge	①固定费用 ②固定消耗量	固定費用
fixed cost	固定成本	固定成本
fixed-growth system	固定成长系统	固著生長系統
fixed nitrogen	固定氮	固定氮
fixed oil	不挥发油	不揮發油
fixing agent	固定剂	固定劑
flagella	鞭毛	鞭毛
flame	火焰	火焰
flame reactor	火焰反应器	火焰反應器
flame resisting agent	阻燃剂	耐燃劑
flame spectrometer	火焰分光仪	火焰分光計
flame stability	火焰稳定性	火焰穩定性
flame temperature	火焰温度	火焰溫度
flammability	可燃性	可燃性
flange	法兰	法蘭,凸緣
flange connection	法兰连接	凸緣連接
flange coupling	法兰接头	凸緣接頭
flange joint	法兰联管节	凸緣接頭
flange tap	法兰旋塞	凸緣分接頭
flapper	挡板	擋板
flapper-nozzle mechanism	挡板喷嘴机理	擋板噴嘴機構
flash calculation	闪蒸计算	驟沸計算
flash chamber	闪蒸室	驟沸室
flash chilling	骤冷	驟冷
flash chlorination	骤氯化	驟氯化［反應］
flash concentrator	闪蒸浓缩器	驟沸濃縮器
flash distillation	闪速蒸馏	驟餾

英 文 名	大 陆 名	台 湾 名
flash distillation column	闪蒸塔	驟餾塔
flash drier	急骤干燥器	驟乾器
flash drum	闪蒸槽	驟沸桶
flash drying	闪速干燥	急驟乾燥
flash evaporation	闪蒸	驟沸蒸發
flash film evaporator	薄膜闪蒸器	薄膜閃蒸器
flash mixer	骤混器	驟混器
flash mixing	骤混	驟混
flash pasteurizer	瞬时杀菌器	瞬間殺菌器
flash point	闪点	閃[火]點
flash point apparatus	闪点测定器	閃[火]點測定器
flash roaster	闪速焙烧炉	急驟焙燒爐
flash tower	闪蒸塔	驟餾塔
flash vapor	闪蒸气	驟沸蒸气
flash vaporization	闪蒸	驟沸蒸發
flash vaporization curve	闪蒸曲线	驟沸蒸發曲線
flask	烧瓶	燒瓶
flat paddle	平板桨	平板槳
flat plate	平板	平板
flat-plate approximation	平板近似法	平板近似法
flat-plate flow	平板流动	平板流動
flavoring materials	调味剂	調味料
flax	亚麻	亞麻
flax fiber	亚麻纤维	亞麻纖維
flax-seed oil	亚麻子油	亞麻仁油
flexural stress	弯曲应力	彎曲應力
flickering flame	闪烁火焰	閃爍火焰
flight conveyor	刮板输送机	梯板運送機,梯運機
flint	燧石	燧石
flint clay	硬质黏土	燧石黏土
flint fireclay	硬质耐火黏土	燧石耐火黏土
flint glass	火石玻璃	燧石玻璃
flint pebble	燧卵石	燧卵石
float	浮标,浮子	浮標,浮子,浮筒
float glass	浮法玻璃	浮法玻璃
float indicator	浮标指示计	浮標指示計
floating action	浮动作用	浮動作用
floating control	浮动控制	浮動控制

英　文　名	大　陆　名	台　湾　名
floating-cover digester	浮盖消化槽	浮[動]蓋消化槽
floating head heat exchanger	浮头换热器	浮[動]頭熱交換器
floating pressure control	浮动压力控制	浮動壓力控制
floating pressure operation	浮动压力操作	浮動壓力操作
floating rate	浮动速率	浮移率
floating roof tank	浮顶罐	浮頂槽
floating soap	浮水皂	浮水皂
floating speed	浮动速率	浮動速率
floating type bell gage	浮式钟型压力计	浮式鐘形壓力計
floating zone method	悬浮区法	浮動帶域[純化]法
float valve	浮阀	浮閥
float-valve tray	浮阀塔板	浮閥板
floc	絮凝物	絮凝體
flocculant	絮凝剂	絮凝劑
flocculate	絮凝物	絮凝物
flocculating mechanism	絮凝机理	絮凝機構
flocculation	絮凝	絮凝[作用]
flocculation reaction	絮凝反应	絮凝反應
flocculation tank	絮凝槽	絮凝槽
flocculator	絮凝器	絮凝器
flocculent particle	絮凝粒子	絮凝粒子
flocculent settling	絮凝沉降	絮凝沈降
flocculent suspension	絮凝悬浮液	絮凝懸浮液
floc growth	絮凝物生长	絮凝體成長
floc particle	絮凝粒子	絮凝粒子
floc size	絮凝物尺寸	絮凝體大小
floc test	絮凝测试	絮凝試驗
flooded suction line	溢流吸入管线	溢流吸入管線
flooding	液泛	溢流
flooding point	泛点	溢流點
flooding velocity	泛点速度	溢流速度
Flory-Huggins theory	弗洛里-哈金斯理论	弗[洛里]-哈[金斯]理論
flotation	浮选	浮選[法]
flotation process	浮选过程	浮選法
flotation tank	浮选槽	浮選槽
flow	流动	流動
flow behavior index	流动行为指数	流動行為指數

英　文　名	大　陆　名	台　湾　名
flow calorimeter	流动量热计	流動卡計
flow control	流量控制	流量控制
flow diagram	流程图	流程圖
flow diagram symbol	流程图符号	流程圖符號
flow gage	流量计	流量計
flow indicator	流量指示计	流量指示器
flow integration	流量积分	流量積分
flow integrator	流量积分器	流量積分器
flow lag	流动落后	流動滯延
flow line	流动线	流動線
flow-line interception	流线截取	流線截取
flow measuring equipment	流量测量设备	流量測量裝置
flowmeter	流量计	流量計
flow method	流动法	流動法
flow mixer	流动混合器	流動混合器
flow net	流动网	流動網
flow nozzle	流量喷嘴	流體噴嘴
flow number	[搅拌]流量数	[攪拌]流量數
flow pattern	流型	流動型式
flow process	流水作业	流動程序
flow rate	流率	流率
flow reactor	流动反应器	流動反應器
flow resistance	流动阻力	流動阻力
flowsheet	流程图	流程圖
flowsheet design	流程图设计	流程圖設計
flowsheet symbol	流程图符号	流程圖符號
flowsheet synthesis	流程综合	流程組合
flow system	流动系统	流動系統
flow visualization	流动显示	流動顯示
flow work	流动功	流動功
fluctuating velocity	①涨落速度 ②脉动流速	波動速度
fluctuation	①涨落 ②脉动	①漲落 ②波動
flue	烟道	煙道
flue gas	烟道气	煙道氣
fluid	流体	流體
fluid dynamics	流体动力学	流體動力學
fluid energy mill	流能磨	流體能量研磨機
fluid flow	流体流动	流體流動

英　文　名	大　陆　名	台　湾　名
fluid flow resistance	流体流动阻力	流體流動阻力
fluid-fluid reaction	流体–流体反应	流體–流體反應
fluid friction	流体摩擦	流體摩擦
fluid head	流体高差	流體高差
fluidics	流控技术	流控技術
fluidizable particle size	可流态化颗粒大小	可流體化粒子大小
fluidization	流态化	流體化
fluidization number	流化数	流[體]化數
fluidized bed	流化床	流[體]化床
fluidized bed adsorber	流化床吸附器	流[體]化床吸附器
fluidized bed boiler	流化床锅炉	流[體]化床鍋爐
fluidized bed combustion	流化床燃烧	流[體]化床燃燒
fluidized bed combustor	流化床燃烧器	流[體]化床燃燒器
fluidized-bed conversion	流化床转化	流[體]化床轉化
fluidized bed dryer	流化床干燥器	流[體]化床乾燥器
fluidized bed gasifier	流化床气化器	流[體]化床氣化器
fluidized bed reactor	流化床反应器	流[體]化床反應器
fluidized catalyst bed	流化催化剂床	流[體]化觸媒床
fluidized distillation	流化蒸馏	流[體]化蒸餾
fluidized mixing	流化混合	流[體]化混合
fluidized reactor	流化反应器	流[體]化反應器
fluidized roaster	流化焙烧炉	流[體]化焙燒爐
fluidizing velocity	流化速度	流[體]化速度
fluid kinematics	流体运动学	流體運動學
fluid mechanics	流体力学	流體力學
fluid phase	流体相	流體相
fluid-solid reaction	流体–固体反应	流體–固體反應
fluid velocity	流体速度	流體速度
fluorapatite	氟磷灰石	氟磷灰石
fluorescein	荧光黄	螢光黃
fluorescence	荧光	螢光
fluorescence spectroscope	荧光分光镜	螢光譜儀
fluorescent crystal	荧光晶体	螢光晶體
fluorescent dye	荧光染料	螢光染料
fluorescent spectrum	荧光光谱	螢光光譜
fluorescent whitening agent	荧光增白剂	螢光增白劑
fluorocarbon	氟碳化物	氟碳化物
fluorocarbon resin	氟碳树脂	氟碳樹脂

英文名	大陆名	台湾名
fluorspar	萤石	螢石,氟石
flush filter plate	平槽压滤板	平槽壓濾板
flushing	涌料	沖洗
flush plate	平槽滤板	平槽濾板
flux	通量	通量
flux density	通量密度	通量密度
foam	泡沫	泡沫
foam agent	发泡剂	發泡劑
foam breaker	消泡器	消泡機
foam concrete	泡沫混凝土	泡沫混凝土
foam fractionation	泡沫分离	泡沫分餾
foaming	发泡	發泡
foam plastic	泡沫塑料	發泡塑膠
foam rubber	泡沫橡胶	泡沫橡膠
foam separation	泡沫分离	泡沫分離
foam stabilizer	稳泡剂	泡沫穩定劑
focus	焦点	焦點
food processing	食品加工	食品加工
food-to-microorganism ratio	食物-微生物比	食物-微生物比
foot valve	底阀	底閥,腳閥
force balance	力平衡	力均衡
force-current analogy	力-电流类比	力-電流類比
forced circulation evaporation	强制循环蒸发	強制循環蒸發
forced circulation evaporator	强制循环蒸发器	強制循環蒸發器
forced convection	强制对流	強制對流
forced convection evaporation	强制对流蒸发	強制對流蒸發
forced convection heat transfer	强制对流传热	強制對流熱傳
forced diffusion	强制扩散	強制擴散
forced draft	强制通风	強制通風,鼓風
forced draft aerator	鼓风通气机	鼓風通氣機
forced draft cooling tower	强制通风凉水塔	鼓風冷卻塔
forced draft fan	鼓风机	鼓風扇
forced oscillation	强制振荡	強制振盪
forced vortex	强制涡旋	強制漩渦
force-mass conversion constant	力-质量换算常数	力-質量換算常數
force potential	力势	力勢
force-voltage analogy	力-电压类比	力-電壓類比
forcing function	强制函数	強制函數

英　文　名	大　陆　名	台　湾　名
formalin	福尔马林	福馬林,甲醛水[溶液]
formal report	正式报告	正式報告
formamide	甲酰胺	甲醯胺
formamide process	甲酰胺法	甲醯胺法
form drag	形状阻力	形狀阻力
formula	①[化学]式 ②公式	①化學式 ②[公]式
formulation	配方	①配方 ②公式化
forsterite	镁橄榄石	矽酸镁石,鎂橄欖石
forsterite brick	镁橄榄石砖	鎂橄欖石磚,矽酸鎂石 磚
forsterite porcelain	镁橄榄石瓷	鎂橄欖石瓷,矽酸鎂石 瓷
forward feed	顺向进料	順向進料
forward feeding	顺向进料	順向進料
forward reaction	正向反应	正反應
fossil	化石	化石
fossil fuel	化石燃料	化石燃料
fouling	结垢	積垢
fouling coefficient	污垢系数	積垢係數
fouling factor	污垢系数	積垢因數
fourdrinier machine	长网成型机	長綱[製紙]機
Fourier number	傅里叶数	傅立葉數
four-way valve	四通阀	四通閥
fractal	分形	碎形
fraction	①分数 ②分率	①分數 ②分率 ③分式
fractional absorption	分级吸收	部分吸收
fractional adsorption	分级吸附	部分吸附
fractional analysis	分段分析	分段分析,分餾分析
fractional column	分馏塔	分餾塔,分餾柱
fractional condensation	分凝作用	分級冷凝
fractional condenser	分凝器	分凝器
fractional condensing tube	分凝管	分凝管
fractional crystallization	分步结晶	分段結晶
fractional dialysis	分级渗析	分級透析
fractional distillation	分馏	分餾
fractional efficiency	分级效率	分級效率
fractional expansion	分步膨胀	分步膨脹
fractional extraction	分馏萃取	分步萃取

英 文 名	大 陆 名	台 湾 名
fractional life	部分寿命	部分壽命
fractional-life method	部分衰减法	部分衰减法
fractional-order reaction	分数级反应	分數級反應,分數階次反應
fractional precipitation	分级沉淀	分級沈澱
fractional yield	产量分率	產量分率
fractionating column	分馏柱	分餾柱,分餾塔,分餾管
fractionating tower	分馏塔	分餾塔
fractionation	分馏	分餾
fractionation assembly	分馏装置	分餾裝置
fractionator	分馏器	分餾器,分餾塔
fraction of coverage	覆盖率	覆蓋率
fracture mechanics	断裂力学	破壞力學
fracture strength	断裂强度	破裂強度
fracture stress	断裂应力	破裂應力
fragility	易碎性	易碎性
fragmentation	碎裂	碎斷[作用]
frame	框架	框,架
frame and plate filter	板框过滤机	框板過濾機
frame filter press	框式压滤机	框式壓濾機
frame of reference	参考坐标	參考坐標
franklinite	锌铁尖晶石	鋅鐵礦
free acid	游离酸	自由酸
free alkali	游离碱	自由鹼
free-centered lattice	面心晶格	面心晶格
free convection	自由对流	自由對流
free crushing	自由压碎	自由壓碎
free energy	自由能	自由能
free energy-composition diagram	自由能组成图	自由能-組成圖
free enzyme	游离酶	游離酶
free falling velocity	自由沉降速度	自由沈降速度
free flow	自由流动	自由流動
free flow head	自由流动压头	自由流動高差
free moisture	游离水分	自由水分
free moisture content	游离水含量	自由水含量
free path	自由程	自由徑
free radical	自由基	自由基
free radical mechanism	自由基机理	自由基機構

英　文　名	大　陆　名	台　湾　名
free radical polymerization	自由基聚合	自由基聚合[反應]
free radical reaction	自由基反应	自由基反應
free settling	自由沉降	自由沈降
free site concentration	自由部位浓度	自由部位濃度
free stream	自由流	自由流
free surface	自由表面	自由表面
free vortex	自由涡旋	自由漩渦
free water	自由水	自由水
freeze coagulation	冷冻凝聚作用	冷凍凝聚[作用]
freeze drier	冷冻干燥机	冷凍乾燥機
freeze drying	冷冻干燥	冷凍乾燥
freezer	冷冻器	冷凍器
freezing point	凝固点	凝固點
freezing point depression	凝固点降低	凝固點下降
freezing process	冷冻过程	冷凍程序
freon	氟氯烷	氟氯烷
frequency	频率	頻率
frequency band	频带	頻帶
frequency domain	频域	頻域
frequency factor	频率因子	頻率因數
frequency response	频率响应	頻率應答
frequency response analysis	频率响应分析	頻率應答分析
frequency shift	频移	頻移
frequency test	频率测试	頻率試驗[法]
frequency tracer	频率示踪器	頻率示蹤器
friability	脆性	易碎性
friction	摩擦	摩擦[力]
frictional resistance	摩擦阻力	摩擦阻力
friction coefficient	摩擦系数	摩擦係數
friction corrosion	摩擦腐蚀	摩擦腐蝕
friction coupling	摩擦接头	摩擦接頭
friction drag	摩擦阻力	摩擦阻力
friction drop	摩擦压力降	摩擦壓力降
friction factor	摩擦因子	摩擦因數
friction head	摩擦压头	摩擦高差
frictionless flow	无摩擦流动	無摩擦流動
friction loss	摩擦损失	摩擦損失
friction loss factor	摩擦损失因子	摩擦損失因數

英　文　名	大　陆　名	台　湾　名
friction separator	摩擦分离器	摩擦分離器
friction test	摩擦试验	摩擦試驗
frit	玻璃料	玻璃料
fritted glass filter	烧结玻璃过滤器	燒結玻璃濾器
frontal analysis	锋面分析	鋒面分析
frosted glass	磨砂玻璃	磨砂玻璃,毛玻璃
froth	泡沫	泡沫
froth depth	泡沫深度	泡沫深度
froth flotation	泡沫浮选	泡沫浮選[法]
fructose	果糖	果糖
F-test	F 试验	F 檢定
fuel	燃料	燃料
fuel-air ratio	燃料–空气比	燃料–空氣比
fuel alcohol	燃料酒精	燃料酒精
fuel bed	燃料床	燃料床
fuel cell	燃料电池	燃料電池
fuel cost	燃料费	燃料成本
fuel industry	燃料工业	燃料工業
fuel oil	燃料油	燃料油
fuel ratio	燃料比	燃料比
fugacity	逸度	逸壓
fugacity coefficient	逸度系数	逸壓係數
fugitive accelerator	短效加速剂	短效催速劑
fuller's earth	漂白土	漂白土
fully developed flow	充分发展流	完全發展流動
fumagillin	烟曲霉素	煙黴素
fume	烟	煙
fumigant	熏蒸剂	燻蒸劑
functional equation	函数方程	泛函方程式
functional group	官能团	官能基
functional polymer	功能高分子	功能性聚合物
functional relationship	函数关系	函數關係[式]
function generator	函数发生器	函數波產生器
fundamental principle	基本原理	基本原理
fundamental property relation	基本性质关系式	基本性質關係[式]
fundamental unit	基本单位	基本單位
fungi	真菌	真菌
fungicide	杀菌剂	殺真菌劑,殺黴劑

英　文　名	大　陆　名	台　湾　名
furan plastic	呋喃塑料	呋喃塑膠
furan resin	呋喃树脂	呋喃樹脂
furnace	［窑］炉	爐
furnace black	炉黑	爐黑
fusel oil	杂醇油	雜醇油
fusion	熔融	熔化
fusion curve	熔化曲线	熔化曲線
fusion point	熔点	熔點

G

英　文　名	大　陆　名	台　湾　名
gage number	规号	規號
gain	增益	增益
gain compensation	增益补偿	增益補償
gain crossover	增益越界	增益交越
gain crossover frequency	增益越界频率	增益交越頻率
gain crossover point	增益越界点	增益交越點
gain margin	增益裕量	增益裕量
gain matrix	增益矩阵	增益矩陣
gain-phase plot	增益相图	增益相位圖
galvanic action	电流作用	迦凡尼效應
galvanic cell	原电池	電池
galvanic corrosion	电偶腐蚀	電流腐蝕
galvanic protection	镀锌保护	鍍鋅保護
galvanization	镀锌	鍍鋅
galvanometer	检流计	電流計
gangue	煤矸石	脈石
gas	气［体］	氣
gas analyzer	气体分析仪	氣體分析儀
gas burner	气体燃烧器	氣體燃燒器
gas calorimeter	气体量热计	氣體卡計
gas chromatogram	气相色谱图	氣相層析圖
gas chromatograph	气相色谱仪	氣相層析儀
gas chromatography	气相色谱法	氣相層析法
gas constant	气体常数	氣體常數
gas distributor	气体分布器	氣體分配器
gaseous emission standard	气体排放标准	氣體排放標準

英 文 名	大 陆 名	台 湾 名
gaseous floatation	气体浮选	氣體浮選[法]
gaseous fuel	气体燃料	氣體燃料
gaseous solution	气体溶液	氣體溶液
gas field	气田	天然氣田
gasfilled tube	充气管	通氣管
gas film	气膜	氣膜
gas film contactor	气膜接触器	氣膜接觸器
gas film control	气膜控制	氣膜控制
gas-forming reaction	成气反应	成氣反應
gas holdup	持气率	持氣量
gasification	气化	氣化
gasification process	气化过程	氣化程序
gasification reaction	气化反应	氣化反應
gasification technology	气化技术	氣化技術
gasifier	气化炉	氣化爐,氣化器
gasifying reaction	气化反应	氣化反應
gasket groove	垫圈沟	墊圈溝
gasket materials	垫片材料	墊片材料
gas law	气体定律	氣體定律
gas-liquid absorption	气液吸收	氣液吸收
gas-liquid contacting	气液接触	氣液接觸
gas-liquid contactor	气液接触器	氣液接觸器
gas-liquid equilibrium	气液平衡	氣液平衡
gas-liquid equilibrium process	气液平衡过程	氣液平衡程序
gas-liquid mass transfer equipment	气液传质设备	氣液質傳裝置
gas-liquid reaction	气液反应	氣液反應
gas-liquid reactor	气液反应器	氣液反應器
gas-liquid separation	气液分离	氣液分離
gas-liquid-solid reaction	气液固反应	氣液固反應
gas-liquid-solid reactor	气液固反应器	氣液固反應器
gas main	气体干管	氣體幹管
gas measuring tube	量气管	量氣管
gas meter	煤气表	氣量計
gasohol	酒精汽油	酒精汽油
gas-oil ratio	气油比	氣-油比
gasoline	汽油	汽油
gasoline engine	汽油机	汽油引擎
gasoline trap	汽油阱	汽油阱

英　文　名	大　陆　名	台　湾　名
gas permeability	透气性	透氣性
gas permeation	气体渗透	氣體滲透
gas phase	气相	氣相
gas phase control	气相控制	氣相控制
gas phase mass transfer coefficient	气相传质系数	氣相質傳係數
gas phase reaction	气相反应	氣相反應
gas producer	煤气发生炉	煤氣發生爐
gas receiver	气体收集器	集氣器
gas reforming	气体重整	[烴]氣重組
gas regulator	气体调节器	氣體調節器
gas scrubber	气体洗涤器	洗氣器
gas shift	水煤气变换	水煤氣轉化
gas shift process	水煤气变换过程	水煤氣轉化程序
gas shift reaction	水煤气变换反应	水煤氣轉化反應
gas-solid equilibrium	气固平衡	氣固平衡
gas-solid reaction	气固反应	氣固反應
gas-solid reactor	气固反应器	氣固反應器
gas stripping	气提	氣提
gas thermometer	气体温度计	氣體溫度計
gas trap	气阱	氣阱
gas turbine	燃气轮机	燃氣渦輪機
gas turbine engine	燃气轮发动机	燃氣渦輪引擎
gas venting system	排气系统	排氣系統
gate	闸门	閘,澆口
gate valve	闸阀	閘閥
gauge	①表 ②计	①表 ②計
gauge pressure	表压	錶壓
Gaussian elimination	高斯消元法	高斯消去法
gauze catalyst	网状催化剂	網狀觸媒
gaylussite	单斜钠钙石	單斜鈉鈣石
gear	齿轮	齒輪
gear mixer	齿轮式混合器	齒輪式混合器
gear pump	齿轮泵	齒輪泵
gear train	齿轮组	齒輪組
gel	凝胶	凝膠
gelatin	明胶	明膠
gelatination	胶凝作用	膠凝[作用]
gelatin dynamite	胶质炸药	膠質炸藥

英　文　名	大　陆　名	台　湾　名
gelatinizing agent	胶凝剂	膠凝劑
gelation	胶凝作用	膠凝[作用],膠化[作用]
gel chromatography	凝胶色谱[法]	凝膠層析法
gel entrapment	胶体包埋	膠體包埋
gel filtration	凝胶过滤	凝膠過濾
gel filtration [chromatography]	凝胶过滤色谱[法]	凝膠過濾層析法
gel immunoelectrophoresis	凝胶免疫电泳	凝膠免疫電泳
gene	基因	基因
generalized coordinates	广义坐标	廣義坐標
generalized equation	普适方程	通式
generalized Newtonian fluid	广义牛顿流体	廣義牛頓流體
generalized Z chart	通用压缩因子图	壓縮因數通用圖
general purpose detergent	通用清洁剂	通用清潔劑
general reduced gradient	通用简约梯度法	通用簡約梯度法
generation time	世代时间	世代時間
genetically engineered cell	基因工程细胞	基因[工程]改造細胞
genetically modified cell	基因改造细胞	基因改造細胞
genetic engineering	基因工程	基因工程
geometric factor	几何因数	幾何因數
geometric isomer	几何异构物	幾何異構物
geometric mean	几何平均	幾何平均
geometric mean area	几何平均面积	幾何平均面積
geometric mean temperature difference	几何平均温差	幾何平均溫差
geometric programming	几何规划	幾何規畫
geometric similarity	几何相似	幾何相似性
geometric surface area	几何表面积	幾何表面積
germicide	杀菌剂	殺菌劑
Gibbs-Duhem equation	吉布斯–杜安方程	吉布斯–杜安方程
Gibbs free energy	吉布斯自由能,自由焓	吉布斯自由能,自由焓
gibbsite	三水铝石	水礬土,水鋁氧
gibbsite fireclay	三水铝石耐火土	水礬土耐火土
glacial acetic acid	冰醋酸	冰醋酸
glass	玻璃	玻璃
glass blowing	玻璃吹制	玻璃吹製
glass ceramics	玻璃陶瓷	玻璃陶瓷
glass drawing	玻璃拉制	玻璃拉製
glass electrode	玻璃电极	玻璃電極

英 文 名	大 陆 名	台 湾 名
glass enamel	玻璃搪瓷	玻璃搪瓷
glass fiber	玻璃纤维	玻璃纖維
glass fiber reinforced plastics	玻璃纤维强化塑料	玻璃纖維強化塑膠
glass filter	玻璃过滤器	玻璃[過]濾器,玻璃濾 光片
glass level gage	玻璃液位计	玻璃液位計,玻璃量規
glass-lined pipe	衬玻璃管	玻璃襯管
glass lining	玻璃衬里	玻璃襯裏
glass pipe	玻璃管	玻璃管
glass pipe and fitting	玻璃管及配件	玻璃管子及配件
glass reinforced plastics	玻璃强化塑料	玻璃強化塑膠
glass stem thermometer	玻璃温度计	玻璃溫度計
glass transition temperature	玻璃化温度	玻璃轉移溫度
glass tube	玻璃管	玻璃管
glass wool	玻璃棉	玻璃綿,玻璃絨
glassy state	玻璃态	玻璃態
glassy zone	玻璃区	玻璃區
glauberite	钙芒硝	鈣芒硝
glaze	釉	釉
glazing	上釉	上釉
global optimum	总体最优[值]	總體最佳[值]
global rate	总体速率	總速率
global reaction	总体反应	總體反應
global reaction rate	总体反应速率	總反應速率
global stability	总体稳定性	總體穩定性
globe valve	球阀	球閥
glost kiln	釉烧窑	釉窯
glucan	葡聚糖	葡聚糖,葡[萄]聚糖
glucosan	葡聚糖	葡聚糖,葡[萄]聚糖
glucose	葡萄糖	葡萄糖
glue	胶	膠
gluten	面筋	麵筋
glutin	明胶蛋白	明膠蛋白
glyceride	甘油酯	甘油酯
glycerine	甘油	甘油
glycogen	糖原	肝醣
glycol	乙二醇	乙二醇
glycolysis	糖酵解	醣解[作用]

英 文 名	大 陆 名	台 湾 名
glycolytic pathway	糖酵解途径	醣解途徑
glyptal resin	甘酞树脂	甘[油]酞[酸]樹脂
golden section searching method	黄金分割法	黃金分割搜尋法
Gouy-Stodola theorem	古伊-斯托多拉定理	古依-斯託多拉定理
governing equation	支配方程	統御方程式
gradient	梯度	梯度
gradientless contactor	无梯度接触器	無梯度接觸器
gradientless reactor	无梯度反应器	無梯度反應器
gradient method	梯度法	梯度法
Graetz number	格雷茨数	格雷茲數
graft copolymer	接枝共聚物	接枝共聚物
graft polymer	接枝聚合物	接枝聚合物
graft polymerization	接枝聚合	接枝聚合[反應]
grain	晶粒	晶粒
grain growth	晶粒长大	晶粒生長
grain shape	[晶]粒形[状]	粒形
grain size	晶粒大小	晶粒大小,粒徑
grain size analysis	粒度分析	晶粒分析
grain size analyzer	粒度分析仪	晶粒分析儀
gramicidin	短杆菌肽	短桿菌素
grand canonical ensemble	巨正则系综	大正則系綜
grand canonical partition function	巨正则配分函数	大正則分配函數
granite	花岗岩	花崗岩
granular bed filter	颗粒层过滤器	顆粒層過濾器
granular materials	颗粒材料	粒狀物質
granular solid filter	粒状固体过滤器	粒狀固體過濾機
granulation	造粒	造粒
granule	颗粒体	顆粒
granulometer	粒度分析仪	粒度計
granulometry	粒度测量	粒度分析
graphical analysis	图解分析	圖解分析法
graphical calculation	图解计算	圖解計算
graphical integration	图解积分	圖解積分法
graphical method	图解法	圖解法
graphical solution	图解	圖解
graphic panel	图示板	圖示面板
graphite	石墨	石墨
graphite crucible	石墨坩埚	石墨坩鍋

英　文　名	大　陆　名	台　湾　名
graphite electrode	石墨电极	石墨電極
graphite fiber reinforced plastics	石墨纤维强化塑料	石墨纖維強化塑膠
graphite grease	石墨润滑脂	石墨[潤]滑脂
graphite refractory	石墨耐火材料	石墨耐火物
graphitic fiber	石墨纤维	石墨纖維
graphitization	石墨化	石墨化
graph theory	图论	圖論
Grashof number	格拉斯霍夫数	葛瑞斯何夫數
gravimetric analysis	重量分析[法]	重量分析[法]
gravitation	重力	重力
gravitational constant	重力常数	重力常數
gravitational conversion factor	重力换算因数	重力換算因數
gravitational field	引力场	重力場
gravitational filter	重力过滤器	重力[過]濾器
gravitational force	重力	重力
gravitational separation	重力分离	重力分離
gravitation settling	重力沉降	重力沈降
gravity	重力	重力
gravity cell	重力电池	重力電池
gravity circulation	重力循环	重力循環
gravity filter	重力式过滤器	重力過濾器
gravity filtration	重力过滤	重力過濾
gravity flow	重力流动	重力流動
gravity head	重力压头	重力高差
gravity meter	重力计	比重計
gravity segregation	重力分凝	重力離析
gravity separation	重力分离	重力分離
gravity separator	重力分离器	重力分離器
gravity settler	重力沉降器	重力沈降器
gravity settler separator	重力沉降分离器	重力沈降分離器
gravity settling	重力沉降	重力沈降
gravity thickener	重力增稠器	重力增稠器
gravity thickening	重力增稠	重力增稠
gray body	灰体	灰體
gray surface	灰面	灰面
gray system	灰色系统	灰色系統
grease	润滑脂	[潤]滑脂
greenhouse effect	温室效应	溫室效應

英 文 名	大 陆 名	台 湾 名
green liquor	绿液	綠液
grey body	灰体	灰體
grid agitator	框式搅拌器	框式攪拌器
grind	研磨	研磨
grindability	可磨性	可磨性
grindability index	可磨性指数	可磨性指數
grinder	磨床	磨床,研磨機
grinding	研磨	研磨
grinding additive	研磨辅料	研磨添加劑
grinding aid	助磨剂	助磨劑
grinding ball	磨球	磨球
grinding machine	磨床	磨床,研磨機
grinding medium	研磨介质	研磨介質,磨媒
grizzly	格筛	柵篩
gross earnings cost	毛利	毛利
gross error	过失误差	總誤差,毛誤差
gross error identification	过失误差检出	總誤差鑑定
gross heating value	总热值	總熱值
groundwater	地下水	地下水
groundwater pollution	地下水污染	地下水污染
groundwater treatment	地下水处理	地下水處理
groundwood pulp	磨木浆	磨木漿,細木漿
group activity coefficient	基团活度系数	基團活性係數
group fraction	基团分率	基團分率
group specificity	基团专一性	基特異性
growth curve	生长曲线	生長曲線
growth factor	生长因子	生長因子
growth parameter	生长参数	生長參數
growth rate	生长率	生長速率
growth yield	生长收率	生長收率
growth yield coefficient	生长收率系数,增殖收率系数	生長收率係數
guanase	鸟嘌呤酶	鳥嘌呤酶
guanine	鸟嘌呤	鳥嘌呤
guano	海鸟粪	[海]鳥糞[石]
guano phosphate	鸟粪磷酸盐	[海]鳥糞磷礦
guaranteed quality	保证质量	保證品質
guard bed	护床	護床

英 文 名	大 陆 名	台 湾 名
guard catalyst	防护催化剂	防護觸媒
gum	树胶	膠
gun cotton	硝棉	[強]硝化棉,火棉
gypsite	土石膏	土[狀]石膏
gypsum	石膏	石膏
gyrating screen	回转振动筛	偏旋篩
gyratory break	回转轧碎机	偏旋破碎機
gyratory crusher	回转破碎机	偏旋壓碎機
gyroscopic flowmeter	回转流量计	回轉流量計

H

英 文 名	大 陆 名	台 湾 名
hair hygrometer	毛发湿度计	毛髮濕度計
hair-pin tube	发夹管	髮夾管
hairpin tube heat exchanger	U 型管换热器	U 型管熱交換器
hair spring	发状弹簧	髮狀彈簧
half body	半体	半體
half life	半衰期	半衰期,半生期
half life method	半衰期法	半衰期法,半生期法
half life of enzyme	酶半衰期	酶半衰期,酶半生期
halite	石盐	岩鹽
halogenation	卤化	鹵化[反應]
hammer crusher	锤[式破]碎机	鎚碎機
hammer mill	锤[式]磨[碎]机	鎚磨機
handling facility	装卸设备	裝卸設備
hard detergent	硬性洗涤剂	硬性清潔劑
hardened oil	硬化油	硬化油
hardener	硬化剂	硬化劑
hardening	硬化	硬化
hardness	硬度	硬度
hardness scale	硬度标度	硬度標度
hardness test	硬度试验	硬度試驗
hard rubber	硬橡胶	硬橡膠
hard sphere	硬球	硬球
hard sphere theory	①硬球理论 ②碰撞理论	硬球理論
hardware	硬件	硬體
hard water	硬水	硬水

英　文　名	大　陆　名	台　湾　名
hard wax	硬蜡	硬蠟
harmonic analysis	谐波分析	諧波分析
harmonic cycling	谐和循环	諧和循環
harmonic mean	谐和平均	諧和平均
harmonic oscillator	谐振子	諧波振盪器
harmonic response	谐波应答	諧波應答
hastelloy	哈斯特镍基合金	赫史特合金
Hatta number	八田数	八田數
hay filter	干草过滤器	乾草過濾器
hazard evaluation	危险性评估	危險性評估
hazard index	危险指数	危險指數
hazardous waste	有害性废弃物	有害性廢棄物
head	扬程	揚程,高差
head condenser	顶冷凝器	頂冷凝器
head effect	高差效应	高差效應
header	集管	集管[箱]
head loss	水头损失	壓頭損失,高差損失
heap and dump leaching	堆[积]浸[取]	堆積浸取
hearth	炉缸	爐膛,腔,爐床
heat	热	熱
heat accumulator	蓄热器	蓄熱器
heat balance	热量衡算	熱量均衡
heat capacity	热容	熱容[量]
heat capacity at constant pressure	等压热容	等壓熱容[量]
heat capacity at constant volume	等容热容	定容熱容
heat capacity flow rate	热容流率	熱容流率
heat carrier	载热体	熱載體
heat conduction	热传导	熱傳導
heat conductivity	热导率	熱傳導係數
heat content	热含量	熱含量
heat convection	热对流	熱對流
heat dissipation	热耗散	熱散逸
heat duty	热负载	熱負載
heat effect	热效应	熱效應
heat efficiency	热效率	熱效率
heat emission	热发射	熱發射
heat energy	热能	熱能
heat engine	热机	熱機

英　文　名	大　陆　名	台　湾　名
heater	加热器	加熱器,加熱爐
heat exchange	换热	熱交換
heat exchanger	换热器,热交换器	熱交換器
heat exchanger network	换热器网路	熱交換器網路
heat flow	热流	熱流動
heat flux	热[流]通量	熱通量
heat generation	热生成	熱生成
heat generation curve	热生成曲线	熱生成曲線
heat generation line	热生成线	熱生成線
heating	加热	加熱
heating area	加热面积	加熱面積
heating coil	加热盘管	加熱旋管
heating cost	加热成本	加熱成本
heating rate	加速速率	加熱速率
heating time	加热时间	加熱時間
heating unit	加热单元	加熱單元,加熱裝置
heating value	热值	熱值
heat insulating materials	绝热材料,保温材料	隔熱材料,保溫材料
heat insulation	绝热	隔熱
heat integration	热集成	熱整合
heat interchanger	热交换器	熱交換器
heat liberating reaction	放热反应	放熱反應
heat loss	热损失	熱損失
heat of absorption	吸收热	吸收熱
heat of adhesion	附着热	附著熱
heat of adsorption	吸附热	吸附熱
heat of coagulation	凝结热	凝聚熱
heat of combination	化合热	化合熱
heat of combustion	燃烧热	燃燒熱
heat of compression	压缩热	壓縮熱
heat of condensation	冷凝热	冷凝熱,凝結熱
heat of conversion	转化热	轉化熱
heat of crystallization	结晶热	結晶熱
heat of decomposition	分解热	分解熱
heat of dilution	稀释热	稀釋熱
heat of evaporation	蒸发热	蒸發熱
heat of formation	生成热	生成熱
heat of fusion	熔化热	熔化熱,熔合熱

英　文　名	大　陆　名	台　湾　名
heat of hydration	水合热	水合熱
heat of liquefaction	液化热	液化熱
heat of mixing	混合热	混合熱
heat of neutralization	中和热	中和熱
heat of reaction	反应热	反應熱
heat of solidification	固化热	固化熱
heat of solution	溶解热	溶解熱
heat of sublimation	升华热	升華熱
heat of transition	转变热	［相］轉變熱
heat of vaporization	汽化热	汽化熱
heat pipe	热管	熱管
heat pipe exchanger	热管换热器	熱管交換器
heat pollution	热污染	熱污染
heat pump	热泵	熱泵
heat rate	热功转化率	熱功轉化率
heat recovery	热量回收	熱回收
heat regeneration	加热再生	熱再生
heat regenerator	蓄热器	蓄熱器
heat removal curve	热量去除曲线	熱量去除曲線
heat removal line	热量去除曲线	熱量去除曲線
heat reservoir	储热器	貯熱器
heat resistance	耐热性	耐熱性
heat sealing adhesive	热封黏合剂	熱封黏合劑
heat sink	热阱	熱匯
heat source	热源	熱源
heat transfer	热量传递,传热	熱傳
heat transfer area	传热面积	熱傳面積
heat transfer coefficient	传热系数	熱傳係數
heat transfer equipment	传热设备	熱傳裝置
heat transfer medium	传热介质	熱傳介質
heat transfer rate	传热速率	熱傳速率
heat transfer resistance	传热阻力	熱傳阻力
heat transport	热量输送	熱輸送
heat treatment	热处理	熱處理
heat value	热值	熱值
heavy chemical	重化学品	重化學品
heavy duty detergent	高效型洗涤剂	強效清潔劑
heavy ends	重馏分	重餾分

英 文 名	大 陆 名	台 湾 名
heavy fuel oil	重燃料油	重燃料油
heavy hydrocarbon	重烃	重烴
heavy hydrogen	重氢	重氫
heavy oil	重质油	重油
heavy straight run	重直馏油	重直餾油
heavy water	重水	重水
height equivalent of a theoretical plate	等[理论]板高度,理论板当量高度	理論板相當高度
height of a [heat] transfer unit	传热单元高度	熱傳單元高度
height of a [mass] transfer unit(HTU)	传质单元高度	質傳單元高度
helical agitator	螺旋搅拌器	螺旋攪拌器
helical coil	螺旋线圈	螺旋管
helical fin	螺旋翅片	螺旋鰭片
helical pressure spring	螺旋压力弹簧	螺旋壓力彈簧
helical ribbon agitator	螺带搅拌器	螺帶攪拌器
helical slidewire	螺旋滑线	螺旋滑線
heliotrope oil	葵花油	葵花油
Helmholtz free energy	[亥姆霍兹]自由能	亥姆霍茲自由能
hematite	赤铁矿	赤鐵礦
hemicellulose	半纤维素	半纖維素
hemp seed oil	麻油	木麻油
Henry law	亨利定律	亨利定律
herbicide	除草剂	除草劑
Hess law	赫斯定律	赫斯定律
heteroazeotrope	非均相共沸混合物	異相共沸[混合]液
heterofermentative bacteria	异型发酵菌	異型發酵菌
heterogeneity	①不均匀性 ②多相性	不匀性
heterogeneous azeotrope	非均相共沸混合物	異相共沸液
heterogeneous catalysis	非均相催化	非均相催化,異相催化
heterogeneous equilibrium	多相平衡	多相平衡,異相平衡
heterogeneously fluidized bed	非均一流化床	非均匀流[體]化床
heterogeneous nucleation	异相成核	異相成核
heterogeneous polymerization	非均相聚合	非均相聚合
heterogeneous population	异质群体	雜菌群,雜族群
heterogeneous reaction	非均相反应	非均相反應
heterogeneous reactor	非均相反应器	非均相反應器
heterogeneous state	非均相状态	非均匀狀態
heterogeneous system	非均相系统	非均匀系統

英 文 名	大 陆 名	台 湾 名
heteropolyacid catalyst	杂多酸催化剂	異聚酸觸媒
heuristic method	启发式方法	經驗推斷法
heuristic rule	启发式规则	試探規則
hexagonal closest packed lattice	六方最密堆积晶格	六方最密堆積
hexose phosphate shunt pathway	磷酸己糖旁路途径	磷酸己糖旁路途徑
hidden layer	隐蔽层	隱藏層
hierarchical control	递阶控制	階層控制
high alumina brick	高铝砖	高鋁磚
high alumina brick	高铝砖	高鋁磚
high alumina cement	高铝水泥	高鋁水泥
high alumina cement	高铝水泥	高鋁水泥
high alumina clay	高铝黏土	高鋁黏土
high alumina glass	高铝玻璃	高鋁玻璃
high alumina refractory	高铝耐火材料	高鋁耐火物
high antiknock fuel	高抗爆性燃料	高抗震爆燃料
high-BTU gas	高热值气体	高熱量氣體
high density polyethylene	高密度聚乙烯	高密度聚乙烯
high heating value	高热值	高熱值
high impact polystyrene	高抗冲聚苯乙烯	耐衝擊聚苯乙烯
high-iron Portland cement	高铁含量[波特兰]水泥	高鐵[卜特蘭]水泥
high lift device	增升装置	高升裝置
high-pass filter	高通滤波器	高通濾波器
high performance liquid chromatography	高效液相色谱法	高效能液相層析儀
high polymer	高聚物	聚合物,高分子
high-rate digester	高速率消化槽	高速率消化槽
high selector switch	高选择器开关	高選擇器開關
high silica iron	高硅铁	高矽鐵
high-speed carbon steel	高速碳钢	高速碳鋼
high-speed evaporator	高速蒸发器	高速蒸發器
high strength portland cement	高强硅盐水泥	高強度[卜特蘭]水泥
high styrene rubber	高苯乙烯橡胶	高苯乙烯橡膠
high tensile strength	高抗拉强度	高抗拉強度
high test molasses	高级糖蜜	高級糖蜜
high transmission glass	高透光玻璃	高透光玻璃
high vacuum distillation	高真空蒸馏	高真空蒸餾
high vacuum operation	高真空操作	高真空操作
hindered sedimentation	受阻沉降	受阻沈降

英　文　名	大　陆　名	台　湾　名
hindered settling	受阻沉降	受阻沈降
histogram	直方图	直方圖
hoist	起重机	起重機,吊車
holding period	保持期间	貯留期間
holding pond	滞留池	貯留池
holding tank	滞留槽	貯留槽
holding time	保持时间	貯留時間
hold-up	滞留量	貯留量,滯留量
hold-up time	滞留时间	貯留時間
hollander	打浆机	打漿機
hollow agitator	自吸搅拌器,空心叶轮 　搅拌器	空心自吸攪拌器
hollow-fiber module	中空纤维组件	中空纖維組件
holocellulose	全纤维素	全纖維素
homoazeotrope	均相共沸混合物	均相共沸液
homofermentative bacteria	同型发酵菌	同型發酵菌
homogenate	匀浆	匀漿,均質[物]
homogeneity	均匀性	均匀性
homogeneous azeotrope	均相共沸液	均相共沸液
homogeneous body	均匀体	均匀體
homogeneous catalysis	均相催化	均相催化[作用]
homogeneous combustion	均相燃烧	均相燃燒
homogeneous equilibrium	均相平衡	均相平衡
homogeneously fluidized bed	均匀流化床	均匀流[體]化床
homogeneous nucleation	均相成核	均相成核
homogeneous phase	均相	均相
homogeneous reaction	均相反应	均相反應
homogeneous reactor	均相反应器	均相反應器
homogeneous system	均相系统	均匀系統
homogeneous turbulence	均匀湍流	均匀紊流
homogenization	匀化	均質化
homogenizer	匀化器	均質機
homolog	同系物	同系物
homologous compound	同系化合物	同系物
homopolymer	均聚物	均聚物,同元聚合物
homopolymerization	均聚反应	均聚合[反應],同元聚 　合[反應]
honeycomb catalyst	蜂窝状催化剂	蜂巢狀觸媒

英　文　名	大　陆　名	台　湾　名
hood	防护罩	排氣罩
hoop stress	环向应力	環應力
horizontal centrifugal screen	卧式离心筛	臥式離心篩
horizontal digester	卧式蒸煮器	臥式蒸煮器
horizontal furnace	卧式炉	臥式爐
horizontal mixer	卧式混合器	臥式混合器
horizontal spray chamber	横式喷雾室	橫式噴霧室
horizontal tube bank	平管排	［水］平管排
horizontal tube evaporator	平管式蒸发器	平管式蒸發器
horsepower	马力	馬力
hose	软管	軟管
hot blast heater	热风炉	熱風爐
hot dipping	热浸	熱浸［漬］
hot floor drier	热坑干燥室	熱炕乾燥室
hot gas recycle process	热气再循环过程	熱氣循環程序
hot junction	热接点	熱接點
hot lime process	热石灰法	熱石灰法
hot liming	热汁加灰法	熱汁加灰法
hot molding	加热成型	加熱成型
hot press	热压机	熱壓機
hot pressing	热压	熱壓
hot process	高温法	高溫法
hot reflux	热回流	熱回流
hot rubber	高温橡胶	熱製橡膠
hot runner	热流道	熱澆道
hot spot	热点	熱點
hot vapor	热汽	熱汽
hot vulcanization	热硫化	熱硫化
hp（＝horse power）	马力	馬力
HTU（＝height of a［mass］transfer unit）	传质单元高度	質傳單元高度
humectant	保湿剂	保濕劑
humid chart	湿度表	濕度表
humid heat	湿热	濕［潤］比熱
humidification	增湿	增濕
humidification process	增湿过程	增濕程序
humidifier	增湿器	增濕器
humidifying	增湿	增濕
humidifying chart	增湿图	增濕圖

英　文　名	大　陆　名	台　湾　名
humidity	湿度	濕度
humidity chart	湿度表	濕度表
humidity ratio	湿度比	濕度比
humid volume	湿体积	濕［潤］比容
humus	腐殖质	腐植質
hybrid computer	混合计算机	併合計算機
hybrid tumor	杂交瘤	雜交瘤
hydrant water	消防水	消防水
hydrase	水化酶	水合酶
hydrate	水合物	水合物
hydrated lime	熟石灰	熟石灰
hydrated salt	水合盐	水合鹽
hydration	水合	水合
hydrator	水合器	水合器
hydraulically smooth pipe	水力平滑管	水力平滑管
hydraulically smooth wall	水力平滑壁	水力平滑壁
hydraulic brake	液压制动器	液壓致動器
hydraulic cement	水硬水泥	水硬水泥
hydraulic centrifuge	液压离心机	液壓離心機
hydraulic classification	水力分类	水力類析［法］
hydraulic control	液压控制	液壓控制
hydraulic controller	液压控制器	液壓控制器
hydraulic conveying	液压运送	液壓運送
hydraulic cylinder	液压筒	液壓筒
hydraulic diameter	水力直径	水力直徑
hydraulic efficiency	水力效率	水力效率,液壓效率
hydraulic fluid	液压流体	液壓流體
hydraulic fracturing	水力劈裂	液壓破裂［法］
hydraulic head	水力压头	水力高差
hydraulic jig	水簸机	水簸機
hydraulic jump	水跃	水躍
hydraulic lime	水硬石灰	水硬石灰
hydraulic load	水力负载量	水力負載量
hydraulic mean diameter	水力平均直径	水力平均直徑
hydraulic overload	水力超载量	水力超載量
hydraulic power	水力功率	水力功率
hydraulic press	水压机	液壓機
hydraulic pressure	液压	液壓

英 文 名	大 陆 名	台 湾 名
hydraulic pump	液压泵	水力泵
hydraulic radius	水力半径	水力半径
hydraulic radius for heat transfer	传热水力半径	水力半径
hydraulic resistance	水力阻力	水力阻力
hydraulic resonance	水力共振	水力共振
hydraulics	水力学	水力學
hydraulic sensor	水力传感器	水力感測器
hydraulic separation	水力分离	水力分離
hydraulic separator	水力分离器	水力分離器
hydraulic tachometer	水力转速计	水力轉速計,液壓轉速計
hydraulic turbine	水轮机	水力渦輪機
hydraulic valve	水力阀	水力閥
hydraulic water	水压用水	液壓用水
hydrazine	肼	肼,聯胺
hydrocarbon	碳氢化合物	碳氫化合物
hydrocarbon black	烃黑	烴黑
hydrocarbon fermentation	烃发酵	烴發酵
hydrochloride rubber	氢氯化橡胶	橡膠氫氯化物
hydrocolloid	水胶体	水膠體
hydrocracker	加氢裂化反应器	加氫裂解工場,加氫裂解爐
hydrocracking	加氢裂化	加氫裂解[反應]
hydrocracking unit	加氢裂化单元	加氫裂解單元,加氫裂解工場
hydrocyclone	旋液分离器	旋液分離器
hydrodealkylation	加氢脱烷基化	加氫脱烷[反應]
hydrodemetallization catalyst	加氢脱金属催化剂	加氫脱金屬觸媒
hydrodenitrogenation	加氢脱氮	加氫脱氮
hydrodesulfurization	加氢脱硫	加氫脱硫
hydrodynamically smooth surface	流体力学平滑表面	流力平滑表面
hydrodynamic boundary layer	流体动力边界层	流[體動]力邊界層
hydrodynamic effect	流体动力效应	流[體動]力效應
hydrodynamic interaction	流体动力相互作用	流[體動]力相互作用
hydrodynamic pressure	动水压力	流體壓力
hydrodynamic stability	流体动力稳定性	流[體動]力穩定性
hydroformylation	氢甲酰化	氫甲醯化[反應]
hydrogasification	加氢气化	加氫氣化

英　文　名	大　陆　名	台　湾　名
hydrogasifier	加氢气化器	加氢氯化器
hydrogenase	氢化酶	氢化酶
hydrogenated fat	氢化脂	氢化脂肪
hydrogenated oil	氢化油	氢化油
hydrogenation	加氢	氢化
hydrogenation catalyst	氢化催化剂	氢化觸媒
hydrogenation reaction	氢化反应	氢化反應
hydrogen bomb	氢弹	氢彈
hydrogen bubble technique	氢气泡法	氢氣泡法
hydrogen discharge tube	氢放电管	氢放電管
hydrogen electrode	氢电极	氢電極
hydrogen explosion	氢气爆炸	氢氣爆炸
hydrogen fuel	氢燃料	氢燃料
hydrogen fuel cell	氢燃料电池	氢燃料電池
hydrogen ion comparator	氢离子比色计	氢離子比色計
hydrogenolysis	氢解	氢解
hydrokinematics	流体运动学	流體運動學
hydrolase	水解酶	水解酶
hydroliquefaction	加氢液化	加氢液化
hydrolysate	水解产物	水解產物
hydrolysis	水解	水解[作用]
hydrolytic adsorption	水解吸附	水解吸附
hydrolytic enzyme	水解酶	水解酶
hydromechanics	流体力学	流體力學
hydrometallurgy	湿法冶金[学]	濕法冶金[學]
hydrometer	液体密度计	比重計
hydrophile	亲水物	親水物
hydrophilic colloid	亲水胶体	親水膠體
hydrophilic fiber	亲水纤维	親水纖維
hydrophilicity	亲水性	親水性
hydrophilic-lipophilic balance(HLB)	亲水亲油平衡	親水性–親油性均衡
hydrophilic particle	亲水性颗粒	親水顆粒
hydrophobe	疏水物	疏水物
hydrophobic chromatography	疏水色谱[法]	疏水層析法
hydrophobic colloid	疏水胶体	疏水膠體
hydrophobic fiber	疏水纤维	疏水纖維
hydrophobicity	疏水性	疏水性
hydrophobic particle	疏水性颗粒	疏水性粒子

英　文　名	大　陆　名	台　湾　名
hydropyrolysis	加氢热解	加氫高溫裂解
hydroreforming	加氢重整	加氫重組
hydroreforming catalyst	加氢重整催化剂	加氫重組觸媒
hydrosol	水溶胶	水溶膠
hydrostatic condition	流体静压条件	流體靜力狀態
hydrostatic force	流体静力	流體靜力
hydrostatic force meter	流体静力计	流體靜力計
hydrostatic head	液柱静压头	流體靜力高差
hydrostatic pressure	液体静压[强]	液體靜壓[力]
hydrostatics	液体静力学	液體靜力學
hydrosulfite	亚硫酸氢盐	亞硫酸氫鹽
hydrosulfite bleaching	亚硫酸氢盐漂白	亞硫酸氫鹽漂白
hydrotreater	加氢处理装置	加氫處理器
hydrotreating	加氢处理	加氫處理
hydrotreatment	加氢处理	加氫處理
hydrotrope	水溶助剂	水溶助劑
hydrotropic agent	水溶助剂	水溶助劑
hydrotropy	水溶助长性	水溶助長性
hydroxide ion	氢氧根离子	氫氧離子
hydroxyl group	羟基	羥基
hydroxyl value	羟值	羥值
hygrometer	湿度计	濕度計
hygrometry	测湿法	測濕法
hygromycin	潮霉素	潮黴素
hygroscopicity	吸湿性	吸濕性
hygroscopic water	湿存水	吸濕[水]分
hyperbolic type [kinetic] equation	双曲线型[动力学]方程	雙曲線型[動力學]方程
hyperelastic materials	超弹性材料	超彈性物質
hyperfiltration	超滤	超微過濾
hypersorber	超吸附装置	超吸附裝置
hypersorption	超吸附	超吸附[法]
hypertonic solution	高渗溶液	高滲壓溶液
hypochlorite bleaching	次氯酸盐漂白	次氯酸鹽漂白
hypothesis	假说	假說
hypotonic solution	低渗溶液	低滲壓溶液
hysteresis	滞后	遲滯

I

英 文 名	大 陆 名	台 湾 名
ice bath	冰浴	冰浴
ice point	冰点	冰點
I control（＝integral control）	积分控制	積分控制
ideal contact stage	理想接触级	理想接觸階
ideal cycle	理想循环	理想循環
ideal flow	理想流动	理想流動
ideal fluid	理想流体	理想流體
ideal gas	理想气体	理想氣體
ideal gas constant	理想气体常数	理想氣體常數
ideal gas enthalpy	理想气体焓	理想氣體焓
ideal gas law	理想气体定律	理想氣體定律
ideal gas state	理想气体状态	理想氣體狀態
ideal gas temperature scale	理想气体温标	理想氣體溫標
ideal mixture	理想混合物	理想混合物
ideal plate	理想板	理想板
ideal reactor	理想反应器	理想反應器
ideal solution	理想溶液	理想溶液
ideal stage	理想级	理想階
ideal tray	理想板	理想板
ideal tubular reactor	理想管式反应器	理想管式反應器
ideal work	理想功	理想功
identification	①鉴定 ②辨识	①鑑定,鑑別 ②識別
igneous quartz	火成石英	火成石英
igneous rock	火成岩	火成岩
ignition	点火	點火
ignition point	燃点	燃點
ignition time	点火时间	點火時間
illite	伊利石	伊來石,白雲母石
ilmenite	钛铁矿	鈦鐵礦
immersed surface	浸没表面	浸沒表面
immersion error	浸没误差	浸沒誤差
immiscibility	不互溶性	不互溶性
immiscible flow	不互溶流动	不互溶流動

英　文　名	大　陆　名	台　湾　名
immiscible liquid	不互溶液体	不互溶液體
immiscible phase	不互溶相	不互溶相
immiscible solvent	不互溶溶剂	不互溶溶劑
immiscible system	不互溶系统	不互溶系統
immobilization	固相化	固定化
immobilization technology	固定化技术	固定化技術
immobilized cell reactor	固定化细胞反应器	固定化細胞反應器
immobilized enzyme	固定化酶	固定化酶
immobilized liquid membrane	支撑液膜	固定式液膜
immunoadsorption	免疫吸附	免疫吸附
immuno electrophoresis	免疫电泳	免疫電泳
impact	冲击	衝擊
impact crusher	冲击破碎机	衝擊粉碎機
impact head	冲击压头	衝擊高差
impact number	冲击数	衝擊數
impactor	冲击机	衝擊機
impact pressure	冲击压[力]	衝擊壓[力]
impact strength	冲击强度	衝擊強度
impact test	冲击试验	衝擊試驗
impedance	阻抗	阻抗
impeded settling	阻滞沉降	阻滯沈降
impeller	叶轮	葉輪,槳
impeller agitator	叶轮搅拌器	葉輪攪拌器
impeller diameter	叶轮直径	葉輪直徑
impeller mixer	叶轮混合器	葉輪混合器
impeller shaft	叶轮轴	葉輪軸
impeller type flowmeter	叶轮式流量计	葉輪流量計
impermeability	不渗透性	不透性
impermeable partition	不透性隔板	不透性隔板
impingement	冲击	衝擊
impingement centrifugal separator	冲击[式]离心分离器	衝擊離心分離器
impingement corrosion	冲击腐蚀	衝擊腐蝕
impingement scrubber	冲击[式]涤气器	衝擊洗氣器
implicit enumeration method	隐枚举法	隱枚舉法
implicit function	隐函数	隱函數
implicit method	隐式法	隱式法
impregnation	浸渍	浸漬
impulse	冲量	①脈衝 ②衝量

英　文　名	大　陆　名	台　湾　名
impulse disturbance	脉冲搅动	脈衝擾動
impulse function	脉冲函数	脈衝函數
impulse response	脉冲响应	脈衝應答
impulse tachometer	脉冲转速计	脈衝轉速計
impulse type	脉冲型	脈衝型
impurity	杂质	雜質
inactivation	失活	去活[作用]
inactivator	惰化剂	惰化劑
incandescent light	白炽灯光	白熱燈光
incidence matrix	关联矩阵	關聯矩陣
incineration	焚化	焚化
incinerator	焚化炉	焚化爐
incipient fluidization	起始流化态	起始流體化
incipient fluidizing velocity	起始流化速度	起始流體化速度
inclined grate cooler	斜栅冷却器	斜柵冷卻器
inclined manometer	斜管压力计	斜管壓力計
inclined plate	斜板	斜板
inclined tube evaporator	斜管蒸发器	斜管蒸發器
income tax	所得税	所得稅
incompressibility	不可压缩性	不可壓縮性
incompressible filter cake	不可压缩滤饼	不可壓縮濾餅
incompressible flow	不可压缩流动	不可壓縮流動
incompressible fluid	不可压缩流体	不可壓縮流體
incondensable gas	不冷凝气体	不冷凝氣體
incongruent melting point	分熔点	分熔點
incremental investment	增资	增資
incubation period	潜伏期	培養期
incubator	恒温箱	培養器,孵化器,保溫箱
indanthrene blue	阴丹士林蓝	陰丹士林藍
indanthrene dye	阴丹士林染料	陰丹士林染料
independent reaction	独立反应	獨立反應
independent variable	自变量	自變數
indicator	①指示剂 ②指示计	①指示劑 ②指示器
indigo blue	靛蓝	靛藍
indigo brown	靛棕	靛棕
indigo carmine	靛洋红	食用藍色二號,靛胭脂
indigo dyeing	靛染	靛染
indigo dye vat	靛染缸	靛染缸

英　文　名	大　陆　名	台　湾　名
indigo white	靛白	靛白
indirect condenser	间接式冷凝器	間接[式]冷凝器
indirect cost	间接成本	間接成本
indirect diaphragm valve	间接隔膜阀	間接隔膜閥
indirect evaporation	间接蒸发	間接蒸發
indirect heat exchange	间接热交换	間接熱交換
indirect heating	间接加热	間接加熱
indirect hydrogenation	间接氢化	間接氫化
indirect liquefaction	间接液化	間接液化
indirect liquid level measurement	间接液位测量	間接液位量測
indirect method	间接法	間接法
individual heat transfer coefficient	传热分系数	個別熱傳係數
individual mass transfer coefficient	传质分系数	個別質傳係數
induced dipole	诱导偶极	誘發偶極
induced draft	引风	抽氣通風
induced draft cooling tower	引风式凉水塔	抽氣通風冷卻塔
induced drag	诱导发力	誘發阻力
induced fit theory	[酶]诱导契合学说	[酶]誘導契合學說
induced reaction	诱导反应	誘發反應
induction furnace	感应炉	感應爐
induction period	诱导期	誘導期
industrial alcohol	工业酒精	工業酒精
industrial chemistry	工业化学	工業化學
industrial gas	工业煤气	工業用氣體
industrial instrument	工业仪器	工業儀器
industrial process	工业过程	工業程序
industrial reactor	工业反应器	工業反應器
industrial stoichiometry	工业化学计量学	工業化學計量學
industrial-use detergent	工业用洗涤剂	工業用清潔劑
industrial waste	工业废料	工業廢棄物
industrial wastewater	工业废水	工業廢水
industrial wastewater treatment	工业废水处理	工業廢水處理
industrial water	工业用水	工業用水
industrial water treatment	工业水处理	工業用水處理
industry	工业	工業,產業
inert	惰性物	惰性物
inert gas	惰性气体	惰性氣體
inertia	惯性	慣性

英　文　名	大　陆　名	台　湾　名
inertial centrifugal separator	惯性离心分离器	慣性離心分離器
inertial device	惯性装置	慣性裝置
inertial force	惯性力	慣性力
inertial settling	惯性沉降	慣性沈降
inferential control	推理控制	推算控制
infiltration	渗滤	滲濾
infinite dilution	无限稀释	無限稀釋
infinite reflux	无限回流	無限回流
inflammability	可燃性	可燃性
inflating agent	发泡剂	充氣劑
inflection point	拐点	拐點,反曲點
influent	流入物	流入物
informal report	非正式报告	非正式報告
information	信息	資訊
informational polymer	信息聚合物	資訊聚合物
information board	信息板	資訊板
information flow diagram	信息流图	資訊流程圖
information processing	信息处理	資訊處理
information storage	信息储存	資訊儲存
infrared analysis	红外分析	紅外線分析
infrared analyzer	红外线分析仪	紅外線分析儀
infrared drier	红外线干燥器	紅外線乾燥器
infrared drying	红外干燥	紅外線乾燥
infrared spectrophotometer	红外分光光度计	紅外線光譜儀
infrared spectrophotometry	红外光谱	紅外線光譜法
infrared spectroscopy	红外光谱法	紅外線光譜法
ingot iron	铁锭	熟鐵
inherent viscosity	比浓对数黏度	比濃對數黏度
inhibition	抑制	抑制[作用]
inhibitor	抑制剂	抑制劑
inhomogeneity	非均匀性	不匀性
initial boiling point	初沸点	初沸點
initial condition	初始条件	初始條件
initial rate	初速率	初速率
initial relaxed system	初始松弛系统	初始鬆弛系統
initial state	初[始状]态	初[始狀]態
initial value theorem	初值定理	初值定理
initial velocity	初速[度]	初速[度]

英 文 名	大 陆 名	台 湾 名
initiator	引发剂	引發劑,起始劑
injection molding	注射成型	射出成型
injector	注射器	注射器
inlet	入口	入口
in-line filter	管线过滤器	管線過濾器
in-line mixer	管路混合器	管内混合器,沿線混合器
in-line tube arrangement	直列管排	直列管排
inner tube	内胎	内胎
inner valve	内阀	内閥
inoculation	接种	接種
inoculum	接种物	接種液
input	输入	輸入
input disturbance	输入搅动	輸入擾動
input function	输入函数	輸入函數
input node	输入节点	輸入節點
input-output	投入产出	輸入輸出,投入產出
input signal	输入信号	輸入信號,輸入訊號
input variable	输入变量	輸入變數
insecticide	杀虫剂	殺蟲劑
inside diameter	内径	内徑
in-situ combustion	就地燃烧	就地燃燒
in-situ processing	就地加工	就地加工
in-situ recovery	就地回收	就地回收
in-situ regeneration	就地再生	就地再生
insolubilization	不溶解化	不溶解化
instability	不稳定性	不穩定性,不安定性
instability constant	不稳定常数	不穩定常數,不安定常數
installation	安装	安裝
installation cost	安装成本	安裝成本
instantaneous fractional yield	瞬时产量分率	瞬間產量分率
instantaneous reaction	瞬时反应	瞬間反應
instantaneous reaction rate	瞬时反应速率	瞬間反應速率
instantaneous system	瞬时系统	瞬間系統
instrument	仪器	儀器
instrumental analysis	仪器分析	儀器分析
instrumentation	仪表	儀表

英　文　名	大　陆　名	台　湾　名
instrumentation diagram	仪表配置图	儀表配置圖
instrument flowsheet	仪表流程图	儀器流程圖
instrument servomechanism	仪器伺服机构	儀器伺服機制
insulated column	隔热柱	隔熱柱
insulated pipe	保温管	隔熱管
insulated runner	隔热绕道	隔熱澆道
insulating brick	保温砖	隔熱磚
insulating efficiency	保温效率	絕緣效率,隔熱效率
insulating firebrick	隔热耐火砖	隔熱耐火磚
insulating lining	隔热衬里	絕緣襯裏,隔熱襯裏
insulating materials	隔热材料	絕緣材料,隔熱材料
insulating paper	绝缘纸	絕緣紙
insulating refractory	隔热耐火材料	隔熱耐火物
insulation	①保温 ②绝缘	絕緣
insulator	绝缘体	絕緣體
insulin	胰岛素	胰島素
insurance	保险	保險
intake pressure	进气压力	進氣壓[力]
intake screen	进料滤网	進料篩網
intake stroke	进气行程	進氣衝程
intalox saddle	矩鞍形填料	矩鞍形填料
integer programming	整数规划	整數規劃
integral	积分	積分[式]
integral action	积分作用	積分作用
integral control	积分控制	積分控制
integral control action	积分控制作用	積分控制作用
integral distribution curve	积分分布函数	累積分佈曲線
integral energy equation	积分能量方程	積分能量方程式
integral feedback	积分反馈	積分回饋
integral heat of solution	积分溶解热	積分溶解熱
integral method	积分法	積分法
integral momentum equation	积分动量方程	積分動量方程式
integral of absolute error	绝对误差积分	絕對誤差積分
integral of square error	误差平方积分	誤差平方積分
integral reactor	积分反应器	整體反應器
integral test	积分检验[法]	積分檢驗[法]
integral time	积分时间	積分時間
integral time constant	积分时间常数	積分時間常數

英　文　名	大　陆　名	台　湾　名
integral windup	积分饱卷	積分飽和
integral yield	累积产率	累積產率
integrated error	累积误差	累積誤差
integrated process	集成过程	整合程序
integration	①集成 ②整合	①積分 ②整合
integrator	积分仪	積分器
intensity	强度	強度
intensity of turbulence	湍流强度	紊流強度
intensive property	强度性质	強度性質
intensive state variable	强度状态变量	內含狀態變數
interacting loops	相互作用环路	相互作用環路
interacting system	相互作用系统	相互作用系統
interaction	相互作用	相互作用
interaction coefficient	相互作用系数	交互作用係數
interaction energy	相互作用能	相互作用能
interaction factor	相互作用因数	相互作用因數
interaction force	相互作用力	相互作用力
interaction index	相互作用指数	相互作用指數
interaction of control systems	控制系统相互作用	控制系統相互作用
interaction potential	相互作用势	相互作用勢
intercoagulation	相互凝聚	相互凝聚[作用]
intercondenser	中间冷凝器	中間冷凝器
intercooler	中间冷却器	中間冷卻器
interdiffusion	相互扩散	相互擴散
interest	利息	利息
interest recovery period	利息回收期	利息回收期
interface	界面	界面
interface composition	界面组成	界面組成
interface control	界面控制	界面控制
interfacial agent	界面活性剂	界面活性劑
interfacial area	界面面积	界面面積
interfacial concentration	界面浓度	界面濃度
interfacial energy	界面能	界面能
interfacial engineering	界面工程	界面工程
interfacial film	界面薄膜	界面薄膜
interfacial phenomenon	界面现象	界面現象
interfacial potential	界面势	界面勢,界面電位
interfacial property	界面性质	界面性質

英　文　名	大　陆　名	台　湾　名
interfacial resistance	界面阻力	界面阻力
interfacial shape	界面形状	界面形狀
interfacial temperature	界面温度	界面溫度
interfacial tensimeter	界面张力计	界面張力計
interfacial tension	界面张力	界面張力
interfacial turbulence	界面湍流	界面紊流
interference pattern	干涉图样	干涉圖型
interferon	干扰素	干擾素
intergranular corrosion	晶间腐蚀	粒間腐蝕
interlaminar stress	层间应力	層間應力
intermediate	中间体	中間體
intermediate clarifier	中间澄清器	中級澄清器,中級沈降槽
intermediate crusher	中间压碎机	中級壓碎機
intermediate product	中间产物	中間產物
intermediate storage	中间产物储存	中間產物儲存
intermetallic compound	金属间化合物	介金屬[化合物]
intermittent drier	间歇干燥室	間歇乾燥室
intermixing	互混	互混
intermolecular force	分子间力	分子間力
internal adsorption	内吸附	內吸附
internal capacity	内在容量	內在容量
internal combustion	内燃	內燃
internal combustion engine	内燃机	內燃機
internal coordinate	内坐标	內坐標
internal diffusion	内扩散	內擴散
internal disturbance	内扰	內在擾動
internal energy	内能	內能
internal flow	内流	內部流動
internal friction	内耗	內摩擦
internal heating	内部加热	內部加熱
internal mixer	密炼机	密閉混合器
internal pressure	内压力	內[部]壓[力]
internal rate of return	内部收益率	內部收益率
internal recirculation reactor	内循环反应器	內循環反應器
internal recycle	内循环	內循環
internal reflux	内回流	內回流
internal reflux control	内回流控制	內回流控制

英　文　名	大　陆　名	台　湾　名
internals	内[部]构件	內部零件
internal surface	内表面	內表面
international practical temperature scale	国际实用温标	國際實用溫標
interparticle diffusion	粒间扩散	粒間擴散
interphase diffusion	相间扩散	相間擴散
interphase effectiveness	相内有效性	相內有效度因數
interphase exchange coefficient	相间交换系数	相間交換係數
interphase-intraphase diffusion	相间相内扩散	相間相內擴散
interphase-intraphase effectiveness	相间相内有效性	相間相內有效度因數
interphase mass transfer	相间传质	相間質傳
interphase reaction	相间反应	相間反應
interphase temperature	相间温度	相間溫度
interphase temperature	相内温度	相內溫度
interphase yield	相内产率	相內產率
interpolation	内插	內插法
interstage cooling	级间冷却	階間冷卻
interstage feed injection	级间注射进料	階間[注射]進料
interstage heating	级间加热	階間加熱
interstage recirculation	级间循环	階間循環
interstitial velocity	空隙速度	間隙速度
interval halving	半区间法	半區間法
intracellular enzyme	胞内酶	[細]胞內酵素
intramolecular force	分子内力	分子內力
intraparticle diffusion	粒内扩散	粒內擴散
intraparticle diffusivity	粒内扩散系数	粒內擴散係數
intrinsic activation energy	固有活化能	固有活化能
intrinsic energy	固有能	固有能
intrinsic kinetics	本征动力学	本徵動力學
intrinsic pressure	蕴压	蘊壓[力]
intrinsic rate	固有速率	固有速率
intrinsic safety	本征安全	固有安全
intrinsic viscosity	特性黏度	本質黏度,極限黏度
invariant	不变式	不變式
invention	发明	發明
inventory	①存量 ②目录	存量
inventory control	库存控制	存量控制
inverse derivative control	反微分控制	反微分控制
inversed solubility curve	逆溶度曲级	反轉溶解度曲線

英　文　名	大　陆　名	台　湾　名
inverse emulsion	反相乳液	反相乳液
inverse lever arm rule	反杠杆法则	反槓桿法則
inverse response	反应答	反應答
inverse transformation	逆变换	逆變換
inversion point	转化点	反轉點
inversion temperature	反转温度	反轉溫度
invertase	转化酶	轉化酶
inverted manometer	反向压力计	反向壓力計
invert sugar	转化糖	轉化糖
investment capital	投资资本	投資資本
investment recovery	投资回收	投資回收
inviscid flow	非黏流动	非黏性流動
in vitro	体外	[生]體外
in vivo	体内	[生]體內
iodine value	碘值	碘值
ion-beam deposition	离子束沉积法	離子束沈積[法]
ion-beam etching	离子束蚀刻法	離子束蝕刻[法]
ion exchange	离子交换	離子交換
ion exchange agent	离子交换剂	離子交換劑
ion exchange capacity	离子交换容量	離子交換容量
ion exchange chromatography	离子交换色谱[法]	離子交換層析法
ion exchange equilibrium	离子交换平衡	離子交換平衡
ion exchange membrane	离子交换膜	離子交換膜
ion exchange process	离子交换过程	離子交換法
ion exchanger	离子交换剂	離子交換劑
ion exchange resin	离子交换树脂	離子交換樹脂
ion exchange separation	离子交换分离	離子交換分離
ion exclusion	离子排斥	離子排斥
ion flotation	离子浮选	離子浮選[法]
ionic diffusion	离子扩散	離子擴散
ionic migration	离子迁移	離子遷移
ionic mobility	离子迁移率	離子移動率
ionic polymerization	离子聚合	離子聚合[作用]
ionic potential	离子电位	離子電位
ionic reaction	离子反应	離子反應
ionic strength	离子强度	離子強度
ionic surfactant	离子型表面活性剂	離子[型]界面活性劑
ion implantation	离子注入	離子植入[法]

英　文　名	大　陆　名	台　湾　名
ionization	电离	游離
ionization constant	电离常数	游離常數
ionization energy	电离能	游離能
ionization potential	电离电位	游離電位
ionomer	离子聚合物	離子聚合物
iron-alumina catalyst	铁铝氧催化剂	鐵鋁氧觸媒
iron bacteria	铁细菌	鐵細菌
iron mordant	铁媒染剂	鐵媒染劑
iron tannage	铁鞣	鐵鞣
irreversibility	不可逆性	不可逆性
irreversible catalysis	不可逆催化	不可逆催化[作用]
irreversible cell	不可逆电池	不可逆電池
irreversible chemical reaction	不可逆化学反应	不可逆化學反應
irreversible process	不可逆过程	不可逆程序,不可逆過程
irreversible reaction	不可逆反应	不可逆反應
irreversible thermodynamics	不可逆[过程]热力学	不可逆熱力學
irreversible work	不可逆功	不可逆功
irritant	刺激物	刺激物
irrotational flow	无旋流	無旋轉流
irrotational motion	无旋运动	無旋轉運動
isenthalpic compression	等焓压缩	等焓壓縮
isenthalpic expansion	等焓膨胀	等焓膨脹
isenthalpic process	等焓过程	等焓過程,等焓程序
isentropic compression	等熵压缩	等熵壓縮
isentropic expansion	等熵膨胀	等熵膨脹
isentropic flow	等熵流	等熵流動
isentropic process	等熵过程	等熵過程,等熵程序
isobar	等压线	等壓線
isobaric process	等压过程	等壓過程,等壓程序
isobaric superheating	等压过热	等壓過熱
isochore	等容线	等容線
isochoric process	等容过程	等容過程,等容程序
isocline	等斜线	等斜率線
isoconcentration curve	等浓度曲线	等濃度曲線
isocyanate adhesive	异氰酸酯黏合剂	異氰酸酯黏合劑
isocyanate polymer	异氰酸酯聚合物	異氰酸酯聚合物
isoelectric focusing	等电聚焦	等電聚焦

英 文 名	大 陆 名	台 湾 名
isoelectric point	等电点	等電點
isoelectric precipitation	等电沉淀	等電沈澱
isoelectric separation	等电点分离	等電分離
isoforming	异构重整	異構重組
isoionic point	等离点	等離子點,等電點
isokinetic point	等动力学点	等動力點
isokinetic temperature	等动力学温度	等動力溫度
isolated system	隔离系统,孤立系统	隔離系統
isolating oil	绝缘油	隔離油
isomer	异构体	異構物
isomerization	异构化	異構化,異構作用
isometric process	等容过程	等容過程,等容程序
isometrics	等容线	等容線,等量線
isopiestic method	等压法	等壓法
isoprene rubber	异戊[二烯]橡胶	異戊二烯橡膠,異平橡膠
isopropylbenzene	异丙苯	異丙苯
isostere	同电子排列体	同電子排列體
isotachophoresis	等速电泳	等速電泳
isotactic polymer	全同立构聚合物,等规聚合物	同排聚合物
isotactic polypropylene	等规聚丙烯	同排聚丙烯
isotherm	等温线	等溫線
isothermal adsorption	等温吸附	等溫吸附
isothermal compressibility	等温压缩性	等溫壓縮性
isothermal compression	等温压缩	等溫壓縮
isothermal diffusion	等温扩散	等溫擴散
isothermal distillation	等温蒸馏	等溫蒸餾
isothermal effectiveness	等温有效因子	等溫有效度因數
isothermal environment	等温环境	等溫環境
isothermal expansion	等温膨胀	等溫膨脹
isothermal flow	等温流动	等溫流動
isothermal line	等温线	等溫線
isothermal multiplicity	等温多重性	等溫多重性
isothermal operation	等温操作	恆溫操作
isothermal process	等温过程	等溫過程,等溫程序
isothermal reaction	等温反应	恆溫反應
isothermal reactor	等温反应器	恆溫反應器

英 文 名	大 陆 名	台 湾 名
isothermal thermogravimetry	等温热重量分析法	等溫熱重量法
isotherm migration	等温线迁移	等溫線遷移
isotonic concentration	等渗压浓度	等滲壓濃度
isotonic solution	等渗溶液	等滲壓溶液
isotope	同位素	同位素
isotope separation	同位素分离	同位素分離
isotopic abundance	同位素丰度	同位素豐度
isotopic method	同位素法	同位素法
isotopic tracer	同位素示踪剂	同位素示蹤劑
isotropic flow	各向同性流动	各向同性流動,等向性流動
isotropic solid	各向同性固体	各向同性固體,等向性固體
isotropic turbulence	各向同性湍流	各向同性紊流
isotropy	各向同性	各向同性,等向性
iterative method	迭代法	疊代法

J

英 文 名	大 陆 名	台 湾 名
jacket	夹套	夾套
jacketed autoclave	夹套加压釜	夾套高壓釜
jacketed crystallizer	夹套结晶器	夾套結晶器
jacketed evaporator	夹套蒸发器	夾套蒸發器
jacketed kettle	夹套锅	[夾]套鍋
jacketed seamless kettle	无缝夹套锅	無縫[夾]套鍋
jacketed still	夹套蒸馏器	夾套蒸餾器
jacketed wall	夹套壁	夾套壁
jacket heating	夹套加热	夾套加熱
Jacobian matrix	雅可比矩阵	賈可比矩陣
japan	日本[天然]漆	日本[天然]漆
japan black	黑漆	黑漆
jar fermenter	缸式发酵槽	缸式發酵槽
jaw breaker	颚碎机	顎碎機
jaw crusher	颚式破碎机	顎碎機
j_D-factor	传质j因子	j_D-因數
jet compression	喷射压缩	噴射壓縮
jet condenser	喷射冷凝器	噴射冷凝器

英 文 名	大 陆 名	台 湾 名
jet drier	喷射干燥器	噴射乾燥器
jet engine	喷气式发动机	噴射引擎
jet exit pressure	喷射出口压	噴射出口壓[力]
jet flooding	喷射液泛	噴射溢流
jet fuel	喷气燃料	噴射機燃料
jet mill	气流粉碎机	噴射磨機
jet penetration length	射流穿透长度	噴射穿透長度
jet pump	喷射泵	噴射泵
jet reactor	射流反应器	噴射反應器
jet tray	舌形板	噴射塔板
jet type washer	喷射式洗涤器	噴射式洗滌器
j-factor	j 因子	j 因數
j_H-factor	传热 j 因子	j_H 因數,熱傳 j 因數
jigged fluidized bed	跳汰流化床	跳汰流[體]化床
jigger	辘轳	簸選機
jigging	簸动	簸選
jigging screen	簸动筛	簸選篩
joint	接头	接頭
joint efficiency	接合效率	接合效率
joule	焦[耳]	焦耳
Joule-Thomson coefficient	焦耳–汤姆孙系数	焦[耳]–湯[姆森]係數
Joule-Thomson effect	焦耳–汤姆孙效应	焦[耳]–湯[姆森]效應
junction	接合	接點,接面
junction box	接线盒	接線盒
junction point	接点	接點
junction potential	接点电位	接合電位

K

英 文 名	大 陆 名	台 湾 名
kainite	钾盐镁矾	鉀鎂礬,鉀瀉鹽
kali salt	钾盐	鉀鹽
kaolin	高岭土	高嶺土
kaolinite	高岭石	高嶺石
kapok oil	木棉油	木棉子油
karbate	无孔碳	高密度碳
kaya oil	榧子油	榧子油
kephalin	脑磷脂	腦磷脂

英　文　名	大　陆　名	台　湾　名
keratin	角蛋白	角蛋白
kerosene	煤油	煤油
kerosene cracking	煤油裂解	煤油裂解
kerosene degreasing	煤油脱脂	煤油脱脂
kerosene emulsion	煤油乳剂	煤油乳劑
kerosene fuel	煤油燃料	煤油燃料
kerosine	煤油	煤油
kettle	①锅 ②釜	鍋
key component	关键组分	關鍵成分
key word	关键字	關鍵字
kid-skin	猾子皮	羔羊皮
kieselguhr	硅藻土	矽藻土
kieselguhr dynamite	硅藻土炸药	矽藻土炸藥
kieserite	水镁矾	硫酸鎂石
killed steel	镇静钢	全靜碳鋼,脫氧鋼
kiln	窑	窯
kinase	激酶	激酶
kinematic property	运动性质	運動性質
kinematics	运动学	運動學
kinematic similarity	运动相似	運動相似性
kinematic viscometer	运动黏度计	動黏度計
kinematic viscosity	运动黏度	動黏度
kinetic analysis	动力分析	動力分析
kinetic control	动力学控制	動力學控制
kinetic energy	动能	動能
kinetic expression	动力学式	動力學式
kinetic head	动压头	動高差
kinetic model	动力模式	動力模式
kinetic parameter	动力学参数	動力參數
kinetic property	动力性质	動力性質
kinetics	动力学	動力學
kinetic theory	动力学理论	動力論
Kirchhoff law	基尔霍夫定律	克希何夫定律
kneader	捏合机	捏合機
kneading	捏合	捏合
knife barker	刀式去皮机	刀式去皮機
knock characteristic test	爆震试验	爆震試驗
knockout drum	分离鼓	液氣分離器

英　文　名	大　陆　名	台　湾　名
knot screen	结筛	結篩
knowledge base	知识库	知識庫
knowledge engineering	知识工程	知識工程
Knudsen diffusion	克努森扩散	努生擴散
Knudsen number	克努森数	努生數
Kolmogorov scale	科尔莫戈罗夫尺度	科摩哥洛夫尺度
konimeter	测尘器	測塵計
kraft paper	牛皮纸	牛皮紙
kraft process	牛皮纸浆法	牛皮紙漿法
kraft pulp	硫酸盐浆	牛皮紙漿
Kremser diagram	克伦舍尔图	克倫舍爾圖
Kühni extractor	屈尼萃取塔	屈尼萃取塔
kurtosis	峰度	峰態
K value	*K* 值	汽液平衡比值,*K* 值

L

英　文　名	大　陆　名	台　湾　名
labile region	不稳定区	不穩定區
labile zone	不稳定区	不穩定區
laboratory manual	实验室手册	實驗室手冊
laboratory reactor	实验室反应器	實驗室反應器
labor cost	人工成本	人工成本
lac	紫胶	蟲漆
lacquer	漆	漆
lactalbumin	乳清蛋白	乳白蛋白
lactam	内酰胺	内醯胺
lactase	乳糖酶	乳酸酶
lactic acid	乳酸	乳酸
lactobiose	乳二糖	乳糖
lactone	内酯	内酯
lactose	乳糖	乳糖
ladle	钢包	澆斗
lag	滞后	滞延
lag compensation	滞后补偿	滞延補償
lag element	落后元件	滞延元件
lagged pipe	保温管	保溫管
lagging	隔热层	保溫層

英 文 名	大 陆 名	台 湾 名
lag phase	滞后期	滞延期
lake	色淀	色澱,沈澱色料
lake pigment	色淀颜料	色澱,沈澱色料
lamina	薄层	薄層
laminar boundary layer	层流边界层	層流邊界層
laminar flow	层流	層流
laminar flow reactor	层流反应器	層流反應器
laminar jet absorber	层流式喷流吸收器	層流式噴射吸收器
laminar region	层流区	層流區[域]
laminar sublayer	层流底层	層流次層
laminate	层压材料	積層板
laminated safety glass	夹层安全玻璃	積層安全玻璃
laminated wall	层压板壁	積層板壁
lamination	层压	積層
lamination press	层压压制机	積層壓製機
lamp black	灯黑	燈黑,燈煙
land disposal	废弃物陆上处理	[廢棄物]陸上處理[法]
landfill	废渣埋填	[廢棄物]掩埋場
land pollution	土地污染	土地污染
lanolin	羊毛脂	羊毛脂
lantern gland	灯笼填函盖	燈籠填函蓋
lap joint	搭接接头	搭接接頭
lard	猪脂	豬油
lard oil	猪油	豬油
latent energy	潜能	潛能
latent heat	潜热	潛熱
latent heat of vaporization	汽化潜热	汽化潛熱
lateral stress	侧向应力	側向應力
latex	胶乳	乳膠
latex paint	乳胶漆	乳膠漆
lattice	①点阵 ②晶格	晶格
laundry detergent	衣物洗涤剂	洗衣清潔劑
laundry soap	洗衣皂	洗衣皂
laundry soda	洗衣碱	洗衣鹼
laurel oil	月桂油	月桂油
law	定律	定律
law of corresponding state	对应态律	對應狀態定律

英 文 名	大 陆 名	台 湾 名
law of definite proportions	定比定律	定比定律
law of mass action	质量作用定律	質量作用定律
layout	布置	布置
layout check list	布置检查表	布置檢查表
layout diagram	布置图	布置圖
leachate	浸取液	瀝取液
leaching	浸取	瀝取
leaching process	浸取法	瀝取法
leaching tank	浸取槽	瀝取槽
lead	铅	鉛
lead alkali glass	铅碱玻璃	鉛鹼玻璃
lead chamber	铅室	鉛室
lead compensation	超前补偿	超前補償
leaded gasoline	含铅汽油	加鉛汽油
leaded zinc	含铅氧化锌	含鉛氧化鋅
lead glass	铅玻璃	鉛玻璃
leading edge	前沿	前緣
lead pipe	铅管	鉛管
lead soap	铅皂	鉛皂
lead storage battery	铅蓄电池	鉛蓄電池
lead white	铅白	鉛白
lead wire	导线	導線,引線
lead wire error	导线误差	引線誤差
leaf filter	叶滤机	葉濾器
leakage	渗漏	漏洩
leakage test	泄漏试验	漏洩試驗
leak detector	检漏器	測漏器
lean gas	贫气	貧氣
lean oil	贫油	貧油
lean solution	贫溶液	貧溶液
learning control system	学习控制系统	學習控制系統
least squares analysis	最小二乘分析	最小平方[分析]法
least squares method	最小二乘法	最小平方法
leather	皮革	皮革
leathery state	皮革态	皮革態
lecithin	卵磷脂	卵磷脂
legal liability	法定责任	法定責任
lehr	退火炉	退火窯

英 文 名	大 陆 名	台 湾 名
lemon-grass oil	柠檬草油	檸檬草油
leucite	白榴石	白榴石
leuco base	隐色基	隱色鹼
leuco compound	隐色体	隱色化合物
leucomycin	柱晶白霉素	白黴素
level control	液位控制	液位控制
level controller	物位控制器	液位控制器
level gauge	液位计	液位計
level transmitter	物位传送器	液位傳送器
lever principle	杠杆原理	槓桿原理
lever rule	杠杆法则	槓桿法則
lever safety valve	杠杆安全阀	槓桿安全閥
levo-rotatory sugar	左旋糖	左旋醣
levulose	果糖	果糖
lift	扬程	上升距離
lift coefficient	上升系数	上升係數
light duty detergent	轻垢洗涤剂	輕污清潔劑
light ends	轻馏分	輕餾分
light fuel oil	轻质燃料油	輕燃料油
light hydrocarbon	轻烃	輕烴
light key component	轻关键组分	輕關鍵成分
light naphtha	轻石脑油	輕石油腦
light oil	轻质油	輕質油
light soda ash	轻灰	輕鈉鹼灰
light straight-run gas oil	轻直馏油	輕直餾油
lignification	木质化	木質化
lignin	木[质]素	木質素
ligninase	木质酶	木質酶
lignin sulphonic acid	木质磺酸	木質磺酸
lignite	褐煤	褐煤
lignitic bituminite	褐烟煤	褐煙煤
lignocellulose	木素纤维素	木質纖維素
lime	石灰	石灰
lime feldspar	钙长石	鈣長石
lime glass	钙玻璃	鈣玻璃
lime kiln	石灰窑	石灰窯
lime milk	石灰乳	石灰乳
lime soap	石灰皂	鈣皂

英　文　名	大　陆　名	台　湾　名
lime-soda process	石灰苏打法	石灰蘇打法
limestone	石灰石	石灰石
liming	浸灰	浸灰法
liming process	浸灰法	浸灰法
limit control	限度控制	限度控制
limit current	极限电流	極限電流
limit cycle	极限环	極限循環
limiter	限制器	限制器
limiting batch size	极限批量	極限批式產能
limiting component	限量成分	限量成分
limiting cycle time	极限周期	極限周期
limiting factor	极限因子	限制因素
limiting reflux condition	极限回流条件	極限回流條件
limiting viscosity	极限黏度	極限黏度
limonite	褐铁矿	褐鐵礦
linear absorption coefficient	线性吸收系数	線性吸收係數
linear analysis	线性分析	線性分析
linear azeotrope	线性共沸液	線性共沸液
linear control	线性控制	線性控制
linear control valve	线性控制阀	線性控制閥
linearity	线性	線性
linearization	线性化	線性化
linearized system	线性化系统	線性化系統
linear lead	线性超前	線性超前
linear low density polyethylene	线型低密度聚乙烯	線型低密度聚乙烯
linear model	线性模型	線性模型
linear momentum principle	线性动量原理	線性動量原理
linear notch	线性凹槽	線性凹槽
linear programming（LP）	线性规划	線性規畫
linear regression analysis	线性回归分析	線性回歸分析[法]
linear system	线性系统	線性系統
linear valve	线性阀	線性閥
linear velocity	线速度	線速度
linear viscoelasticity	线性黏弹性	線性黏彈性
lined pipe	衬里管	襯管
line loss	线路损失	線路損失
linen	亚麻布	亞麻布
line sink	线汇座	線匯座

英　文　名	大　陆　名	台　湾　名
line source	线源	線源
lining	衬里	襯裏
linoleum	油毡	油氈, 油布
linseed oil	亚麻子油	亞麻仁油
linter	精制棉短绒	棉絨
lipase	脂肪酶	脂肪酶
liquefaction	液化	液化
liquefaction point	液化点	液化點
liquefaction process	液化过程	液化程序
liquefaction reactor	液化反应器	液化反應器
liquefied fuel	液化燃料	液化燃料
liquefied gas	液化气体	液化氣[體]
liquefied natural gas	液化天然气	液化天然氣
liquefied petroleum gas	液化石油气	液化石油氣
liquefier	①液化剂 ②液化器	液化器
liquid	液体	液體
liquid air	液态空气	液態空氣
liquid ammonia	液氨	液氨
liquid chromatogram	液相色谱图	液相層析圖
liquid chromatograph	液相色谱仪	液相層析儀
liquid chromatography	液相色谱法	液相層析[法]
liquid column gage	液柱压力计	液柱壓力計
liquid crystal	液晶	液晶
liquid crystal display	液晶显示	液晶顯示器
liquid crystal polymer	液晶高分子	液晶聚合物
liquid crystal transition	液晶转变	液晶轉變
liquid cyclone	水力旋流器	液體旋流器
liquid cyclone separator	水力旋流分离器	液體旋流分離器
liquid extract	液体萃取物	液體萃取物
liquid film	液膜	液膜
liquid film coefficient	液膜系数	液膜係數
liquid film contactor	液膜接触器	液膜接觸器
liquid film resistance	液膜阻力	液膜阻力
liquid flotation	液体浮选	液體浮選[法]
liquid fluidized bed	液体流化床	液體流[體]化床
liquid fuel	液体燃料	液體燃料
liquid-gas reaction	液气反应	液氣反應
liquid head	液体压头	液體高差

英 文 名	大 陆 名	台 湾 名
liquid hourly space velocity	液时空速	液體空間時速
liquid interface control	液体界面控制	液體界面控制
liquid level	液位	液位
liquid level control	液面控制	液位控制
liquid level gage	液面计	液位計
liquid level indicator	液位指示器	液位指示器
liquid-liquid equilibrium	液体平衡	液液平衡
liquid-liquid extraction	液液萃取	液液萃取
liquid-liquid reaction	液液反应	液液反應
liquid-liquid separator	液液分离器	液液分離器
liquid membrane	液膜	液[態薄]膜
liquid nitrogen	液氮	液[態]氮
liquid oxygen	液氧	液[態]氧
liquid paraffin	液体石蜡	液態石蠟
liquid phase	液相	液相
liquid phase cracking	液相裂解	液相裂解
liquid phase reaction	液相反应	液相反應
liquid phase reactor	液相反应器	液相反應器
liquid propellant	液体推进剂	液體推進劑
liquid reaction	液态反应	液態反應
liquid relief valve	安全泄液阀	液體釋放閥
liquid ring compressor	液环压缩机	液環壓縮機
liquid seal	液封	液封
liquid-solid cyclone	液固旋风器	液固旋風器
liquid-solid reaction	液固反应	液固反應
liquid solution	液体溶液	液態溶液
liquid thermometer	液体温度计	液體溫度計
liquid trap	捕液器	液阱
liquidus	液相线	液相線
liquid-vapor equilibrium	液汽平衡	液汽平衡
literature	文献	文獻
literature survey	文献调查	文獻調查
lithium-base grease	锂基润滑脂	鋰基潤滑脂
lithography	平版印刷术	微影術,蝕刻法
lithopone	锌钡白	鋅鋇白
litmus paper	石蕊试纸	石蕊試紙
live steam	新鲜蒸汽	活蒸汽
lixiviation	浸滤	瀝取

英 文 名	大 陆 名	台 湾 名
load	负载	負載,負荷
load change	负荷变动	負載變化,負荷變化
load disturbance	负荷搅动	負荷擾動
loading effect	负荷效应	負荷效應
loading point	负载点	負載點
local acceleration	局部加速度	局部加速度
local composition	局部组成	局部組成
local equilibrium	局部平衡	局部平衡
local optimum	局部最优[值]	局部最適[值]
local stability	局部稳定性	局部穩定性
lock-and-key theory	锁钥学说	鎖鑰理論
lock hopper	闭锁式料斗	閉鎖式料斗
locus	轨迹	軌跡
locus of constant phase	等相位轨迹	等相位軌跡
locus of critical point	临界点轨迹	臨界點軌跡
loft drying	风干	風乾
logarithmic mean	对数平均	對數平均[數]
logarithmic mean area	对数平均面积	對數平均面積
logarithmic mean mole fraction	对数平均摩尔分率	對數平均莫耳分率
logarithmic mean radius	对数平均半径	對數平均半徑
logarithmic mean temperature difference	对数平均温差	對數平均溫差
logarithmic phase	对数生长期	對數[生長]期
logarithmic plot	对数图	對數圖
log normal distribution function	对数正态分布函数	對數常態分佈函數
longitudinal dispersion coefficient	纵向分散系数	縱向分散係數
longitudinal fin	纵向翅片	縱向鰭片
longitudinal stress	纵向应力	縱向應力
long sweep elbow	长弯头	長彎頭
long tube evaporator	长管蒸发器	長管蒸發器
loop	环路,回路	環路,回圈
loop configuration	环路构型	環形組態
looped pipe system	环状管路系统	環狀管路系統
loop reactor	环流反应器	環流反應器
loop response	环路应答	環路應答
loop transfer function	环路转移函数	環路轉移函數
loss	损耗	損耗
loss tangent	损耗角正切	損耗模數比,損耗正切
lost work	损失功	損失功

英　文　名	大　陆　名	台　湾　名
louver type baffle	百叶窗挡板	百葉窗型擋板
low-BTU gas	低热值煤气	低熱值燃氣
low density polyethylene	低密度聚乙烯	低密度聚乙稀
lower bound	下限	下限
lower limit	下限	下限
low pass filter	低通滤波器	低通濾波器
low selector switch	低选择器开关	低選擇器開關
low temperature carbonization	低温碳化	低溫碳化[作用]
LP(=linear programming)	线性规划	線性規畫
lubricant	润滑剂	潤滑劑
lubricating grease	润滑脂	潤滑脂
lubricating oil	润滑油	潤滑油
lubrication	润滑	潤滑
lubrication mechanics	润滑力学	潤滑力學
lubricity	润滑性	潤滑性
luminescence	①发光 ②冷光	①發光 ②冷光
luminescent dye	发光染料	發光染料
luminous paint	发光涂料	發光漆
lumped parameter model	集总参数模型	集總參數模式
lumped parameter system	集总参数系统	集總參數系統
lumped system	集总系统	集總系統
lumping	参数集总	集總
lumping kinetics	集总动力学	集總動力學
lustering agent	上光剂	亮光劑
lye	碱液	鹼水
lyolysis	液解	溶劑解
lyophilic colloid	亲液胶体	親液膠體
lyophilization	冷冻干燥	冷凍乾燥
lyophobic colloid	疏液胶体	疏液膠體
lyotropic liquid crystal	感胶液晶	溶致液晶
lysine	赖氨酸	離胺酸
lysozyme	溶菌酶	溶菌酶

M

英　文　名	大　陆　名	台　湾　名
maceration	浸渍	浸解[作用]
machine molding	机械造型	機械成型
Mach number	马赫数	馬赫數
macroemulsion	粗滴乳状液	粗滴乳液
macrofluid	宏观流体	巨觀流體
macrokinetics	宏观动力学	宏觀動力學
macromixing	宏观混合	巨觀混合
macromolecule	大分子	巨分子, 大分子
macropore	大孔	大孔
macroscale behavior	宏观规模行为	巨觀行為
macroscopic balance	宏观平衡	巨觀均衡
macroscopic reversibility	宏观可逆性	巨觀可逆性
madder	茜草	茜草
madder root	茜草根	茜草根
magazine grinder	箱式研磨机	箱式研磨機
magma	岩浆	岩漿
magnesia	镁砂	氧化鎂
magnesia brick	镁砖	鎂氧磚
magnesite	菱镁矿	菱鎂礦
magnesite brick	镁砖	鎂磚
magnesite-chrome brick	镁铬砖	鎂鉻磚
magnesite refractory	镁质耐火材料	鎂氧耐火物
magnetically stabilized fluidized bed	磁稳流化床	磁穩定[體]流化床
magnetic data storage	磁式数据储存	磁式資料儲存
magnetic disk	磁盘	磁碟
magnetic field	磁场	磁場
magnetic filtration	磁力过滤	磁力過濾
magnetic flowmeter	磁流量计	磁力流量計
magnetic impurity	磁性不纯物	磁性不純物
magnetic moment	磁矩	磁矩
magnetic separation	磁分离	磁力分離
magnetic separator	磁力分离器	磁力分離器, 磁選機
magnetic stirrer	磁搅拌器	磁攪拌器

英 文 名	大 陆 名	台 湾 名
magnetic susceptibility	磁化率	磁化率
magnetic tape	磁带	磁帶
magnetite	磁铁矿	磁鐵礦
magnetochemistry	磁化学	磁化學
magneto fluidization	磁力流态化	磁力流體化
magnetohydrodynamics	磁流体动力学	磁性流體力學
magnitude ratio	大小比值	大小比值,量比值
magnitude scaling	大小标度	大小量度
maintenance	维修	維護,保養
maintenance cost	维修费	維護成本
maintenance downtime	停修时间	停工維護時間
maintenance management	维修管理	維護管理
malachite	孔雀石	孔雀石
malachite green	孔雀绿	孔雀綠
maldistribution	不良分布	不良分佈
malic acid	苹果酸	蘋果酸
malonic acid	丙二酸	丙二酸
malt	麦芽	麥芽
maltase	麦芽糖酶	麥芽糖酶
maltose	麦芽糖	麥芽糖
manganite	水锰矿	水錳礦
manhole	人孔	人孔
man hour	工时	工時
manifold	管汇	歧管
manipulating variable	操纵变量	操縱變數
man-made fiber	人造纤维	人造纖維
manometer	压力计	[流體]壓力計
manometric fluid	测压液	測壓液
manometric head	压力计压头	壓力計高差
manual operation	人工操作	人工操作
manual reset	手动重设	手動重設[定]
manufacture	制造	製造
manufacturing cost	制造成本	製造成本
manufacturing technology	制造技术	製造技術
mapping	映射	映射
marcasite	白铁矿	白鐵礦
margarine	人造黄油	人造奶油
marginal profit	边际利润	邊際利潤

英　文　名	大　陆　名	台　湾　名
margin gain	边界增益	邊限增益
margin of stability	稳定界限	穩定邊限
margin phase	边界相	邊限相位
Margules equation [of state]	马居尔方程	馬古利斯方程[式]
market cost	市场成本	市場成本
market price	市价	市價
market research	市场研究	市場研究
market value	市值	市價
MARR(=minimum acceptable rate of return)	最低容许收益率	最低容許收益率
marsh gas	沼气	沼氣
Martin-Hou equation	马丁-侯[虞钧]方程	馬[丁]侯[虞鈞]方程
Martin-Hou equation of state	马丁-侯[虞钧]状态方程	馬[丁]侯[虞鈞]狀態方程式
masking power	掩蔽能力	遮蓋[能]力
mass	质量	質量
mass absorption coefficient	质量吸收系数	質量吸收係數
mass action kinetics	质量作用动力学	整體作用動力學
mass average velocity	质量平均速度	質量平均速度
mass balance	①质量衡算 ②质量平衡	質量均衡
mass concentration	质量浓度	質量濃度
mass conservation	质量守恒	質量守恆
mass diffusivity	质量扩散系数	質量擴散係數
mass flow	质量流	質量流
mass flowmeter	质量流量计	質量流量計
mass flow rate	质量流率	質量流率
mass flux	质量通量	質量通量
mass fraction	质量分率	質量分率
mass generation curve	质量生成曲线	質量生成曲線
massive bed reactor	宽床反应器	寬床反應器
mass polymerization	本体聚合	本體聚合[反應]
mass removal curve	质量去除曲线	質量去除曲線
mass separating agent	物质分离剂	物質分離劑
mass spectrometer	质谱仪	質譜儀
mass spectrometry	质谱法	質譜法
mass spectrum	质谱图	質譜
mass transfer	传质,质量传递	質傳,質量傳送
mass transfer coefficient	传质系数	質傳係數

英　文　名	大　陆　名	台　湾　名
mass transfer control	传制控制	質傳控制
mass transfer equipment	传质装置	質傳裝置
mass transfer operation	传质操作	質傳操作
mass transfer rate	传质速率	質傳速率
mass transfer resistance	传质阻力	質傳阻力
mass transfer zone(MTZ)	传质区	質傳區
mass transport	质量迁移	質量輸送
mass velocity	质量流速	質量流速
master controller	主令控制器	主控制器
material	物料	物料
material and energy balance	物料与能量平衡	質能均衡
material balance	物料平衡	物質均衡
material balance flowsheet	物料平衡流程图	物質均衡流程圖
material cost	材料成本	材料成本
material cost index	材料成本指数	材料成本指數
material derivative	物质导数	物質導數
material design	材料设计	材料設計
material function	物质函数	物質函數
material manufacturing	材料制造	材料製造
material processing	材料加工	材料加工
materials engineering	材料工程	材料工程
materials of construction	建筑材料	營建材料
materials science	材料科学	材料科學
material synthesis	材料合成	材料合成
mathematical model	数学模型	數學模式
mathematical modeling	数学模拟	數學模式化
mathematical programming	数学程序	數學規劃
matrix	①矩阵 ②基体	①矩陣 ②基體
matrix arithmetics	矩阵算术	矩陣算術
matrix inversion	矩阵求逆	矩陣求逆
matter	物质	物質
matt glaze	无光釉	無光釉
matt paint	无光油漆	無光油漆
maximax criterion	大中取大判据	大中取大準則
maximin utility criterion	小中取大效用判据	小中取大效用判據
maximum allowable pressure	最高允许压力	最高容許壓力
maximum allowable temperature	最高允许温度	最高容許溫度
maximum boiling azeotrope	最高沸点共沸物	最高沸點共沸液

英 文 名	大 陆 名	台 湾 名
maximum likelihood principle	最大似然原理	最大概似
maximum mixedness	最大混合度	最大混合度
maximum principle	极大值原理	極大原理
maximum specific growth rate	最大比生长速率	最大比生長速率
Maxwell relation	麦克斯韦关系	馬克士威關係[式]
McCabe-Thiele diagram	M-T图	M-T圖
McMahon packing	网鞍填料	麥氏網鞍填料
MD method(=molecular dynamic method)	分子动态法	分子動力法
MD tray(=multidowncomer tray)	多降液管塔板	多降液管塔板
mean	均值	平均[值]
mean deviation	平均偏差	平均偏差
mean error	平均误差	平均誤差
mean free path	平均自由程	平均自由徑
mean radius	平均半径	平均半徑
mean residence time	平均停留时间	平均滯留時間
mean square error	均方误差	均方誤差
mean temperature difference	平均温差	平均溫差
measurability	可测量性	可量測性
measurable property	可测量性质	可量測性質
measurement	测量	量測,測定
measurement lag	测量滞后	量測滯延
measuring device	测量装置	量測裝置
measuring element	测量元件	量測元件
measuring flask	量瓶	量瓶
mechanical agitation	机械搅拌	機械攪拌
mechanical automation	机械自动化	機械自動化
mechanical centrifugal separator	机械离心分离器	機械離心分離器
mechanical controller	机械式控制器	機械式控制器
mechanical design	机械设计	機械設計
mechanical disintegration	机械破碎	機械粉碎
mechanical draft	机械通风	機械通風
mechanical draft cooling tower	机械通风冷却塔	機械通風冷卻塔
mechanical energy	机械能	機械能
mechanical energy balance	机械能量平衡	機械能均衡
mechanical energy equation	机械能方程式	機械能方程式
mechanical equivalent of heat	热功当量	熱功當量
mechanical filter	机械过滤器	機械過濾器
mechanical flotation cell	机械浮选池	機械浮選池

英 文 名	大 陆 名	台 湾 名
mechanical flow diagram	工程流程图	機械流程圖
mechanical flowsheet	工程流程图	機械流程圖
mechanical loss	机械损失	機械損失
mechanical-physical separation process	机械–物理分离过程	機械–物理分離程序
mechanical plating	机械镀	機械鍍
mechanical pulp	磨木浆	機械紙漿
mechanical pulping process	磨木浆法	機械製漿法
mechanical seal	机械密封	機械密封
mechanical separation	机械分离	機械分離
mechanical separator	机械分离器	機械分離器
mechanical work	机械功	機械功
mechanics	力学	力學
mechanism deficiency	机理缺陷	機構缺陷
mechanism of catalysis	催化作用机理	催化機構
mechanism structure	机理结构	機制結構
mechanistic equation	机械学方程式	機理方程式
mechanochemical pulping process	化学动力过程	機械化學製漿法
median	中位值	中位數, 中值
median selector	中值选择器	中值選擇器
medium	介质	介質, 培養基
medium crushing	中级压碎	中級壓碎
medium filter	介质过滤器	介質過濾器
medium grinding	中级研磨	中級研磨
medium screening	中级筛选	中級篩選
medium sizing	中级筛选	中級篩選
medium sweep elbow	中弯头	中彎頭
melamine adhesive	三聚氰胺黏合剂	三聚氰胺黏合劑, 美耐 皿黏合劑
melamine formaldehyde resin	三聚氰胺甲醛树脂	三聚氰胺甲醛樹脂
melanogen dye	黑素原染料	硫化染料
melanoid	类黑素	類黑素
melanoidin	蛋白黑素	梅納汀
Mellapak packing	板波纹填料	板波紋填料
melt	①熔化 ②熔化物	①熔體 ②熔態
melt flow index	熔体流动指数	熔體[流動]指數
melt index	熔体指数	熔體指數
melting point	熔点	熔點
melting point depression	熔点下降	熔點下降

英 文 名	大 陆 名	台 湾 名
melting range	熔点范围	熔點範圍
melting temperature	熔化温度	熔化溫度,熔融溫度
melt spinning	熔纺	熔紡[絲]
melt viscosity	熔融黏度	熔體黏度
membrane	膜	膜
membrane bioreactor	膜生物反应器	膜生物回應器
membrane distillation	膜蒸馏	[薄]膜蒸餾
membrane extraction	膜萃取	膜萃取
membrane filter	膜滤器	膜濾器
membrane module	膜组件	膜組件
membrane permeability	膜通透性	膜滲透性
membrane permeation	膜渗透	膜滲透
membrane pump	膜泵	膜泵
membrane reactor	膜反应器	[薄]膜反應器
membrane separation	膜分离	[薄]膜分離
membrane separation technology	膜分离技术	[薄]膜分離技術
membrane strength	膜强度	膜強度
membrane support	膜支撑物	膜支撐物
membrane technology	膜技术	膜技術
membrane vesicle	膜囊	膜囊泡
memory cell	①记忆细胞 ②存储单元	①記憶細胞 ②儲存單元
memory fluid	记忆流体	記憶流體
memory function	记忆函数	記憶函數
menthol	薄荷醇	薄荷腦
mercaptan	硫醇	硫醇[類]
mercerization	丝光处理	絲光處理
mercerized cotton	丝光棉	絲光棉
mercerized yarn	丝光纱	絲光紗
mercerizing	丝光处理	絲光處理
mercury cathode cell	汞阴极电解槽	汞陰極電解槽
mercury cell	汞电池	汞電池,水銀電池
mercury decomposer	汞分解器	汞分解器
mercury diffusion pump	汞扩散泵	汞擴散泵
mercury float manometer	汞浮子压力计	汞浮子壓力計
mercury pollution	汞污染	汞污染
mercury thermometer	水银温度计	水銀溫度計,汞溫度計
mesh	①筛目 ②网孔	①篩目 ②篩孔
mesh analysis	筛析	篩析

英　文　名	大　陆　名	台　湾　名
mesh efficiency	筛目效率	篩網效率
MESH equations	MESH 方程组	MESH 方程組
mesh filter	筛网过滤器	篩網過濾器
mesh screen	网筛	網篩
mesh separator	筛网分离器	篩網分離器
mesomorphic state	介晶态	介[晶]態
mesophase	中间相	介相
mesophile	中温菌	嗜中溫菌
mesophilic bacteria	中温菌	嗜中溫菌
mesopore	介孔	中孔
mesoscale behavior	介标行为	介觀行為
metabolic control	代谢控制	代謝控制
metabolic rate	代谢速率	代謝速率
metabolism	代谢	代謝[作用]
metabolite	代谢物	代謝物
metal	金属	金屬
metal complex acid dye	金属络合酸性染料	金屬錯合物酸性染料
metal finishing	金属表面处理	金屬表面處理
metal ion inactivator	金属离子惰化剂	金屬離子惰化劑
metallic catalyst	金属催化剂	金屬觸媒
metallic glass	金属玻璃	金屬玻璃
metallic mordant	金属媒染剂	金屬媒染劑
metallic soap	金属皂	金屬皂
metallized dye	金属配位染料	金屬化染料
metallography	金相学	金相學
metallorganic chemical vapor deposition	金属有机化学蒸气沉积	金屬有機化學氣相沈積法
metallurgical coke	冶金焦	冶金焦
metallurgical process	冶金过程	冶金程序
metallurgy	冶金学	冶金學
metal plating	金属电镀	金屬電鍍
metal spraying	金属喷镀	金屬噴霧
metastability	亚稳定性	介穩定性
metastable limit	亚稳界限	介穩定界限
metastable region	亚稳区	介穩定區[域]
metastable state	亚稳态	介穩定狀態
metathesis	复分解	複分解
metathetical reaction	复分解反应	複分解反應

英文名	大陆名	台湾名
metering	计量	計量
metering pump	计量泵	計量泵
methanation	甲烷化	甲烷化[法]
methanation reaction	甲烷化反应	甲烷化反應
methanator	甲烷化反应器	甲烷化器
methane bacteria	甲烷菌	甲烷菌
methane-forming bacteria	甲烷细菌	甲烷菌
methane hydrate	甲烷水合物	甲烷水合物
methane lean gas	甲烷贫气	甲烷貧氣
methane rich gas	甲烷富气	甲烷富氣
methanol	甲醇	甲醇
method	方法	方法
method of excess	过量法	過量法
method of finite difference	有限差分法	有限差分法
method of initial rate	初速率法	初速率法
method of isolation	隔离法	隔離法,單離法
method of moments	力矩法	矩法
method of moments matching	配矩法	配矩法
methyl alcohol	甲醇	甲醇
methylated spirit	变性酒精	變性酒精
methylating agent	甲基化剂	甲基化劑
methylation	甲基化作用	甲基化[作用]
methyl cellulose	甲基纤维素	甲基纖維素
methylene blue	亚甲蓝	亞甲藍
methylmethacrylate-acrylonitrile-butadiene-styrene copolymer（MABS）	甲基丙烯酸甲酯－丙烯腈－丁二烯－苯乙烯共聚物	甲基丙烯酸甲酯－丙烯腈－丁二烯－苯乙烯共聚物
methyl orange	甲基橙	甲基橙
methyl parathion	甲基对硫磷	甲[基]巴拉松
methyl red	甲基红	甲基紅
methyl rubber	甲基橡胶	甲基橡膠
methyl violet	甲基紫	甲基紫
methyl yellow	甲基黄	甲基黃
metric system	公制	公制,米制
mica	云母	雲母
micellar catalysis	胶束催化	微胞催化[作用]
micellar flooding	胶束驱油	微胞泛流[法]
micelle	胶束	微胞

英　文　名	大　陆　名	台　湾　名
micellization	胶束化	微胞化
Michaelis-Menten constant	米氏常数	米氏常數
Michaelis-Menten equation	米氏方程	米氏方程[式]
Michaelis-Menten kinetics	米氏动力学	米氏動力學
microanalysis	微量分析	微量分析[法]
microbe	微生物	微生物
microbial contamination	杂菌污染	雜菌污染
microbial corrosion	微生物腐蚀	微生物腐蝕
microbial dynamics	微生物动力学	微生物動力[學]
microbial fermentation	微生物发酵	微生物發酵[法]
microbial film	微生物膜	微生物膜
microbial fouling	微生物结垢	微生物積垢
microbial kinetics	微生物动力学	微生物動力學
microbial process	微生物法	微生物法
microbial reaction	微生物反应	微生物反應
microbial reactor	微生物反应器	微生物反應器
microbiological process	微生物过程	微生物法
microbiology	微生物学	微生物學
microcanonical ensemble	微正则系综	微正則系綜
microcanonical partition function	微正则配分函数	微正則配分函數
microcapsule	微胶囊	微膠囊
microcarrier	微载体	微載體
microcatalytic reactor	微型催化反应器	微型觸媒反應器
microchemistry	微量化学	微量化學
microcircuit	微电路	微電路
microcomputer	微型计算机	微電腦
microcrystal	微晶	微晶
microcrystalline structure	微晶结构	微晶結構
microcrystalline wax	微晶蜡	微晶蠟
microemulsion	微乳	微乳液
microencapsulation	微囊化	微膠囊化
microfiltration	微滤	①微量過濾 ②微[孔]過濾
microfluid	微观流体	微觀流體
micromeritics	微粒学	微粒學
micrometer	测微器	測微計
micromixing	微观混合	微觀混合
micron	微米	微米

英　文　名	大　陆　名	台　湾　名
micronizer	微粉磨	微磨機
microorganism	微生物	微生物
microphotometer	显微光度计	微光度計
micropore	微孔	微孔
microporous filter	微孔过滤器	微孔過濾器
microporous membrane	微孔滤膜	微孔透膜
microporous separator	微孔分离器	微孔分離器
microreaction engineering	微型反应工程	微反應工程
microreactor	微型反应器	微型反應器
microscale phenomenon	微观现象	微觀現象
microscopic balance	微观平衡	微觀平衡
microscopic reversibility	微观可逆性	微觀可逆性
microscopic state	微观态	微觀狀態
microscopy	显微术	顯微術
microsphere	微球	微球
microstrainer	微滤器	濾微器
microstraining	微滤	微[過]濾
microstructure	微结构	微結構
microstructured materials	微结构材料	微結構化材料
microwave	微波	微波
microwave drying	微波干燥	微波乾燥
microwave spectroscopy	微波波谱学	微波光譜學
microwave spectrum	微波谱	微波譜
middle oil	中质油	中質油
migration potential	迁移电位	遷移電位
mild steel	软钢	軟鋼
milky glass	乳白玻璃	乳白玻璃
mill capacity	研磨能力	研磨容量
mill efficiency	研磨效率	研磨效率
milligram	毫克	毫克
millimeter	毫米	毫米
milling capacity	研磨能力	碾磨能力
millivoltmeter	毫伏表	毫伏特計
mill roller	研磨辊	碾磨輥
mill room	研磨室	碾磨室
mineral acid	无机酸	礦酸
mineral dye	无机染料	礦物染料
mineral fertilizer	矿物肥料	礦物肥料

英　文　名	大　陆　名	台　湾　名
mineralogy	矿物学	礦物學
mineral oil	矿物油	礦油
mineral resources	矿物资源	礦物資源
mineral wax	矿蜡	礦蠟
mingler	拌和机	拌和機
miniature instrument	微型仪器	微型儀器
minimum acceptable rate of return （MARR）	最低容许收益率	最低容許收益率
minimum boiling azeotrope	最低沸点共沸物	最低沸點共沸液
minimum deviation	最小偏差	最小偏差
minimum fluidization	最小流化态	最低流體化態
minimum fluidization velocity	最小流化速度	最低流體化速度
minimum fluidizing velocity	最小流化速度	最低流體化速度
minimum freezing point	最低凝固点	最低凝固點
minimum principle	最小值原理	極小原理
minimum reflux	最小回流	最小回流
minimum reflux ratio	最小回流比	最小回流比
minimum tray number	最小塔板数	最少板數
mining explosive	矿用炸药	礦用炸藥
minisupercomputer	小型超级计算机	迷你超級電腦
minium	铅丹	鉛丹,鉛紅
MINLP(=mixed integer nonlinear programming)	混合整数非线性规划	混合整數非線性規劃
minor loop	副回路	次環路
mirror colorimeter	镜式比色计	鏡式比色計
mirror filter reflectometer	滤镜反光计	濾鏡反光計
mirror polishing	镜面磨光	鏡面磨光
mirror spectrophotometer	镜式光谱仪	鏡式光譜儀
miscalibration	误校正	誤校正,校正失誤
miscibility	互溶性	互溶性
miscible flooding	互溶泛滥	互溶氾流[法]
miscible system	互溶系统	互溶系統
mist	雾	[煙]霧
mist collector	集雾器	集霧器
mist flow	雾状流	霧沫流動
mist particle	雾粒	霧粒
mist separator	油雾分离器	分霧器
miter elbow	斜弯头	斜彎頭

英　文　名	大　陆　名	台　湾　名
miticide	杀螨剂	殺蟎劑
mixed acid	混合酸	混合酸
mixed catalyst	混合催化剂	混合觸媒
mixed crystal	混晶	混晶
mixed feeding	混合进料	混合進料
mixed fertilizer	混合肥料	混合肥料
mixed flow	混流	混合流動
mixed flow pump	混流泵	混流泵
mixed flow reactor	混合流动反应器	混合流動反應器
mixed integer nonlinear programming （MINLP）	混合整数非线性规划	混合整數非線性規劃
mixed juice	混合计	混合汁
mixed liquor	混合液	混合液
mixed-liquor suspended solid	混合液悬浮固体	混合液懸浮固體
mixed-liquor volatile suspended solid	混合液挥发性悬浮固体	混合液揮發性懸浮固體
mixed node	混合节点	混合節點
mixed order	混合序列	混合階
mixed production	混合生产	混合生產
mixed reactor	混合反应器	混合反應器
mixed suspension	混合悬浮	混合懸浮
mixer	混流器	混合器,混合機
mixer power	混合器功率	混合器功率
mixer-settler	混合澄清槽	混合沈降槽
mixing	混合	混合
mixing agitator	混合搅拌器	混合攪拌機
mixing baffle	混合用挡板	混合用擋板
mixing capacity	混合能力	混合能力
mixing chamber	混合室	混合室
mixing column	混合柱	混合[管]柱
mixing cup temperature	混合杯温	混合杯溫
mixing depth	混合深度	混合深度
mixing device	混合装置	混合裝置
mixing draft tube	混合引管	混合導管
mixing drum	混合罐	混合桶
mixing entrainment	混合雾沫	混合霧沫
mixing index	混合指数	混合指數
mixing intensity	混合强度	混合強度
mixing length	混合长	混合長度

英　文　名	大　陆　名	台　湾　名
mixing paddle	混合桨叶	混合槳葉
mixing pattern	混合型式	混合型式
mixing rate	混合速率	混合速率
mixing rule	混合规则	混合法則
mixing tank	混合槽	混合槽
mixing temperature	混合温度	混合溫度
mixing time	混合时间	混合時間,攪拌時間
mixture	混合物	混合物
mobile phase	流动相	流動相
mobile reactor	流动反应器	流動反應器
mobility	迁移率	遷移率,移動率
mockup experiment	冷模试验	冷模試驗
modacrylic fiber	改性聚丙烯腈纤维	改質[聚]丙烯腈纖維
model	模型	模型
model building	模型建立	模式建立
model identification	模型辨识	模型辨識
modeling	建模	建模
model parameter	模型参数	模型參數
model scale	模型比尺	模型比例
model test	模型试验	模型試驗
modified alkyd resin	改性醇酸树脂	改質醇酸樹脂
modified starch	改性淀粉	改質澱粉
modulation	调制	調變
module	模块	模組,組件
modulus	模量	模數
modulus ratio	模量比	模數比
moist materials	湿物料	濕物料
moisture	水分	水分
moisture content	湿含量	水分含量
moisture determination apparatus	湿度测定器	水分測定器
moisture factor	湿度因数	水分因數
moisture meter	湿度计	濕度計
moisture separator	去湿器	濕氣分離器,除濕器
molal average boiling point	摩尔平均沸点	莫耳平均沸點
molal concentration	质量摩尔浓度	重量莫耳濃度
molality	重量摩尔浓度	重量莫耳濃度
molal solution	摩尔溶液	重量莫耳溶液
molar average boiling point	摩尔平均沸点	莫耳平均沸點

英　文　名	大　陆　名	台　湾　名
molar average diffusivity	摩尔平均扩散系数	莫耳平均擴散係數
molar average velocity	摩尔平均速度	莫耳平均速度
molar concentration	摩尔浓度	[體積]莫耳濃度
molar flux	摩尔通量	莫耳通量
molar heat capacity	摩尔热容	莫耳熱容[量]
molar humidity	摩尔湿度	莫耳濕度
molarity	摩尔浓度	體積莫耳濃度
molar solution	摩尔溶液	莫耳溶液
molar volume	摩尔体积	莫耳體積
molasses	糖蜜	糖蜜
molasses alcohol	糖蜜酒精	糖蜜酒精
molasses graining	糖蜜起晶	糖蜜起晶
molasses medium	糖蜜塔养基	糖蜜培養基
mold	①霉菌 ②模	①黴菌 ②模
mold amylase	霉菌淀粉酶	黴菌澱粉酶
molding	模塑	成型,模製
molding composition	模压成分	模製成分
molding compression	模压	模[製]壓[縮]
molding press	压模机	模壓機
mold-release agent	脱模剂	脱模劑
mole balance	摩尔平衡	莫耳平衡
molecular beam deposition	分子束沉积法	分子束沈積[法]
molecular beam epitaxy	分子束外延	分子束磊晶
molecular biology	分子生物学	分子生物學
molecular configuration	分子构型	分子組態
molecular diffusion	分子扩散	分子擴散
molecular diffusivity	分子扩散系数	分子擴散係數
molecular dispersion	分子分散	分子分散
molecular distillation	分子蒸馏	分子蒸餾
molecular dynamic method（MD method）	分子动态法	分子動力學法
molecular interaction	分子相互作用	分子相互作用
molecularity	反应分子数	分子數,分子性
molecular parameter	分子参数	分子參數
molecular partition function	分子配分函数	分子配分函數
molecular rearrangement	分子重排	分子重排
molecular self-assembly	分子自集	分子自組裝
molecular sieve	分子筛	分子篩
molecular-sieve catalyst	分子筛催化剂	分子篩觸媒

英　文　名	大　陆　名	台　湾　名
molecular-sieve zeolite	分子筛沸石	分子篩沸石
molecular simulation	分子模拟	分子模擬
molecular species	分子物种	分子物種
molecular structure	分子结构	分子結構
molecular symmetry	分子对称性	分子對稱性
molecular thermodynamics	分子热力学	分子熱力學
molecular transformation	分子转换	分子變換
molecular transport	分子输送	分子輸送
molecular velocity	分子速度	分子速度
molecular weight	分子量	分子量
molecular weight distribution（MWD）	分子量分布	分子量分佈
molecule	分子	分子
Mollier diagram	莫利尔图	莫里爾圖
molten-bath gasifier	熔浴气化器	熔浴氣化器
molybdenum catalyst	钼催化剂	鉬觸媒
moment	矩	矩
moment of force	力矩	力矩
moment of inertia	转动惯量	轉動慣量
momentum	动量	動量
momentum balance	动量平衡	動量平衡
momentum conservation	动量守恒	動量守恆
momentum diffusivity	动量扩散系数	動量擴散係數
momentum equation	动量方程	動量方程式
momentum flux	动量通量	動量通量
momentum principle	动量原理	動量原理
momentum separator	动量分离器	動量分離器
momentum transfer	动量传递	動量傳送
momentum transfer coefficient	动量传递系数	動量傳送係數
momentum transport	动量输送	動量輸送
monazite	独居石	獨居石
monitor	监视器	監測器
monitoring	监视	監測
monitoring station	监测站	監測站
monoazo dye	单偶氮染料	單偶氮染料
monochromatic emissive power	单色发射能力	單色發射能力
monoclonal antibody	单克隆抗体	單源抗體,單株抗體
monococcus	单球菌	單球菌屬
Monod growth kinetics	莫诺生长动力学	莫氏生長動力學

英　文　名	大　陆　名	台　湾　名
monodisperse	单分散	單分散
monodispersity	单分散性	單分散性
monofilament	单丝	單絲纖維
monoglyceride	甘油单酯	單甘油酯
monolithic catalyst	整装催化剂	整裝觸媒,蜂巢狀觸媒
monomer	单体	單體
monomolecular reaction	单分子反应	單分子反應
monosaccharide	单糖	單醣
Monte Carlo simulation	蒙特卡罗模拟	蒙地卡羅模擬
monticellite	钙镁橄榄石	鈣橄欖石
montmorillonite	蒙脱石	蒙脱石,蒙脱土
mordant	媒染剂	媒染劑
mordant assistant	媒染助剂	媒染助劑
mordant azo dye	媒染偶氮染料	媒染偶氮染料
mordant color	媒染色料	媒染色料
mordant dye	媒染染料	媒染染料
mordanting process	媒染法	媒染法
mortar	砂浆	灰泥
most probable distribution	最概然分布	最可能分佈
most probable number	最大概率数	最可能數
mother cell	母细胞	母細胞
mother liquor	母液	母液
motion	运动	運動
motionless mixer	无动混合器	無動混合器
motor	电动机	電動機
motor actuator	电动机驱动器	馬達致動器
motor constant	电动机常数	馬達常數
mould	霉菌	黴[菌]
mounting	装配	安裝
moving bed	移动床	移動床
moving bed adsorber	移动床吸附器	移動床吸附器
moving bed gasifier	移动床气化器	移動床氣化器
moving bed process	移动床法	移動床法
moving bed reactor	移动床反应器	移動床反應器
moving boundary	移动界面	移動界面,移動邊界
moving catalyst bed	移动催化剂床	移動觸媒床
moving coil galvanometer	圈转电流计	圈轉電流計
moving phase	移动相	移動相

英 文 名	大 陆 名	台 湾 名
MTZ(=mass transfer zone)	传质区	質傳區
mud press	滤泥机	濾泥機
mud settler	泥浆沉降器	泥漿沈降器
mullite	莫来石	莫來石,富鋁紅柱石
mullite porcelain	莫来石瓷	富鋁紅柱石瓷
mullite refractory	莫来石耐火材料	富鋁紅柱石耐火材
multiaxial stress	多轴应力	多軸應力
multibed reactor	多床反应器	多床反應器
multiblade fan	多叶片风机	多葉[片]風扇
multiblade mixer	多叶片混合器	多葉[片]混合器
multicasing turbine	多室涡轮机	多室渦輪機
multichamber centrifuge	多室离心机	多室離心機
multichannel analyzer	多道分析仪	多通道分析儀
multicomponent	多组分	多成分,多元
multicomponent absorption	多组分吸收	多成分吸收
multicomponent azeotrope	多组分共沸物	多成分共沸液
multicomponent diffusion	多组分扩散	多成分擴散
multicomponent mixture	多元混合物,多组分混合物	多成分混合物
multicomponent separation sequence	多元分离序列	多成分分離順序
multicomponent system	多元系[统],多组分系统	多成分系統
multicone separator	多锥管分离器	多錐管分離器
multicyclone	多管旋风分离器	多[管]旋風分離器
multidowncomer tray(MD tray)	多降液管塔板	多降液管塔板
multifunctional catalyst	多功能催化剂	多功能觸媒
multifunctional enzyme	多功能酶	多功能酶
multilayer adsorption theory	多层吸附理论	多層吸附理論
multilevel method of optimization	多层次优化法	多層次優化法
multiloop control system	多回路控制系统	多環路控制系統
multiloop system	多回路系统	多環路系統
multi-objective programming	多目标规划	多目標規劃
multipass condenser	多程冷凝器	多程冷凝器
multipass exchanger	多程换热器	多程熱交換器
multipass heater	多程加热器	多程加熱器
multiphase flow	多相流	多相流
multiphase reactor	多相反应器	多相反應器
multiphase system	多相系统	多相系統

英　文　名	大　陆　名	台　湾　名
multiple batch extraction	多级批式萃取	多階批式萃取
multiple bed reactor	多床反应器	多床反應器
multiple blade mixer	多叶混合器	多葉混合器
multiple distributed component	多分布组分	多分佈成分
multiple evaporation	多效蒸发	多效蒸發
multiple evaporator	多效蒸发器	多效蒸發器
multiple extraction	多次萃取	多階萃取
multiple feed stream	多重进料流	多重進料流
multiple hearth furnace	①多室反应器 ②多膛炉	複床爐
multiple-input-multiple-output-system	多输入多输出系统	多輸入多輸出系統
multiple-input system	多重输入系统	多[重]輸入系統
multiple loop system	多回路系统	多環路系統
multiple output control	多重输出控制	多重輸出控制
multiple-output system	多重输出系统	多[重]輸出系統
multiple pipe system	多管系统	多管系統
multiple product column	多重产物塔	多重產物塔
multiple reactions	多重反应	多重反應
multiple reactor system	多重反应器系统	多反應器系統
multiple reboiler	多重再沸器	多重再沸器
multiple split point	多分隔点	多分隔點
multiple stability	多重稳态	多重穩定性
multiple stage centrifuge	多级离心机	多階離心機
multiple stage compression	多级压缩	多階壓縮
multiple stage compressor	多级压缩机	多階壓縮機
multiple steady state	多重稳态	多重穩態
multiplet	多重态	多重態,多重[譜]線
multiplicity of steady states	稳态多重性	穩態多重性
multiplier	乘法器	乘法器
multiplying manometer	倍示压力计	倍示壓力計
multipoint recorder	多点记录器	多點記錄器
multiport system	多进出口系统	多進出口系統
multiproduct batch plant	产品批量工厂	多產品批式工廠
multi-region model	多区模型	多區模型
multistage centrifugal pump	多级离心泵	多階離心泵
multistage compression	多级压缩	多階壓縮
multistage compressor	多级压缩机	多階壓縮機
multistage diffusion	多级扩散	多階擴散
multistage distillation	多级蒸馏	多階蒸餾

英 文 名	大 陆 名	台 湾 名
multistage ejector	多级喷射泵	多階噴射器
multistage fluidized bed	多级流化床	多階流[體]化床
multistage grinding	多级研磨	多階研磨
multistage process	多段过程	多階程序
multistage system	多段体系	多階系統
multitube reactor	多管反应器	多管反應器
multivariable control system	多变量控制系统	多變數控制系統
multivariable optimization	多变量最优化	多變數最適化
multivariable system	多变量系统	多變數系統
municipal waste	城市垃圾	都市廢棄物
municipal wastewater	城市污水	都市污水
muriate of potash	氯化钾	氯化鉀
muriatic acid	盐酸	鹽酸
Murphree efficiency	默弗里效率	莫氏效率
muscovado sugar	混糖	黃砂糖
muscovite	白云母	白雲母
mushroom distilling column	蕈状蒸馏塔	蕈狀蒸餾塔
mustard oil	芥子油	芥子油
mutagen	诱变剂	誘變劑
mutation	突变	突變
mutation frequency	突变频率	突變頻率
mutual solubility	互溶度	互溶度
MWD(=molecular weight distribution）	分子量分布	分子量分佈

N

英 文 名	大 陆 名	台 湾 名
nano-crystal	纳米微晶	奈米晶體
nanodroplet	纳米液滴	奈米液滴
nanofiltration	纳米过滤	奈米過濾
nanomaterials	纳米材料	奈米材料
nanometer	纳米	奈米
nanoparticle	纳米粒子	奈米粒子,奈米顆粒
nano-polymer materials	纳米高分子材料	奈米高分子材料
nanoscale science and technology	纳米科学[与]技术	奈米科技
nanostructure	纳米结构	奈米結構
nanotechnology	纳米技术	奈米技術
naphtha	石脑油	石油腦,輕油

英 文 名	大 陆 名	台 湾 名
naphtha cracking	石脑油裂解	輕油裂解
naphtha pyrolysis	石脑油热解	輕油熱解
naphthene	环烷	環烷
naphthol dye	萘酚染料	萘酚染料
Nash pump	纳氏泵	納氏泵
natural aging	自然老化	自然老化
natural circulation evaporator	自然循环蒸发器	自然循環蒸發器
natural convection	自然对流	自然對流
natural cooling	自然冷却	自然冷卻
natural draft	自然通风	自然通風
natural draft cooling tower	自然通风冷却塔	自然通風冷却塔
natural evaporation	自然蒸发	自然蒸發
natural frequency	自然频率	自然頻率
natural gas	天然气	天然氣
natural gas hydrate	天然气水合物	天然氣水合物
natural gas liquefaction	天然气液化	天然氣液化
natural gasoline	天然汽油	天然氣汽油,天然氣凝結油
natural gas reforming	天然气重整	天然氣重組
natural period	自然周期	自然週期
natural resources	自然资源	天然資源
natural rubber	天然橡胶	天然橡膠
natural soda	天然碱	天然鹼
natural varnish	天然清漆	天然清漆
natural ventilation	自然通风	自然通風
needle number	针孔数	針孔數
needle valve	针形阀	針閥
negative adsorption	负吸附	負吸附
negative azeotrope	负共沸混合物	負共沸液
negative catalysis	负催化作用	負催化[作用]
negative catalyst	负催化剂	負觸媒
negative deviation	负偏差	負偏差
negative feedback	负反馈	負回饋
negative feedback	负反馈	負回饋
negative resistance	负电阻	負電阻
nematic liquid crystal	向列型液晶	向列[型]液晶
nematic state	向列态	向列液晶態
neomycin	新霉素	新黴素

英　文　名	大　陆　名	台　湾　名
neoprene	氯丁橡胶	新平橡膠,氯丁二烯橡膠
neoprene adhesive	氯丁橡胶黏合剂	新平黏合劑
neoprene rubber	氯丁橡胶	新平橡膠,氯丁二烯橡膠
nepheline	霞石	霞石
nested control loop	巢式控制环路	巢式控制環路
net flux	净通量	淨通量
net heating value	净热值	淨熱值
net positive suction head（NPSH）	汽蚀余量	淨正吸高差
net present value	净现值	淨現值
net present worth	净现值	淨現值
net profit	净利润	淨利潤
net rate	净速率	淨速率
net weight	净重	淨重
net work	净功	淨功
network	网络	網路
neural network	神经网络	類神經網路
neural network training	神经网络训练	神經網絡訓練
neuron	神经元	神經元
neutral fertilizer	中性肥料	中性肥料
neutral firebrick	中性耐火砖	中性耐火磚
neutral glass	中性玻璃	中性玻璃
neutralization	中和	中和[作用]
neutralization number	中和值	中和值
neutralization point	中和点	中和點
neutralization reaction	中和反应	中和反應
neutralization titration	中和滴定	中和滴定[法]
neutralization value	中和值	中和值
neutralizing agent	中和剂	中和劑
neutral refractory	中和耐火材料	中性耐火材
neutral size	中性施胶	中性膠料
neutral soap	中性肥皂	中性肥皂
neutral sulfite process	中性亚硫酸盐法	中性亞硫酸[鹽]法
neutral sulfite semichemical process	中性亚硫酸盐半化学法	中性亞硫酸[鹽]半化學法
neutral zone	中和区	中和區
news-printing paper	新闻纸	新聞紙

英　文　名	大　陆　名	台　湾　名
Newtonian fluid	牛顿流体	牛頓流體
Newton method for convergence	牛顿收敛法	牛頓[迭代]法
Newton-Raphson method	牛顿–拉弗森法	牛[頓]–拉[福森]法
nichrome	镍铬合金	鎳鉻合金
nickel alumina catalyst	镍铝氧催化剂	鎳鋁氧觸媒
nickel carbonyl catalyst	羰基镍催化剂	鎳羰觸媒
nickel crucible	镍坩埚	鎳坩堝
nickel plating	镀镍	鍍鎳
nickel resistance bulb	镍电阻球	鎳電阻球
nickel steel	镍钢	鎳鋼
nickel storage battery	镍蓄电池	鎳蓄電池
niger	皂脚	皂腳
niger oil	皂脚油	皂腳油
nigrosine	苯胺黑	苯胺黑
niter cake	硝饼	硝餅
niter cake furnace	硝饼炉	硝餅爐
niter oven	硝炉	硝爐
niter pot	硝锅	硝鍋
nitration	硝化作用	硝化
nitration process	硝化法	硝化法
nitrator	硝化器	硝化器
nitrification	硝化作用	[氮]硝化
nitrifying bacteria	硝化菌	[氮]硝化菌
nitrile rubber	丁腈橡胶	腈橡膠
nitrocellulose	硝酸纤维素	硝化纖維素
nitrocellulose powder	硝酸纤维素火药	硝化纖維素火藥,無煙 　火藥
nitro dye	硝基染料	硝基染料
nitroexplosive	硝基火药	硝基炸藥
nitrogen cycle	氮循环	氮循環
nitrogen fixation	固氮[作用]	固氮作用
nitrogen guano	富氮海鸟粪	富氮[海]鳥糞
nitrogen mustard gas	氮芥子气	氮芥子氣
nitrogenous fertilizer	氮肥	氮肥
nitrogen oxygen demand	氮氧需要量	氮氧需要量
nitrogen purge	氮气吹扫	氮氣沖洗
nitroglycerine explosive	硝化甘油炸药	硝化甘油炸藥
nitrosamine red	亚硝胺红	亞硝胺紅

英 文 名	大 陆 名	台 湾 名
nitroso dye	亚硝基染料	亞硝基染料
node	节点	節點
nodulizing kiln	造粒转窑	粒化轉窯
noise	噪声	噪音
noise pollution	噪声污染	噪音污染
noise thermometer	噪声温度计	噪音溫度計
nomenclature	命名法	命名[法]
nominal pipe diameter	公称管径	標稱管徑
nominal size	公称尺寸	標稱尺寸
nominal stress	名义应力	標稱應力
nominal value	标称值	標稱值
nonactivated chemisorption	非活化吸附	非活化化學吸附
nonadiabatic reaction	非绝热反应	非絕熱反應
nonadiabatic reactor	非绝热反应器	非絕熱反應器
non-affinity adsorption	非亲和吸附	非親和吸附
noncatalytic reaction	非催化反应	非催化反應
nonchain reaction	非连锁反应	非連鎖反應
noncoking coal	非炼焦煤	非煉焦煤
noncompetitive inhibition	非竞争性抑制	非競爭性抑制[作用]
noncondensable gas	不凝气体	不冷凝氣體
noncrystalline	非晶性的	非晶性的
nondestructive test	非破坏性试验	非破壞性試驗
nondrying oil	非干性油	非乾性油
nonelementary reaction	非基本反应	非基本反應
non-equilibrium stage model	非平衡级模型	非平衡階模型
non-equilibrium system	非平衡系统	非平衡系統
non-equilibrium thermodynamics	非平衡热力学	非平衡[態]熱力學
nonhomogeneity	非均匀性	非均匀性
nonideal flow	非理想流动	非理想流動
nonideal gas	非理想气体	非理想氣體
nonideal reactor	非理想反应器	非理想反應器
nonideal solution	非理想溶液	非理想溶液
nonideal surface	非理想表面	非理想表面
noninertial coordinate system	非惯性坐标系统	非慣性坐標系統
noninteracting system	无相互作用系统	無相互作用系統
noninvasive instrument	非侵入性仪器	非侵入性儀器
noninvasive sensor	非侵入式传感器	非侵入性感測器
nonionic detergent	非离子洗涤剂	非離子清潔劑

英　文　名	大　陆　名	台　湾　名
nonionic surface active agent	非离子型表面活性剂	非離子界面活性劑
nonionic surfactant	非离子型表面活性剂	非離子界面活性劑
non-isothermal absorption	非等温吸收	非等溫吸收
nonisothermality	非等温性	非等溫性
nonisothermal reactor	非等温反应器	非等溫反應器
nonisotropic turbulence	各向异性湍流	非等向性紊流
nonlinear controller	非线性控制器	非線性控制器
nonlinear control system	非线性控制系统	非線性控制系統
nonlinear control valve	非线性控制阀	非線性控制閥
nonlinearity	非线性	非線性
nonlinear kinetics	非线性动力学	非線性動力學
nonlinear programming	非线性规划	非線性規劃
nonlinear regression analysis	非线性回归分析	非線性回歸分析[法]
nonlinear stability	非线性稳定性	非線性穩定性
nonlinear system	非线性系统	非線性系統
nonlinear viscoelasticity	非线性黏弹性	非線性黏彈性
nonmetal	非金属	非金屬
nonminimum phase lag	非最小相位滞后	非最小相位滯延
non-Newtonian fluid	非牛顿流体	非牛頓流體
nonoverlapping operation	非重叠操作	非重疊操作
nonporous membrane	非多孔膜	無孔透膜
nonporous pellet absorber	非多孔颗粒吸收器	無孔顆粒吸收器
nonproductive period	辅助周期	不生產期
non-random-two-liquid equation(NRTL equation)	非随机两液方程	NRTL 方程
nonreactive component	无反应性组分	無反應性成分
nonreactivity	无反应性	無反應性
nonregular solution	非正规溶液	非正規溶液
nonrenewable energy sources	非再生能源	非再生性能源
nonrepetitive polymer	非重复聚合物	非重複性聚合物
nonsegregated backmixing	非隔离返混	非隔離逆混
nonsegregated mixing	非隔离混合	非隔離混合
nonsegregated reactor	非隔离反应器	非隔離反應器
nonspontaneous process	非自发过程	非自發程序
nonspontaneous reaction	非自发反应	非自發反應
nonsteady state process	非稳态过程	非穩態程序
nontronite	绿脱石	綠脫石,矽鐵石
nonuniform flow	非均匀流	非均匀流動

英　文　名	大　陆　名	台　湾　名
nonvolatility	不挥发性	不揮發性
nonwashing plate	非洗式板框	非洗式板框
nonwetting surface	非润湿表面	不潤濕表面
nonwoven fabrics	无纺布	不織布
normal channel depth	正常渠深	正常渠深
normal concentration	当量浓度	當量濃度
normal condition	正常状态	常態
normal coordinates	法坐标	法坐標,正規坐標
normal distribution	正态分布	常態分佈
normal distribution function	正态分布函数	常態分佈函數
normalization	归一化	正規化
normal linearization	正常线性化	常態線性化
normal operation	正常操作	正常操作
normal reaction kinetics	正常反应动力学	正常反應動力學
normal shock wave	正激波	正震波
normal solution	当量溶液	當量溶液
normal stress	法向应力	法向應力
normal stress difference	法向应力差	法向應力差
no slip condition	无滑动条件	無滑動[邊界]條件
notation	记号	記號,記法
notch	凹槽	凹槽,V 形槽
notched weir	切口堰	V 形堰
novobiocin	新生霉素	新生黴素
noxious gas	有害气体	有害氣體
nozzle	喷嘴	噴嘴
nozzle pressure	喷嘴压力	噴嘴壓力
NPSH(=net positive suction head)	汽蚀余量	淨正吸高差
NRTL equation	NRTL 方程	NRTL 方程
NTU(=number of transfer unit)	传质单元数	質傳單元數
nuclear bomb	核弹	核彈
nuclear chemistry	核化学	核化學
nuclear energy	核能	核能
nuclear fuel	核燃料	核燃料
nuclear fusion	核聚变	核熔合
nuclear physics	核物理学	核子物理學
nuclear radiation	核辐射	核輻射
nuclear reactor	核反应堆	核反應器
nucleate boiling	泡核沸腾	成核沸騰

英　文　名	大　陆　名	台　湾　名
nucleate crystallization	有核结晶	成核結晶
nucleation	成核	成核[作用]
nucleation crystallization	成核结晶	成核結晶
nucleic acid	核酸	核酸
nucleoside	核苷	核苷
nucleotide	核苷酸	核苷酸
nucleus	晶核	晶核
nucleus formation	晶核形成	晶核形成
nuclide	核素	核種
null balance	零点平衡	零點平衡
null hypothesis	零假设	虛無假設
null potentiometer	零电位器	零點電位計
number average molecular weight	数均分子量	數量平均分子量
number of heat transfer unit	传热单元数	熱傳單元數
number of overall transfer units	总传质单元数	總傳送單元數
number of transfer unit（NTU）	传质单元数	質傳單元數
numerical analysis	数值分析	數值分析
numerical control	数字控制	數值控制
numerical method	数值方法	數值方法
numerical solution	数值解	數值解
Nusselt number	努塞特数	那塞數
nylon	尼龙	尼龍,耐綸
nystatin	制霉菌素	奈黴素

O

英　文　名	大　陆　名	台　湾　名
objective control	目标控制	目標控制
objective function	目标函数	目標函數
objective management	目标管理	目標管理
observability	可观测性	可觀測性
observed order	观测次序	測得階
obsolescence	退化	退化
occluded resin	包藏树脂	包藏樹脂
occlusion	包藏	包藏
occurrence matrix	事件矩阵	事件矩陣
ocean pollution	海洋污染	海洋污染
ocher	赭石	赭石

英 文 名	大 陆 名	台 湾 名
octane enhancer	辛烷值增进剂	辛烷值增進劑
octane number	辛烷值	辛烷值
odor	气味	氣味
odor pollutant	恶臭污染物	氣味污染物
off-line	离线	離線
offset	偏离	偏差
offset paper	胶版印刷纸	平板印紙
off-spec product	不合格产品	不合[规]格產品
ohmmeter	欧姆计	歐姆計,電阻計
oil	油	油
oil atomizer	油喷雾器	油霧化器
oil bath	油浴	油浴
oil cake	油饼	油餅
oiled paper	油纸	油紙
oil field	油田	油田
oil flotation	油浮选	油浮選[法]
oil foot	油脚	油腳,油渣
oil gas	油气	油氣
oil in water emulsion	水包油乳状液	水中油乳液
oil mordant	油媒染剂	油媒染劑
oil of vitriol	浓硫酸	濃硫酸,礬油
oil paint	油脂涂料	油性漆
oil recovery	采油	油回收
oil sand	油砂	油砂
oil shale	油页岩	油頁岩
oil tank	储油罐	油箱,油槽
oil tanker	油轮	油輪
oil-water separator	油水分离器	油水分離器
ointment base	软膏基质	軟膏基
oleandomycin	竹桃霉素	安黴素,奧連黴素
oleo-resin	油性树脂	油性樹脂
oleum	发烟硫酸	發煙硫酸
oligoclase	奥长石	鈉鈣長石
oligomer	低聚物	寡聚物
oligomerization	低聚反应	低聚合[反應]
oligosaccharide	寡糖	寡醣
olive-kernel oil	橄榄仁油	橄欖仁油
olive oil	橄榄油	橄欖油

英　文　名	大　陆　名	台　湾　名
olivine	橄榄石	橄欖石
once-through operation	单程操作	單程操作
once-through process	单程过程	單程程序
one-component system	单组分系统	單成分系
one-dimensional model	一维模型	一維模型
on-line	在线	線上
on-line adaptation	在线适应	線上調適
on-line computer control	线上计算机控制	線上控制
on-line tuning	线上调谐	線上調諧
on-off action	开关作用	開關作用
on-off control	通断控制	開關控制
Onsager reciprocal relation	昂萨格倒易关系	翁沙格倒易關係
opacifier	乳浊剂	失透劑
opacifying agent	乳浊剂	失透劑
opacity	不透明度	不透明度
opalescent glass	乳白玻璃	乳光玻璃
opaque glass	乳浊玻璃	不透明玻璃
opaque glaze	不透明釉	不透明釉
open boundary	开式边界	開放邊界
open channel	明渠	明渠
open channel	明槽	明渠
open channel flow	明槽流	明渠流
open circuit	开路	開路,斷路
open circuit grinding	开路研磨	開路研磨
open filter	敞式沙滤器	開敞過濾器
open impeller	开式叶轮	敞動葉輪
open loop	开环	開環
open-loop control	开环控制	開環控制
open-loop pole	开环极点	開環極點
open-loop system	开环系统	開環系統
open-loop test	开环测试	開環測試
open-loop transfer function	开环转移函数	開環轉移函數
open-loop zero	开环零点	開環零點
open nozzle	开敞喷嘴	開敞噴嘴
open reactor	敞式反应器	開敞反應器
open sand filter	敞式沙滤器	開敞砂濾器
open sequence reaction	敞式序列反应	非連鎖反應
open system	敞开系统	開放系統

英　文　名	大　陆　名	台　湾　名
open vessel	开式容器	開放式容器
operability	可操作性	可操作性
operating condition	操作条件	操作條件
operating cost	操作成本	操作成本
operating curve	操作曲线	操作曲線
operating data	操作数据	操作數據
operating diagram	操作图	操作圖
operating flexibility	操作弹性	操作彈性
operating frequency	操作频率	操作頻率
operating limit	操作极限	操作極限
operating line	操作线	操作線
operating manual	操作手册	操作手冊
operating parameter	操作参量	操作參數
operating point	操作点	操作點
operating pressure	操作压力	操作壓力
operating principle	操作原理	操作原理
operating stability	操作稳定性	操作穩定性
operating temperature	操作温度	操作溫度
operating time	操作时间	操作時間
operating variable	操作变量	操作變數
operating window	操作窗口	操作窗
operational variable	操作变量	操作變數
operation amplifier	操作放大器	操作放大器
operation condition	操作条件	操作條件
operation cost	操作费用	操作成本
operation cycle	操作周期	操作周期
operation diagram	操作图	操作圖
operation efficiency	操作效率	操作效率
operation flowsheet	操作流程图	操作流程圖
operation length	操作时间	操作期
operation line	操作线	操作線
operation mode	操作模式	操作方式
operation regeneration cycle	操作再生循环	操作再生循環
operation standard	操作标准	操作標準
opposing reaction	对峙反应	逆向反應
optical compensator	光学补偿器	光學補償器
optical density	①光学密度 ②黑度	光學密度
optical disc	光碟	光碟

英 文 名	大 陆 名	台 湾 名
optical fiber	光纤	光纖[維]
optical glass	光学玻璃	光學玻璃
optical isomer	旋光异构体	光學異構物
optical pyrometer	光学高温计	光學高溫計
optical rotatory dispersion	旋光色散	旋光色散,旋光分散
optimal composition	最优组成	最適組成
optimal condition	最佳工况	最適條件,最適狀況
optimal control	最优控制	最適控制
optimal conversion	最优转化率	最適轉化率
optimal design	最优设计	最適設計
optimality	最优性	最適性
optimal operation condition	最佳操作条件	最適操作條件,最適操作狀況
optimal path	最佳途径	最適途徑
optimal policy	最适策略	最適策略
optimal reflux	最佳回流	最適回流
optimal searching method	最佳搜寻法	最適搜尋法
optimal set point	最优设定点	最適設定點
optimal system	最优系统	最適系統
optimal temperature	最适温度	最適溫度
optimal tuning	最佳调谐	最適調諧
optimal value	最优值	最適值
optimization	最优化	最適化
optimum control	最优控制	最適控制
optimum conversion	最优转化率	最適轉化率
optimum design	最优设计	最適設計
optimum path	最佳途径	最適途徑
optimum policy	最适策略	最適策略
optimum reflux	最佳回流	最適回流
optimum reflux ratio	最佳回流比	最適回流比
optimum set point	最优设定点	最適設定點
optimum switching	最优切换	最適切換
optimum system	最优系统	最適系統
optimum temperature	最适温度	最適溫度
optimum tuning	最佳调谐	最適調諧
optimum value	最优值	最適值
order of deactivation	去活化级	去活化階
order of magnitude	数量级	[數]量級

英　文　名	大　陆　名	台　湾　名
order of reaction	反应级数	反應級數
ordinary differential equation	常微分方程	常微分方程[式]
ore dressing	选矿	選礦
ore flotation	矿石浮选	礦石浮選[法]
organic acid	有机酸	有機酸
organic binder	有机黏合剂	有機黏結劑
organic builder	有机补助剂	有機補助劑
organic carbon	有机碳	有機碳
organic fertilizer	有机肥料	有機肥料
organic load	有机负荷量	有機物負載
organic overload	有机超负载	有機物超負載
organic sludge	有机污泥	有機污泥
organic solid	有机固体	有機固體
organism	生物	生物
organization	组织	組織
organoclay	有机黏土	有機黏土
organomercury	有机汞	有機汞
organometallic compound	有机金属化合物	有機金屬化合物
organosol	有机溶胶	有機溶膠
orientation	①定向 ②取向	①定向 ②位向
orientation distribution	定向分布	位向分佈
orientation flat	定向平面	定向平面
oriented adsorption	定向吸附	位向吸附
orifice	孔口	孔口
orifice coefficient	孔口系数	孔口[計]係數
orifice discharge	孔口排放	孔口排放
orifice flange	孔板法兰	孔口凸緣,孔口法蘭
orifice flowmeter	孔板流量计	孔口[流量]計
orifice meter	孔板流量计	孔口[流量]計
orifice meter coefficient	孔板计系数	孔口計係數
orifice plate	孔板	孔口板
orifice ring	孔口环	孔口環
orifice tap	孔板计接头	孔口計接頭
orifice tube	锐孔管	孔口管
orlon	奥纶	奥綸
orthogonal collocation	正交配置	正交配置[法]
orthogonality	正交性	正交性
oscillating reaction	振荡反应	振盪反應

英　文　名	大　陆　名	台　湾　名
oscillating screen	振动筛	振盪篩
oscillation	振荡	振盪
oscillation frequency	振荡频率	振盪頻率
oscillatory element	振荡元件	振盪元件
oscillograph	示波器	示波器
oscilloscope	示波器	示波器
Oslo evaporative crystallizer	奥斯陆蒸发结晶器	奥斯陸蒸發結晶器
osmometer	渗透压计	滲[透]壓[力]計
osmoscope	渗透试验器	滲壓測定器
osmosis	渗透[作用]	滲透
osmotic coefficient	渗透系数	滲透係數
osmotic diffusion	渗透扩散	滲透擴散
osmotic pressure	渗透压	滲透壓[力]
outgassing	出气	除氣
outlet	出口	出口
outlet temperature	出口温度	出口溫度
outlet valve	出口阀	出口閥
output	①输出 ②产量	①輸出 ②產量
output function	输出函数	輸出函數
output layer	输出层	輸出層
output pressure	输出压力	輸出壓力
output set	输出集	輸出集
output signal	输出信号	輸出信號
output variable	输出[变]量	輸出變數
outside diameter	外径	外徑
outside reflux	外回流	[塔]外回流
outside reflux ratio	外回流比	[塔]外回流比
oven	烘箱	烘箱,爐
overall absorption coefficient	总吸收系数	總吸收係數
overall coefficient	总系数	總係數
overall composition	总组成	總組成
overall efficiency	总效率	總效率
overall energy balance	总能量平衡	總能量均衡
overall entropy balance	总熵平衡	總熵均衡
overall fractional yield	总产量分率	總產量分率
overall heat balance	总热平衡	總熱量均衡
overall heat transfer coefficient	总传热系数	總熱傳係數
overall mass balance	总质量平衡	總質量均衡

英　文　名	大　陆　名	台　湾　名
overall mass transfer coefficient	总传质系数	總質傳係數
overall process	总过程	總程序
overall rate	总速率	總速率
overall recovery	总回收	總回收[率]
overall transfer function	总传递函数	總傳遞函數,總轉移函數
overall transfer unit	总传递单元	總傳送單元
overcapacity	生产能力过剩	生產能力過剩
overdamped response	过阻尼响应	過阻尼應答
overdamped system	过阻尼系统	過阻尼系統
overdamping	过阻尼	過阻尼
over-driven buhrstone mill	过载磨石机	上動石磨機
overexpansion phenomenon	过膨胀现象	過膨脹現象
overflow	溢流	溢流
overflow alarm	溢流警报	溢流警報
overflow pipe	溢流管	溢流管
overflow rate	溢流速率	溢流速率
overflow velocity	溢流速度	溢流速度
overflow weir	溢流堰	溢流堰
over gain	过增益	過增益
overhead condenser	塔顶冷凝器	塔頂冷凝器
overhead cost	管理费用	管理費
overhead distillate	头馏分	塔頂餾出物
overhead vapor	塔顶蒸汽	塔頂蒸氣
overheating	过热	過熱
overlapping operation	重叠操作	重疊操作
overload	过载	超載,超負荷
overpotential	过电位	過電位
overproduction	生产过剩	過度生產
over range protection	超限保护	超限保護
overrelaxation method	超松弛法	過鬆弛法
oversaturation	过饱和	過飽和
overshoot	超调	超越量,超調量
overshooting	过调节	超越,超調
oversize distribution curve	筛上物分布曲线	篩上物分佈曲線
oxidant	氧化剂	氧化劑
oxidase	氧化酶	氧化酶
oxidation	氧化	氧化[作用]

英　文　名	大　陆　名	台　湾　名
oxidation catalyst	氧化催化剂	氧化觸媒
oxidation ditch	氧化沟	氧化渠
oxidation inhibitor	抗氧[化]剂	氧化抑制劑
oxidation period	氧化期	氧化期
oxidation pond	氧化塘	氧化池
oxidation potential	氧化电位	氧化電位
oxidation process	氧化过程	氧化程序
oxidation reaction	氧化反应	氧化反應
oxidation reduction cell	氧化还原电池	氧化還原電池
oxidation reduction potential	氧化还原电位	氧化還原電位
oxidation reduction reaction	氧化还原反应	氧化還原反應
oxidation reduction titration	氧化还原滴定	氧化還原滴定[法]
oxidative degradation	氧化降解	氧化降解
oxidative pyrolysis	氧化热裂化	氧化熱裂解
oxidized oil	氧化油	氧化油
oxidized starch	氧化淀粉	氧化澱粉
oxo process	羰基合成	甲醯化
oxydase	氧化酶	氧化酶
oxydo-reductase	氧化还原酶	氧化還原酶
oxygen analyzer	氧分析仪	氧分析儀
oxygenation	氧合作用	加氧[反應],充氧
oxygenator	充氧器	充氧器
oxygen consumption rate	耗氧速率	耗氧速率
oxygen content	氧含量	含氧量
oxygen demand	需氧量	需氧量
oxygen depletion	氧量耗尽	氧耗盡
oxygenizer	充氧器	充氧器
oxygen supply	供氧	供氧
oxygen transfer	氧传递	氧傳送
oxygen transfer coefficient	传氧系数	氧傳[送]係數
oxygen transfer rate	传氧速率	氧傳[送]速率
oxygen uptake	摄氧	攝氧
oxygen uptake rate	摄氧速率	攝氧速率
oxygen yield coefficient	氧收率系数	氧產率係數
oxyluminescence	氧化发光	氧化發光
ozone	臭氧	臭氧
ozone cracking	臭氧龟裂	臭氧龜裂
ozonization	臭氧化	臭氧化[作用]

英　文　名	大　陆　名	台　湾　名
ozonometer	臭氧计	臭氧計

P

英　文　名	大　陆　名	台　湾　名
packaged equipment	成套设备	套裝設備
package software	软件包	套裝軟體
packaging	包装	包裝
packed bed	填充床	填充床
packed bed reactor	填充床反应器	填充床反應器
packed column	填充柱	①填料塔 ②填料柱
packed column absorber	填料吸收器	填料式吸收塔
packed density	填充密度	填充密度
packed distillation column	填料蒸馏塔	填充蒸餾塔
packed extractor	填料萃取塔	填料式萃取塔
packed height	填充高度	填充高度
packed scrubber	填充式洗涤器	填料式洗滌塔
packed spray tower	填充式喷淋塔	填料式噴霧塔
packed tower	填料塔	填料塔
packing density	填充密度	填充密度
packing effect	装填效应	裝填效應
packing factor	填料因子	填料因子
packing fraction	充填率	裝填分率
packing leather	密封革	墊皮
packing materials	填充物	填充物
packing paper	包装纸	包裝紙
packing ring	①密封圈 ②填料环	墊圈
packing support	填料支承板	填料支架
padding	①垫料 ②浸染	襯墊料
padding liquor	浸染液	浸染液
padding machine	浸染机	浸染機
paddle	平桨	平槳
paddle agitator	桨式搅拌器	槳式攪拌器
paddle flocculator	桨式絮凝器	槳式絮凝器
paddle impeller	桨式叶轮	槳式葉輪
paddle mixer	桨式混合器	槳式混合器
paddle stirrer	桨式搅拌器	槳式攪拌器
paddle wheel	叶轮	槳輪

英 文 名	大 陆 名	台 湾 名
paint	①油漆 ②涂料	①油漆 ②塗料
paint remover	脱漆剂	除漆劑
pair annihilation	[正负电子]对湮没	成對毀滅
pair creation	电子偶产生	成對生成
pair glass	双层中空玻璃	雙層玻璃
pairing variables	对偶变量	變數配對
pair production	电子偶生成	成對產生
palladium catalyst	钯催化剂	鈀觸媒
pallet	平板架	棧板
Pall ring	鲍尔环	鮑爾環
palm kernel oil	棕榈仁油	棕櫚仁油
palm nut oil	棕榈仁油	棕櫚仁油
palm oil	棕榈油	棕櫚油
palm oil grease	油润滑脂	棕櫚油潤滑脂
palm wax	棕榈蜡	棕櫚蠟
panel board	①仪表板 ②配电盘	儀表板
panel spalling test	[耐火砖]格子散裂试验	[耐火磚]屏列剝落試驗
pan feeder	加料盘	加料盤
pan salt	锅盐	鍋鹽
pan seed	缸内晶种	罐內晶種
pan seeding	缸内晶种	罐內種晶
papain	木瓜蛋白酶	木瓜[蛋白]酶
paper chromatography	纸色谱法	紙層析[法]
paper electrophoresis	纸电泳	紙電泳[法]
paper partition chromatography	纸分配色谱	紙分配層析[法]
parabolic flume	抛物线溜槽	抛物線槽
parabolic valve	抛物线阀	抛物線閥
paraboloid condenser	抛物线体聚光器	抛物線體聚光器
paraffin	石蜡	石蠟
paraffin conversion	石蜡转化	石蠟轉化
paraffin distillate	石蜡馏分	石蠟餾出物
paraffin oil	石蜡油	石蠟油
paraffin paper	石蜡纸	[石]蠟紙
paraffin slop	含蜡废油	含蠟污油
paraffin soap	石蜡皂	石蠟皂
paraffin wax	石蜡	石蠟
paraflow	并流,同向流	並流

英　文　名	大　陆　名	台　湾　名
paraflow heat exchange	并流热交换	並流熱交換
parahelium	仲氦	仲氦
parahydrogen	仲氢	仲氫
parallel circuits	并联电路	並聯電路
parallel compensation	并联补偿	並聯補償
parallel compound turbine	并列复式涡轮机	並列複式渦輪機
parallel deactivation	平行失活	平行去活化
parallel feed	平行进料	平行進料
parallel operation	并行操作	並聯操作
parallel processes	并行过程	並聯程序
parallel processing	平行处理	平行處理
parallel reaction	平行反应	平行反應
parallel running	并联运行	並聯運轉
paramagnetic body	顺磁体	順磁體
paramagnetic effect	顺磁效应	順磁效應
paramagnetic oxygen analyzer	顺磁氧分析仪	順磁氧分析儀
paramagnetism	顺磁性	順磁性
parameter	参数	參數
parameter selection	参数选择	參數選擇
parameter sensitivity	参数灵敏度	參數靈敏度
parametric pump	参数泵	參數泵
parametric pumping	参数法分离	參數法分離
parasite	寄生物	寄生物,寄生蟲
parathion	对硫磷	巴拉松
parchment paper	羊皮纸	羊皮紙
parent cell	母细胞	母細胞,親細胞
parent element	母体元素	母元素
parent nuclide	母体核素	母核種
parison	型坯	型坯
parity	奇偶性	奇偶性
paromomycin	巴龙霉素	巴龍黴素
partial combustion	部分燃烧	部分燃燒
partial condensation	部分冷凝	部分冷凝
partial condenser	分凝器	部分冷凝器
partial conversion	部分转化	部分轉化
partial correlation	偏相关	部分相關
partial cracking	部分裂化	部分裂解
partial decomposition	部分分解	部分分解

英　文　名	大　陆　名	台　湾　名
partial decoupling	部分解耦	部分解偶
partial differential equation	偏微分方程	偏微分方程[式]
partial fraction	部分分式	部分分式
partial-fraction expansion	部分分式展开式	部分分式展開[法]
partial ionization	部分游离	部分游離
partially miscible system	部分互溶系统	部分互溶系统
partially segregated flow	部分分隔流动	部分分隔流[動]
partial methylation	部分甲基化	部分甲基化
partial miscibility	部分互溶	部分互溶性
partial molar enthalpy	偏摩尔焓	偏莫耳焓,部分莫耳焓
partial molar excess property	偏摩尔超额物性函数	部分莫耳過剩性質
partial molar Gibbs free energy	偏摩尔吉布斯自由能	偏莫耳吉布斯自由能,部分莫耳吉布斯自由能
partial molar internal energy	偏摩尔内能	部分莫耳内能,偏莫耳内能
partial molar property	偏摩尔物性函数	部分莫耳性質,偏莫耳性質
partial molar quantity	偏摩尔量	偏莫耳[數]量,部分莫耳[數]量
partial molar volume	偏摩尔体积	部分莫耳體積,偏莫耳體積
partial oxidation	部分氧化	部分氧化
partial oxidation reaction	部分氧化反应	部分氧化反應
partial pressure	分压力	分壓
partial reaction	部分反应	部分反應
partial recycle process	部分再循环过程	部分循環程序
partial reflux	部分回流	部分回流
partial separation	部分分离	部分分離
partial vacuum	部分真空	部分真空
particle	粒子	粒子
particle board	碎料板	塑合板,碎屑膠合板
particle characteristics	颗粒特性	粒子特性
particle conversion	粒子转化	粒子轉化
particle cyclone separator	粒子旋风分离器	粒子旋風分離器
particle density	颗粒密度	顆粒密度
particle diameter	粒径	粒[子直]徑
particle kinetics	粒子动力学	粒子動力學

英 文 名	大 陆 名	台 湾 名
particle shape	颗粒形状	粒形
particle shape factor	颗粒形状因子	粒形因數
particle size	粒度	粒度,粒子大小
particle size analysis	粒度分析	粒徑分析
particle size analyzer	粒度分析仪	粒徑析儀
particle size distribution	粒度分布	粒度分佈,粒子大小分佈
particle size separator	粒度分离器	顆粒分離器
particle swarm	颗粒群	粒子群
particulate	微粒	顆粒,粒狀物
particulate bed	颗粒床	顆粒床
particulate fluidization	散式流态化	顆粒流體化
particulate process	颗粒过程	顆粒程序
particuology	颗粒学	顆粒學
parting agent	脱模剂	脱模劑,離型劑
parting line	分界线	分模線
partition chromatography	分配色谱法	分配層析[法]
partition coefficient	分配系数	分配係數
partition constant	分配常数	分配常數
partitioning	分隔	分隔
pascal	帕[斯卡](压力单位)	帕[斯卡](壓力單位)
passivation	钝化	鈍化
passive iron	钝化铁	鈍化鐵
passive state	钝态	鈍態
paste	糊剂	糊
paste paint	厚漆	厚漆
paste resin	糊状树脂	糊狀樹脂
pasteurization	巴氏消毒法	低溫殺菌法,巴氏殺菌法
pasteurizer	巴氏灭菌器	低溫殺菌器
pastille	锭剂	錠劑
pat	试饼	試餅
patch board	接线板	接線板
patch cord	转接线	插線
patching	①插线 ②修补	①插線 ②修補
patch test	斑贴试验	斑貼試驗
patent	专利	專利
patent leather	漆革	漆革,黑漆皮

英 文 名	大 陆 名	台 湾 名
path function	路径函数	路徑函數
path length	程长	路徑長度
path line	迹线	徑線,路線
pathogen	病原体	病原體
path tracing	路径追踪	路徑追蹤
pattern	模式	圖型,圖樣
pattern recognition	模式识别	圖型辨識
pattern search	模式搜索	樣式搜尋[法]
payback period	投资回收期	回收期
payout	支付	支付
payroll	工资单	薪資單
peach bloom glaze	桃红釉	桃紅釉
peach-kernel oil	桃仁油	桃仁油
peak	峰值	尖峰
peak flow	峰流量	尖峰流[量]
peak flow rate	最高流速	尖峰流量
peak gain	峰增益	尖峰增益
peak gain ratio	峰增益比	尖峰增益比
peak load	峰负荷	尖峰負載
peak resonance	峰共振	尖峰共振
peak time	峰时间	尖峰時間
peanut oil	花生油	花生油
peat	泥炭	泥煤
peat bog	泥炭沼地	泥煤田,泥煤沼
pebble	卵石	卵石
pebble bed	卵石层	卵石床
pebble heater	卵石加热器	卵石加熱器
pebble mill	砾磨机	卵石磨
pecan oil	胡桃油	胡桃油
Peclet number	佩克莱数	佩克萊數
pectin	果胶	果膠
pectinase	果胶酶	果膠酶
pecto-cellulose	果胶纤维素	果膠纖維素
pedestal	底座	底座
peeling	剥离	剝離
peep hole	观察孔	視孔
pegmatite	伟晶岩	偉晶岩
pelleted catalyst	粒状催化剂	錠粒觸媒

英　文　名	大　陆　名	台　湾　名
pelletizer	造粒机	造粒機
pelletizing	造粒	造粒
penalty function	罚函数	處罰函數
pendant drop test	悬滴试验	懸滴試驗
pendulum	摆	[鐘]擺
pendulum-type tension testing machine	摆式张力试验机	擺式張力試驗機
pendulum viscometer	摆式黏度计	擺式黏度計
penetrability	穿透性	穿透性
penetrant	渗透剂	滲透劑
penetrating agent	渗透剂	滲透劑
penetration index	针入度指数	針入[度]指數
penetration probability	穿透概率	穿透機率
penetration test	渗入测试	針入試驗{煤}
penetration theory	穿透理论	穿透理論
penetrometer	针入度测定计	針入計
Peng-Robinson equation	PR[状态]方程	彭[定宇]-羅[賓遜] 狀態方程
penicillin	青霉素	青黴素,盤尼西林
penicillinase	青霉素酶	青黴素酶
peppermint oil	薄荷油	薄荷油
peptidase	肽酶	肽酶
peptide	肽	[胜]肽
peptization	胶溶	解膠[作用]
peptizing agent	①胶溶剂 ②塑解剂	解膠劑
peptone	胨	腖
peptonization	胨化	腖化[作用]
peracetic acid	过乙酸	過醋酸
percentage absolute humidity	百分绝对湿度	百分絕對濕度
percentage humidity	百分湿度	百分濕度
percolate	渗滤液	滲漉液
percolation	渗滤	滲濾,浸透
percolation extractor	渗滤器	滲漉[萃取]器
perdistillation	透析蒸馏	透析蒸餾
perfect control	完全控制	理想控制
perfect decoupling	完全解偶	完全解偶
perfect fluid	理想流体	理想流體
perfect gas	理想气体	理想氣體
perfectly mixed reactor	完全混合反应器	完全混合反應器

英 文 名	大 陆 名	台 湾 名
perfect mixing	全混	完全混合
perfect solution	理想溶液	理想溶液
perforated brick	多孔砖	多孔磚
perforated false bottom	多孔假底	多孔假底
perforated pipe	多孔管	多孔管
perforated plate	多孔板	多孔板
perforated plate column	孔板塔	孔板塔
perforated plate distillation column	孔板蒸馏塔	孔板蒸餾塔
perforated plate tower	孔板塔	孔板塔
perforated screen	多孔板筛	多孔[板]篩
perforated tray	多孔塔板	穿孔塔板
perforation	穿孔	穿孔,打洞
performance	性能	性能
performance characteristics	运行特性	性能特徵
performance chart	[操作]性能图	性能圖
performance control	[操作]性能控制	性能控制
performance criterion	特性判据	性能準則
performance curve	特性曲线	性能曲線
performance equation	特性方程式	性能方程式
performance index	性能指标	性能指數
performance number	特性数	性能值
performance test	性能试验	性能試驗
performance test report	性能试验报告	性能試驗報告
perform tray	网孔塔板	網孔塔板
perfume	香料	香料
perfume fixative	定香剂	香料固定劑
perfume fixing agent	定香剂	香料固定劑
perfume oil	芳香油	香料油
perfusion culture	灌注培养	灌注培養
periclase	方镁石	方鎂石
perilla oil	紫苏子油	紫蘇油
period	周期	週期
periodic coagulation	周期凝聚	週期凝聚[作用]
periodic flow	周期流	週期流[動]
periodic flow reactor	周期流反应器	週期流[動]反應器
periodicity	周期性	週期性
periodic kiln	间歇窑	間歇窯,週期窯
periodic load	周期负荷	週期負載

英　文　名	大　陆　名	台　湾　名
periodic operation	周期操作	週期操作
periodic oscillation	周期振荡	週期振盪
periodic phenomenon	周期现象	週期現象
periodic solution	周期解	週期解
periodic table of elements	元素周期表	週期表
peripheral speed	圆周速度	周邊速率
peripheral velocity	圆周速度	周邊速度
periscope	潜望镜	潛望鏡
perlite	珍珠岩	珠岩
permanent deformation	永久变形	永久變形
permanent disturbance	永久扰动	永久擾動
permanent hardness	永久硬度	永久硬度
permanent loss	永久损失	永久損失
permanganate number	高锰酸盐值	過錳酸[鹽]值
permanganate value	高锰酸盐[滴定]值	過錳酸[鹽]值
permeability	①渗透性 ②渗透率	滲透率,磁導率
permeability apparatus	渗透仪	滲透儀
permeability test	渗透试验	滲透[性]試驗
permeate	渗透物	滲透物
permeation flux	渗透通量	滲透通量
permutite	人造沸石	人造沸石
peroxide	过氧化物	過氧化物
peroxide decomposer	过氧化物分解剂	過氧化物分解劑
persistent agent	长效剂	長效劑,持久劑
pertraction	渗透萃取	透[過]萃[取]法
perturbation	微扰	微擾
perturbation method	微扰法	微擾法
perturbation theory	微扰理论,摄动理论	微擾理論
perturbation variable	微扰变量	微擾變數
perturbed hard chain theory(PHC theory)	微扰硬链理论	微擾硬鏈理論
pervaporation	渗透蒸发	滲透蒸發
pesticide	杀虫剂	殺蟲劑
petrochemical	石油化学品	石[油]化[學]品
petrochemical complex	石油化工厂	石[油]化[學]工業區
petrochemical industry	石油化学工业	石[油]化[學]工業
petrochemistry	石油化学	石油化學
petrol	车用汽油	車用汽油
petrolatum	矿脂	石蠟脂

英　文　名	大　陆　名	台　湾　名
petroleum	石油	石油
petroleum coke	石油焦	石油焦
petroleum engineering	石油工程	石油工程
petroleum ether	石油醚	石油醚
petroleum gas	石油气	石油氣
petroleum naphtha	石脑油	石油腦
petroleum processing engineering	石油加工工程	石油煉製工程
petroleum processing technology	石油加工技术	石油煉製技術
petroleum refinery	炼油厂	煉油廠
petroleum refining	石油炼制	石油煉製
phage	噬菌体	噬菌體
pharmaceutical chemistry	药物化学	藥物化學
pharmaceutical industry	制药工业	製藥工業
pharmaceutical preparation	药物制剂	藥物製劑
pharmaceutics	药剂学	藥劑學
pharmacokinetics	药动学	藥物動力學
pharmacology	药理学	藥理學, 藥物學
pharmacopoeia	药典	藥典
phase	①相 ②相位	①相 ②相位
phase advance	相位超前	相位超前
phase angle	相角	相角
phase angle locus	相角轨迹	相角軌跡
phase behavior	相特性	相行為
phase change	相变	相變化
phase crossover	相位交叉	相位交叉
phase crossover frequency	相位交叉频率	相位交叉頻率
phase crossover point	相位交叉点	相位交叉點
phase diagram	相图	相圖
phase difference	相[位]差	相位差
phase equilibrium	相平衡	相平衡
phase lag	相位滞后	相位滯延
phase lead	相位超前	相位超前
phase margin	相位裕量	相位裕量
phase plane	相平面	相平面
phase plane analysis	相平面分析	相平面分析
phase plane plot	相平面标绘	相平面圖
phase plane portrait	相平面图	相平面圖
phase relation	位相关系	相關係

英 文 名	大 陆 名	台 湾 名
phase rule	相律	相律
phase shift	相移	相位位移
phase space	相空间	相[位]空間
phase transfer	相转移	相[間]轉移
phase transfer catalysis	相转移催化	相[間]轉移催化[作用]
phase transfer catalyst	相转移催化剂	相[間]轉移觸媒
phase transfer catalyzed reaction	相转移催化反应	相[間]轉移催化反應
phase transformation	相变	相變換
phase velocity	相速度	相速度
pH control	pH 控制	pH 控制
PHC theory(=perturbed hard chain theory)	微扰硬链理论	微擾硬鏈理論
pH electrode	pH 电极	pH 電極
phenol formaldehyde resin	酚醛树脂	酚[甲]醛樹脂
phenolic resin	酚醛树脂	酚[醛]樹脂
phenolics	酚醛塑料	酚醛樹脂,含酚物
phenolphthalein	酚酞	酚酞
phenol red	酚红	酚紅
phenol resin adhesive	酚树脂黏合剂	酚樹脂黏合劑
phenomenological coefficient	唯象系数	現象學係數
phenomenological model	唯象模型	現象學模式
pH meter	pH 计	pH 計,酸度計
phosgenation process	光气法	光氣法
phosgene	光气	光氣,二氯化羰
phosgenite	角铅矿	角鉛礦
phosphate buffer	磷酸盐缓冲剂	磷酸鹽緩衝劑
phosphate builder	磷酸盐增效助剂	磷酸鹽增滌劑
phosphate fertilizer	磷肥	磷肥
phosphate glass	磷酸盐玻璃	磷酸鹽玻璃
phosphate rock	磷酸盐岩	磷鹽岩
phospholipid	磷脂	磷脂[質]
phosphorescence	磷光	磷光,燐光
phosphorylation	磷酸化[作用]	磷酸化[反應]
phosphosilicate glass	磷硅酸盐玻璃	磷矽酸鹽玻璃
photoactivation	光活化	光活化[作用]
photocatalysis	光催化	光催化[作用]
photocatalyst	光催化剂	光觸媒

英　文　名	大　陆　名	台　湾　名
photocell	光电池	光電池
photochemical absorption law	光化学吸收律	光化學吸收律
photochemical activity	光化学活性	光化[學]活性
photochemical chlorination	光化学氯化	光氯化
photochemical decomposition	光化学分解	光化[學]分解
photochemical degradation	光化学降解	光化[學]降解
photochemical efficiency	光化学效率	光化[學]效率
photochemical equivalent	光化学当量	光化[學]當量
photochemical excitation	光化学激化	光化[學]激化
photochemical induction	光化学诱导	光化[學]誘發
photochemical kinetics	光化学动力学	光化學動力學
photochemical ozonization	光化学臭氧化	光化[學]臭氧化
photochemical process	光化学过程	光化學程序
photochemical reaction	光化学反应	光化學反應
photochemical reactivity	光化学反应性	光化學反應性
photochemical reactor	光化反应器	光化學反應器
photochemical threshold	光化学低限	光化[學]低限
photochemical yield	光化学产量	光化學產率
photochemistry	光化学	光化學
photochlorination	光氯化	光氯化
photocolorimetry	光比色法	光比色法
photoconductive materials	光导材料	光導電材料
photoconductivity	光电导性	光導電性
photodecomposition	光解	光分解
photodegradation	光降解	光降解
photodensitometer	光密度计	光密度計
photodetector	光电检测器	光偵檢計
photodissociation	光致离解	光解離
photoelasticity	光弹性	光彈性
photoelectric cell	光电池[管]	光電池, 光電管
photoelectric colorimeter	光电比色	光電比色計
photoelectric effect	光电效应	光電效應
photoelectricity	光电学	光電學
photoelectric polarimeter	光电旋光计	光電旋光計
photoelectric spectrophotometer	光电分光光度计	光電光譜儀
photoelectrometer	光电计	光電計
photoetching process	光刻蚀过程	光蝕刻程序
photogenic bacteria	发光菌	發光菌

英 文 名	大 陆 名	台 湾 名
photographic desensitizer	感光减敏剂	感光减敏劑
photographic emulsion	感光乳剂	感光乳液
photographic paper	照相纸	感光紙
photographic sensitizer	照相增感剂	感光敏化劑
photolithography	光刻	光蚀刻法，微影術
photoluminescence	光致发光	光致發光
photolysis	光解	光解[作用]
photometer	光度计	光度計
photometric analysis	光度分析	光度分析
photometric method	光度测定法	光度測定法
photometry	光度学	光度學
photon	光子	光子
photoneutron	光中子	光中子
photonic materials	光子材料	光子材料
photonuclear reaction	光[致]核反应	光[致]核反應
photooxidation	光氧化	光氧化
photooxidative aging	光氧化老化	光氧化老化
photopolymer	光[致]聚合物	光聚合物
photopolymerization	光致聚合	光聚合[反應]
photoreaction	光反应	光反應
photoreceptor	光感受体	光受器，光受體
photoresist	光致抗蚀剂	光阻劑
photorespiration	光呼吸	光呼吸
photosensitivity	光敏性	光敏性
photosensitized oxidation	光敏氧化	光敏氧化[反應]
photosensitizer	光敏剂	光敏劑
photostabilization	耐光作用	耐光作用
photosynthesis	光合作用	光合作用
photothermal analysis	光热分析	光熱分析[法]
phototube	光电管	光電管
photoyellowing	光黄化	光黃化
pH-sensitive electrode	pH 敏感电极	pH 敏感電極
phthalocyanine blue	酞菁蓝	酞[花]青藍
phthalocyanine dye	酞菁染料	酞[花]青染料
phthalocyanine green	酞菁绿	酞[花]青綠
phthalocyanine pigment	酞菁颜料	酞[花]青顏料
pH value	pH 值	pH 值，酸度值
physical absorbent	物理吸收剂	物理吸收劑

英　文　名	大　陆　名	台　湾　名
physical absorption	物理吸收	物理吸收
physical adsorption	物理吸附	物理吸收
physical chemistry	物理化学	物理化學
physical constant	物理常数	物理常數
physical constraint	物理约束	物理限制
physical exergy	物理㶲	物理可用能
physical law	物理定律	物理定律
physical model	物理模型	物理模式
physical parameter	物理参数	物理参數
physical process	物理过程	物理程序
physical property	物理性质	物理性質
physical separation	物理分离	物理分離
physical system	物理系统	物理系統
physical test	物理试验	物理試驗
physical tracer	物理示踪剂	物理示蹤劑
physical transport step	物理传递步骤	物理輸送步驟
physical treatment	物理处理	物理處理
physical treatment process	物理处理过程	物理處理法
physical vapor deposition	物理气相沉积	物理氣相沈積
physicochemical absorption	物理化学吸收	物理化學吸收
physisorption	物理吸附	物理吸附
phytosterol	植物甾醇	植固醇
pickle bath	酸洗槽	酸洗槽
pickle liquor	酸洗液	酸洗液
pickling	酸洗	酸洗
pickling bath	酸洗槽	酸洗槽
pickling process	酸洗法	酸洗法
pictorial flowsheet	图形流程图	圖形流程圖
PID（=piping and instrument diagram）	管路-仪表流程图	管線及儀器圖
piezoelectric crystal	压电晶体	壓電晶體
piezometer ring	环形流压计	測壓環
pig iron	生铁	生鐵
pigment	颜料	颜料
pilot plant	中间试验装置	先導工廠,試驗工廠
pilot valve	导向阀	導引閥
pinch	箍缩	狹點
pinch effect	箍缩效应	狹點效應
pinch point	夹点	夾點

英 文 名	大 陆 名	台 湾 名
pinch technology	夹点技术	夾點技術
pine oil	松油	松油
pine seed oil	松子油	松子油
pin-point gate	针点浇口	針點澆口
pipe	管	管
pipe bend	管肘	管肘
pipe bending machine	弯管机	彎管機
pipe diameter	管径	管徑
pipe filter	管过滤器	管過濾器
pipe fitting	管件	管接頭
pipe flow	管流	管流
pipe friction	管路摩擦	管路摩擦
pipe joint	管接头	管接頭
pipeline	管路	管線
pipe line list	管线表	管線表
pipeline network	管路网络	管線網路
pipe reducer	异径管	漸縮管
pipe relative roughness	管子相对粗糙度	管子相對粗糙度
pipe roughness	管子粗糙度	管粗糙度
pipe schedule number	管壁厚度系列号	管壁厚度系列號
pipe sealing	管接头密封	管密封
pipe size	管子尺寸	管子尺寸
pipe still	管式蒸馏器	管餾器
pipe tap	锥管螺纹	攻牙器
piping	配管,管路	配管,管路
piping and instrument diagram(PID)	管路-仪表流程图	管線及儀器圖
piping flowsheet	管道流程图	配管流程圖
piping standard	配管标准	配管標準
piping stress	管线应力	管線應力
piping system	管路系统	管路系統,配管系統
piston	活塞	活塞
piston actuator	活塞执行机构	活塞致動器
piston flow reactor	活塞流反应器	塞流反應器
piston gauge	活塞式压力计	活塞壓力計
piston meter	活塞流量计	活塞流量計
piston pump	活塞泵	活塞泵
piston valve	活塞阀	活塞閥
pit	凹坑	坑,槽,池

英　文　名	大　陆　名	台　湾　名
pitch	①节距 ②硬沥青	①節距 ②瀝青
pitching	纵摇	添加酵母[釀酒]
pitching tank	加酵母槽	種母槽[釀酒]
pi-theorem	π定理	pi 定理,π定理
Pitot tube	皮托管	皮托管
pitting	剥蚀	麻點,凹痕
pitting corrosion	点蚀	麻點腐蝕
pivot	枢轴	樞軸
plait	褶	褶,疊
plait point	褶点	褶點
plane stress	平面应力	平面應力
planet agitator	行星式搅拌器	行星式攪拌器
planimeter	求积仪	面積計
plant asset	工厂资产	工廠資產
plant cost estimate	工厂成本估算	工廠成本估計值
plant growth hormone	植物生长激素	植物生長激素
plant growth regulator	植物生长调节剂	植物生長調節劑
plant hormone	植物激素	植物激素
plant layout	工厂布置	工廠布置
plant location	厂址	廠址
plant site selection	厂址选择	廠址選擇
plasma	①等离子体 ②血浆	電漿,血漿
plasma-enhanced etching	等离子[体]增强刻蚀	電漿加強蝕刻[法]
plasma etching	等离子[体]刻蚀	電漿蝕刻[法]
plaster	硬石膏	熟石膏,硬膏劑
plaster mold	石膏模	石膏模
plastic	塑料	塑膠
plastic binder	塑料黏合剂	塑膠黏結劑
plastic cement	塑料黏结剂	塑膠膠合劑
plastic engineering	塑料工程	塑膠工程
plastic fluid	塑性流体	塑性流體
plastic foam	塑料泡沫	塑膠泡體,塑膠泡沫
plasticimeter	塑料计	塑性試驗計
plastic index	塑性指数	塑性指數
plasticity	可塑性	塑性
plasticization	增塑作用	塑化
plasticizer	增塑剂	塑化劑,可塑劑
plastic materials	塑性材料	塑性材料,塑膠材料

英　文　名	大　陆　名	台　湾　名
plastic packing	塑料填料	塑膠填充物
plastic processing	塑料成形加工	塑膠加工
plastic state	黏流态	可塑狀態
plastic sulfur	弹性硫	彈性硫
plastic yield	塑性屈伏	塑性降伏
plastic yield point	塑性屈伏点	塑性降伏點
plastigel	塑性凝胶	塑性凝膠
plastimeter	塑度计	塑性計
plastisol	增塑溶胶	塑料溶膠
plastograph	塑性计	塑性測定器,塑性儀
plastometer	塑度计	塑性計
plate	塔板	塔板
plate aerator	板式通气器	板式通氣器
plate and frame filter	板框式过滤器	板框式過濾器
plate-and-frame filter press	板框式压滤机	板框壓濾機
plate-and-frame module	板框组件	板框組件
plateau	高原	[曲線]平台區
plateau effect	平坦效应	平台效應
plate column	板式塔	板式塔
plate culture	平板培养	平板培養
plate efficiency	[塔]板效率	板效率
plate-fin heat exchanger	板翅换热器	板翅熱交換器
plate-frame filter press	板框式压滤机	板框壓濾機
plate heat exchanger	板式换热器	板式熱交換器
plate number	板数	板數
plate tower	板式塔	板式塔
plate-type evaporator	板式蒸发器	板式蒸發器
plate type heat exchanger	平板式换热器	板式熱交換器
platform	平台	平臺
platformate	铂重整产品	鉑媒重組油
platform balance	台式天台	檯秤
platforming	铂重整	鉑媒重組
platforming process	铂重整过程	鉑媒重組程序
plating industry	电镀工业	電鍍工業
platinized asbestos	附铂石棉	附鉑石綿
platinized asbestos catalyst	附铂石棉催化剂	附鉑石綿觸媒
platinized carbon electrode	镀铂碳电极	鍍鉑碳[電]極
platinized pumice	附铂浮石	附鉑浮石

英 文 名	大 陆 名	台 湾 名
platinized silica gel catalyst	附铂硅胶催化剂	附鉑矽膠觸媒
platinum-alumina catalyst	铂氧化铝催化剂	鉑鋁氧觸媒
platinum black catalyst	铂黑催化剂	鉑黑觸媒
platinum contact process	铂接触过程	鉑接觸法
platinum crucible	铂坩埚	鉑坩堝,白金坩堝
platinum gauze	铂网	鉑網
platinum nickel catalyst	铂-镍催化剂	鉑鎳觸媒
platinum oxide catalyst	氧化铂催化剂	氧化鉑觸媒
platinum resistance bulb	铂电阻球	鉑電阻球
platinum resistance thermometer	铂电阻温度计	鉑電阻溫度計
platinum sponge	铂海绵	鉑海綿
plenum chamber	充气室	充氣室
plot	[绘]图	圖
plug	插塞	塞[子]
plug flow	平推流,活塞流	塞流,栓流
plug flow backmixing	平推流返混	塞流逆混
plug flow reactor	活塞流反应器	塞流反應器
plugging	堵塞	阻塞
plug valve	旋塞阀	塞閥
plumb bob float	铅锤式浮标	錘式浮標
plum kernel oil	梅仁油	梅仁油
plunger	柱塞	柱塞
plunger pump	柱塞泵	柱塞泵
plutonium reactor	钚反应器	鈽反應器
plywood	胶合板	合板
pneumatic actuator	气动执行机构	氣動致動器
pneumatic analog	气动模拟	氣動類比計算機
pneumatic classifier	气力分级器	氣動類析器
pneumatic control	气动控制	氣動控制
pneumatic controller	气动调节器	氣動控制器
pneumatic control valve	气动调节阀	氣動控制閥
pneumatic conveying	气力输送	氣動運送
pneumatic conveyor	气流输送器	氣動運送機
pneumatic device	气动装置	氣動裝置
pneumatic dryer	气流干燥器	氣流乾燥機
pneumatic electric relay	气动电流继电器	氣動繼電器
pneumatic flow transmitter	气动式流动传送器	氣動式流動傳送器
pneumatic force meter	气动测力仪	氣動測力計

英 文 名	大 陆 名	台 湾 名
pneumatic liquid density gage	气动式液体密度计	氣動液體密度計
pneumatic liquid level gage	气动式液面计	氣動液位計
pneumatic mechanism	气动机理	氣動機構
pneumatic sensor	气动传感器	氣動感測器
pneumatic set controller	气动设定控制器	氣動設定控制器
pneumatic signal	气动信号	氣動信號
pneumatic stirring	气动搅拌	氣動攪拌
pneumatic system	气动系统	氣動系統
pneumatic transmission	气动传动	氣動傳送
pneumatic transmission line	气动传动线	氣動傳送線路
pneumatic transmission receiver	气动传动接受器	氣動傳送接受器
pneumatic transmitter	气动传动器	氣動傳送器
pneumatic transport	气动输运	氣動運送
pneumatic valve	气动阀	氣動閥
pocket grinder	袋式研磨机	袋式研磨機
point efficiency	点效率	點效率
point of contact	接触点	接觸點
point of steepest ascent	最陡上坡点	最陡上坡點
point of steepest descent	最陡下坡点	最陡下坡點
point selectivity	点选择性	點選擇性
point source	点源	點源
poise	泊	泊
poison	毒物	毒物,毒劑
poisoning	中毒	中毒
polar coordinates	极坐标	極坐標
polarimeter	旋光计	旋光計,偏光計
polariscope	偏光镜	偏光計
polarity	极性	極性
polarizability	极化率	極化率,極化性
polarization	①极化 ②偏振	①極化 ②偏光
polarization factor	①极化因子②偏振化因子	偏光因子
polarograph	极谱仪	極譜儀
polarographic method	极谱法	極譜法
polarography	极谱法	極譜法
polaroid	偏振片	偏光玻璃
polar plot	极坐标图	極坐標圖
polar solvent	极性溶剂	極性溶劑

英 文 名	大 陆 名	台 湾 名
polishing	抛光	磨光
polishing pan	上光盘	上光盤
pollutant	污染物	污染物
pollutant standards index	污染物标准指数	污染物標準指數
polluted water	污染水	污水
pollution	污染	污染
pollution abatement	污染减少	污染减量
pollution control	污染控制	污染控制,污染防治
pollution engineering	污染工程	污染工程
pollution source	污染源	污染源
polyacrylate	聚丙烯酸酯	聚丙烯酸酯
polyacrylate resin	聚丙烯酸酯树脂	聚丙烯酸酯樹脂
polyacrylonitrile	聚丙烯腈	聚丙烯腈
polyacrylonitrile fiber	聚丙烯腈纤维	聚丙烯腈纖維
polyalcohol	多元醇	多元醇
polyalkylation	多烷基化作用	多烷基化[作用]
polyamide	聚酰胺	聚醯胺
polyamide fiber	聚酰胺纤维	聚醯胺纖維
polyamide resin	聚酰胺树脂	聚醯胺樹脂
polyblend	高分子共混物	高分子摻合物
polycaprolactam	聚己内酰胺	聚己內醯胺
polycarbonate	聚碳酸酯	聚碳酸酯
polycarbonate resin	聚碳酸酯树脂	聚碳酸酯樹脂
polycondensation	缩聚反应	聚縮[作用]
polyester	聚酯	聚酯
polyester fiber	聚酯类纤维	聚酯纖維
polyester resin	聚酯类树脂	聚酯樹脂
polyester rubber	聚酯类橡胶	聚酯橡膠
polyether	聚醚	聚醚
polyethylene	聚乙烯	聚乙烯
poly-fluid theory	多流体理论	多流體理論
polyformaldehyde	聚甲醛	聚甲醛
polyforming	叠合重整	疊合重組[作用]
polyisobutylene	聚异丁烯	聚異丁烯
polyisoprene rubber	聚异戊二烯橡胶	聚異戊二烯橡膠,異平橡膠
polymer	聚合物	聚合物,高分子,聚合體
polymer blend	共混聚合物	高分子摻合物

英 文 名	大 陆 名	台 湾 名
polymer degradation	聚合物降解	聚合物降解
polymer flooding	聚合物驱	聚合物氾流[法]
polymer gasoline	叠合汽油	疊合汽油
polymerization	聚合[反应]	聚合[反應]
polymerization kinetics	聚合动力学	聚合[反應]動力學
polymerization mechanism	聚合机理	聚合反應機構
polymerization reaction	聚合反应	聚合反應
polymerization reaction engineering	聚合反应工程	聚合反應工程
polymerization reactor	聚合反应器	聚合反應器
polymerized oil	聚合油	聚合油
polymer processing	聚合物加工	聚合物加工,高分子加工
polymer processing engineering	聚合物加工工程	聚合物加工工程
polymer processing technology	聚合物加工技术	聚合物加工技術
polymer stabilization	聚合物稳定剂	聚合物穩定化
polymetallic catalyst	多金属催化剂	多金屬觸媒
polymorphism	多态[现象]	多形性
polymyxin	多黏菌素	多黏菌素
polyolefin resin	聚烯烃树脂	聚烯烴樹脂
polypropylene	聚丙烯	聚丙烯
polypropylene fiber	聚丙烯纤维	聚丙烯纖維
polysaccharidase	多糖酶	多醣分解酶
polysaccharide	多糖	多醣
polystyrene	聚苯乙烯	聚苯乙烯
polysulfide polymer	多硫聚合物	多硫聚合物
polysulfide rubber	聚硫橡胶	多硫橡膠
polysulfone	聚砜	聚碸
polytetrafluoroethylene	聚四氟乙烯	聚四氟乙烯
polytrifluorochloroethylene	聚三氟氯乙烯	聚三氟氯乙烯
polytropic constant	多变常数	多變常數
polytropic expansion	多变膨胀	多變膨脹
polytropic index	多变指数	多變指數
polytropic process	多变过程,多方过程	多變程序
polytropic specific heat	多变比热	多變比熱
polytropism	多变性	多變性
polyurethane	聚氨基甲酸酯	聚胺甲酸酯
polyurethane rubber	聚氨酯橡胶	聚胺[甲酸]酯橡膠
polyvinyl acetate	聚乙酸乙烯酯	聚乙酸乙烯酯

英 文 名	大 陆 名	台 湾 名
polyvinyl alcohol	聚乙烯醇	聚乙烯醇
polyvinyl butyral resin	聚乙烯酯缩丁醛树脂	聚乙烯醇縮丁醛樹脂
polyvinyl chloride	聚氯乙烯	聚氯乙烯
polyvinyl chloride resin	聚氯乙烯树脂	聚氯乙烯樹脂
polyvinyl ether	聚乙烯基醚	聚乙烯醚
polyvinyl ether adhesive	聚乙烯醚黏合剂	聚乙烯醚黏合劑
polyvinylidene chloride	聚偏氯乙烯	聚偏二氯乙烯
polyvinylidene fluoride	聚偏氟乙烯	聚偏二氟乙烯
polyvinyl pyrrolidone	聚乙烯吡咯烷酮	聚乙烯氫吡咯酮
Ponchon-Savarit diagram	P–S 图	焓–濃圖,P–S 圖
pool boiling	池沸腾	池沸騰
pool boiling curve	池沸腾曲线	池沸騰曲線
population balance	种群平衡	種群均衡,粒數衡算
population density function	种群密度函数	種群密度函數
population dynamics	种群动态	種群動態
population parameter	总付参数	種群參數
porcelain	瓷	瓷
porcelain crucible	瓷坩埚	瓷坩堝
porcelain enamel	搪瓷	搪瓷
porcelain enameled steel	搪瓷钢	搪瓷鋼
porcelain glaze	瓷釉	瓷釉
porcelain insulator	瓷绝缘子	瓷絕緣體
pore	孔隙	孔隙,[細]孔
pore diffusion	微孔扩散	孔[隙]擴散
pore diffusion resistance	微孔扩散阻力	孔[隙]擴散阻力
pore entrance	细孔入口	孔隙入口
pore geometry	孔隙结构	孔隙[幾何]形狀
pore-mouth poisoning	孔口中毒	孔口中毒
pore orientation distribution	细孔方向分布	孔隙方向分佈
pore radius	孔半径	孔[隙]半徑
pore shape factor	细孔形状因子	孔[隙]形狀因數
pore size	孔径	孔徑,孔隙大小
pore size distribution	孔径分布	孔徑分佈
pore structure	孔隙结构	孔[隙]結構
pore volume	孔体积	孔隙體積
porosimeter	孔隙计	孔隙計
porosity	孔隙率	孔隙率
porosity apparatus	孔隙仪	孔隙儀

英　文　名	大　陆　名	台　湾　名
porosity effect	孔效应	孔隙效應
porous barrier	孔膜隔板	多孔隔板
porous catalyst	多孔催化剂	多孔觸媒
porous materials	多孔材料	多孔[性]材料
porous medium	多孔介质	多孔介質
porous membrane	多孔膜	多孔膜
porous packing	多孔填料	多孔填料
porous pellet	多孔粒	多孔粒
porous pellet pulse reactor	多孔粒脉冲反应器	多孔粒脈衝反應器
porous plate	密孔板	多孔板
porous solid	多孔固体	多孔固體
porous-wall tube	多孔壁管	多孔壁管
positioner	焊接变位机	定位器
positive azeotrope	正共沸混合物	正共沸液
positive deviation	正偏差	正偏離
positive displacement	正位移	正排量
positive displacement flowmeter	容积式流量计	正排量流量計
positive displacement meter	正位移计	正排量[流量]計
positive displacement pump	容积式泵	正排量泵
positive feedback	正反馈	正回饋
positive-negative azeotrope	正负共沸物	正負共沸液
postbaking	后烘	後焙
post vulcanization	后硫化	後硫化
potable water	饮用水	飲用水
potash feldspar	钾长石	鉀長石
potash fertilizer	钾肥	鉀肥
potash glass	钾玻璃	鉀玻璃
potash-lead glass	钾铅玻璃	鉀鉛玻璃
potash-lime glass	钾钙玻璃	鉀鈣玻璃
potash soap	钾皂	鉀皂
potash superphosphate	含钾过磷酸钙	含鉀過磷酸鈣
potential	①势 ②电势 ③位势	①勢 ②電位 ③位能
potential barrier	势垒	勢障,位能障壁
potential correction	势修正	勢修正
potential deviation	势偏差	勢偏差
potential difference	势差	勢差,[電]位差
potential drop	[电]位降	位降
potential energy	势能	位能

英　文　名	大　陆　名	台　湾　名
potential energy surface	势能面	位能面
potential flow	势流	勢流
potential gradient	电势梯度	[電]位梯度
potential head	位头	勢差,位能差
potential line	势线	勢線
potentiometer	电位器	電位計
potentiometric analysis	电位分析[法]	電位分析
potentiometric method	电位法	電位法
potentiometric titration	电位滴定[法]	電位滴定[法]
poundal	磅达	磅達
pouring glazing	浇注釉	澆釉
powder	粉[体]	粉[體]
powder density	粉体密度	粉體密度
powder metallurgy	粉末冶金	粉末冶金[學]
powder technology	粉体技术	粉[粒]體技術
power	功率	功率,能力
power consumption	功率消耗	功率消耗[量]
power efficiency	功率效率	功率效率
power factor	功率因数	功率因數
power function type equation	幂函数型方程	冪函數型方程
power generator	动力发动机	發電機
power-law fluid	幂律流体	冪次律流體
power loss	功率损耗	功率損失
power number	功率数	功率數
power plant	①动力装置 ②发电站	動力廠,發電廠
power plant cycle	发电厂循环	發電廠循環
power requirement	功率需要量	功率需要量
Poynting correction	坡印亭校正	波印亭[校正]因子
pozzolana	火山灰	火山灰
practical stability	实际稳定性	實用穩定性
Prandtl number	普朗特数	卜朗特數
prebaking	预焙	預焙
precedence ordering	排序	排序
precipitate	沉淀物	沈澱物
precipitation	①沉淀 ②析出	沈澱
precipitation polymerization	沉淀聚合	沈澱聚合[反應]
precipitation reaction	沉淀反应	沈澱反應
precipitation titration	沉淀滴定[法]	沈澱滴定[法]

英　文　名	大　陆　名	台　湾　名
precipitator	①沉淀器 ②[静电]除尘器	①沈澱器 ②集塵器
precise fractionation	精密分馏	精密分餾
precision	精[密]度	精密度,精確度
precision fractional distillation	精密蒸馏	精密分餾
precocity of accelerator	促进剂的早熟性	催速劑早熟性
precooler	预冷器	預冷器
precursor	前体	前驅物
predesign cost estimate	预设计成本估算	預設計成本估算
prediction	预测	預測
predictive control	预测控制	預測控制
preevaporator	预蒸发器	預蒸發器
preexponential factor	指前因子	指數前因數,頻率因數
preferential solubility	有择溶解度	優先溶解度
prefiltration	预滤	預濾
preflash tower	预闪蒸塔	預閃蒸塔
prefractionation	预分馏	預分餾
preheater	预热器	預熱器
preheat furnace	预热炉	預熱爐
preheating	预热	預熱
preheating evaporator	预热蒸发器	預熱蒸發器
preheating period	预热期	預熱期
preheating zone	预热区	預熱區
preliminary design	初步设计	初步設計,前期設計
preliminary estimate	初步概算	初估值
preliminary filter	初滤机	初濾機
preliminary sizing	粗筛选	初級篩選,初上漿,尺寸初估
preliminary treatment	预处理	初步處理
preload	预载荷	預負載
premix molding	预混模制物	預混成型
present value	现值	現值
present worth	现值	現值
present worth factor	现值因子	現值因數
present worth method	现值法	現值法
preservative	防腐剂	防腐劑
press mold	压模	壓模
pressure balanced valve	压力平衡阀	壓力均衡閥

英 文 名	大 陆 名	台 湾 名
pressure calibration	压力校正	壓力校正
pressure coefficient	压力系数	壓力係數
pressure compensation	压力补偿	壓力補償
pressure-composition diagram	压力－组成图	壓力-組成圖
pressure control	压力控制	壓力控制
pressure controller	压力控制器	壓力控制器
pressure cooker	加压蒸煮器	加壓蒸煮器,壓力鍋
pressure correction	压力校正	壓力修正
pressure difference	压差	壓[力]差
pressure diffusion	压力扩散	壓力擴散
pressure distillate	加压馏出物	加壓餾出物
pressure distillation	加压蒸馏	加壓蒸餾
pressure distribution	压力分布	壓力分佈
pressure drop	压降	壓[力]降
pressure effect	压力效应	壓力效應
pressure efficiency	压力效率	壓力效率
pressure energy	压力能	壓[力]能
pressure-enthalpy diagram	压焓图	壓力-焓圖
pressure equalizer	压力平衡器	均壓器
pressure filter	压滤机	壓濾機
pressure filtration	加压过滤	加壓過濾
pressure flotation	加压浮选	加壓浮選[法]
pressure fractionation	加压分馏	加壓分餾
pressure gage	压力计[表]	壓力計
pressure gradient	压力梯度	壓力梯度
pressure head	压头	壓力高差
pressure indicator	压力指示计	壓力指示器
pressure leaf filter	加压叶滤机	加壓葉濾機
pressure loss	压力损失	壓力損失
pressure measurement	压力测量	壓力量測
pressure meter	压力计	壓力計
pressure oxidation	加压氧化	加壓氧化[反應]
pressure range	压力范围	壓力範圍
pressure recorder	压力记录器	壓力記錄器
pressure reducing valve	减压阀	減壓閥
pressure regulating valve	调压阀	壓力調節閥
pressure regulator	调压器	壓力調節器
pressure sand filter	加压砂滤器	加壓砂濾器

英　文　名	大　陆　名	台　湾　名
pressure screen	加压筛	壓力篩
pressure sensitive adhesive(PSA)	压敏胶	壓敏膠,感壓膠
pressure signal	压力信号	壓力信號
pressure swing adsorption	变压吸附	變壓吸附
pressure system	气压系统	壓力系統
pressure tap	测压孔	測壓分接頭
pressure-temperature diagram	压力–温度图	壓力–溫度圖
pressure transducer	压力传感器	壓力傳感器
pressure transmitter	压力变送器	壓力傳送器
pressure type thermometer	压力式温度计	壓力式溫度計
pressure-vacuum valve	压力－真空阀	壓力真空閥
pressure vessel	压力容器	壓力容器,壓力槽
pressure-volume diagram	压容图	壓力–體積圖
pressure-volume work	压力–容积功	壓力–體積功
pretreatment	预处理	預處理
preventative treatment	预防处理	預防處理
prevention	预防	預防
prilling	成球	製粒
prilling tower	造粒塔	製粒塔
primary clarifier	初级澄清器	初級澄清器
primary design variable	主要设计变量	主要設計變數
primary dimension	主要尺寸	主因次
primary distillation	初[级蒸]馏	初[級蒸]餾
primary element	主要元素	主要元素
primary feeder	主加料机	主飼機,主進料器
primary impact	主要冲击	主要衝擊
primary liquid	原液	原液
primary loop	主环路	主環路
primary normal stress coefficient	一级正应力系数	一次法向應力係數
primary plasticizer	主增塑剂	主塑化劑
primary product	主产品	主產物
primary purification	初级净化	初級純化
primary reaction	主反应	主要反應,初步反應
primary settler	初级沉降槽	初級沈降槽
primary standard	一级标准	一級標準,原級標準
primary tower	初馏塔	初餾塔
primary treatment	一级处理	初級處理
primary variable	主变量	主要變數

英　文　名	大　陆　名	台　湾　名
primer	底漆	①底漆 ②雷管 ③起動注給器
principal axis	主轴	主軸
principal fermentation	主发酵	主發酵
principal stress	主应力	主應力
principle	原理	原理
principle of conservation of energy	能量守恒原理	能量守恆原理
principle of conservation of mass	质量守恒原理	質量守恆原理
principle of conservation of moment of momentum	动量矩守恒原理	角動量守恆原理
principle of conservation of momentum	动量守恒原理	動量守恆原理
principle of corresponding state	对比状态原理	對應狀態原理
principle of detailed balance	精细平衡原理	細部平衡原理
principle of dimensional homogeneity	因次均一性原理	因次均一性原理
principle of entropy increase	熵增原理	熵增原理
principle of material objectivity	物质客观性原理	物質客觀性原理
principle of microscopic reversibility	微观可逆性原理	微觀可逆性原理
probabilistic process	随机过程	隨機程序
probability	概率	機率
probability curve	概率曲线	機率曲線
probability density	概率密度	機率密度
probability density function	概率密度函数	機率密度函數
probability distribution	概率分布	機率分佈
probability distribution function	概率分布函数	機率分佈函數
probability of acceptance	认可概率	允收機率
probability plot	概率标绘图	機率圖
probe	探头	探頭
process analysis	过程分析	程序分析
process automation	过程自动化	程序自動化
process behavior	过程行为特性	程序行為
process calculation	工艺计算	程序計算
process capacity	过程能力	製程能力指數
process characteristics	过程特性	程序特性
process chart	工艺流程图	程序圖
process condition	工艺条件	程序條件
process control	过程控制	程序控制
process control pattern	过程控制模式	程序控制模式
process decomposition	过程分解	程序分解

英　文　名	大　陆　名	台　湾　名
process design	工艺过程设计	程序設計
process design engineer	工艺设计工程师	程序設計工程師
process designer	工艺设计师	程序設計師
process development	过程开发	程序開發
process dynamics	过程动态[学]	程序動態學
process element	过程组成元件	程序元件
process engineer	工艺工程师	製程工程師,方法工程師
process engineering	过程工程	程序工程,方法工程
process engineering check list	过程工程核查表	程序工程查核表
process equipment	工艺设备	製程設備
process evaluation	过程评价	程序評估
process feed	过程进料	程序進料
process flow diagram	工艺流程图	程序流程圖,方法流程圖
process flowsheet	工艺流程图	程序流程圖,方法流程圖
process gain	过程增益	程序增益
process identification	过程辨识	程序辨識
process industry	过程工业	程序工業
processing	加工	加工,處理
processing engineering	加工工程	加工工程
processing industry	加工工业	加工業
processing technology	加工技术	加工技術
processing variable	加工变量	加工變數
process instability	过程非定态性	程序不穩定性
process integration	过程集成	程序整合
process layout	工艺布置图	程序配置[圖]
process load	工艺负荷	程序負載
process load change	工艺负荷变化	程序負載變化
process modeling	过程模型化	程序模式化
process optimization	过程优化	程序最適化
process overload	过程超载	程序超載
process parameter	工艺参数	程序參數
process performance	工艺性能	程序性能
process planning	工艺规划	程序規畫
process reaction curve	过程反应曲线	程序反應曲線
process reliability	过程可靠性	程序可靠性

英 文 名	大 陆 名	台 湾 名
process resilience	过程反弹	程序韌性
process response	过程响应	程序應答
process response curve	过程响应曲线	程序應答曲線
process safety	工艺安全	程序安全[性]
process-scale chromatography	工业色谱	工業級層析法
process sensitivity	过程灵敏度	程序靈敏度
process simulation	过程模拟	程序模擬
process stability	过程稳定性	程序穩定性
process steam	工艺蒸气	製程蒸汽
process stream	工艺物流	程序流
process system	工艺系统	程序系統
process system analysis	过程系统分析	程序系統分析
process system engineering	过程系统工程	程序系統工程
process technology	工艺技术	程序技術
process time lag	过程时间滞后	程序時延
process unit	工艺单元	程序單元
process upset	过程失稳	程序失穩
process vacuum	工艺用真空[系统]	程序真空[系統]
process variability	过程变异性	程序變異性
process variable	过程变量	程序變數
process water	生产用水	製程用水
producer	煤气发生炉	發生爐
producer gas	发生炉煤气	發生爐氣
product cost	产品成本	產品成本
product design	产品设计	產品設計
product development	产品开发	產品開發
product distribution	产物分布	產物分佈
product distribution distortion	产品分布变异	產品分佈變異
product engineering	产品工程	產品工程
product inhibition	产物抑制	產物抑制[作用]
production	①生产 ②产量	①生產 ②產量
production capacity	生产能力	生產能力
production control	生产控制	生產管控
production cost	生产成本	生產成本
production data	生产数据	生產數據
production line	生产线	生產線
production rate	生产率	生產速率
production schedule	生产进度表	生產時程表

英　文　名	大　陆　名	台　湾　名
productivity	生产力	生產力
product planning	产品规划	產品規畫
product quality	产品质量	產品品質
product recovery	产品回收	產物回收
product separation	产品分离	產物分離
product separator	产品分离器	產物分離器
product shipment check list	产品船运清单	產品運送清單
product specification	产品规格	產品規格
product storage	产品储存	產品儲存
product stream	产品物流	產物流
profile	剖面[图]	剖面[圖]
profit	利润	利潤
profitability	可获利程度	獲利能力
profitability factor	利润因子	利潤因數
profitability index	利润指数	利潤指數
profit function	利润函数	利潤函數
profit margin	利润限度	利潤率
program evaluation and review technique	①计划评审法 ②统筹法	計畫評審法
programmable calculator	可编程计算器	可程式計算器
programmable controller	可编程调制器	可程式控制器
programmable logic control	可编程逻辑控制	可程式邏輯控制
programmed control	程序控制	程式化控制
programmed controller	程序控制器	程式化控制器
programming	编程	规畫
progressive conversion model	渐进转化模型	漸進轉化模型
progressive freezing	逐步冷冻法	漸進冷凍
project	项目	專案,計畫
project engineer	设计主管工程师	專案工程師
project engineering	项目工程	專案工程
promoted catalyst	促进化催化剂	促進化觸媒
promoter	助催化剂	促進劑
propagation	传播	傳播
propagation of chain reaction	链增长反应	連鎖反應傳播,鏈[鎖]反應傳播
propagation of error	误差传递	誤差傳播
propagation reaction	链增长反应	增長反應
propagator	传播函数	傳播函數
propellant	推进剂	推進劑

英 文 名	大 陆 名	台 湾 名
propeller	螺旋桨	螺旋槳,推进器
propeller agitator	螺旋桨式搅拌器	螺旋槳攪拌器
propeller impeller	螺旋桨式叶轮	螺旋槳式葉輪
propeller type fan	螺旋桨风扇	螺旋槳型風扇
property change	性质变化	性質變化
property relationship	性质关联式	性質關係[式]
property tax	财产税	財產稅
proportional action	比例作用	比例作用
proportional band	比例度	比例帶
proportional band adjustment	比例带调整	比例帶調整
proportional control	比例控制	比例控制
proportional controller	比例控制器	比例控制器
proportional counter	正比计数器	比例計數器
proportional-derivative action	比例–微分作用	比例–微分作用
proportional-derivative control	比例–微分控制	比例–微分控制
proportional element	比例元件	比例元件
proportional gain	比例增益	比例增益
proportional-integral action	比例–积分作用	比例–積分作用
proportional-integral control	比例–积分控制	比例–積分控制
proportional-integral-derivative action	比例–积分–微分作用	比例–積分–微分作用
proportional plus reset action	比例–重调作用	比例–重設作用
proportional plus reset plus rate action	比例–重调–比率作用	比例–重設–速率作用
proportional sensitivity	比例灵敏度	比例靈敏度
propylene oligomer	丙烯低聚物	丙烯寡聚物
propylene tetramer	四聚丙烯	四聚丙烯
propylene trimer	三聚丙烯	三聚丙烯
protease	蛋白酶	蛋白酶
protein	蛋白质	蛋白質
proteinase	蛋白酶	蛋白酶
protein engineering	蛋白质工程	蛋白質工程
protein fractionation	蛋白质分级	蛋白質份化
proteolysis	蛋白酶解	蛋白質水解
proteolytic enzyme	蛋白[水解]酶	蛋白[水解]酶
protic solvent	质子溶剂	質子性溶劑
protoplasm	原生质	原生質
protoplast	原生质体	原生質體
prototype	原型	原型,雛型
prototype experiment	原型试验	原型實驗,雛型實驗

英　文　名	大　陆　名	台　湾　名
protozoa	原生动物	原生動物
protruded corrugated sheet packing	压延孔板波纹填料	壓延孔板波紋填料
Prussian blue	普鲁士蓝	普魯士藍
PSA (=pressure swing adsorption)	变压吸附	變壓吸附
pseudocritical constant	假临界常数	擬臨界常數
pseudocritical method	假临界点法	擬臨界點法
pseudocritical pressure	假临界压力	擬臨界壓[力]
pseudocritical temperature	假临界温度	擬臨界溫度
pseudo-homogeneous model	拟均相模型	擬均相模型
pseudokinetics	假动力学	擬動力學
pseudomechanism	虚拟机理	虛擬機構
pseudo-parameter	虚拟系数	虛擬參數
pseudo plastic fluid	假塑性流体	擬塑性流體
pseudo plasticity	假塑性	假塑性,擬塑性
pseudo reaction	虚拟反应	虛擬反應
pseudo steady state	假稳态	擬穩態
pseudo variable	虚拟变量	虛擬變量
p-styrene sulfonic acid	对苯乙烯磺酸	對苯乙烯磺酸
psychrometer	干湿表	[乾濕球]濕度計
psychrometric chart	湿度图	濕度圖
psychrometric difference	湿度差	乾濕球溫度差
psychrometric line	湿度线	濕度線
psychrometric ratio	干湿比	濕度係數
psychrometry	湿度测定法	濕度測定法
psychrophile	嗜冷微生物	嗜冷菌
psychrophilic bacteria	嗜冷菌	嗜冷菌
pug mill	揉捏机	捏泥機
pulp	纸浆	紙漿
pulp beater	打浆机	打漿機
pulping	制浆	製漿
pulping process	制浆法	製漿法
pulsating fluidized bed	脉动流化床	脈動流[體]化床
pulsation	脉动	脈動
pulsation damper	脉动阻尼器	脈動阻尼器
pulsation velocity	脉动速度	脈動速度
pulse	脉冲	脈衝
pulse disturbance	脉冲扰动	脈衝擾動
pulsed sieve plate column	脉冲筛板塔	脈衝篩板塔

英　文　名	大　陆　名	台　湾　名
pulse flow technique	脉冲流技术	脈衝流動法
pulse function	脉冲函数	脈衝函數
pulse input	脉动输入	脈動輸入
pulse method	脉冲法	脈衝法
pulse reactor	脉冲反应器	脈衝反應器
pulse response	脉冲响应	脈衝響應
pulsing flow	脉冲流	脈衝流動
pulverization	粉碎	粉碎
pulverized coal	粉煤	粉煤
pulverized fuel	粉状燃料	粉狀燃料
pulverizer	粉磨机	粉碎機
pulverizing mill	粉磨机	粉碎機
pumice	浮石	浮石
pump	泵	泵
pump characteristics	泵特性	泵特性
pump efficiency	泵效率	泵效率
pump horsepower	泵功率	泵馬力
pump impeller	泵轮	泵葉輪
pumping loss	泵送损耗	泵抽損失
pumping work	泵送功	泵功
pump pressure drop	泵压降	泵壓[力]降
pump rotor	泵转子	泵轉子
pump series	串联泵组	串聯泵組
pump shaft	泵轴	泵軸
pure component	纯组分	純成分
purge gas	吹扫气体	吹掃氣,排放氣
purification	提纯	純化
purification process	净化过程	純化程序
purified water	净化水	純化水
purity	纯度	純度
puzzolana cement	火山灰水泥	火山灰水泥
pycnometer	比重瓶	比重瓶
pyrazolone dye	吡唑啉酮染料	吡唑啈染料
pyrethrin	除虫菊酯	除蟲菊酯
pyrite	黄铁矿	黃鐵礦
pyrogenic decomposition	高温分解	高溫分解,熱解
pyrolusite	软锰矿	軟錳礦
pyrolysis	热解	熱[裂]解

英　文　名	大　陆　名	台　湾　名
pyrolysis furnace	热解炉	熱[裂]解爐
pyrolysis oil	热解油	熱解油
pyrolysis process	热解过程	熱[裂]解程序
pyrometallurgical process	火法冶金过程	火法冶金法,高溫冶金法
pyrometer	高温计	高溫計
pyrophyllite	叶蜡石	葉蠟石
pyrrhotite	磁黄铁矿	磁黃鐵礦

Q

英　文　名	大　陆　名	台　湾　名
q-line	进料状态线	進料狀態線,q線
quadratic lag	二次滞后	二次滯延
qualitative analysis	定性分析	定性分析
quality control	质量控制	品質管制
quality specification	质量规格	品質規格
quantimet	图像分析仪	圖像分析儀
quantitative analysis	定量分析	定量分析
quantitative flow diagram	物量流程图	計量流程圖
quantity meter	[累计]总量表	總量表
quantum effect	量子效应	量子效應
quartz	石英	石英
quartz glass	石英玻璃	石英玻璃
quartzite	石英岩	石英石
quartz tube	石英管	石英管
quasi-chemical approximation	准化学近似	準化學近似
quasi-chemical solution model	准化学溶液模型	準化學溶液模型
quasi-linearization	拟线性化	準線性化
quasi-static process	准静态过程	準靜態過程
quasi-static thermogravimetry	准静态热重量分析法	似靜態熱重量法
quasi-stationary state	准静态	準靜態
quasi-steady state	拟稳态	似穩態
quenched system	急冷系统	[已]抑止系統
quenching	淬火	淬火,驟冷
quenching oil	淬火油	淬火油,驟冷油
quenching water	淬火水	淬火水,驟冷水
quick lime	生石灰	生石灰

英 文 名	大 陆 名	台 湾 名
quick-opening valve	速启阀	速啟閥
quiescent bed	静止床	靜止床

R

英 文 名	大 陆 名	台 湾 名
racemic mixture	外消旋混合物	[外]消旋混合物
racemization	外消旋化	[外]消旋反應
radial bearing	径向轴承	徑向軸承
radial dispersion	径向分散	徑向分散
radial dispersion coefficient	径向分散系数	徑向分散係數
radial distribution function	径向分布函数	徑向分佈函數
radial flow reactor	径向反应器	徑向流反應器
radial flow turbine	辐流式汽轮机	徑向流渦輪
radial grooved filter plate	辐射沟槽滤板	輻射溝濾板
radial mixing	径向混合	徑向混合
radial stress	径向应力	徑向應力
radial velocity	径向速度	徑向速度
radiant beam	辐射束	輻射束
radiant emissivity	辐射系数	輻射發射係數
radiant energy	辐射能	輻射能
radiant energy emission	辐射能发射	輻射能發射
radiation	辐射	輻射
radiation bonnet	辐射帽	輻射帽
radiation chemistry	辐射化学	輻射化學
radiation constant	辐射常数	輻射常數
radiation damage	辐射损伤	輻射損害
radiation density	辐射密度	輻射密度
radiation dose	辐射剂量	輻射劑量
radiation dosimeter	辐射剂量计	輻射劑量計
radiation energy	辐射能	輻射能
radiation hazard	辐射危害	輻射危害
radiation heat	辐射热	輻射熱
radiation heat transfer	辐射传热	輻射熱傳
radiation intensity	辐射强度	輻射強度
radiation law	辐射定律	輻射定律
radiation level	辐射能级	輻射級位
radiation loss	辐射损耗	輻射損失

英　文　名	大　陆　名	台　湾　名
radiation meter	辐射测量仪	輻射計
radiation of heat	热辐射	熱輻射
radiation penetration	辐射穿透	輻射穿透
radiation pressure	辐射压[强]	輻射壓力
radiation protection	辐射防护	輻射防護
radiation pyrometer	辐射高温计	輻射高溫計
radiation safety	辐射安全	輻射安全
radiation screen	辐射隔屏	輻射屏
radiation section	辐射段	輻射段
radiation shield	辐射屏蔽	輻射屏蔽
radiation sterilization	辐射消毒	輻射滅菌
radiator	①辐射器 ②散热器	①輻射器 ②散熱器
radical scavenger	自由基捕获剂	自由基捕獲劑
radioactive contamination	放射性污染	放射性污染
radioactive decay	放射性衰变	放射性衰變
radioactive disintegration	放射性蜕变	放射性蛻變
radioactive dust	放射性尘	放射[性]塵
radioactive element	放射性元素	放射元素
radioactive fallout	放射性沉降物	放射性落塵
radioactive fission	放射性裂变	放射性分裂
radioactive indicator	放射性指示剂	放射性指示劑
radioactive isotope	放射性同位素	放射性同位素
radioactive nucleus	放射性核	放射性核
radioactive nuclide	放射性核素	放射性核種
radioactive poison	放射性毒物	放射性毒
radioactive pollutant	放射性污染物	放射性污染物
radioactive source	放射源	放射[性]源
radioactive tracer	放射性示踪物	放射性示蹤劑
radioactive waste	放射性废物	放射性廢棄物
radioactivity	放射性	放射性
radiochemistry	放射化学	放射化學
raffinate	萃余物	萃餘物
raining solid reactor	淋粒反应器	淋粒反應器
rammer process	夯锤法	夯鎚法
ramp disturbance	斜坡扰动	斜坡擾動
ramp function	斜坡函数	斜坡函數
ramp response	斜坡响应	斜坡應答
random copolymer	无规共聚物	隨機共聚物

英　文　名	大　陆　名	台　湾　名
random degradation	无规降解	隨機降解
random diffusion	无规扩散	隨機擴散
random distribution	随机分布	隨機分佈
random disturbance	随机扰动	隨機擾動
random error	随机误差	隨機誤差
random fluctuation	随机波动	隨機起伏
random load change	随机负荷变化	隨機負載變化
randomness	随机性	隨機性
random number	随机数	隨機數
random packing	散堆填料	隨機填料
random process	随机过程	隨機程序
random sampling	随机抽样	隨機取樣
random search	随机搜索	隨機搜尋
random signal	随机信号	隨機信號
random surface renewal	随机表面更新	隨機表面更新
random variable	随机变量	隨機變數
rangeability of control valve	控制阀量程	控制閥可調範圍
Rankine cycle	兰金循环	郎肯循環
Raoult law	拉乌尔定律	拉午耳定律
rapid filter	快速过滤器	快速過濾器
rapid sand filtration	快速砂滤速率	快速砂濾[法]
rate of evaporation	蒸发率	蒸發速率
rate of return on investment	投资收益率	投資收益率
ratio-flow control	流量比例调节	流量比控制
rational function	有理函数	有理函數
rational number	有理数	有理數
raw gas	原料气	原料氣
raw material	原料	原料
raw material check list	原料检查报表	原料檢查報表
raw material cost	原料成本	原料成本
raw material storage	原料储罐	原料儲存
raw sewage	原污水	原污水
raw sludge	原污泥	原污泥
raw wastewater	原废水	原廢水
raw water	原水	原水
Rayleigh number	瑞利数	瑞立數
Raymond mill	雷蒙磨	雷蒙磨
Rayon	人造丝	嫘縈

英　文　名	大　陆　名	台　湾　名
rayon pulp	人造丝浆	嫘縈紙漿
rayon staple	人造丝短纤维	嫘縈棉
rayon tow	人造丝束	嫘縈束
reachability matrix	可及矩阵	可達性矩陣
reactant	反应物	反應物
reactant ratio	反应物比例	反應物比例
reacting phase	反应相	反應相
reaction	反应	反應
reaction affinity	反应亲和力	反應親和力
reaction chamber	反应室	反應室
reaction coordinate	反应坐标	反應坐標
reaction curve	反应曲线	反應曲線
reaction curve method	反应曲线法	反應曲線法
reaction cycle	反应循环	反應循環
reaction cycle time	反应周期	反應週期
reaction diagram	反应图	反應圖
reaction diffusion network	反应扩散网络	反應擴散網[路]
reaction engineering	反应工程	反應工程
reaction kettle	反应釜	反應釜
reaction kinetics	反应动力学	反應動力學
reaction mechanism	反应机理	反應機構
reaction network	反应网络	反應網路
reaction network analysis	反应网络分析	反應網路分析
reaction order	反应级数	反應級數
reaction pair	反应对	反應對
reaction path	反应途径	反應途徑
reaction probability	反应概率	反應機率
reaction process	反应过程	反應程序
reaction progress variable	反应过程变量	反應進展變數
reaction rate	反应速率	反應速率
reaction rate constant	反应速率常数	反應速率常數
reaction rate theory	反应速率理论	反應速率理論
reaction sensitivity	反应灵敏度	反應靈敏度
reaction specificity	反应专一性	反應特異性,反應特定性
reaction step	反应步骤	反應步驟
reaction system	反应系统	反應系統
reaction temperature	反应温度	反應溫度

英　文　名	大　陆　名	台　湾　名
reaction time	反应时间	反應時間
reaction tower	反应塔	反應塔
reaction-type distributor	反应式分配器	作用式分配器
reaction velocity	反应速度	反應速度
reaction zone	反应区	反應區
reactivation	复活	再活化
reactive center	反应中心	反應中心
reactive distillation	反应蒸馏	反應[性]蒸餾
reactive extraction	反应萃取	反應萃取
reactive ion etching	反应离子蚀刻	反應離子蝕刻[法]
reactive process	反应过程	反應程序
reactive sputtering etching	反应溅射蚀刻法	反應濺射蝕刻[法]
reactivity	反应性	反應性
reactor	反应器	反應器
reactor analysis	反应器分析	反應器分析
reactor capacity	反应器规模	反應器容量
reactor configuration	反应器构型	反應器構型
reactor control	反应器控制	反應器控制
reactor core	堆芯活性区	[核]反應器爐心
reactor design	反应器设计	反應器設計
reactor dynamics	反应器动态学	反應器動力學
reactor engineering	反应器工程	反應器工程
reactor kinetics	反应器动力学	反應器動力學
reactor model	反应器模型	反應器模式
reactor network	反应器网络	反應器網路
reactor operation	反应器操作	反應器操作
reactor optimization	反应器优化	反應器最適化
reactor parameter	反应器参数	反應器參數
reactor performance	反应器操作特性	反應器性能
reactor poisoning	反应器(催化剂)中毒	反應器中毒
reactor safety	反应器安全	反應器安全性
reactor setup	反应器装配	反應器設置
reactor stability	反应器稳定性	反應器穩定性
reactor system	反应器系统	反應器系統
reactor theory	反应器理论	反應器理論
reactor train	反应器组	反應器組
ready-mixed paint	调合漆	調和漆
reaeration	再曝气	再曝氣

英　文　名	大　陆　名	台　湾　名
reaggregation	再聚集	再聚集[作用]
real composition	真实组成	真實組成
real gas	真实气体	真實氣體
realizability	可实现性	可實現性
real stage	实际阶段	實際階
rear bearing	后轴承	後軸承
reboiler	再沸器,重沸器	再沸器
reboiler condenser	再沸冷凝器	再沸冷凝器
reboiler dynamics	再沸器动力学	再沸器動力學
recalcination	再锻烧	再燃燒[反應]
recalcining process	再锻烧过程	再煅燒程序
receptor	受体	受體,受器
recessed plate	凹槽式滤板	凹槽式濾板,帶框濾板
reciprocal plot	倒数标绘	倒數圖
reciprocal relation	倒数关系	倒數關係
reciprocal rule	倒数法则	倒數法則
reciprocating	往复式	往復運動
reciprocating compressor	往复式压缩机	往復[式]壓縮機
reciprocating feeder	往复加料器	往復進料器
reciprocating piston compressor	往复式活塞压缩机	往復式活塞壓縮機
reciprocating plate column	往复板式塔	往復板萃取塔
reciprocating plate extraction column	往复板式萃取塔	往復板萃取塔
reciprocating pump	往复泵	往復泵
reciprocating rake	往复耙	往復耙
reciprocating reactor	往复循环反应器	往復反應器
reciprocating screen	往复筛	往復篩
reciprocating vacuum pump	往复式真空泵	往復式真空泵
recirculating pump	往复泵	循環泵
recirculating ratio	循环比	循環比
recirculation	再循环	循環
recirculation factor	循环因子	循環因數
recirculation flow	循环流动	循環流動
recirculation pattern	循环模式	循環模式
recirculation ratio	循环比	循環比
recirculation reactor	循环反应器	循環反應器
reclaimed gypsum	回收石膏	再生石膏
reclaimed materials	再生材料	再生材料
reclaimed rubber	再生胶	再生橡膠

英 文 名	大 陆 名	台 湾 名
reclaiming tank	回收槽	回收槽
reclaiming unit	回收装置	回收裝置
reclamation	回收	回收
recombinant DNA	重组 DNA	重組 DNA
recombination	重组	重組
recorder	记录仪	記錄器
recorder-controller	记录控制器	記錄控制器
recording barometer	记录式气压计	記錄氣壓計
recording chart	记录图表	記錄圖表
recording equipment	记录设备	記錄裝置
recording materials	记录材料	記錄材料
recoverable stress	可恢复应力	可復應力
recovery	回收[率]	回收
recovery curve	回收率曲线	回收曲線
recovery factor	回收系数	回收因數
recovery system	回收系统	回收系統
recovery time	恢复时间	回收時間
recovery tower	回收塔	回收塔
recrystallization	再结晶	再結晶
rectangular coordinate	直角坐标	直角坐標
rectangular pulse	矩形脉冲	矩形脈衝
rectangular weir	矩形堰	矩形堰
rectification	精馏	精餾
rectification section	精馏段	精餾段
rectifier	①精馏器 ②整流器	①精餾器 ②整流器
rectifier column	精馏塔	精餾塔
rectifying column	精馏塔	精餾塔
rectifying plate	精馏塔板	精餾板
rectifying section	精馏段	精餾段
rectifying still	精馏釜	精餾釜
rectifying tower	精馏塔	精餾塔
rectifying tray	精馏板	精餾板
recuperator	蓄热器	復熱器
recursion	递归	遞回
recursion formula	递推公式	遞回公式
recursive function	递归函数	遞回函數
recursive relation	递推关系式	遞回關係[式]
recycle	循环	循環

英　文　名	大　陆　名	台　湾　名
recycle feedstock	循环料	回收[进]料
recycle oil	循环油	循環油
recycle process	循环过程	循環程序
recycle ratio	循环比	循環比
recycle stream	循环物流	循環流
recycle synthesis gas	循环合成气	循環合成氣
recycle system	循环系统	循環系統
recycling	循环	循環
redeposition	再沉积	再沈積
red lead	红铅	紅鉛,紅丹
Redlich-Kwong equation (RK equation)	RK 方程	RK 方程
redox potential	氧化还原电位	氧化還原電位
redox reaction	氧化还原反应	氧化還原反應
red phosphorus	赤磷	赤磷,紅磷
reduced coordinates	对比坐标	對比坐標
reduced density	对比密度	對比密度
reduced equation of state	对比状态方程	約化狀態方程式
reduced mass	约化质量	約化質量
reduced model	降阶模型	降階模式
reduced molecular weight	对比分子量	約化分子量
reduced pressure	对比压力	對比壓力
reduced property	对比性质	約化性質
reduced saturated vapor pressure	对比饱和蒸汽压	對比飽和蒸汽壓
reduced temperature	对比温度	對比溫度
reduced turbidity	对比浊度	對比濁度
reduced variable	对比变量	約化變數
reduced volume	对比体积	對比體積
reducer	异径管	漸縮管
reducer resistance	大小头阻力	漸縮管阻力
reducing agent	还原剂	還原劑
reducing coupling	缩径管接头	漸縮接頭
reducing elbow	异径弯头	漸縮彎頭
reducing fitting	异径管件	漸縮管件
reducing gear	减速齿轮	減速齒輪
reducing tee	异径三通管	漸縮三通管,漸縮 T 型管
reducing valve	减压阀	減壓閥
reductant	还原剂	還原劑

英　文　名	大　陆　名	台　湾　名
reduction	还原	還原［反應］
reduction potential	还原电位	還原電位
reduction ratio	破碎比	磨碎比
reduction reaction	还原反应	還原反應
reduction value	还原值	還原值
reduction zone	还原区	還原帶
reductive bleaching	还原性漂白	還原漂白
reductive dehalogenation	还原性脱卤	還原性脱鹵［反應］
redundancy	冗余［度］	冗餘［度］
re-entrainment	二次夹带	二次夾帶
reference condition	参比条件	參考條件
reference electrode	参比电极	參考電極
reference input	参考输入	參考輸入
reference junction	参比接点	參考接點
reference standard	参考标准	參考標準
reference temperature	参考温度	參考溫度
reference value	参考值	參考值
refill	再装满,回填	再裝滿
refinement	精炼	精煉
refiner	精制机	精製機
refinery	炼油厂	煉油廠
refinery gas	炼厂气	煉油氣
refining	精炼	精煉
reflectance spectrophotometer	反射分光光度计	反射分光光度針
reflectance spectroscopy	反射光谱法	反射光譜法
reflecting condenser	反射聚光器	反射聚光器
reflection spectrophotometer	反射分光光度计	反射分光光度針
reflectivity	反射率	反射率
reflectometer	反射计	反射計
reflux	回流	回流
reflux accumulator	回流罐	回流貯槽
reflux column	回流塔	回流塔
reflux condenser	回流冷凝器	回流冷凝器
reflux distillation	回流蒸馏	回流蒸餾
reflux drum	回流槽	回流槽
reflux operation	回流操作	回流操作
reflux ratio	回流比	回流比
reflux tube	回流管	回流管

英　文　名	大　陆　名	台　湾　名
reformed gasoline	重整汽油	重組汽油
reformer	重整器	重組器
reforming	重整	重組
reforming catalyst	重整催化剂	重組觸媒
reforming reaction	重整反应	重組反應
refraction temperature coefficient	折射温度系数	折射溫度係數
refractive index	折射率	折射率
refractometer	折射计	折射計
refractory brick	耐火砖	耐火磚
refractory porcelain	耐火瓷	耐火瓷
refractory shale	耐火页岩	耐火頁岩
refractory surface	耐火表面	耐火表面
refrigerant	冷冻剂	冷凍劑,冷媒
refrigerating coil	冷冻盘管	冷凍旋管
refrigerating fluid	冷冻液	冷凍流體
refrigerating ton	冷冻吨	冷凍噸
refrigeration	冷冻	冷凍
refrigeration capacity	冷冻能力	冷凍能力
refrigeration cycle	制冷循环	冷凍循環
refuse	矸石	廢棄物
refuse pit	废物坑	廢棄物坑
regain	回收	回收
regenerated cellulose	再生纤维素	再生纖維素
regenerated fiber	再生纤维	再生纖維
regenerated rubber	再生橡胶	再生橡膠
regeneration	再生	再生
regeneration furnace	再生炉	再生爐,蓄熱爐
regenerative cycle	再生循环	再生循環
regenerative heat	再生热	再生熱
regenerative heat exchanger	蓄热式换热器	蓄熱式熱交換器
regenerative system	再生系统	再生系統
regenerator	再生器	再生器
region	区[域]	區[域]
region of asymptotic stability	渐近稳定区	漸近穩定區
region of attraction	吸引力区[域]	吸引區
region of practical stability	实际稳定区	實用穩定區
regression	回归	回歸
regression analysis	回归分析	回歸分析

英　文　名	大　陆　名	台　湾　名
regression coefficient	回归系数	回歸係數
regression equation	回归方程	回歸方程式
regression line	回归线	回歸線
regular solution	正规溶液	正規溶液
regular solution theory	正规溶液理论	正規溶液理論
regulating box	调节箱	調節箱
regulator	①调节器 ②调节剂	①調節器 ②調節劑
regulatory control	调节控制	調節控制
reheater	再热器	再熱器
reinforced plastic	增强塑料	強化塑膠
reinforcing filler	增强填充料	增強填料
reinforcing materials	增强材料	增強材料
relative absorptivity	相对吸收率	相對吸收率
relative abundance	相对丰度	相對豐度
relative adsorptivity	相对吸附能力	相對吸附度
relative blackness	相对黑度	相對黑度
relative deviation	相对偏差	相對偏差
relative error	相对误差	相對誤差
relative gain	相对增益	相對增益
relative humidity	相对湿度	相對濕度
relative rate constant	相对速率常数	相對速率常數
relative roughness	相对粗糙度	相對粗糙度
relative saturation	相对饱和度	相對飽和度
relative stability	相对稳定性	相對穩定性
relative supersaturation	相对过饱和度	相對過飽和度
relative vapour pressure	相对蒸汽压	相對蒸氣壓
relative velocity	相对速度	相對速度
relative velocity factor	相对速度因子	相對速度因數
relative viscosity	相对黏度	相對黏度
relative volatility	相对挥发度	相對揮發度
relaxation method	松弛法,弛豫法	鬆弛法
relaxation modulus	弛豫模量	鬆弛模數
relaxation phenomenon	松弛现象	鬆弛現象
relaxation test	松弛试验	鬆弛試驗
relaxation time	弛豫时间	鬆弛時間
relaxed state	松弛状态	鬆弛狀態
relay	继电器	繼電器
relay valve	继动阀	繼動閥

英　文　名	大　陆　名	台　湾　名
release agent	脱模剂	脱模劑,離型劑
release valve	放气阀	洩氣閥
reliability	可靠性	可靠性
relief device	泄放装置	洩放裝置
relief paint	浮雕漆	浮雕漆
relief valve	调压阀	洩氣閥
remote control	遥控	遙控
renewable energy source	可再生能源	[可]再生能源
renewable resources	可再生资源	[可]再生資源
renormalization	再归一化	再歸一化,重整
reoxygenation	再充氧作用	再充氧[作用]
repellent	防水剂	撥水劑
repetitive operation	重复操作	重複操作
repetitive polymer	重复性聚合物	重複性聚合物
replacement	置换	置換
replacement cost	更换费用	置換成本
replacement evaluation	更换评估	置換評估
replacement value	更换价值	置換價值
repolymerization	再聚合	再聚合[反應]
report	报告	報告
reprocessing	后处理	再處理,再加工
reprocessing plant	[核燃料]后处理工厂	再處理廠
reproducibility	再现性	再現性
reproduction	复制	繁殖
reproduction factor	再现因素	繁殖因數
reproduction rate	繁殖率	繁殖速率
reproductive cycle	繁殖周期	繁殖週期
repulsive potential	推斥势	推斥勢
rerun column	再蒸馏塔	重餾塔
rerunning tower	再蒸馏塔	重餾塔
rerun yield	再蒸馏收率	重餾產率
research	研究	研究
reservoir	蓄水池	貯槽
reset	复位	重設
reset action	复位作用	重設作用
reset corner	复位角	重設角
reset rate	复位速率	重設速率
reset response	复位响应	重設應答

英　文　名	大　陆　名	台　湾　名
reset time	返位时间	重設時間,積分時間
reset windup	复位终结	重設繞緊
residence time	停留时间	滯留時間
residence time distribution	停留时间分布	滯留時間分佈
residual analysis	残差分析	殘差分析
residual contribution	残余贡献	剩餘貢獻
residual enthalpy	残余焓	剩餘焓
residual entropy	残余熵	剩餘熵
residual error	残差	殘餘誤差
residual fraction	残余馏分	殘留分率
residual gas	残余气体	殘留氣體
residual product	残油	殘留產物
residual property	残余性质	剩餘性質
residual strain	残余应变	殘留應變
residual term	残余项	殘留項
residual volume	残余体积	剩餘體積
residue	渣油	殘渣,殘留物
resilience	弹性	彈[性]能
resin	树脂	樹脂
resin soap	树脂皂	樹脂皂
resin tannage	树脂鞣法	樹脂鞣法
resistance	①阻力 ②电阻	①阻力 ②電阻
resistance coefficient	阻力系数	阻力係數
resistance furnace	电阻炉	電阻爐
resistance thermometer	电阻温度计	電阻溫度計
resolution	分辨率	解析度
resolving power	分辨能力	解析能力
resonance	共振	共振
resonance frequency	共振频率	共振頻率
resonance integral	共振积分	共振積分
resonance peak	共振峰	共振峰
resonant frequency	共振频率	共振頻率
resonant peak	共振峰	共振峰
resources cost	资源成本	資源成本
respiration	呼吸	呼吸
respiratory chain	呼吸链	呼吸鏈
response curve	响应曲线	應答曲線
response speed	响应速度	應答速率

英　文　名	大　陆　名	台　湾　名
response surface	响应面	應答面
response time	响应时间	應答時間
restrainer	抑制剂	抑制劑,制止器
resuspension	再悬浮	再懸浮
retardation factor	阻滞因子	阻滯因數
retardation time	推迟时间	阻滯時間
retarded motion	阻迟运动	阻滯運動
retarder	阻滞剂	阻滯劑
retarding force	减速力	減速力
retarding torque	减速扭矩	減速扭矩
retentate	渗余物	滲餘物
retention	截留,保留	滯留
retention period	保留周期	滯留期間
retention time	保留时间	滯留時間
retention volume	保留体积	滯留體積
retort	甑,干馏釜	甑,蒸餾罐
retort process	甑馏法	甑[蒸]餾法
retrogradation	逆反作用	[澱粉]老化,衰退[作用]
retrograde condensation	逆反冷凝	降壓冷凝
return action	反回作用	返回動作
return bend	回弯头	回彎管
return on investment	投资收益率	投資報酬率
return rate	返回率	報酬率
return trap	回流阱	回流阱
return tube	溢流管	回流管
reverse current	反向电流	反向電流,逆流
reverse diffusion	反扩散	逆擴散
reversed micelle	可逆胶囊	逆微胞
reverse extraction	反萃取	逆萃取
reverse flow	反流	逆向流動
reverse fractionation	逆分馏法	逆相分餾[法]
reverse micelle extraction	反胶团萃取	逆微胞萃取
reverse osmosis	反渗透	逆滲透
reverse reaction	逆反应	逆反應
reverse-running pump	反向泵	反向泵
reversibility	可逆性	①可逆度 ②可逆性
reversible adiabatic flow	可逆绝热流动	可逆絕熱流動

英 文 名	大 陆 名	台 湾 名
reversible cell	可逆电池	可逆電池
reversible change	可逆变化	可逆變化
reversible chemical reaction	可逆化学反应	可逆化學反應
reversible decomposition potential	可逆分解电位	可逆分解電位
reversible diffusion	可逆扩散	可逆擴散
reversible mechanism	可逆机理	可逆機構
reversible precipitation	可逆沉淀	可逆沈澱
reversible process	可逆过程	可逆程序
reversible reaction	可逆反应	可逆反應
reversible work	可逆功	可逆功
revolution per minute(rpm)	每分钟转数	分鐘轉數
revolving screen	回转筛	迴轉篩
Reynolds number	雷诺数	雷諾數
rheological property	流变性质	流變性質
rheology	流变学	流變學
rheometer	流变仪	流變儀
rheometry	流变测量学	流變測定法
rheopectic fluid	震凝性流体	搖變增黏流體
ribbon mixer	螺带混合机	螺條混合機
ribonucleic acid(RNA)	核糖核酸	核糖核酸
ribose	核糖	核糖
rice bran oil	米糠油	米糠油
rich phase	富相	富相
rich solution	浓溶液	濃溶液
Rideal mechanism	里迪尔机理	里迪爾機構
rigidity	刚度	剛性
rigidity modulus	刚性模数	剛性模數
rigorous method	严格法	嚴格法
ringing pole	振铃极点	振鈴極點
ring joint	环形接头	環形接頭
ring piezometer	环形压强计	環形測壓計
ring roll mill	环滚磨机	環輥磨
ring roll pulverizer	环滚粉碎机	環輥粉碎機
ring sprayer	环形喷雾器	環狀噴霧器
ring type grinder	环形研磨机	環形研磨機
ring type manometer	环形压强计	環形壓力計
ripple tray	波纹塔板	波紋塔板
rippling	波动的	梳麻

英　文　名	大　陆　名	台　湾　名
riser	提升管	上升管,竖板
riser reactor	提升管反应器	上升管反應器
risk	风险	風險
risk analysis	风险分析	風險分析
risk earning rate	风险收益率	風險收益率
risk factor	风险因子	風險因素
risk function	风险函数	風險函數
ristocetin	瑞斯托菌素	瑞斯特黴素
RK equation (= Redlich-Kwong equation)	RK 方程	RK 方程
RNA (= ribonucleic acid)	核糖核酸	核糖核酸
roaster	焙烧炉	焙燒爐
robot	机器人	機器人
robustness	①鲁棒性 ②稳健性	韌性
robust process control	鲁棒过程控制	韌性程序控制
robust stabilizability	鲁棒稳定性	韌性穩定化度
robust stabilization	鲁棒稳定化	韌性穩定化
rocket	火箭	火箭
rocket engine	火箭发动机	火箭引擎
rocket fuel	火箭燃料	火箭燃料
rock sugar	冰糖	冰糖
rodenticide	杀鼠剂	殺鼠劑
rod mill	棒磨机	桿磨
roll crusher	辊式破碎机	輥[壓]碎機
roller	滚柱	輥,滚筒
roller bearing	滚轴支座	滚筒軸承,輥軸承
roller mill	①开[放式]炼胶机 ②辊[式研]磨机	輥磨機
roller mill press	辊压机	輥壓機
roller printing machine	滚筒印花机	滚筒印花機
rolling mill	[辊]轧机	輥軋機
root locus	根轨迹	根軌跡
root locus angle	根轨迹角	根軌跡角
root locus method	根轨迹法	根軌跡法
root locus plot	根轨迹图	根軌跡圖
root locus technique	根轨迹方法	根軌跡法
root-mean-square deviation	均方根偏差	均方根偏差
root-mean-square error	均方根误差	均方根誤差
root-mean-square fluctuating velocity	均方根振荡速度	均方根波動速度

英 文 名	大 陆 名	台 湾 名
Roots blower	罗茨鼓风机	鲁氏鼓風機
rosin	松香	松香, 松脂
rosined soap	松香皂	松香皂
rosin oil	松香油	松香油
rosin size	松香胶	松香膠料
rosin sizing	松香施胶	松香上膠
rotameter	转子流量计	浮子流量計
rotary blower	回转鼓风机	旋轉鼓風機
rotary compressor	回转压缩机	旋轉壓縮機
rotary continuous filter	旋转连续过滤机	連續旋濾機
rotary cooler	旋转冷却器	旋轉冷卻器
rotary crusher	旋转压碎机	旋轉壓碎機
rotary crystallizer	旋转结晶器	旋轉結晶器
rotary disc filter	转盘过滤机	旋盤濾機
rotary disc meter	转盘流量计	轉盤流量計
rotary displacement pump	旋转排代泵	旋轉排量泵
rotary distributor	旋转分配器	旋轉分配器
rotary drum filter	转鼓过滤机	旋桶濾機
rotary dryer	回转干燥器	旋轉[式]乾燥器
rotary efficiency	旋转效率	旋轉效率
rotary feeder	旋转进料器	旋轉飼機
rotary filter	旋转过滤机	旋濾機
rotary filter press	旋转压滤机	旋轉壓濾機
rotary kiln	回转窑	旋[轉]窯
rotary pump	回转泵	旋轉泵
rotary roaster	回转焙烧炉	旋轉焙燒爐
rotary screen	旋转筛	旋轉篩
rotary shaker	旋转摇床	旋轉搖動器
rotary sieve	旋转筛	旋轉篩
rotary sieve shaker	旋转筛摇床	旋轉搖篩器
rotary vacuum drum filter	转筒真空过滤机	真空旋桶濾機
rotary vacuum pump	旋转真空泵	旋轉真空泵
rotary valve	旋转阀	旋轉閥
rotary viscometer	旋转黏度计	旋轉黏度計
rotary washer	旋转洗涤机	旋轉洗滌機
rotating-basket reactor	旋筐反应器	旋籃反應器
rotating disc	转盘	旋轉盤
rotating disc contactor	转盘塔	轉盤塔

英　文　名	大　陆　名	台　湾　名
rotating-disc process	生物转盘法	[生物]旋轉盤法
rotating-drum absorber	转筒吸收器	轉桶吸收器
rotating drum dryer	滚筒干燥器	旋桶乾燥器
rotating extractor	旋转萃取器	旋轉萃取器
rotating sprinkler	旋转式喷头	旋轉噴灑器
rotational energy	转动能	旋轉能
rotational energy level	转动能级	旋轉能階
rotational flow	旋转流	旋轉流動
rotational partition function	转动配分函数	轉動配分函數
rotation viscometer	旋转黏度计	旋轉黏度計
rotatory power	旋光能力	旋光能力
rotenone	鱼藤酮	魚藤酮
rotifer	轮形动物	輪蟲
rotor	转子	轉子
rotor blade	转子叶片	轉子葉片
roughing bed	粗滤床	粗濾床
roughing filter	粗滤床	粗濾床
roughness	粗糙度	粗糙度
roughness factor	粗糙因子	粗糙因數
roughness scale	粗糙标度	粗糙標度
rounded orifice	圆孔口	圆孔口
route	途径	途徑
royalty	专利权使用费	權利金
rpm(= revolution per minute）	每分钟转数	每分轉數
rubber	橡胶	橡膠
rubber hose	胶管	橡皮軟管
rubber latex	橡胶胶乳	橡膠乳膠
rubber-like liquid	橡胶态液体	橡膠態液體
rubber-lined pipe	橡胶衬里管	橡膠襯管
rubber mill	橡胶磨	橡膠磨碾機
rubber tube	橡皮管	橡皮管
rubbery state	橡胶状态	橡膠狀態
run	运转	運轉,操作
runaway	失控	失控
run-down tank	溢流槽	接收槽
runner	①浇道 ②碾碎机	澆道
rupture stress	断裂应力	斷裂應力
rust	锈	鏽

英 文 名	大 陆 名	台 湾 名
rust remover	除锈剂	除鏽劑
rutile	金红石	金紅石

S

英 文 名	大 陆 名	台 湾 名
saccharide	糖类	醣
saccharification	糖化作用	糖化[作用]
saccharifying agent	糖化剂	糖化劑
saccharimeter	糖量计	糖量計
saccharin	糖精	糖精
saccharomyces	酵母菌属	酵母菌
saccharose	蔗糖	蔗糖
saddle azeotrope	鞍点共沸物	鞍點共沸液
safety check list	安全检查清单	安全檢查報表
safety device	安全装置	安全裝置
safety door	安全门	安全門
safety explosive	安全炸药	安全炸藥
safety factor	安全系数	安全因數
safety glass	安全玻璃	安全玻璃
safety margin	安全裕度	安全裕量
safety-relief area	安全泄压面积	安全釋放面積
safety-relief valve	安全泄压阀	安全洩壓閥
safety testing	安全性试验	安全性試驗法
safety valve	安全阀	安全閥
safety vent	安全放空	安全排氣孔
safe working pressure	安全工作压强	安全操作壓力
safe working stress	安全工作应力	安全工作應力
sag flow	淌流	淌流
sag flow board	淌流板	淌流板
salicylic acid	水杨酸	水楊酸,柳酸
salmon oil	鲑鱼油	鮭魚油
salt	盐	鹽
salt bridge	盐桥	鹽橋
salt catcher	盐捕集器	鹽接受器
salted hide	盐皮	鹽[醃]皮
salt effect	盐效应	鹽效應
salt-fog resistance	防盐雾性	抗鹽霧性

英　文　名	大　陆　名	台　湾　名
salt-free polyelectrolyte solution	无盐聚电解质溶液	無鹽高分子電解質溶液
salt glaze	盐釉	鹽釉
salting	腌制	鹽漬
salting in	盐溶	鹽溶
salting out	盐析	鹽析
salting out agent	盐析剂	鹽析劑
saltpeter	硝石	硝石
salvage value	残值	殘餘價值
sampled data	采样数据	採樣數據
sampled data control	采样数据控制	採樣數據控制
sampled data control system	采样数据控制系统	採樣數據控制系統
sampled data signal	采样数据信号	採樣數據信號
sampled data system	采样数据系统	採樣數據系統
sample system	样品系统	樣品系統
sampling	①取样 ②采样	採樣
sampling delay	取样滞后	採樣延遲
sampling error	取样误差	取樣誤差
sampling system	采样系统	採樣系統
sampling tube	取样管	取樣管
sandal oil	檀香油	檀香油
sandarac	桧树胶	檜樹膠
sand-bed filter	砂滤器	砂濾器
sand-blast	喷砂	噴砂
sand filtration	砂滤法	砂濾[法]
sand mill	砂磨	砂磨
sand paper	砂纸	砂紙
sandwich structure	夹层结构	夾層結構
sanitary paper	卫生纸	衛生紙
sanitary wastewater	生活污水	生活污水
SAN（＝styrene-acrylonitrile）	苯乙烯－丙烯腈[共聚物]	苯乙烯丙烯腈
saponification	皂化	皂化[作用]
saponification number	皂化值	皂化值
saponified acetate rayon	皂化乙酸人造丝	皂化乙酸嫘縈
saponified cellulose acetate	皂化纤维素乙酸酯	皂化乙酸纖維素
saponified polymer	皂化聚合物	皂化聚合物
saponifying agent	皂化剂	皂化劑
sapphire	蓝宝石	藍寶石

英 文 名	大 陆 名	台 湾 名
saran	莎纶	紗綸,聚偏二氯乙烯
saran pipe	莎纶管	紗綸管
sardine oil	沙丁鱼油	沙丁魚油
SAS（=small angle scattering）	小角散射	小角[度]散射
satellite plant	卫星工厂	衛星工廠
satin weave	纤维编织	緞紋組織
saturated air	饱和空气	飽和空氣
saturated humidity	饱和湿度	飽和濕度
saturated hydrocarbon polymer	饱和烃聚合物	飽和碳氫聚合物
saturated liquid	饱和液体	飽和液
saturated paper	防潮纸	油紙
saturated polyester	饱和聚酯	飽和聚酯
saturated solution	饱和溶液	飽和溶液
saturated state	饱和态	飽和態
saturated steam	饱和水蒸气	飽和蒸汽
saturated vapor	饱和蒸汽	飽和蒸氣
saturated vapor pressure	饱和蒸汽压	飽和蒸氣壓
saturating paper	浸渍纸	吸油紙
saturation	饱和	飽和
saturation curve	饱和曲线	飽和曲線
saturation limit	饱和极限	飽和極限
saturation pressure	饱和压力	飽和壓力
saturation temperature	饱和温度	飽和溫度
saturator	饱和器	飽和器
SAXD（=small angle X-ray diffraction）	小角 X 射线衍射	小角度 X 射線繞射
SAXS（=small angle X-ray scattering）	小角 X 射线散射	小角度 X 射線散射
scalar	标量	純量
scale	污垢	[污]垢
scale down	缩小	等比縮小
scale factor	标度因子	比例因數
scale fiber	有鳞[片]纤维[黏胶]	有鱗[片]纖維[黏膠]
scale formation	结垢形成	積垢
scale indicator	标度指示器	標度指示器
scale up	放大	等比增大
scale-up problem	放大问题	放大問題
scale-up rule	放大法则	放大法則
scaling factor for heat transfer	传热污垢因子	熱傳積垢因數
scaling law	标度律	標度律

英　文　名	大　陆　名	台　湾　名
scanning calorimetry	扫描量热法	掃描量熱法
scanning electron microscope（SEM）	扫描电子显微镜	掃描電子顯微鏡
scanning electron microscopy	扫描电子显微镜学	掃描電子顯微法
scanning rate	扫描速率	掃描速度
scarf joint	嵌接	嵌接頭
scattering	散射	散射
scattering angle	散射角	散射角
scattering cell	散射池	散射池
scattering coefficient	散射系数	散射係數
scattering of light	光散射	光散射
scattering pattern	散射形式	散射圖樣
scattering volume	散射体积	散射體積
scavenger	①净化剂 ②捕获剂	①去除劑 ②捕獲劑
schedule number	管壁厚度系列号	管壁厚度系列號
schematic diagram	示意图	示意圖
Schiff base	席夫碱	希夫鹼
Schiff base polymer	席夫碱聚合物	希夫鹼聚合物
Schmidt number	施密特数	施密特數
Schob elastometer	肖伯弹力试验机	肖伯彈性計
Schob resilience	肖伯回弹性	肖伯回彈性
schonite	软钾镁矾	硫酸鉀鎂礦
Schopper tensile tester	朔佩尔张力试验机	肖珀張力試驗機
Schulz-Zimm distribution	舒尔茨－齐姆分布	舒爾茨－齊姆分佈
scission	断键	斷裂
SCLC（＝space- charge-limited current）	空间电荷限制电流	空間電荷限制電流
sclerometer	硬度计	硬度計
scope	范围	範圍
scorching	焦烧	焦化［現象］
scorching test	焦烧试验	焦化試驗
scorch-resisting treatment	抗焦处理	防焦處理
scorch retarder	防焦剂	防焦劑
Scott flexometer	斯科特挠度计	斯科特撓度計
scouring	洗毛	洗滌
scraper	刮板	刮刀
scraper chiller	刮刀式冷却器	刮刀冷凍器
scraper conveyor	刮板输送机	刮運機
scraper heat exchanger	刮板式换热器	刮刀熱交換器
scrap rubber	废橡胶	廢橡膠

英　文　名	大　陆　名	台　湾　名
scratch	刮痕	刮痕
scratch hardness	刮痕硬度	刮痕硬度
scratch resistance	抗刮性	抗刮性
screen	筛网	篩網
screen analysis	筛析	篩析
screen aperture	筛孔	篩孔
screen bank	筛组	篩組
screen changer	过滤网调换装置	換濾網裝置
screen conveyor	筛网输送机	篩運機
screen effectiveness	屏蔽有效度	篩有效度
screen filter	筛滤器	篩濾機
screening	筛选	篩選
screening device	筛选装置	篩選裝置
screening efficiency	筛选效率	篩選效率
screening test	筛选测验	篩選試驗
screening trommel	转筒筛	轉筒篩
screen mill	筛磨	篩磨機
screen opening	筛孔	篩孔
screen pack	过滤网组[合]	篩組件
screen plate	筛板	篩板
screw	螺杆	螺桿
screw core pin	螺杆芯孔销	螺桿蕊栓
screw dislocation	螺型位错	螺旋差排
screw efficiency	螺杆效率	螺桿效率
screw extractor	螺杆出料机	拔螺釘器
screw extruder	螺杆压出机	螺桿擠出機
screw extruder reactor	螺杆挤出反应器	螺桿擠出反應器
screw extrusion	螺旋挤出	螺桿擠出成形
screw preplasticizing type injection molding machine	螺杆式预塑化注压成型机	螺桿式預塑化射出成型機
screw pump	螺杆泵	螺桿泵
screw structure	螺旋结构	螺旋結構
screw type injection molding machine	螺杆式注塑成型机	螺桿式射出成型機
scroop	丝鸣[绸料摩擦声]	絲鳴
scrubber	洗涤器	洗滌器
scrubbing tower	洗涤塔	洗滌塔
scum	浮渣	浮渣
scum-collecting device	浮渣收集装置	浮渣收集設備

英　文　名	大　陆　名	台　湾　名
sea island cotton	海岛型棉	海島棉
seal	密封	密封
sealant	密封胶	密封劑
seal cement	密封胶合剂	密封膠合劑
sealed tube	密封管	密封管
sealer	①密封器 ②封堵剂	封口機
seal gum	密封胶	封口膠
sealing	①热合 ②封口	封口
sealing compound	密封胶	密封劑
sealing joint	热合接头	密封接頭
sealing liquid	密封液	密封液
sealing machine	封口机	封口機
sealing oil	密封油	密封油
sealing property	密封性能	密封性質
sealing wax	火漆	火漆,封蠟
seal oil	海豹油	海豹油
seal pot	密封罐	密封罐
seamless pipe	无缝管	無縫管
seam welder	缝焊机	縫焊機
seam welding	缝焊	縫焊
seaweed fiber	海藻纤维	海藻纖維
sebacic acid	癸二酸	癸二酸
secant modulus	正割模量	正割模數
secondary accelerator	助促进剂	助促進劑
secondary amine	二级胺	二級胺
secondary antioxidant	助抗氧剂	助抗氧劑
secondary cell wall	次生细胞壁	次生細胞壁
secondary clarifier	二级澄清器	二級澄清器
secondary creep	蠕变恒速区	二期潛變
secondary crystallization	后期结晶	二次結晶
secondary dimension	次要因次	次要因次
secondary dispersion	二次色散	次生分散
secondary feeder	二级加料器	二級加料器
secondary gluing	二次胶合	二次膠合
secondary impact	二次撞击	次要衝擊
secondary loop	副环路	副環路
secondary normal stress coefficient	次正应力系数	次正應力係數
secondary nucleation	二次成核	次成核

英　文　名	大　陆　名	台　湾　名
secondary nucleation	二次成核作用	二次成核
secondary oil recovery	二次采油	二次石油回收
secondary plasticizer	辅助增塑剂	助塑化劑
secondary polymerization reaction	二级聚合反应	二級聚合反應
secondary product	次级产品	次要產物
secondary reaction	二次反应	次要反應
secondary relaxation	次级弛豫	次級鬆弛
secondary relaxation temperature	次级弛豫温度	次級鬆弛溫度
secondary settler	二级沉降槽	二級沈降槽
secondary standard	二级标准	二級標準
secondary structure	次生结构	二級結構
secondary transition	次级转变	次級轉變
secondary treatment	二级处理	二級處理
secondary trickling filter	二级滴滤池	二級滴濾池
second moment	二次矩	二次[力]矩,二次動差
second order crystallization	二级结晶	二級結晶
second-order fluid	二阶流体	二階流體
second order ionization potential	二级电离电位	二級游離電位
second order lag	二级滞后	二級落後
second order moment	二次矩	二次[力]矩
second order nucleation	二级成核作用	二級成核作用
second-order response	二阶响应	二階應答
second-order system	二阶系统	二階系統
second order transition	二级转变	二級[相]轉變
second order transition temperature	二级转变温度	二級[相]轉變溫度
second virial coefficient	第二位力系数	位力係數
sectioning	切片	剖切
sediment	沉积物	沈積物
sedimentary clay	沉积黏土	沈積黏土
sedimentary kaolin	沉积高岭土	沈積高嶺土
sedimentary rock	沉积岩	沈積岩
sedimentation	沉降	沈降
sedimentation average molecular weight	沉降平均分子量	沈降平均分子量
sedimentation basin	沉淀池	沈降池
sedimentation bottle	沉淀瓶	沈降瓶
sedimentation chamber	沉淀槽	沈降槽
sedimentation coefficient	沉降系数	沈降係數
sedimentation constant	沉降常数	沈降常數

英　文　名	大　陆　名	台　湾　名
sedimentation equilibrium	沉降平衡	沈降平衡
sedimentation equilibrium method	沉降平衡法	沈降平衡法
sedimentation method	沉降法	沈降法
sedimentation pan	沉降盘	沈降盤
sedimentation potential	沉积电势	沈降電位
sedimentation rate	沉积速率	沈積速率
sedimentation tank	沉积池	沈積槽
sedimentation velocity	沉降速度	沈降速度
sedimentation velocity method	沉降速度法	沈降速度法
seed cotton	籽棉	籽棉
seed crystal	晶种	晶種
seeded crystallization	晶种结晶	晶種結晶
seed grain	晶种粒	晶種粒
seeding	加晶种	播晶種
seeding polymerization	种子聚合	核種聚合
seed mixer	晶种混合槽	晶種混合槽
segment	链节	［鏈］段
segmental-bed reactor	分段床反应器	分段床反應器
segmental Brownian motion	链段布朗运动	鏈段布朗運動
segmental copolymer	多嵌段共聚物	鏈段共聚物
segmental downtake calandria	弓形降液管	弓形下導排管
segmental friction factor	链段摩擦因子	鏈段摩擦因子
segmental jump frequency	链段跃迁频率	鏈段躍遷頻率
segmental motion	①链段运动 ②摆动	鏈段運動
segmental orifice	弓形孔口	弓形孔口
segmental orifice plate	弓形孔口板	弓形孔口板
segment anisotropy	链段各向异性	鏈段各向異性
segment copolymer	多嵌段共聚物	鏈段共聚物
segment copolymerization	嵌段共聚合	鏈段共聚合
segment-density distribution	链段密度分布	鏈段密度分佈
segmented copolymer	多嵌段共聚物	鏈段共聚物
segment-interaction parameter	链段相互作用参数	鏈段交互作用參數
segment rotation	链段旋转	鏈段旋轉
segregated flow	隔离流动	隔離流動
segregated reactor	隔离流反应器	隔離流反應器
segregation	离析	偏析
selected-area diffraction	选区衍射	選區繞射
selection rule	选择定则	選擇法則

英　文　名	大　陆　名	台　湾　名
selective absorbent	选择吸收剂	選擇[性]吸收劑
selective absorption	选择吸收	選擇[性]吸收
selective control	选择性控制	選擇[性]控制
selective controller	选择性控制器	選擇[性]控制器
selective cracking	选择裂化	選擇[性]裂解
selective extraction	选择性萃取	選擇[性]萃取
selective fermentation	选择性发酵	選擇[性]發酵
selective flotation	选择性浮选	選擇[性]浮選
selective hydrogenation	选择加氢	選擇[性]氫化
selective leaching	选择性浸取	選擇[性]瀝取
selective poisoning	选择性中毒	選擇[性]中毒
selective polymerization	选择聚合	選擇[性]聚合
selective rectification	选择精馏	選擇[性]精餾
selective solvent	选择性溶剂	選擇性溶劑
selectivity	选择性	選擇性
selectivity coefficient	选择性系数	選擇性係數
selectivity dispersion curve	选择性分散曲线	選擇性分散曲線
selectivity ratio	选择性比	選擇性比
selector	选择器	選擇器
selenium polymer	硒聚合物	硒聚合物
self-adaptive control	自适应控制	自適應控制
self-adhesive	自黏合剂	自黏著劑
self-adhesive tape	自黏胶带	自黏膠帶
self ageing	自老化	自老化
self-balancing bridge	自平衡电桥	自均衡電橋
self-balancing potentiometer	自平衡电位计	自均衡電位計
self-bonded fiber	自黏合纤维	自黏合纖維
self-crosslinking acrylate rubber	自交联丙烯酸酯橡胶	自交聯丙烯酸酯橡膠
self-crosslinking acrylic resin	自交联丙烯酸树脂	自交聯丙烯酸樹脂
self-cure	自固化	自硬化
self-cure epoxy resin	自固化环氧树脂	自硬化環氧樹脂
self-curing	①自固化 ②自硫化	自硬化
self-curing adhesive	自固化黏合剂	自固化黏合劑
self-diffusion	自扩散	自擴散
self extinguish ability	自熄性	自熄性
self-extinguishing	自熄	自熄
self-heating property	自动加热性	自熱性
self-ignition	自燃	自燃

英　文　名	大　陆　名	台　湾　名
self-ignition temperature	自燃温度	自燃溫度
self-initiation	自引发	自引發
self-oligomerization	自身低聚化	自低聚合[反應]
self-polymerization	自聚合	自聚合[作用]
self-priming pump	自起动泵	自起動泵
self-propagation	自增长	自增長
self purification	自净化	自淨化[作用]
self regulation	自调整	自調節
self-reinforcing polymer	自增强聚合物	自強化聚合物
self-sustained reaction	自持续反应	自持續反應
self sustaining	自持续	自持續
self-termination	自终止	自終止[作用]
self-tuning controller	自校正控制器	自調諧控制器
self-tuning regulator	自校正调节器	自調諧調節器
self-vulcanizing	自硫化	自硫化
semi-anthracite	半无烟煤	半無煙煤
semi-automatic press	半自动压机	半自動壓機
semi-automatic system	半自动系统	半自動系統
semi-batch operation	半分批式操作	半批式操作
semi-batch process	半分批法	半分批法
semi-batch reactor	半分批反应器	半批次反應器
semi-batch selectivity	半分批选择性	半批式選擇性
semi-bituminous coal	半烟煤	半煙煤
semi-boiled soap	含水皂	含水皂
semi-boiling process	半煮法	半煮法
semi-chemical pulp	半化学纸浆	半化學紙漿
semi-chemical pulping process	半化学制浆法	半化學製漿法
semiconducting polymer	高分子半导体	半導體聚合物
semiconductive coating	半导体涂料	半導電塗料
semiconductive polymer	半导体聚合物	半導電聚合物
semiconductor	半导体	半導體
semiconductor catalyst	半导体催化剂	半導體觸媒
semi-continuous kiln	半连续窑	半連續窯
semi-continuous polymerization	半连续聚合	半連續聚合
semi-continuous process	半连续过程	半連續程序
semi-crystal	半晶体	半晶體
semi-crystalline polymer	半结晶聚合物	半結晶聚合物
semi-cure	半硫化	半硬化

英 文 名	大 陆 名	台 湾 名
semi-curing	半硫化	半交聯
semidrying oil	半干性油	半乾性油
semi-dull	半无光	半消光織物
semi-durable adhesive	半耐久性黏合剂	半耐久性黏合劑
semi-enclosed impeller	半封闭叶轮	半封閉葉輪
semi-flexible chain	半挠性链	半可撓鏈
semiflow reactor	半流动式反应器	半流動式反應器
semigloss coating	半光涂料	半光塗層
semi-interpenetrating polymer network	半互穿透聚合物网络	半互穿聚合物網絡
semi-logarithmic paper	半对数坐标纸	半對數[坐標]紙
semi-microanalysis	半微量分析	半微量分析
semi-opaque	半透明的	半透明的,半蔽光的
semi-paste paint	半糊状漆	去漆膏
semipermeable membrane	半透膜	半透膜
semipolar bond	半极性键	半極性鍵
semi-reinforcing furnace black	半补强炉黑	半強化爐黑
semi-rigid foam	半硬泡沫	半硬發泡體
semisynthetic fiber	半合成纤维	半合成纖維
semisynthetic oil	半合成[机]油	半合成[機]油
SEM（=scanning electron microscope）	扫描电子显微镜	掃描電子顯微鏡
sensible heat	显热	顯熱
sensing element	敏感元件	感測元件
sensitivity	灵敏度	靈敏度
sensitivity analysis	灵敏度分析	靈敏度分析
sensitivity estimate	灵敏度估计	靈敏度估計
sensitized paper	感光纸	感光紙
sensitizer	敏化剂	敏化劑
sensitizing dye	增感染料	敏化染料
sensor	传感器	感測器
separated application adhesive	组分分涂黏合剂	雙組分黏合劑
separated flow	分离流	分離流動
separation	分离	分離
separation cell	分离池	分離池
separation cyclone	旋风分离器	旋風分離器
separation factor	分离因子	分離因數
separation point	分离点	分離點
separation process	分离过程	分離程序
separation specification table	分离技术规格表	分離規格表

英 文 名	大 陆 名	台 湾 名
separation technology	分离技术	分離技術
separative membrane	分离膜	分離膜
sephadex gel	交联葡聚糖[凝胶]	交聯葡凝膠
septum	隔膜	隔膜
sequence	序列	序列
sequence control	顺序控制	順序控制
sequence length	序列长度	序列長度
sequence-length distribution	序列长度分布	序列長度分佈
sequencing on-off operation	逐次开闭操作	逐次開關操作
sequencing valve	顺序阀	順序閥
sequential analysis	序贯分析	順序分析
sequential copolymer	序列共聚物	序列共聚物
sequential interpenetrating network	序列互穿网络	時序[形成]互穿網結構
sequential operation	顺序操作	順序操作
sequential polymerization	序列聚合	序列聚合
sequential sampling	序贯抽样	順序取樣
serial processes	串行过程	串聯程序
sericin	丝胶蛋白	絲膠
series compensation	串联补偿	串聯補償
series flow	串联流动	串流
series-parallel reactions	串-并联反应	串並聯反應
series reactions	串联反应	串聯反應
series reactors	串联反应器	串聯反應器
serration	锯齿形	鋸齒形
serum	①血清 ②乳清[橡胶]	血清
serum albumin adhesive	血清白蛋白黏合剂	血清白蛋白黏合劑
serum globulin	血清球蛋白	血清球蛋白
serviceable life	使用期	使用期
servo control	伺服控制	伺服控制
servo mechanism	伺服机构	伺服機構
servomechanism compensation	伺服机构补偿	伺服機構補償
servomechanism-type problem	伺服问题	伺服問題
servo problem	伺服问题	伺服問題
sesame oil	芝麻油	芝麻油
set point	①设定值 ②沉淀点	設定點
set point control	设定值控制	設定點控制
set point disturbance	设定值扰动	設定點擾動

英　文　名	大　陆　名	台　湾　名
set point response	设定值响应	設定點應答
set point tracking	设定值跟踪	設定點追蹤
set point variation	设定值变动	設定點變動
set time	凝固时间	設定時間
setting temperature	凝固温度	凝固溫度
setting time	凝固时间	凝結時間,固化時間
settler	沉降器	沈降器,沈降槽
settling	沉降	沈降
settling basin	沉降池	沈降池
settling chamber	沉降室	沈降室
settling column	沉降柱	沈降柱
settling column analysis	沉降柱分析	沈降柱分析
settling filtration	沉降过滤	沈降過濾
settling tank	沉降槽	沈降槽
settling time	沉降时间	沈降時間,安定時間
settling velocity	沉降速度	沈降速度
settling zone	沉降区	沈降區
setup cost	装置建立成本	配置成本
set-up effect	开始[硫化]效应	開始[硫化]效應
sewage disposal	污水处理	污水處理
sewer pipe	污水管	污水管
sexamer	六聚物	六聚物
shadowing	屏蔽	影屏
shaft	轴	軸
shaft brake horsepower	轴制动马力	軸制動馬力
shaft work	轴功	轉軸功
shaker	震动器	搖動器
shaking conveyor	振动输送机	搖運機
shaking screen	摇动筛	搖動篩
shale	页岩	頁岩
shale gas	页岩气	頁岩氣
shale oil	页岩油	頁岩油
shallow-bed reactor	浅床反应器	淺床反應器
shape birefringence	形状双折射	形狀雙折射
shape factor	形状系数	形狀因數
shape-selective catalyst	择形催化剂	形狀選擇觸媒
shape stability	形状稳定性	形狀穩定性
shared electron pair	共价电子对	共用電子對

英　文　名	大　陆　名	台　湾　名
shark skin	鲨鱼皮	鯊魚皮
sharp edged orifice	锐缘孔口	銳緣孔口
sharp separation	清晰分离	清晰分離
shear	剪切	剪切,切變
shear band	剪切带	剪切帶
shear banding	剪切带	剪切帶
shear compliance	剪切柔量	剪切柔量,剪力順性
shear creep	剪切蠕变	剪切潛變
shear deformation	剪切变形	剪切變形
shear flexure	剪切挠曲	剪切撓曲
shear force	剪切力	剪[切]力
shearing	剪切	剪切
shearing disc viscometer	剪切圆盘式黏度计	剪切盤式黏度計
shearing stiffness	剪切刚度	剪切勁度
shearing strength	剪切强度	剪切強度
shear modulus	剪切模量	剪切模數,切變模數
shear rate	剪切率	剪[切速]率
shear relaxation	剪切松弛	剪切鬆弛
shear storage modulus	剪切储能模量	剪切儲存模數
shear strain	剪切应变	剪應變
shear strength	剪切强度	剪切強度
shear stress	剪应力	剪[切]應力
shear test	剪切试验	剪切試驗
shear thickening	剪切致稠	剪切增黏
shear-thickening fluid	剪切增稠流体	剪切增黏流體
shear thinning	剪切致稀	剪切減黏
shear-thinning fluid	剪切稀化流体	剪切減黏流體
shear velocity	剪切速度	剪速度
shear viscosity	剪切黏度	剪切黏度
shear viscosity coefficient	剪切黏滞系数	切變黏滯係數
sheathing paper	绝热[沥青]纸	絕熱紙
sheet blowing method	片料吹塑法	薄膜吹製法
sheeting	①压片 ②挡板	壓片
sheet molding	片状成型	片狀成型
sheet molding compound	片状成型料	片狀成型塑料
sheet polymer	片型聚合物	片型聚合物
shelf dryer	柜式干燥机	箱形乾燥器
shelf life	储存期限	儲存期限,儲存壽命

英　文　名	大　陆　名	台　湾　名
shellac	紫胶	蟲膠
shellac varnish	紫胶清漆	蟲膠清漆
shellac wax	紫胶蜡	蟲膠臘
shell-and-tube heat exchanger	列管换热器	殼管熱交換器
shell-and-tube reactor	列管式反应器	殼管反應器
shell balance	壳平衡	殼均衡
shell molding	壳模铸造	殼狀成型
shell molding resin	壳模铸造[用]树脂	殼狀成型樹脂
shell-side heat transfer coefficient	壳程传热系数	殼側熱傳係數
shell still	简单[壳式]蒸馏釜	[殼式]蒸餾釜
shielding	屏蔽	屏蔽,屏護
shielding length	屏蔽长度	屏蔽長度
shift converter	变换炉	[水煤氣]轉化器
shift factor	平移因子	平移因子
shifting-order reaction	变阶反应	變階反應
shift reaction	变换反应	[水煤氣]轉化反應
shift reactor	[水煤气]变换反应器	[水煤氣]轉化反應器
shikimic acid	莽草酸	莽草酸
ship bottom paint	船底漆	船底漆
shish-kebab	串晶	串晶
shish-kebab crystalline structure	串晶结构	串晶結構
shock coagulation	冲击凝固	衝擊凝聚[作用]
shock elasticity	冲击弹性	衝擊彈性
shock load	冲击载荷	衝擊負載
shock proof lacquer	防震涂料	防震塗料
shock wave	冲击波	震波
shock wave polymerization	冲击波聚合	震波聚合[作用]
shock wave reactor	冲击波反应器	震波反應器
Shore durometer	肖氏硬度计	蕭氏硬度計
Shore elastometer	肖氏弹性计	蕭氏彈性計
Shore hardness	肖氏硬度	蕭氏硬度
short-chain branch	短支链	短支鏈
short-chain branching	短链支化	短鏈支化
shortcut design method	简捷设计法	簡捷設計法
shortcut method	简捷法	簡捷法
shortening oil	起酥油	酥脆油
short fiber	短纤维	短纖維
short fiber composite	短纤维复合材料	短纖維複合材料

英　文　名	大　陆　名	台　湾　名
short fiber reinforced plastic	短纤维增强塑料	短纖強化塑膠
short fiber reinforcement	短纤维增强	短纖強化
short nipple	双头螺纹短接头	雙頭螺紋短接頭
short range	近程	近程,短距離
short-range interaction	近程相互作用	短程交互作用
short-range intrachain crankshaft movement	近程链内曲柄运动	近程鏈內曲柄運動
short-range intramolecular interaction	近程分子内相互作用	分子內短程[相互]作用
short-range order	近程有序	短程有序
short-range structure	近程结构	短程結構
short ripening	短期熟成	短期熟成
short stopped polymerization	速止聚合	速止聚合
short stopper	速止剂	速止劑
short stopping agent	速止剂	速止劑
shot capacity	注射量	射出容量
shot cycle	注射周期	射出週期
shot rate	注射速率	射出速率
shot size	注射尺寸	射出尺寸
shot volume	注射体积	射出體積
shot volume controller	注射体积控制器	射出體積控制器
shot weight	注射量	射出[重]量
shredder	撕碎机	撕碎機
shredding	水淬	水淬
shredding device	撕碎装置	撕碎裝置
shrink	收缩	收縮
shrinkage	收缩	收縮
shrinkage crack	收缩裂纹	縮裂
shrinkage curve	收缩曲线	收縮曲線
shrinkage jigs	防缩模	防縮模
shrinkage rate	收缩率	收縮率
shrinkage tension tester	收缩张力试验机	收縮張力試驗機
shrink fixture	防缩器	防縮裝置
shrinking agent	收缩剂	收縮劑
shrinking machine	[热]收缩机	收縮機
shrinking power	收缩能力	收縮能力
shrinking stress	收缩应力	收縮應力
shrink mark	缩痕	縮陷處

英 文 名	大 陆 名	台 湾 名
shrink resistant finish	防缩整理	防縮加工
shrunk glass	高硅氧玻璃	耐熱玻璃
shutdown control	停车控制	停機控制
shutdown inspection	停堆检查	停工檢查
shutdown period	停工期	停工期
shutdown schedule	停工日程表	停機時程
shutdown time	停工时间	停工時間
shut-off head	关闭高差	關閉高差
side branch	侧[链]支化	側支,側鏈
side capacitance	旁容量	旁容量
side chain motion	侧链运动	側鏈運動
side chain radical	侧链基	側鏈[自由]基
side cooler	中间冷却器	側流冷卻器
sidecut distillate	侧馏分	側流餾出物
side effect	副作用	副效應
side group	侧基	側基
side reaction	副反应	副反應
side reflux	侧回流	側回流
siderite	菱铁矿	菱鐵礦
side stream	侧流	側流,支流
sidestream column	侧线塔	側流塔
side stream stripper	侧线[馏分]汽提塔	側流汽提塔
side stripper	侧线汽提塔	側流汽提塔
side tube	支管	支管
sieve	筛	篩
sieve analysis	筛析	篩析
sieve plate	筛板	篩板
sieve plate column	筛板塔	篩板塔
sieve plate tower	筛板塔	篩板塔
sieve shaker	摇筛器	搖篩器
sieve test	筛分试验	篩分試驗
sieve tray	筛板	篩板,篩盤
sieving	筛分	篩[分]
sifter	筛	篩
sifting machine	筛选机	篩粉機
sight box	窥箱	窺箱
sight glass	视镜	窺鏡
sight hole	视孔	視孔,窺孔

英　文　名	大　陆　名	台　湾　名
signal flow diagram	信号流程图	信號流程圖
signal flow graph	信号流程图	信號流程圖
signal smoke	信号烟	信號煙
signal transfer lag	信号传递滞后	信號傳送落後
sign convention	符号规定	符號規定
silane	硅烷	矽烷
silane adhesion promoter	硅烷类增黏剂	矽烷助黏著劑
silane coupling agent	硅烷偶联剂	矽烷偶合劑
silanol	甲硅烷醇	矽醇
silazane polymer	硅氮烷聚合物	矽氮烷聚合物
silica	二氧化硅	二氧化矽,矽石
silica aerogel	硅气凝胶	矽氣凝膠
silica brick	硅砖	矽磚
silica flour	石英细粉	矽砂粉
silica gel	硅胶	矽膠
silica glass	石英玻璃	矽玻璃
silica refractory materials	硅质耐火材料	矽石耐火物
silica rock	硅岩	矽岩
silicate adhesive	硅酸盐类黏合剂	矽酸鹽黏合劑
siliceous clay	硅质黏土	矽質黏土
siliceous earth	硅藻土	矽藻土
siliceous fireclay	硅质耐火黏土	矽質耐火黏土
siliceous fireclay brick	硅质耐火黏土砖	矽質耐火黏土磚
siliceous limestone	硅石灰石	矽質石灰石
silicon carbide	碳化硅	碳化矽
silicon carbide fiber	碳化硅纤维	碳化矽纖維
silicone adhesive	有机硅黏合剂	聚矽氧黏合劑
silicone elastomer	有机硅弹性体	聚矽氧彈性體
silicone enamel	有机硅瓷漆	聚矽氧瓷漆
silicone grease	硅[润滑]脂	聚矽氧潤滑脂
silicone nitrile rubber	氰硅橡胶	[聚]矽[氧]腈橡膠
silicone oil	硅油	[聚]矽[氧]油
silicone plastic	有机硅塑料	矽氧塑膠
silicone polymer	有机硅聚合物	矽氧聚合物
silicone release	有机硅脱模剂	矽氧脱膜劑
silicone resin	有机硅树脂	矽氧樹脂
silicone rubber	硅橡胶	矽氧橡膠
silicone rubber adhesive	硅橡胶黏合剂	矽氧橡膠黏著劑

英　文　名	大　陆　名	台　湾　名
silicone rubber compound	硅橡胶混炼胶	矽氧橡膠聚合物
silicone rubber foam	泡沫硅橡胶	發泡矽氧橡膠
silicon nitride	氮化硅	氮化矽
silicon-nitrogen polymer	硅氮聚合物	矽氮聚合物
silicon rectifier	硅整流器	矽整流器
silicon rubber	硅橡胶	矽氧橡膠
silicon wafer	硅片	矽晶片
silk	蚕丝	蠶絲
silk degumming	生丝精练	生絲精練,絲綢精練
silk fiber	丝纤维	絲纖維
silk rubber	绢丝橡胶	絲狀橡膠
silk scouring	生丝精练	絲洗練
silk spinning	丝纺	絹紡
silkworm gut	蚕胶丝	蠶膠
sillimanite	硅线石	矽線石
siloxane	硅氧烷	矽氧烷
silver necking	银颈	銀頸
silver number	银值	銀值
silvichemical	林产化学品	木材化學品
similarity	相似性	相似性
similarity solution	相似性解	相似解
similarity theory	相似理论	相似性理論
similarity transformation	相似变换	相似變換
similar law	相似定律	相似定律
simile paper	模造纸	模造紙
simple cubic lattice	简单立方晶格	簡單立方晶格
simple elongation	简单伸长	簡單伸長
simple fluid	简单流体	簡單流體
simple interest	单利	單利
simple pole	单极点	單極點
simple reaction	简单反应	簡單反應
simple shear	简单剪切	簡單剪切
simple shear deformation	简单剪切形变	簡單剪切變形
simple shear flow	简单剪流动	簡單剪切流
simplex algorithm	单纯形算法	單體演算法
simple xanthating machine	简式黄原酸化机	簡式黄[原酸]化機
simplex method	单纯形法	單體[演算]法
simplex reciprocating pump	单缸往复泵	單缸往復泵

英　文　名	大　陆　名	台　湾　名
simple zero	单零点	單零點
simulation	模拟	模擬
simulation model	模拟模型	模擬模型,模擬模式
simulator	模拟器	模擬器
simultaneous absorption	同时吸收	同時吸收
simultaneous interpenetrating network	同步互穿网络	同步[形成]互穿網結構
simultaneous operation	同时操作	同時操作
simultaneous polymerization	同时聚合	同時聚合
simultaneous reactions	同时反应	併發反應
sine response	正弦响应	正弦應答
sine wave	正弦波	正弦波
sine wave generator	正弦波发生器	正弦波發生器
single acting pump	单动泵	單動泵
single-bridged polymer	单桥聚合物	單橋高分子
single cavity mold	单腔模具	單穴模具
single cell protein	单细胞蛋白	單細胞蛋白
single channel analyzer	单道分析仪	單道分析儀
single component	单组分	單成分
single crystal	单晶	單晶
single-crystal fiber	单晶纤维	單晶纖維
single distributed component	单分布组分	單分佈成分
single effect evaporator	单效蒸发器	單效蒸發器
single impression mold	单型腔模	單穴模
single-input-single-output system	单输入单输出系统	單輸入單輸出系統
single liquid approximation	单一液体近似法	單一液體近似法
single-loop control	单环路控制	單環路控制
single pass conversion	单程转化率	單程轉化率
single-pass heater	单程加热器	單程加熱器
single-phase reactor	单相反应器	單相反應器
single-phase system	单相系统	單相系
single-point determination	单点测定法	單點測定[法]
single reaction	单反应	單一反應
single roll crusher	单辊压碎机	單輥壓碎機
single-screw extruder	单螺杆挤出机	單螺桿擠出機
single-screw mixer	单螺杆混合机	單螺桿混合器
single-site mechanism	单部位机理	單部位機構
single-split point	单分隔点	單分隔點

英　文　名	大　陆　名	台　湾　名
single stage centrifugal pump	单级离心泵	單級離心泵
single-stage centrifuge	单级离心机	單階離心機
single stage distillation	单级蒸馏	單階蒸餾
single stage extraction	单级萃取	單階萃取
single stage operation	单级操作	單階操作
single stage pilot valve	单级导向阀	單階導引閥
single-stage process	单级过程	單階程序
single-stage resin	一步法[酚醛]树脂	一步法[酚醛]樹脂
single-strand polymer	单股聚合物	單股聚合物
single-stroke preforming press	单冲程预塑机	單衝程預塑機
single substrate reaction	单底物反应	單受質反應
single thread tester	单纱强力试验机	單紗強度試驗機
single twist	单丝加捻	單撚
single unit depreciation	单元折旧率	單元折舊率
single yarn	单纱	單紗
singular point	奇点	奇點
sink	①汇 ②汇点	匯座
sink mark	凹痕	縮痕
sink point	汇壑点	匯座點
sinter	烧结	燒結
sintered glass	烧结玻璃	燒結玻璃
sintered glass filter	垂熔玻璃滤器	燒結玻璃過濾器
sintered materials	烧结料	燒結材料
sinter forming	烧结成型	燒結成型
sintering	烧结	燒結
sintering zone	烧结带	燒結區
sinter membrane	烧结膜	燒結膜
sinusoidal analysis	正弦分析	正弦分析
sinusoidal deformation	正弦式变形	正弦形變
sinusoidal disturbance	正弦扰动	正弦擾動
sinusoidal response	正弦响应	正弦應答
sinusoidal signal	正弦信号	正弦信號
sinusoidal wave	正弦波	正弦波
siphon	①虹吸管 ②虹吸	①虹吸管 ②虹吸
siphon barometer	虹吸气压表	虹吸氣壓計
siphon gauge	虹吸压力计	虹吸表
siphon oiler	虹吸加油器	虹吸加油器
site-model theory	位置模型理论	位置模型理論

英　文　名	大　陆　名	台　湾　名
size analysis	粒度分析	粒度分析,粒径分析
size distribution	粒度分布	粒度分佈,粒径分佈
size enlargement	粒度增大	尺寸增大
size factor	尺寸因子	尺寸因子
size ratio	尺寸比	尺寸比
size reduction	粉碎	粉碎
size reduction ratio	粉碎比	粉碎比
size separation	粒度分离	粒度分離,篩析
sizing agent	①上胶剂 ②胶黏剂 ③ 浆料	上漿劑
sizing machine	上浆机	上漿機
skein	绞丝	絞紗
skeining	绕成绞	成絞
skeletal vibration	骨架振动	骨架振動
skewness	扭曲度	偏斜度
skim	撇去浮渣	[去]浮沫
skim milk	脱脂奶	脱脂乳
skimming machine	乳油分离机	乳油分離機
skim rubber	胶清橡胶	膠清橡膠
skin	皮	皮,表層
skin and core effect	皮心效应	皮芯效應
skin-core structure	皮心结构	皮芯結構
skin effect	表皮效应	表皮效應,集膚效應
skin friction	壁剪应力	表面摩擦
skinning	结皮[现象]	結皮
skin packaging	表皮型包装	密著包裝
skin temperature	表面温度	表面溫度
skiving machine	削皮机	削片機
slabber	切块机	切塊機
slab glass	光学玻璃板	[光學]玻璃板
slab grating	木浆除滓机	木漿除滓機
slack coal	末煤	碎煤
slacking test	风蚀试验	風蝕試驗
slack mercerization	松弛丝光	鬆弛絲光加工
slack variable	松弛变量	鬆弛變量,附加變數
slack wax	含残油软石蜡	含油石蠟,鬆蠟
slag	炉渣	熔渣
slag action	熔蚀作用	熔蝕作用

英　文　名	大　陆　名	台　湾　名
slag cement	矿渣水泥	熔渣水泥
slag wool	渣棉	渣棉
slaking	湿化	水化
slat conveyor	板式输送机	板式運送機,板運機
slat packed tower	百叶板填充塔	填板塔
sleeve coupling	套筒联轴器	套筒連結器
sleeve joint	套管接头	套筒接頭
slenderness	细长度	細長度
slenderness ratio	高径比	細長比
slice	①薄片 ②平切	片
sliced film	平切薄膜	切片薄膜
slicer	切片机	切片機
slicing	切片	切片
slide core	滑动型芯	滑動芯
slide valve	滑阀	滑閥
slide vane pump	滑板式泵	滑葉泵
slide wire	滑线	滑線
slime	黏泥	黏泥,黏液
slime control	腐浆防治	黏泥控制
slimy fermentation	黏滞发酵	黏液發酵
sling psychrometer	手摇干湿表	搖轉濕度計
slip	滑动	滑動
slip band	滑移带	滑移帶
slip casting process	注浆成型过程	注漿成型法
slip factor	滑移系数	滑動因數
slip friction	滑动摩擦	滑動摩擦
slip glaze	泥釉	泥釉
slippage factor	滑动因子	滑動因子
slip velocity	滑移速度	滑動速度
slit	割缝	狹縫
slitter	切条机	切割機
slop	污油	污油
sloppy separation	粗略分离	粗略分離
slow reaction	缓慢反应	慢反應
slow release	缓释	緩釋
slow release capsule	缓释胶囊	緩釋膠囊
slow release control	缓释控制	緩釋控制
slow release device	缓释器件	緩釋裝置

英　文　名	大　陆　名	台　湾　名
slow release drug	缓释药物	緩釋葯物
slub yarn	竹节丝	粗節紗
sludge blanket	污泥层	污泥層
sludge bulking	污泥蓬松现象	污泥蓬鬆[現象]
sludge density index	污泥密度指数	污泥密度指數
sludge digestion	污泥消化	污泥消化
sludge disposal	污泥处理	污泥處理
sludge disposal process	污泥处理过程	污泥處理法
sludge filtration	污泥过滤	污泥過濾
sludge furnace	污泥焚烧炉	污泥焚燒爐
sludge incinerator	污泥焚烧炉	污泥焚化爐
sludge process	淤泥法	污泥法
sludge promoter	淤渣生成促进剂	污泥促進劑
sludge recycle	淤渣再循环	污泥循環
sludge thickener	污泥浓缩器	污泥濃縮器
sludge thickening	污泥浓缩	污泥增黏
sludge volume index	污泥容积指数	污泥體積指數
slug flow	节涌流	氣包流動
slug flow reactor	节涌流反应器	氣包流反應器
slugging tablet making	压片	乾壓製錠
slurry	浆料	漿料
slurry coating	水浆涂料	漿料塗布
slurry piping system	浆料配管系统	漿料配管系統
slurry polymerization	淤浆聚合	漿料聚合[反應]
slurry process	淤浆法	漿料法
slurry reaction	浆料反应	漿料反應
slurry reaction kinetics	浆料反应动力学	漿料反應動力學
slurry reactor	浆料反应器	漿料反應器
slush molding	①中空铸型法 ②搪塑	中空鑄型[法]
slush pulp	纸浆[粕]液	紙漿[粕]液
small angle scattering(SAS)	小角散射	小角[度]散射
small angle X-ray diffraction (SAXD)	小角 X 射线衍射	小角度 X 光繞射
small angle X-ray scattering (SAXS)	小角 X 射线散射	小角 X 射線散射[法]
small angular light scattering	小角光散射	小角光散射[法]
smectic	层状的	層列狀
smectic phase	层状相	[液晶]層列狀相
smectic state	层状态	[液晶]層列狀態
smectic structure	近晶型结构	[液晶]層列狀結構

英　文　名	大　陆　名	台　湾　名
smelter	冶炼厂	冶煉廠,熔爐
smelter gas	冶炼炉气	熔煉爐氣
smelter hearth	冶炼炉	熔煉爐
smeltery	冶炼厂	熔煉廠
smelting	熔炼	熔煉,冶煉
smelting furnace	熔炼炉	熔煉爐
smelting pot	熔炼坩埚	熔煉[坩]堝
smelting process	冶炼过程	熔煉法
smog	烟雾	煙霧
smoke agent	烟雾剂	發煙劑
smoke bomb	烟幕弹	煙幕彈
smoke candle	烟雾烛[缸]	發煙罐
smoked sheet	烟片[橡胶]	煙片[橡膠]
smoke generator	烟雾发生器	發煙器
smokeless fuel	无烟燃料	無煙燃料
smokeless powder	无烟火药	無煙火藥
smokeless propellant	无烟推进剂	無煙推進劑
smoke point	烟点	發煙點
smoke pot	发烟罐	發煙罐
smoke shell	发烟弹	煙幕彈
smoke tanning	烟鞣制	煙鞣
smoking	熏制	燻製
smoking test	发烟试验	發煙試驗
smoothed signal	修匀信号	修匀信號
smoothing constant	修匀常数	修匀常數
smooth surface	平滑表面	平滑表面
snail	蜗结	小渦輪,蛇輪
snake-cage polyelectrolyte	蛇笼型聚电解质	蛇籠型聚電解質
snapback fiber	弹性纤维	彈性纖維
snarl	卷缩	扭結
snubber roll	①缓冲辊 ②减振器	緩衝輥
soakage	浸湿法	浸漬[液]
soaking factor	均热因子	均熱因數
soak test	浸泡试验	浸泡試驗
soap content	含皂量	皂含量
soap extraction	皂液萃取	皂萃取
soap fastness	耐皂洗[色]牢度	耐皂洗性
soap flake	皂片	皂片

英　文　名	大　陆　名	台　湾　名
soap micelle	皂胶束	皂微胞
soap paste	皂糊	皂糊
soap resistance	耐皂性	耐皂性
social cost	社会成本	社會成本
SOC（＝stress optical coefficent）	应力光学系数	應力光學係數
soda ash	苏打灰	蘇打灰,鈉鹼灰
soda cellulose	碱纤维素	鹼化纖維素
soda feldspar	钠长石	鈉長石
soda glass	钠玻璃	鈉玻璃
soda lime	碱石灰	鹼石灰
soda-lime feldspar	钠钙长石	鈉鈣長石
soda-lime glass	钠钙玻璃	鈉鈣玻璃
soda process	烧碱法	鹼法
soda pulp	烧碱法浆	鈉鹼紙漿
soda soap	钠[硬]皂	鈉皂
sodium alginate	褐藻酸钠	藻酸鈉
sodium-butadiene rubber	丁钠橡胶	丁二烯鈉橡膠
sodium cellulose xanthate	纤维素黄酸钠	黃酸纖維素鈉鹽
sodium ethylene sulfonate polymer	乙二磺酸钠聚合物	乙烯基磺酸鈉聚合物
sodium iron tartrate	酒石酸铁钠	酒石酸鐵鈉
sodium lamp	钠灯	鈉燈
sodium polyacrylate	聚丙烯酸钠	聚丙烯酸鈉
sodium polymerization	钠[引发]聚合作用	鈉聚合[作用]
soft acid	软酸	軟酸
soft coal	烟煤	煙煤
soft detergent	软性洗涤剂	軟性清潔劑
softened water	软化水	軟化水
softener	软化剂	[水]軟化劑
softening	软化	軟化
softening agent	软化剂	軟化劑
softening point	软化点	軟化點
softening temperature	软化温度	軟化溫度
soft fiber	软纤维	軟纖維
soft glass	软玻璃	軟玻璃
soft grease	软润滑脂	軟潤滑脂
soft-mud process	软泥法	軟泥法
softness	①柔软度 ②柔软性	[柔]軟度
soft paraffin	软石蜡	軟石蠟

英 文 名	大 陆 名	台 湾 名
soft polymer	软质聚合物	軟質聚合物
soft poly(vinyl chloride) film	软聚氯乙烯薄膜	軟聚氯乙烯薄膜
soft rubber	软质橡胶	軟質橡膠
soft segment	软[链]段	軟段
software	软件	軟體
soft water	软水	軟水
soft wax	软质蜡	軟蠟
softwood	软木材	軟木
softwood flour	软木粉	軟木粉
soft X-ray	软性X射线	弱[穿]X射線
Sohio process	索亥俄法	蘇黑澳法
soil	土壤	土壤
soil acidity	土壤酸度	土壤酸度
soil bearing pressures	土壤承受压力	土壤承受壓力
soil cement	土水泥	土壤水泥
soil conditioner	土壤改良剂	土壤改良劑
soil contamination	土壤污染	土壤污染
soil corrosion	土壤腐蚀	土壤腐蝕
soil fertility	土壤肥力	土壤肥度
soil mechanics	土壤力学	土壤力學
soil organism	土壤微生物	土壤微生物
soil pollution	土壤污染	土壤污染
soil sol	土壤溶胶	土壤溶膠
soil stabilizer	土壤稳定剂	土壤安定劑
soil sterilant	土壤消毒剂	土壤消毒劑
soil treatment	土壤处理	土壤處理
soil yeast	土壤酵母	土壤酵母
sol	溶胶	溶膠
solar cell	太阳能电池	太陽能電池
solar collector	太阳能集热器	太陽能收集器
solar constant	太阳常数	太陽常數
solar distillation	太阳能蒸馏	太陽能蒸餾
solar energy	太阳能	太陽能
solar engine	太阳能发动机	太陽能引擎
solar evaporation	暴晒蒸发	太陽能蒸發
solar oil	太阳[索拉]油	太陽油
solar radiation	太阳辐射	太陽輻射
solar spectrum	太阳光谱	太陽光譜

英　文　名	大　陆　名	台　湾　名
solation	溶胶化	溶膠化[作用]
solder	[低温]焊料	軟焊料,焊錫
soldering	钎焊	軟焊
soldering paste	钎焊膏	焊膏
solenoid valve	电磁阀	電磁閥
sol fraction	溶胶部分	溶膠部分
sol-gel transformation	溶胶-凝胶转换	溶凝膠轉換
solid	固体	固體
solid-catalyzed reaction	固体催化反应	固體催化反應
solid concentration	固体浓度	固體濃度
solid content	固相含量	固[體]含量
solid filled plastic	固体填充塑料	固體填充塑料
solid flux	固体通量	固體通量
solid flux theory	固体通量理论	固體通量理論
solid fuel	固体燃料	固體燃料
solid handling	固体输送	固體輸送
solidification	凝固	固化[作用]
solidification heat	固化热	固化熱
solidification point	凝固点	固化點
solidification process	凝固过程	固化程序
solidification rate	固化速率	固化速率
solidification rate parameter	固化速率参数	固化速率參數
solidified alcohol	固化酒精	固化酒精
solidified gasoline	固化汽油	固化汽油
solidifying point	凝固点	固化點
solid-liquid separator	固液分离器	固液分離器
solid lubricant	固体润滑剂	固體潤滑劑
solid mechanics	固体力学	固體力學
solid overload	固体超载	固體超載
solid particle	固体粒子	固體粒子
solid phase	固相	固相
solid phase polycondensation	固相缩聚	固相縮聚
solid phase polymerization	固相聚合	固相聚合
solid phase reaction	固相反应	固相反應
solid phase synthesis	固相合成	固相合成
solid polymerization	固体聚合	固態聚合[反應]
solid propellant	固体推进剂	固體推進劑
solid-solid reaction	固-固相反应	固固相反應

英　文　名	大　陆　名	台　湾　名
solid-solid transition	固-固相转变	固固相轉變
solid solution	固溶体	固溶體
solid solution alloy	固溶体合金	固溶體合金
solid state	固态	固態
solid state chemistry	固态化学	固態化學
solid state diffusion	固态扩散	固態擴散
solid state physics	固态物理学	固態物理學
solid state polyelectrolyte	固态高分子电解质	固態聚[合物]電解質
solid state polymerization	固相聚合	固態聚合
solid suspension	固体悬浮体	固體懸浮物
solid transport	固体输送	固體輸送
solidus	固相线	固相線
solid waste	固体废物	固體廢棄物
solid waste disposal	固体废物处理	固體廢棄物處理
solid waste landfill	固体废物掩埋场	固體廢棄物掩埋場
sol rubber	溶胶橡胶	溶膠橡膠
solubility	①溶解度 ②可溶性	溶解度
solubility analysis	溶度分析	溶解度分析
solubility barrier	溶解度势垒	溶解度障壁
solubility curve	溶解度曲线	溶解度曲線
solubility fractionation	溶度分级	溶解度分餾
solubility limit	溶度极限	溶解極限
solubility parameter	溶度参数	溶解度參數
solubility product	溶度积	溶度積
solubility product constant	溶[解]度积常数	溶[解]度積常數
solubility-temperature curve	溶解度-温度曲线	溶解度-溫度曲線
solubility test	溶解度试验	溶[解]度測定
solubilization	增溶作用	溶解化[作用]
solubilizer	增溶剂	助溶劑
solubilizing agent	增溶剂	助溶劑
solubilizing reaction	增溶反应	助溶反應
soluble gum	可溶性胶	可溶[性]膠
soluble polymer	可溶性聚合物	可溶性聚合物
soluble resin	可溶性树脂	可溶性樹脂
soluble RNA（sRNA）	可溶性核糖核酸	可溶性 RNA
soluble starch	可溶性淀粉	可溶澱粉
solute	溶质	溶質
solution	溶液	溶液

英　文　名	大　陆　名	台　湾　名
solution birefringence	溶液双重折射	溶液雙折射
solution casting	溶液浇铸	溶液澆鑄[法]
solution chemistry	溶液化学	溶液化學
solution polymerization	溶液聚合	溶液聚合[作用]
solution pressure	溶解压	溶解壓力
solution spinning	溶液纺丝	溶液紡絲
solution styrene-butadiene rubber	溶液丁苯橡胶	溶液[聚合]苯乙烯丁二烯橡膠
solvability	可解性	可溶解性,溶劑合性
solvate	溶剂合物	溶劑合物
solvating effect	溶剂化效应	溶劑合效應
solvating plasticizer	溶剂化增塑剂	溶劑合塑化劑
solvating power	溶剂化能力	溶劑合能力
solvation	溶剂化	溶劑合[作用]
solvation effect	溶剂化效应	溶劑合效應
solvation energy	溶剂化能	溶劑合能
solvent	溶剂	溶劑
solvent assisted dyeing	溶剂助染	溶劑助染法
solvent based adhesive	溶剂基黏合剂	溶劑基黏合劑
solvent based coating	溶剂基涂料	溶劑基塗料
solvent cast process	溶剂流铸法	溶液鑄造法
solvent cement	溶剂胶浆	溶劑膠合劑
solvent cracking	溶剂裂化	溶劑裂解
solvent crazing	溶剂银纹化	溶劑細紋
solvent deashing	溶剂去灰分[法]	溶劑去灰分[法]
solvent deashing process	溶剂去灰分过程	溶劑去灰分程序
solvent dyeing	溶剂染色	溶劑染色
solvent effect	溶剂效应	溶劑效應
solvent etching	溶剂蚀刻	溶劑蝕刻
solvent extraction	溶剂萃取	溶劑萃取[法]
solvent-free basis	无溶剂基准	無溶劑基準
solvent front	溶剂锋面	溶劑鋒面
solvent-gradient elution	溶剂梯度洗脱	溶劑梯度溶解
solvent hold-up	溶剂滞留量	溶劑滯留量
solvent impregnated resin	浸渍树脂	溶劑浸漬樹脂
solvent interaction parameter	溶剂相互作用参数	溶劑相互作用參數
solvent lamination	溶剂复合	溶劑積層
solventless adhesive	无溶剂黏合剂	無溶劑黏合劑

英 文 名	大 陆 名	台 湾 名
solventless coating	无溶剂涂料	無溶劑塗料
solventless paint	无溶剂漆	無溶劑漆
solvent naphtha	溶剂滤油	溶劑油
solvent process	溶剂法	溶劑法
solvent-refined coal	溶剂精炼煤	溶劑精煉煤
solvent refining	溶剂精制	溶劑精煉[法]
solvent resistance	耐溶剂性	抗溶劑性
solvent-segment interaction	溶剂-链节相互作用	溶劑鏈節作用
solvent separator	溶剂分离器	溶劑分離器
solvent spinning	溶剂纺丝	溶劑紡絲
solvent spun fiber	溶纺纤维	溶紡纖維
solvent strength	溶剂浓度	溶劑強度
solvent test	溶剂试验	溶劑試驗
solvent tolerance	溶剂最大容限	溶劑容許度
solvent type adhesive	溶剂型黏合剂	溶劑型黏合劑
solvent type plasticizer	溶剂型增塑剂	溶劑型塑化劑
solvent welding	溶剂黏接	溶劑熔接
solvolysis	溶剂解	溶劑解[作用]
solvolysis reaction	溶剂化反应	溶劑分解[作用]反應
solvus	固溶线	固溶線
sonic flow	声速流动	音速流動
sonic velocity	声速	音速
sonoluminescence	声致发光	聲致發光
sonolysis	超声波分解	超音波分解
sorbent	①吸着剂 ②吸附剂	吸著劑
sorption	①吸着 ②吸附	吸著
sorption hysteresis	吸附滞后现象	吸附遲滯現象
sorption isotherm	吸附等温线	吸附等溫線
sorption of water	吸湿	吸濕
sorter	[纤维长度]分析器	選別器
sorting	[纤维]分级	揀選,分類
sound-insulating adhesive	隔音黏合剂	隔音黏合劑
sound-insulating materials	隔音材料	隔音材料
sound-proof coating	隔音涂料	隔音塗料
sour fermentation	酸败发酵	酸敗發酵
sour gas	酸气	酸氣
souring	酸败	酸化
sour water	酸水	酸水

英　文　名	大　陆　名	台　湾　名
sour-water stripping	酸水汽提	酸水汽提
soybean fiber	大豆蛋白纤维	大豆蛋白纖維
soybean oil	大豆油	大豆油
soybean protein	大豆蛋白质	大豆蛋白[質]
space charge	空间电荷	空間電荷
space charge limited current	空间电荷限制电流	空間電荷限制電流
space conditioning	空间空调	空調
space cooling	空间冷却	空間冷卻
space factor	空间系数	空間因數
space group	空间群	空間群
space heating	环流空间供暖	空間加熱
space lattice	空间点阵	空間晶格
space network polymer	立体网状聚合物	立體網形聚合物
space polymer	立体聚合物	立體聚合物
space time	时空	空間時間
space velocity	空间速率	空間速度
spacing	间隔	間距
spall	剥落	剝落物
spallation	剥落	剝落
spalling	剥落	剝落
spalling resistance	抗剥落性	抗散裂性
spalling test	剥落试验	散裂試驗
spalling test panel	剥落试验板	散裂試驗板台
span	跨度	跨距
spandex fiber	弹力纤维	彈性纖維
spar	晶石	晶石
sparger	鼓泡器	鼓泡器
sparging gas	鼓泡气	鼓泡氣
sparging reactor	鼓泡反应器	鼓泡反應器
spark discharge	火花放电	火花放電
spark discharge detector	火花放电检测器	火花放電偵檢器
spark erosion	火花电蚀	電火花腐蝕
sparse matrix	稀疏矩阵	稀疏矩陣
spar varnish	桅杆清漆	桅桿清漆
spatial configuration	立体排列	空間組態
spatial oscillation	空间振荡	空間振盪
spatial structure	立体结构	空間結構
spatial velocity	空间速度	空間速度

英　文　名	大　陆　名	台　湾　名
special adhesive	特种黏合剂	特殊黏合劑
speciality chemical	专用化学品	特用化學品
speciality polymer	特殊性能高分子	特用高分子
species	物种	物種,種
specification	规格	規格
specific cake resistance	比滤饼阻力	比濾餅阻力
specific damping capacity	比阻尼容量	比制震能
specific enthalpy	比焓	比焓
specific entropy	比熵	比熵
specific extinction coefficient	比消光系数	比消光係數
specific gravity	比重	比重
specific gravity bottle	比重瓶	比重瓶
specific growth rate	比生长速率	比生長速率
specific heat capacity	比热容	比熱容
specific humidity	比湿	比濕度
specific impulse	比冲量	比衝量
specific inductive capacity	介电常数	介電常數,比電容量
specific insulation resistance	比绝缘电阻	比絕緣電阻
specificity	特异性	特異性
specific rate	比速率	比速率
specific reaction rate	比反应速率	比反應速率
specific reaction rate constant	比反应速率常数	比反應速率常數
specific resistance	比电阻	比阻力,電阻率
specific rigidity	比刚性	比剛性
specific rotation	旋光率	比旋光[度]
specific rotatory power	比旋光度	比旋光度,旋光率
specific speed	比转速	比轉速
specific strength	比强度	比強度
specific surface	比表面	比表面積
specific surface area	比表面积	比表面積
specific thrust	比推力	比推力
specific viscosity	比黏度	比黏度
specific volume	比体积	比容
specific weight	比重	比重
speck	污点	斑點
speck dyeing	斑染	斑染
spectral analysis	光谱分析	光譜分析
spectral transmittance	透光率	透光率

英　文　名	大　陆　名	台　湾　名
spectrochemistry	光谱化学	光譜化學
spectrogram	光谱图	光譜圖
spectrograph	摄谱仪	攝譜儀
spectrometer	光谱仪	光譜儀
spectrometry	光谱测定法	光譜測定法
spectrophotometer	分光光度计	分光光度計,光譜儀
spectroscope	分光镜	分光鏡
spectroscopic analysis	光谱分析	光譜分析
spectroscopy	光谱学	光譜學
spectroturbidimetric titration	分光浊度滴定	分光濁度滴定
spectrum	谱	光譜
speed	速率	速率
spent caustic	废碱	廢鹼
spent liquor	废液	廢液
spent lye	废碱液	廢鹼液
spermaceti oil	鲸蜡油	鯨蠟油
spermaceti wax	鲸蜡	鯨蠟
sperm oil	鲸蜡油	鯨蠟油,抹香鯨油
spew	①溢料缝 ②毛刺	溢料
spew area	溢料面	溢料面
spew groove	溢料槽	溢料槽
spew line	溢料线	溢料線
sphere	球面	球體
spherical cluster	球状团簇	球狀團簇
spherical coordinate	球面坐标	球[形]坐標
spherical polar coordinates	球面极坐标	球極坐標
spherical tank	球形油罐	球形槽
sphericity	球形度	球形度
spherulite	球晶	球晶
spherulite structure	球晶结构	球晶結構
spin coating	旋涂	旋轉塗布
spin coupling	自旋耦合	自旋耦合
spin coupling constant	自旋耦合常数	自旋耦合常數
spin decoupling	自旋去耦	自旋去耦合
spindle oil	锭子油	錠子油
spin-drawing	旋转延伸	紡伸製程
spin echo	自旋回波	自旋回波
spin echo method	自旋回波法	自旋回波法

英　文　名	大　陆　名	台　湾　名
spin-lattice relaxation	自旋晶格弛豫	自旋晶格弛豫
spin-lattice relaxation time	自旋晶格弛豫时间	自旋晶格弛豫時間
spinnability	可纺性	可紡性
spinneret	喷丝头	紡嘴
spinning	①纺丝 ②抽丝	紡絲
spinning acid	纺丝酸	紡絲酸
spinning bath	纺丝浴	紡絲浴
spinning bath stretch	纺丝浴拉伸	紡絲浴拉伸
spinning cake	[纺丝]丝饼	紡絲餅
spinning can	纺丝罐	紡絲罐
spinning cell	纺丝仓	紡絲槽
spinning channel	纺丝甬道	紡絲通道
spinning die	喷丝头	紡絲頭
spinning draft	纺丝头拉伸	紡絲頭拉伸
spinning head	喷丝头	紡絲頭
spinning jet	喷丝头	噴絲頭
spinning machine	纺丝机	紡絲機
spinning nozzle	纺丝头	紡絲嘴
spinning oil	纺丝油	紡絲油
spinning pot	纺丝罐	紡絲罐
spinning solution	纺丝溶液	紡絲液
spinodal	亚稳单相极限线	離相曲線
spin-spin coupling	自旋-自旋耦合	自旋-自旋耦合
spin-spin relaxation	自旋-自旋弛豫	自旋間弛豫
spin-spin relaxation time	自旋-自旋弛豫时间	自旋間弛豫時間
spin welding	旋转焊接	摩擦焊
spiral	螺旋	螺旋
spiral coil heater	螺旋管加热器	螺旋管加熱器
spiral condenser	螺旋板式冷凝器	螺旋冷凝器
spiral conveyor	螺旋运输机	螺旋運送機
spiral flow	螺旋流	螺旋流
spiral gear pump	螺旋齿轮泵	螺旋齒輪泵
spiral plate heat exchanger	螺旋板换热器	螺旋板熱交換器
spiral polymer	螺旋形高分子	螺旋形聚合物
spiral pressure spring	螺旋压力弹簧	螺旋壓力彈簧
spiral ring	螺旋环	螺旋環
spiral separator	螺旋形分离器	螺旋分離器
spiral structure	螺旋形结构	螺旋結構

英 文 名	大 陆 名	台 湾 名
spirit	酒精	酒精
spirit varnish	醇质清漆	酒精清漆
spiropolymer	螺旋聚合物	螺旋聚合物
spitzkasten	角锥[沉淀]池	錐形選粒器
splash disc	防溅盘	防濺盤
splash plate	防溅挡板	防濺板
split feed adsorption	分馈吸附	分饋吸附
split fibre	膜裂纤维	[撕]裂[薄]膜纖維
split flow	分流	分叉流
split mold	分瓣模	對合鑄模
split-range control	分程调节	分程控制
split-tear	剥撕	剝撕
splitter	分流器	分流器,分離器
splitting yarn	裂膜纱	[撕]裂[薄]膜紗
sponge	海绵	海綿
sponge gum	海绵胶	海綿膠
sponge iron	海绵铁	海綿鐵
sponge nickel	海绵镍	海綿鎳
sponge plastic	多孔塑料	海綿塑膠
sponge platinum	海绵铂	海綿鉑
sponge rubber	泡沫橡胶	海綿橡膠
spontaneous coagulation	自然凝结	自發凝聚[作用]
spontaneous combustion	自燃	自燃
spontaneous crystallization	自发结晶	自發結晶
spontaneous emulsification	自发乳化	自發乳化
spontaneous evaporation	自然蒸发	自然蒸發
spontaneous nucleation	自发形核	自發成核
spontaneous polarization	自发极化	自發極化
spontaneous process	自发过程	自發程序
spontaneous reaction	自发反应	自發反應
spontaneous termination	自发终止	自發終止[作用]
spool	卷筒	有邊筒子,線軸
sporadic nucleation	自发成核[作用]	突發成核[作用]
spot	斑点	斑點
spot analysis	斑点分析	斑點分析
spot gluing	点胶合	點膠合
spot reaction	斑点反应	斑點反應
spot test	斑点试验	斑點試驗

英 文 名	大 陆 名	台 湾 名
spot welder	点焊机	點焊機
spot welding	点焊	點銲
spouted bed	喷动床	噴流床
spouting velocity	喷流速度	噴流速度
spray	喷雾	噴霧,噴霧劑
spray atomizer	喷雾雾化器	噴霧[霧化]器
spray chamber	喷雾室	噴霧室
spray chamber contactor	喷雾室接触器	噴霧室接觸器
spray coating process	喷涂法	噴塗法
spray column	喷雾塔	噴霧塔
spray condenser	喷雾冷凝器	噴霧冷凝器
spray cooled crystallizer	喷雾冷却结晶器	噴霧冷卻結晶器
spray cooling	喷雾冷却	噴霧冷卻
spray drier	喷雾干燥器	噴霧乾燥器
spray drying	喷雾干燥	噴霧乾燥
spray drying process	喷雾干燥法	噴霧乾燥法
sprayer	喷雾器	噴霧器
spray gun	雾化器	噴槍
spraying	喷涂	噴霧
spraying drier	喷雾干燥器	噴霧乾燥器
spraying glazing	喷釉	噴釉
spray nozzle	雾化喷嘴	噴嘴
spray paint	喷漆	噴漆
spray pond	喷水池	噴水池
spray separator	喷雾分离器	噴霧分離器
spray tower	喷粉塔	噴霧塔
spreadable life	可涂期	可塗期
spread coater	刮涂机	刮塗機
spreading calender	涂胶压延机	塗膠壓延機
spring actuator	弹簧执行机构	彈簧致動器
spring balance	弹簧秤	彈簧秤
spring balanced type bell gauge	弹簧平衡型钟式压力计	鐘型彈簧壓力計
spring constant	弹簧常量	彈簧常數
spring force	弹簧力	彈簧力
spring hanger	弹簧吊架	彈簧吊架
springless actuator	无弹簧执行机构	無彈簧致動器
spring loaded area meter	弹簧载荷面积计	彈簧負載可變面積流量計

英 文 名	大 陆 名	台 湾 名
spring loaded pressure regulator	弹簧载荷调压器	彈簧負載壓力調節器
spring loaded reducing valve	弹簧载荷减压阀	彈簧負載減壓閥
spring-mass-damper system	弹簧–质量–阻尼系统	彈簧–質量–阻尼系統
springness	弹力性	彈性
spring safety valve	弹簧式安全阀	彈簧安全閥
spring scale	弹簧秤	彈簧秤
spring type hardness tester	弹簧式硬度试验仪	彈簧式硬度試驗儀
sprinkler	洒水器	灑水器
sprinkler head	水喷头	灑水頭
sprue	直浇道	澆口
sprue bush	浇道套	[壓鑄]澆口套筒
sprue ejector	注残料顶杆	殘料頂桿
spun-dyed yarn	纺前染色丝	色紡紗
spun yarn	短纤纱	機紡紗,短纖維加撚紗
spur gear pump	正齿轮泵	正齒輪泵
sputtering	溅射	濺射,濺鍍
sputtering deposition	溅射沉积	濺鍍[法]
sputtering etching	溅射蚀刻	濺射蝕刻[法]
square edge orifice	直边孔口	方[緣]孔[口]
square wave	方波	方波
squeeze pump	挤压泵	擠壓泵
squeezer	压榨机	擠壓機
sRNA（=soluble RNA）	可溶性核糖核酸	可溶性核糖核酸
stabiliser	稳定剂	安定劑,穩定器
stability	稳定性	穩定性,穩定度,安定性
stability analysis	稳定性分析	穩定性分析
stability condition	稳定条件	穩定條件
stability criterion	稳定性判据	穩定準則
stability limit	稳定极限	穩定性極限
stability of steady state	稳态稳定性	穩態穩定性
stability region	稳定区域	穩定區[域]
stability test	稳定性试验	安定性試驗
stabilizability	可稳性	可穩[定]性
stabilization	稳定[作用]	穩定化,安定化
stabilization basin	稳定化池	穩定化池,穩定化槽
stabilization pond	稳定化池	穩定化池
stabilization tank	稳定化槽	穩定化槽
stabilized gasoline	稳定汽油	穩定汽油

英 文 名	大 陆 名	台 湾 名
stabilized reformate	稳定重整油	穩定重組油
stabilizer	稳定塔	穩定器
stabilizing agent	稳定剂	穩定劑
stabilizing ingredient	稳定成分	安定成分
stable emulsion	稳定乳液	穩定乳液
stable isotope	稳定同位素	穩定同位素
stable operating condition	稳定操作条件	穩定操作條件
stable region	稳定区域	穩定區[域]
stable response	稳定响应	穩定應答
stable solution	稳定溶液	穩定溶液
stable state	稳态	穩定狀態
stable steady state	稳定稳态	穩定穩態
stack gas	烟道气	煙道氣
stacking	堆垛	堆垛
stacking effect	堆垛效应	堆垛效應
stack-up reactor	迭式反应器	疊式反應器
stage calculation	逐级计算	分階計算
stage contacting	逐级接触	分階接觸
stage contactor	逐级接触器	分階接觸器
stage efficiency	级效率	階效率
stage ejector	多级喷射器	多階射出器
stage model	分级模型	分階模式
stage operation	分段操作	分階操作
stagewise model	多级模型	分階模式
staggered air heater	拐折空气加热器	交錯空氣加熱器
staggered form	交叉式构象	相錯式
staggered tubes	交错管排	交錯管
stagnant boundary layer	静止边界层	[停]滯邊界層
stagnant film	滞止膜	[停]滯膜
stagnant phase	滞止相	[停]滯相
stagnation point	驻点	停滯點
stagnation pressure	驻点压力	停滯壓力
stagnation property	驻点性质	停滯性質
stagnation temperature	驻点温度	停滯溫度
staining	染色	著色
staining method	着色法	著色法
staining test	着色试验	著色試驗
stainless steel	不锈钢	不鏽鋼

英　文　名	大　陆　名	台　湾　名
stainless-steel fiber	不锈钢纤维	不鏽鋼纖維
staircase reaction	逐级反应	逐次反應,階段反應
stalactite	钟乳石	鐘乳石
stamp battery	捣矿杵组	捣礦杵組
stamp mill	捣磨机	捣碎機
stamp pad ink	打印墨	印墨
standard	标准	標準
standard atmospheric pressure	标准大气压	標準大氣壓力
standard cell	标准电池	標準電池
standard cellulose	标准纤维素	標準纖維素
standard compound	标准化合物	標準化合物
standard consistency	标准稠度	標準稠度
standard deviation	标准差	標準[偏]差
standard enthalpy change of formation	标准生成焓变化	標準生成焓變
standard enthalpy change of reaction	标准反应焓变化	標準反應焓變
standard enthalpy of formation	标准生成焓	標準生成焓
standard error	标准误差	標準誤差
standard gage	标准量规	標準規
standard gate	标准浇口	標準澆口
standard heat of combustion	标准燃烧热	標準燃燒熱
standard heat of formation	标准生成热	標準生成熱
standard heat of reaction	标准反应热	標準反應熱
standardization	标准化	標準化
standardization flow-sheet	标准化流程图	標準化流程圖
standard linear solid	标准线性固体	標準線性固體
standard nomenclature	标准命名法	標準命名法
standard nozzle	标准喷嘴	標準噴嘴
standard operation	标准操作	標準操作
standard operation procedure	标准操作步骤	標準操作步驟
stalandard orifice	标准孔口	標準孔口
standard reaction condition	标准反应条件	標準反應條件
standard screwed elbow	标准螺纹弯头	標準螺旋彎頭
standard sieve	标准筛	標準篩
standard solution	标准溶液	標準溶液
standard state	标准状态	標準狀態
standard state fugacity	标准态逸度	標準狀態逸壓
standard substance	基准物	標準物質
standard symbol	标准符号	標準符號

英　文　名	大　陆　名	台　湾　名
standard temperature	标准温度	標準溫度
standard temperature and pressure	标准温度和压力	標準溫壓
standard test	标准试验	標準試驗
standard testing sieve	标准试验筛	標準試驗篩
standard thermometer	标准温度计	標準溫度計
staple	人造短纤维	短纖維,棉狀纖維,絨[棉]
staple diagram	[纤维]长度分布图	絲毛長度圖
staple fiber	切段[定长]纤维	短纖維
staple glass fiber	定长玻璃短纤维	玻璃短纖維
staple length	纤维长度	纖維長度
staple yarn	棉状纱	棉狀紗
starch	淀粉	澱粉
starch adhesive	淀粉黏合剂	澱粉接著劑
starch based polymer	淀粉基聚合物	澱粉基聚合物
starch iodide paper	淀粉碘化物试纸	澱粉碘試紙
starch iodide test	淀粉碘化物试验	澱粉碘試驗
starch iodine paper	淀粉碘试纸	澱粉碘試紙
starch paper	淀粉试纸	澱粉試紙
starch paste	浆糊	漿糊,澱粉糊
star polymer	星形聚合物	星形聚合物
starter	起动器	起動器,起動機
starved line	缺胶层	缺膠層
state	状态	狀態
state function	状态函数	狀態函數
state of cure	硫化状态	硬化程度
state property	物态参量	狀態參量
state variable	状态变量	狀態變數
static dielectric constant	静电介电常数	靜介電常數
static electrification	带静电[作用]	靜電起電
static eliminator	静电消除器	靜電消除器
static error	静态误差	靜態誤差
static error coefficient	静态误差系数	靜態誤差係數
static fatigue	静态疲劳	靜態疲勞
static friction	静摩擦	靜摩擦
static head	静压头	靜高差
static mixer	静态混合器	靜態混合器
static modulus	静态模量	靜模數

英 文 名	大 陆 名	台 湾 名
static optimization	静态优化	静態最適化
static pressure	静压	靜壓[力]
statics	静力学	靜力學
static state	静态	靜態
static test	静态试验	靜態試驗,靜力試驗
static thermogravimetry	静态热重分析法	靜態熱重量法
stationary growth phase	生长静止期	生長靜止期
stationary phase	静止期	靜止期
stationary platen	固定台	固定平台
stationary point	驻点	駐定點
stationary state	定态	定態
stationary state approximation	稳态近似	穩態近似法
statistical analysis	统计分析	統計分析
statistical chain	统计链	統計鏈
statistical copolymer	统计[结构]共聚物	統計[結構]共聚物
statistical design	统计设计	統計設計
statistical error	统计误差	統計誤差
statistical estimation	统计估计	統計估計
statistical homogeneity	统计均匀性	統計均勻性
statistical mechanics	统计力学	統計力學
statistical segment	统计链段	統計鏈段
statistical theory	统计理论	統計理論
statistical thermodynamics	统计热力学	統計熱力學
statistical unit	统计单元	統計單位
statistical weight matrix	统计权重矩阵	統計權重矩陣
statistics	统计学	統計學
statistics of rupture	断裂统计学	斷裂統計學
stator	定子	定子
Staudinger viscosity law	施陶丁格黏性定律	史陶丁格黏性定律
steady flow	稳定流	穩流
steady motion	定常运动	穩定運動
steady periodic operation	稳态周期操作	穩定週期操作
steady-state	稳态,定常态	穩態
steady state analysis	稳态分析	穩態分析
steady state approximation	稳定近似	穩態近似法
steady-state compliance	稳态柔量	穩態柔量
steady-state creep	稳态蠕变	穩態潛變
steady-state diffusion rate	稳态扩散速率	穩態擴散速率

英　文　名	大　陆　名	台　湾　名
steady state error	稳态误差	穩態誤差
steady state flow	稳态流动	穩態流
steady state gain	稳态增益	穩態增益
steady-state model	稳态模型	穩態模式
steady-state multiplicity	多重稳态性	多重穩態
steady state operation	稳态运行	穩態操作
steady-state performance	稳态性能	穩態效能
steady state process	稳态过程	穩態程序
steady state response	稳态响应	穩態應答
steam	①蒸汽 ②水蒸气	①蒸汽 ②水蒸氣
steam bath	蒸汽浴	蒸汽浴
steam boiler	蒸汽锅炉	蒸汽鍋爐
steam chamber	蒸汽室	蒸汽室
steam coil	蒸汽旋管	蒸汽旋管,蒸汽盤管
steam condensate	蒸汽冷凝水	蒸汽冷凝液
steam condenser	蒸汽冷凝器	蒸汽冷凝器
steam cracking	蒸汽裂解	蒸汽裂解
steam cure	①蒸汽熟化 ②蒸汽硫化	①蒸汽熟化 ②蒸汽硫化
steam distillation	水蒸气蒸馏	蒸汽蒸餾
steam-driven pump	蒸汽驱动泵	蒸汽驅動泵
steam drum	汽包	蒸汽鼓
steam economizer	蒸汽省热器	蒸汽省熱器
steam economy	蒸汽经济	蒸汽經濟
steam ejector	蒸汽喷射器	蒸汽噴射器
steam engine	蒸汽机	蒸汽機
steam hammer	汽锤	蒸汽鎚
steam header	蒸汽总管	蒸汽集管[箱]
steaming	汽蒸	[汽]蒸
steam jacket	蒸汽夹套	蒸汽夾套
steam jet ejector	蒸汽喷射泵	蒸汽噴射器
steam jet pump	蒸汽喷射泵	蒸汽噴射泵
steam power plant	蒸汽动力发电厂	蒸汽發電廠
steam purge	蒸汽吹洗	蒸汽吹驅
steam quality	蒸汽品质	蒸汽乾度
steam reforming	蒸汽转化	蒸汽重組
steam regeneration	蒸汽再生法	蒸汽再生[法]
steam relief valve	蒸汽释放阀	蒸汽釋放閥
steam separator	蒸汽水分离器	[蒸]汽水分離器

英 文 名	大 陆 名	台 湾 名
steam sterilization	蒸汽灭菌	蒸汽滅菌
steam sterilizer	蒸汽消毒器	蒸汽滅菌器
steam stripping	汽提	蒸汽汽提
steam table	水蒸气图表	蒸汽表
steam trap	疏水器	祛水器,蒸汽阱
steam tube	蒸汽管	蒸汽管
steam turbine	蒸汽涡轮	蒸汽渦輪
steam valve	蒸汽阀	蒸汽閥
steam valve capacity	水蒸气阀容量	蒸汽閥容量
steel	钢	鋼
steel fiber	钢纤维	鋼纖維
steel glass	强力玻璃	強力玻璃
steel pipe	钢管	鋼管
steepest ascent method	最速上升法	最陡上升法
steepest descent method	最速下降法	最陡下降法
steeping	浸染	浸透
steeping press	浸渍压榨机	浸[漬]壓[榨]機
stem	主干	針桿,莖
stem cell	干细胞	幹細胞
stem fiber	韧皮纤维	莖纖維
stenter	展幅机	擴幅機
step addition polymer	逐步加成聚合物	逐步加成聚合物
step analysis	阶梯分析	階梯分析
step copolymerization	逐步共聚合	逐步共聚[作用]
step disturbance	阶跃干扰	階梯擾動
step function	阶梯函数	階梯函數
step function response	阶跃函数响应	階梯函數應答
step input	阶跃输入	階梯輸入
stepladder polymer	分段梯形聚合物	立梯型聚合物
step motor	步进电机	步進馬達
step reaction polymerization	逐步聚合	逐步聚合[反應]
step response	阶跃响应	階梯應答
stepwise addition polymerization	逐步加成聚合	逐步加成聚合[反應]
stepwise polymerization	逐步聚合	逐步聚合
stepwise reaction	逐步反应	逐步反應
stereoblock copolymer	立构嵌段共聚物	立構嵌段共聚物
stereocenter	立构中心	立構中心
stereochemistry	立体化学	立體化學

英 文 名	大 陆 名	台 湾 名
stereohybridization	立构杂化作用	立構混成作用
stereoisomer	立体异构体	立體異構物
stereomicrography	立构显微照相法	立體顯微照相法
stereorandom copolymer	立构无规共聚物	隨機立構共聚物
stereoregularity	立构有规性	立體規則性
stereoregular polymer	有规立构聚合物	立體規則性聚合物
stereoregular polymerization	立构规整聚合	立體規則聚合[反應]
stereorepeating unit	立构重复单元	立構重複單元
stereo rubber	立构规整橡胶	立構橡膠
stereoselective polymerization	立构有择聚合	立構選擇聚合
stereoselectivity	立体选择性	立體選擇性
stereosequence	立体序列	立體序列
stereosequence distribution	立体序列分布	立體序列分佈
stereospecific catalyst	立体有择催化剂	立構特定觸媒
stereospecific configuration	立体有规构型	立構特定組態
stereospecific copolymerization	有规立构共聚	立構特定共聚
stereospecificity	立体专一性	立體特定性
stereospecific polymer	立构规整聚合物	立體特定聚合物
stereospecific polymerization	立构规整聚合	立體特定聚合[反應]
stereospecific reaction	立体定向反应	立體特定反應
stereospecific rubber	有规立构橡胶	立構特定橡膠
stereosymmetrical homopolymer	立体对称均聚物	立體對稱均聚物
stereotactic polymer	立构规整聚合物	立體異構聚合物
stereotactic polymerization	立构规整聚合	立體異構聚合
steric effect	空间效应	位阻效應,立體效應
steric factor	空间因子	位阻因素,立體因素
steric hindrance	位阻	位阻,立體阻礙
steric regularity	立构规整性	立體規律性
steric restriction	空间障碍	空間障礙
steric strain	空间张力	立體應變
sterilization	灭菌	滅菌
sterilizer	灭菌器	滅菌器
sterol	甾醇,固醇	固醇類
stickiness	黏性	黏性
sticky stage	黏态	黏性階段
stiff chain	硬性链	剛性鏈
stiffener	硬化剂	加強材,防撓材
stiffening agent	硬化剂	硬挺劑

英　文　名	大　陆　名	台　湾　名
stiff equation	刚性方程	剛性方程
stiff flow	难流动[性]	僵硬流
stiff-mud process	硬泥法	硬泥法
stiffness modulus	刚性模量	剛性模量
stiffness test	刚度试验	剛性試驗
stiffness-weight ratio	比刚度	比剛度
stilbene	二苯乙烯	二苯乙烯
stilbene dye	染料	二苯乙烯染料
still	蒸馏釜	蒸餾釜
stimulus-response technique	受激响应技术	刺激應答法
stirred flow reactor	搅拌流反应器	攪拌流動反應器
stirred fluidized bed	搅拌流化床	攪拌流[體]化床
stirred reactor	搅拌反应器	攪拌反應器
stirred tank	搅拌罐	攪拌槽
stirred-tank contactor	搅拌槽接触器	攪拌槽接觸器
stirred tank reactor	搅拌罐式反应器	攪拌反應槽
stirrer	搅拌器	攪拌器
stirring	搅拌	攪拌
stitch bonded fabric	缝编织物	縫編織物
stochastic analysis	随机分析	隨機分析
stochastic feature	随机特性	隨機特質
stochastic process	随机过程	隨機程序
stochastic stability	随机稳定性	隨機穩定性
stock	原料	[儲]存料
stock dyeing	散纤维染色	纖維染色
stock pile	储料堆	儲料[堆]
stoichiometric calculation	化学计量计算	化學計量計算
stoichiometric coefficient	化学计量系数	化學計量係數
stoichiometric compatibility	化学计量兼容性	化學計量相容性
stoichiometric mixture	化学计量混合物	化學計量混合物
stoichiometric number	化学计量数	化學計量數
stoichiometry	化学计量	化學計量學
stokes	斯托克斯(动力黏度单位)	斯托克(動黏度單位)
Stokes approximation	斯托克斯近似法	斯托克近似法
Stokes line	斯托克斯线	斯托克斯[譜]線
stoneware	炻器	陶石器
stopped flow method	停止流动法	停流法

英　文　名	大　陆　名	台　湾　名
stop valve	截止阀	截流閥
storage	存储器	儲存
storage battery	蓄电池	蓄電池
storage capacity	储存量	儲存容量
storage cell	蓄电池	蓄電池
storage compliance	储能柔量	儲存柔量
storage equipment	储存设备	儲存裝置
storage equipment flowsheet	储存装置流程图	儲存裝置流程圖
storage facility	储存设施	儲存設施
storage life	储存寿命	保存期限
storage loss	储存损失	儲存損失
storage modulus	储能模量	儲存模數
storage stability	储存稳定性	儲存穩定性
storage tank	储槽	儲槽
storage temperature	储存温度	儲存溫度
stored energy function	储能函数	貯能函數
Stormer viscometer	斯氏黏度计	史托馬黏度計
stoving	烘干	烘乾
straight-chain polymer	直链聚合物	直鏈聚合物
straight dipping process	单纯浸渍法	直浸漬過程
straight-line depreciation	直线折旧	直線折舊[法]
straight run	直馏油	直餾
straight-run distillation	直馏[法]	直餾
strain adhesive	应变黏合剂	應變黏合劑
strain birefringence	应变双折射	應變雙折射
strain ellipsoid	应变椭球	應變橢圓體
strain energy	应变能	應變能
strain energy function	应变能函数	應變能函數
strain gauge	应变仪	應變[測量]計
strain hardening	应变硬化	應變硬化
straining	粗滤	粗濾
straining process	粗滤过程	粗濾程序
strain optical coefficient	应变光学系数	應變光學係數
strain softening	应变软化	應變軟化
strain tensor	应变张量	應變張量
stratification	分层	分層[作用]
stratified flow	分层流	分層流
stratified plastic	层压塑料	分層塑膠

英　文　名	大　陆　名	台　湾　名
stratum reticular	网状层	網狀層
straw pulp	草纸浆	草[類]紙漿
stray light	杂散光	雜散光
streak line	条纹线	條紋線
stream function	流函数	流線函數
streaming birefringence	流动双折射	流動雙折射
streaming potential	冲流电压	沖流電壓,沖流電位
streamline	流线	流線
streamline coordinates	流线坐标	流線坐標
streamlined body	流线型物体	流線形物體
streamlined filter	流线式过滤器	流線式濾器
streamlined valve	流线型阀	流線形閥
streamline flow	层流	流線流
streamline motion	流线运动	流線運動
streamlining	流线型化	流線化
strength	强度	強度
strengthening agent	补强剂	強化劑
strength-to-weight ratio	比强度	強度重量比
streptomycin	链霉素	鏈黴素
stress	应力	應力
stress analysis	应力分析	應力分析
stress birefringence	应力双折射	應力雙折射
stress concentration	应力集中	應力集中
stress concentration factor	应力集中系数	應力集中因數
stress corrosion	应力腐蚀	應力腐蝕
stress cracking	应力开裂	應力開裂
stress crazing	应力银纹	應力細紋
stress crystallinity	应力结晶性	應力結晶性
stress-deformation curve	应力－形变曲线	應力－形變曲線
stress distribution	应力分布	應力分佈
stressed shell	预应力外壳	預應力外殼
stress ellipsoid	应力椭球体	應力椭球
stress graphitization	应力石墨化	應力石墨化
stress history	应力史	應力歷程
stress-induced crystallization	应力诱导结晶	應力誘發結晶
stress-induced growth	应力诱导生长	應力誘發生長
stress-induced orientation	应力诱导取向	應力誘發定向
stress-induced polarization	应力诱导极化	應力誘發極化

英 文 名	大 陆 名	台 湾 名
stress intensity factor	应力强度因子	應力強度因子
stress optical coefficient	应力光学系数	應力光學係數
stress power	应力功率	應力功率
stress relaxation	应力松弛	應力弛豫
stress relaxation curve	应力松弛曲线	應力弛豫曲線
stress relaxation modulus	应力松弛模量	應力弛豫模數
stress relaxometer	应力松弛计	應力弛豫計
stress relief test	应力消除试验	應力釋放試驗
stress softening	应力软化	應力軟化
stress-strain behavior	应力－应变行为	應力－應變行為
stress-strain curve	应力－应变曲线	應力－應變曲線
stress-strain curve relation	应力－应变曲线关系式	應力應變曲線關係
stress-strain response	应力－应变响应	應力－應變響應
stress tensor	应力张量	應力張量
stress-transfer mechanism	应力传递机理	應力轉移機制
stress whitening	应力致白	應力致白
stretchability	拉伸性	拉伸性
stretched membrane	拉伸膜	拉伸膜
stretch forming	拉伸成型	拉伸成形
stretching	拉伸	拉伸
stretch orientation	拉伸取向	拉伸定向
stretch rate	拉伸速率	伸展速率
stretch ratio	拉伸比	拉伸比
stretch spinning	拉伸纺丝	拉伸紡絲
stretch yarn	弹力纱	伸縮紗
stringiness	起黏丝性	絲黏性
strip chart	长条记录纸	長條記錄紙
stripe	条纹	條紋,斑紋
strippable coating	可剥性涂料	可剝塗層
stripper	①汽提塔 ②解吸塔 ③脱模机	汽提塔
stripper plate	脱模板	脱模板
stripping	提馏	汽提,脫除
stripping agent	①汽提剂 ②解吸剂 ③褪色剂 ④脱模剂	脫色劑
stripping column	汽提塔	汽提塔
stripping factor	解吸因子	汽提因數
stripping section	提馏段	汽提段

英　文　名	大　陆　名	台　湾　名
stripping steam	汽提用蒸汽	汽提蒸汽
stripping still	汽提器	汽提器
stripping test	①剥离试验 ②褪色试验	剝離試驗
stripping tower	汽提塔	汽提塔
stroboscope	频闪仪	頻閃轉速計
stroke	冲程	衝程
strong viscose rayon	强力黏胶纤维	強黏液嫘縈
structural adhesive	结构黏合剂	結構性接著劑
structural birefringence	结构双折射	結構雙折射
structural composite	结构复合材料	結構複合材料
structural defect	结构缺陷	結構缺陷
structural disorder	结构无序	結構失序
structural fatigue	结构疲劳	結構疲勞
structural foam plastic	结构泡沫塑料	結構泡沫塑膠
structural stability	结构稳定性	結構穩定性
structural unit	结构单元	結構單元
structural viscosity	结构黏度	結構黏性
structure analysis	结构分析	結構分析
structured fluid	有规结构流体	結構化流體
structured model	结构化模型	結構化模式
structured packing	整装填料	結構填充物
stuffing	①填料 ②上脂	加脂［法］
stuffing box	填料箱	填料箱
stuffing box seal	填料函式密封	填料箱密封
styrenated oil	苯乙烯基化油	苯乙烯化油
styrene	苯乙烯	苯乙烯
styrene acrylate copolymer coating	苯乙烯－丙烯酸酯共聚涂料	苯乙烯–丙烯酸酯共聚物塗料
styrene acrylonitrile copolymer	苯乙烯－丙烯腈共聚物	苯乙烯–丙烯腈共聚物
styrene butadiene random copolymer	丁苯无规共聚物	苯乙烯丁二烯隨機共聚物
styrene divinylbenzene copolymer	苯乙烯－二乙烯基苯共聚物	苯乙烯二乙烯苯共聚物
styrene maleic anhydride copolymer	苯乙烯－顺丁烯二酸酐共聚物	苯乙烯–顺丁烯二［酸］酐共聚物
styrene methyl methacrylate resin	苯乙烯－甲基丙烯酸甲酯树脂	苯乙烯–甲基丙烯酸甲酯樹脂
styrene oxide polymer	氧化苯乙烯聚合物	氧化苯乙烯聚合物

英 文 名	大 陆 名	台 湾 名
styrene resin	苯乙烯树脂	苯乙烯樹脂
styrene rubber	苯乙烯橡胶	苯乙烯橡膠
subatmospheric pressure	负压	負壓
subbituminous coal	次烟煤	次煙煤
subcell	亚晶胞	亞晶胞
subcooled boiling	过冷沸腾	過冷沸騰
subcooled liquid	过冷液体	過冷液體
subcooled state	过冷状态	過冷狀態
subcooling	过冷	過冷
sublayer	次层	次層
sublimate	升华物	升華物
sublimate drying	升华干燥法	升華乾燥［法］
sublimation	升华	升華［作用］
sublimation apparatus	升华装置	升華裝置
sublimator	升华器	升華器
sublimer	升华器	升華器
submerged combustion	浸没燃烧	沈沒燃燒
submerged combustion evaporator	浸没燃烧蒸发器	浸沒燃燒蒸發器
submerged condenser	浸没式冷凝器	沈式冷凝器
submerged pump	液下泵	沈式泵
submicrocrack	亚微裂纹	次微米裂紋
submicrofracture	亚微观断裂	次微米斷裂
submicroscopic	亚微观的	亞微觀的
submicroscopic micelle	亚微观胶束	亞微觀微胞
submicroscopic structure	亚微观结构	亞微觀結構
suboptimal control	次优控制	次優控制
suboptimal controller	次优控制器	次優控制器
suboptimality	次优性	次優性
subprogram	子程序	子程式
subroutine	子程序	子程式
subsonic nozzle	亚声速喷嘴	次音速噴嘴
substituent	①取代基 ②替代物	取代基
substituent constant	取代基常数	取代基常數
substituent uniformity	取代基均匀度	取代基均匀度
substitute natural gas	代用天然气	合成天然氣
substitution reaction	取代反应	取代反應
substitution rule	取代法则	代換法則
substrate	底物	受質，基質

英　文　名	大　陆　名	台　湾　名
substrate inhibition	底物抑制	受質抑制[作用]
substrate specificity	基质专一性	受質專一性
subunit	①亚单元 ②亚基	次單元
subunit structure	①亚单元结构 ②亚基结构	次單元結構
successive polymerization	逐次聚合	逐次聚合
successive reaction	逐次反应	逐次反應
successive tear strength	逐次撕裂强度	逐次撕裂強力
sucrose	蔗糖	蔗糖
suction filter	吸滤器	吸濾器
suction head	吸入压头	吸取高差
suction lift	吸引升液器	吸入升力
suction line	吸引管线	吸入管線
suction potential	吸引势	吸入勢
suction pressure	吸入压力	吸入壓力
suction pump	抽水泵	吸取泵
suction velocity	吸入速度	吸入速度
sugar	糖	糖
sugarcane	甘蔗	甘蔗
sugarcane wax	甘蔗蜡	蔗蠟
suitability	适合性	適合性
sulfated oil	硫酸化油	硫酸化油
sulfate process	硫酸盐制纸浆法	硫酸鹽方法,硫酸鹽處理
sulfate pulp	硫酸盐纸浆	硫酸鹽紙漿
sulfate resistant cement	耐硫酸水泥	抗硫酸鹽水泥
sulfating	硫酸化	硫酸化
sulfation	硫酸盐化	硫酸化[反應]
sulfidation	硫化过程	硫化作用
sulfidity	硫化度	硫化度
sulfite process	亚硫酸盐法	亞硫酸鹽法
sulfite pulp	亚硫酸盐浆	亞硫酸[鹽]紙漿
sulfonate	磺酸盐	磺酸鹽
sulfonated oil	磺化油	磺[酸]化油
sulfonating agent	磺化剂	磺[酸]化劑
sulfonation	磺化	磺[酸]化[反應]
sulfonator	磺化器	磺[酸]化器
sulfur	硫	硫

英　文　名	大　陆　名	台　湾　名
sulfur bleach	硫漂白	硫[黄]漂白
sulfur chloride vulcanization	氯化硫溶液硫化	氯化硫[溶液]硫化
sulfur-containing polymer	含硫聚合物	含硫聚合物
sulfur crosslinking	硫交联	硫交聯
sulfur donor	给硫体	硫施體
sulfur donor agent	给硫剂	給硫劑
sulfur dye	硫化染料	硫化染料
sulfur elimination	脱硫	脱硫
sulfuric acid ester	硫酸酯	硫酸酯
sulfuric acid process	硫酸法	硫酸法
sulfurization	硫化	硫化
sulfur vulcanization	硫[磺]硫化	[硫黄]硫化
sulphonic acid ionomer	磺酸型离子交联聚合物	磺酸離子聚合物
sump	污水池	機油箱,廢油池
sun-checking agent	抗日光龟裂剂	抗晒劑
sun crack	晒裂	晒裂
sun cracking	晒裂	晒裂
sun crazing	晒裂	晒裂
sun-discoloration	日晒变色	日晒變色
sunflower oil	向日葵油	葵花[子]油
supercalender	多辊压延机	多輥壓延機
supercentrifuge	超速离心机	高速離心機
supercomputer	超级计算机	超級計算機
superconducting alloy	超导合金	超導合金
superconducting carbonitride	超导碳氮化物	超導碳氮化物
superconducting characteristic	超导特性	超導特性
superconducting generator	超导发电机	超導發電機
superconducting state	超导态	超導態
superconducting thin film	超导薄膜	超導薄膜
superconducting transition temperature	超导转变温度	超導轉變溫度
superconductive polymer	超导聚合物	超導聚合物
superconductivity	超导性	超導性
superconductor	超导体	超導體
supercooled liquid	过冷液	過冷液體
supercooled vapor	过冷蒸气	過冷蒸氣
supercooling	过冷	過冷
supercritical extraction	超临界萃取	超臨界萃取
supercritical flow	超临界流	超臨界流動

英　文　名	大　陆　名	台　湾　名
supercritical gas	超临界气体	超臨界氣體
super drawing	超拉伸	超延伸
super-duty refractory	特级耐火材料	特級耐火物,超強耐火物
super-elasticity	超弹性	超彈性
superficial velocity	表观速度	表觀速度
superfluid	超流体	超流體
superheat	过热[量]	過熱[量]
superheated state	过热状态	過熱狀態
superheated water	过热水	過熱水
superheater	过热器	過熱器
superheating	过热	過熱
super ion-conductive polymer	超离子导电聚合物	超離子導電聚合物
superlattice	超晶格	超晶格
supermolecular order	超分子有序	超分子次序
supermolecular structure	超分子结构	超分子結構
supermolecular texture	超分子织态结构	超分子紋理
supermolecular transition	超分子转变	超分子轉變
supermolecule	超分子	超分子
supernatant	上清液	上澄液
superoxide	超氧化物	超氧化物
superphosphate	过磷酸钙	過磷酸鹽
superpolymer	超高聚物	超聚合物
superposition	叠加	疊加,疊合
superposition principle	叠加原理	疊加原理
supersaturated solution	过饱和溶液	過飽和溶液
supersonic nozzle	超声速喷嘴	超音速噴嘴
supervision	监督	監督
supervisory control	监督控制	監督控制
supply	供应	供應
supply pressure	供给压力	供給壓力
supported bimetallic catalyst	有载体双金属催化剂	負載型雙金屬觸媒
supported catalyst	有载体催化剂	負載型觸媒
supported metal catalyst	有载体金属催化剂	負載型金屬觸媒
supporting film	支持膜	支持膜
support plate	支承板	支撐板
supramolecular structure	超分子结构	超分子結構
supramolecule	超分子	超分子

英 文 名	大 陆 名	台 湾 名
surface abrasion	表面磨蚀	表面磨耗
surface absorber	表面吸收器	表面吸收器
surface acid site	表面酸性部位	表面酸性部位
surface-active agent	表面活性剂	表面活性劑
surface activity	表面活性	表面活性
surface adsorption	表面吸附	表面吸附
surface area	表面积	表面積
surface bonding	表面黏接	表面黏接
surface catalysis	表面催化	表面催化[作用]
surface charge	表面电荷	表面電荷
surface chemistry	表面化学	表面化學
surface coating	表面涂层	表面塗裝,表面塗層
surface complex	表面络合物	表面錯合物
surface concentration	表面浓度	表面濃度
surface condenser	表面式冷凝器	表面冷凝器
surface condition	表面状况	表面狀態
surface conductance	表面电导	表面電導
surface coverage	表面覆盖度	表面覆蓋率
surface crack	表面裂纹	表面裂紋
surface cultivation	表面培养	表面培養
surface density	表面密度	表面密度
surface diffusion	表面扩散	表面擴散
surface dislocation	表面位错	表面移位
surface drag	表面曳引	表面阻力
surface emissivity	表面发射率	表面發射率
surface energy	表面能	表面能
surface energy of fracture	断裂表面能	斷裂表面能
surface engineering	表面工程	表面工程
surface fermentation	表面发酵	表面發酵
surface flow	表面流动	表面流動
surface force	表面力	表面力
surface free energy	表面自由能	表面自由能
surface grafting	表面接枝	表面接枝
surface hardening	表面硬化	表面硬化
surface imperfection	表面缺陷	表面缺陷
surface layer	表面层	表層
surface modification	表面改性	表面改質
surface modified fiber	表面改性纤维	表面改質纖維

英　文　名	大　陆　名	台　湾　名
surface modifier	表面改性剂	表面改質劑
surface moisture	表面水分	表面水分
surface morphology	表面形态	表面形態
surface phenomenon	表面现象	表面現象
surface polymerization	界面聚合	界面聚合[法]
surface porosity	表面孔隙度	表面孔隙度
surface potential	表面电位	表面電位
surface preparation	表面制备	表面製備
surface pressure	表面压力	表面壓力
surface profile	表面轮廓	表面輪廓
surfacer	二道底漆	光面漆
surface rate	表面速率	表面速率
surface reaction resistance	表面反应阻力	表面反應阻力
surface renewal factor	表面更新因数	表面更新因數
surface renewal theory	表面更新理论	表面更新理論
surface resistance	表面电阻	表面電阻,表面阻力
surface resistivity	表面电阻率	表面電阻率
surface rheology	表面流变学	表面流變學
surface roughening treatment	表面粗化处理	表面粗化處理
surface roughness	表面粗糙度	表面粗糙度
surface science	表面科学	表面科學
surface sizing	表面施胶	表面上膠
surface structure	表面结构	表面結構
surface temperature	表面温度	表面溫度
surface tensiometer	表面张力计	表面張力計
surface tension	表面张力	表面張力
surface transcrystallinity	表面横向结晶	表面橫列結晶
surface treatment	表面处理	表面處理
surface viscometer	表面黏度计	表面黏度計
surface viscosity	表面黏度	表面黏度
surface winder	表面卷取机	表面捲取機
surfactant	表面活性剂	界面活性劑
surfactant flooding	表面活性剂泛点	界面活性劑氾流[法]
surge tank	缓冲罐	緩衝槽
surge wave	涌波	湧波
surging	①喘振 ②脉动	①衝擊 ②突波
surroundings	[热力学]环境	環境
surrounding temperature	环境温度	環境溫度

英　文　名	大　陆　名	台　湾　名
survival rate	生存率	生存率
survival theory	生存理论	生存理論
suspended catalyst	悬浮催化剂	懸浮觸媒
suspended growth	悬浮生长	懸浮生長
suspended-growth system	悬浮生长系统	懸浮生長系統
suspended solid	悬浮固体	懸浮固體
suspending agent	助悬剂	懸浮劑
suspension	悬浮	懸浮
suspension adhesive	悬浮黏合剂	懸浮黏合劑
suspension colloid	悬浮胶体	懸[浮]膠體
suspension load	悬浮物负载	懸浮物負載
suspension polymerization	悬浮聚合	懸浮聚合
suspension stabilizer	悬浮稳定剂	懸浮安定劑
suspension system	悬浮系统	懸浮系統
suspension wire	悬浮线	懸線
suspensoid	悬浮体	懸[浮]膠體
sustained oscillation	自持振荡	持續振盪
swabbing process	擦涂法	擦抹取樣過程
sweating	发汗	發汗法
sweating process	发汗工艺	發汗法
sweep diffusion	吹扫扩散	掃掠擴散
sweetener	甜味剂	甜味料
sweetening process	①脱臭过程 ②脱硫过程	[天然氣]脱臭程序
sweet gas	低硫天然气	脱臭天然氣
swell	溶胀	膨潤
swelling agent	溶胀剂	膨潤劑
swelling capacity	溶胀量	膨潤能力
swelling power	溶胀能力	膨潤能力
swelling pressure	①溶胀压 ②膨胀压力[模具]	膨潤壓[力]
swelling ratio	①溶胀比 ②膨胀比[挤塑]	膨潤比
swelling value	溶胀值	膨潤值
switch	开关	開關
switchboard model	接线板模型	插線板模型
switching line	切换线	交換線
symbiosis	共生	共生
symbol	符号	符號

英　文　名	大　陆　名	台　湾　名
symmetric membrane	对称膜	對稱透膜
symmetry	对称性	對稱
symmetry center	对称中心	對稱中心
synchro	同步机	同步器
synchro-control transformer	同步控制变压器	同步控制變壓器
synchro-transmitter	同步传送器	同步傳送器
synchrotron	同步加速器	同步加速器
syncrude	合成原油	合成原油
syndiotactic addition	间周立构加成	對排立構加成[反應]
syndiotacticity	间同立构规整度	對排立構度
syndiotactic placement	间同立构键接	對排立構鍵接
syndiotactic polymer	间同立构聚合物	對排立構聚合物
syndiotactic polymerization	间同立构聚合	對排立構聚合[作用]
syndiotactic sequence	间同立构序列	對排立構序列
syndiotactic unit	间同立构单元	對排立構單元
syneresis	脱水收缩	[膠體]脱水收縮
synergism	增效作用	增效作用
synergist	增效剂	增效劑
synergistic additive	增效添加剂	增效添加劑
synergistic effect	协同效应	增效作用
synergistic flame retardant	增效性阻燃剂	增效阻燃劑
synergistic mechanism	增效机理	增效機構
synergistic mixture	增效混合剂	增效混合物
synergistic stabilizer	增效性稳定剂	增效穩定劑
synfuel	合成燃料	合成燃料
syngas	合成气	合成氣
syntan	合成鞣剂	合成單寧
synthesis	合成	合成
synthesis gas	合成气	合成氣
synthesis process	合成过程	合成程序
synthesis reaction	合成反应	合成反應
synthesis reactor	合成反应器	合成反應器
synthesizer	合成器	合成器
synthetic ammonia	合成氨	合成氨
synthetic chemistry	合成化学	合成化學
synthetic coating materials	合成涂布材料	合成塗料
synthetic coloring agent	合成色素	合成色素
synthetic crude	合成原油	合成原油

英　文　名	大　陆　名	台　湾　名
synthetic crude oil	合成原油	合成原油
synthetic detergent	合成洗涤剂	合成清潔劑
synthetic drying oil	合成干性油	合成乾性油
synthetic elastomer	合成弹性体	合成彈性體
synthetic fat	合成脂肪	合成脂肪
synthetic fatty acid	合成脂肪酸	合成脂肪酸
synthetic fiber	合成纤维	合成纖維
synthetic fiber paper	合成纤维纸	合成纖維紙
synthetic foam	合成泡沫塑料	合成發泡材
synthetic fuel	合成燃料	合成燃料
synthetic gas	合成气	合成氣
synthetic gasoline	合成汽油	合成汽油
synthetic gem	合成宝石	合成寶石
synthetic glycerol	合成甘油	合成甘油
synthetic gypsum	合成石膏	合成石膏
synthetic high polymer	合成高分子	合成高分子
synthetic indigo	合成靛	合成靛
synthetic latex	合成胶乳	合成乳膠
synthetic leather	合成革	合成皮革
synthetic macromolecule	合成高分子	合成大分子
synthetic natural gas	合成天然气	合成天然氣
synthetic natural rubber	合成天然橡胶	合成天然橡膠
synthetic nitrogenous fertilizer	合成氮肥	合成氮肥
synthetic oil	合成油	合成油
synthetic paper	合成纸	合成紙
synthetic plastic	合成塑料	合成塑膠
synthetic plasticizer	合成增塑剂	合成塑化劑
synthetic polymer	合成聚合物	合成聚合物
synthetic polypeptide	合成多肽	合成多肽
synthetic process	合成过程	合成程序
synthetic reaction	合成反应	合作反應
synthetic resin	合成树脂	合成樹脂
synthetic resin adhesive	合成树脂黏合剂	合成樹脂黏合劑
synthetic resin cement	合成树脂胶泥	合成樹脂膠合劑
synthetic resin varnish	合成树脂涂料	合成樹脂清漆
synthetic rubber	合成橡胶	合成橡膠
synthetic rubber adhesive	合成橡胶黏合剂	合成橡膠黏合劑
synthetic rubber latex	合成橡胶胶乳	合成橡膠乳膠

英　文　名	大　陆　名	台　湾　名
synthetic tannin	合成单宁	合成單寧
synthetic tanning materials	合成鞣料	合成鞣料
synthetic wood	合成木材	合成木材
synthetic wool	合成羊毛	合成羊毛
synthon	合成元,合成子	合成組元
syphon	虹吸	虹吸
syphon seal	虹吸式密封	虹吸式密封
syringe	注射器	注射器
syrup	糖浆剂	糖漿
system	系统	系統
system analysis	系统分析	系統分析
system boundary	系统边界	系統邊界
system dynamics	系统动力学	系統動態[學]
system engineering	系统工程	系統工程
system function	①系统函数 ②系统功能	系統函數
system international	国际[单位]系统	國際單位系統
system optimization	系统优化	系統最適化
system parameter	系统参数	系統參數
system performance	系统性能	系統性能
system stability	系统稳定性	系統穩定性

T

英　文　名	大　陆　名	台　湾　名
table salt	食盐	食鹽
tablet	片剂	片劑,錠
tabletting	压片	壓片
tachometer	转速表	轉速計,轉數計
tail gas	尾气	尾氣
tailing	尾料	尾料
tail pipe	尾管	尾管
take off point	接取点	接取點
talc	滑石	滑石
talcum powder	爽身粉	滑石粉
tall oil	妥尔油	松油
tall-oil rosin	妥尔油松香	松香
tall-oil soap	妥尔油皂	松香油皂
tallow	牛脂	牛脂

英　文　名	大　陆　名	台　湾　名
tandem compound turbine	单轴[系]汽轮机	串接雙軸汽輪機
tangential force	切向力	切向力,切線力
tangential stress	切向应力	切線應力
tankcar	槽车	液罐車,油罐車
tank crystallizer	槽式结晶器	槽式結晶器
tanker	油轮	油輪
tank furnace	槽炉	槽爐
tank reactor	槽式反应器	槽式反應器
tanks-in-series model	串联槽模型	串聯槽模式
tannage	鞣制	鞣製
tannery	鞣革厂	製革廠
tannic acid	鞣酸,单宁酸	鞣酸,單寧酸
tannin	鞣质,单宁	鞣酯,單寧
tanning	制革	鞣製
tap density	振实密度	敲緊密度
tapered aeration	渐减曝气	漸減通氣,漸減曝氣
tapered pipe	锥形管	錐形管
tap water	自来水	自來水
tar	焦油	焦油,潻
tar acid	焦油酸	焦油酸,潻酸
tar distillate	焦油馏出液	潻餾出物,焦油馏出物
target efficiency	靶效率	靶效率
target tube	靶管	靶管
tariff	关税	關稅
tar oil	焦油	焦油,潻油
tar sand	沥青砂,焦油砂	潻砂,焦油砂
tartar	酒石	酒石
tartar emetic	吐酒石	吐酒石
tartaric acid	酒石酸	酒石酸
tautomer	互变异构体	互變異構物
tautomerism	互变异构[现象]	互變異構現象
T-die	T 字模	T 字模
tear gas	催泪[毒]气	催淚[毒]氣
tearing	撕裂	裂痕
tearing strength	撕裂强度	撕裂強度
tearing stress	撕裂应力	撕裂應力
technical evaluation	技术评估	技術評估
technical service	技术服务	技術服務

英　文　名	大　陆　名	台　湾　名
technology	技术	技術
technology transfer	技术转移	技術轉移
tee equivalent pipe length	T 型管相当长度	T 形管相當長度
teflon	特氟龙	鐵氟龍,聚四氟乙烯
telemetering	遥测	遥測[術]
temperature	温度	溫度
temperature calibration	温度校正	溫度校正
temperature-composition diagram	温度组成图	溫度-組成圖
temperature control	温度控制	溫度控制
temperature controller	温度控制器	溫度控制器
temperature correction	温度校正	溫度修正
temperature difference	温差	溫差
temperature difference driving force	温差驱动力	溫差驅動力
temperature distribution	温度分布	溫度分佈
temperature-entropy diagram	温熵图	溫熵[關係]圖
temperature gauge	温度计	溫度計
temperature gradient	温度梯度	溫度梯度
temperature-humidity chart	温湿图,T-H 图	溫濕圖
temperature indicating controller	温度指示控制器	溫度指示控制器
temperature indicator	温度指示器	溫度指示器
temperature measuring element	测温组件	測溫元件
temperature profile	温度廓线	溫度剖面圖
temperature programmer	程序变温器	程序變溫器
temperature programming	程序变温	程序變溫
temperature progression	温度历程	溫度歷程
temperature range	温度范围	溫度範圍
temperature recorder	温度记录器	溫度記錄器
temperature regulator	温度调节器	溫度調節器
temperature runaway	飞温,温度失控	溫度失控
temperature scale	温标	溫標
temperature schedule	温度程序	溫度排程
temperature swing adsorption	变温吸附	變溫吸附
temperature transmitter	温度变送器	溫度傳送器
tempered glass	钢化玻璃	強化玻璃
tempering	回火	回火
tempering air	回火空气	回火空氣
tempering coil	调温旋管	調溫旋管
tenacity	韧度	韌度,抗拉強度

英　文　名	大　陆　名	台　湾　名
tenorite	黑铜矿	黑銅礦
tensile strength	抗拉强度	抗拉強度
tensiometer	张力计	張力計
tension	张力	張力
tension testing machine	张力试验机	張力試驗機
tensor	张量	張量
terminal	终端	終端
terminal condition	最终条件	最終條件
terminal falling velocity	终端下降速度	最終下降速度
terminal velocity	终端速度	終端速度
termination agent	终止剂	終止劑
termination reaction	终止反应	終止反應
termolecular reaction	三分子反应	三分子反應
ternary azeotrope	三元共沸物	三元共沸液
ternary system	三元系[统],三组分系统	三成分系[统],三元系[统]
terpene	萜[烯]	[烯]萜[烯]類
terpene oil	萜[烯]油	萜油
terramycin	土霉素	土黴素,地靈黴素
tertiary chemical clarifier	三级化学澄清器	三級化學澄清器
tertiary oil recovery	三次采油	三級採油
tertiary treatment	三级处理	三級處理
tester	测试仪	試驗器
testing	测试	試驗法
test point	测试点	測試點
test procedure	测试过程	測試步驟
test tube	试管	試管
tetracycline	四环素	四環[黴]素
tetraethyl lead	四乙铅	四乙[基]鉛
tetramer	四聚物	四聚物
textile	纺织物,纺织品	織物,紡織品
textile fiber	纺织纤维	紡織纖維
textile finishing	纺织物整理	紡織物整理
thawing	解冻	解凍
theorem	定理	定理
π-theorem	π 定理	π 定理
theoretical air	理论空气量	理論空氣[需要]量
theoretical model	理论模型	理論模式

英　文　名	大　陆　名	台　湾　名
theoretical plate	理论[塔]板	理論板
theoretical plate number	理论板数	理論板數
theoretical stage	理论级	理論階
theoretical tray number	理论板数	理論板數
theory	理论	理論
thermal alkylation	热烷化	熱烷化[反應]
thermal analysis	热分析	熱分析[法]
thermal autoxidation	热自氧化	熱自氧化
thermal balance	热量平衡	熱量均衡
thermal behavior	热行为	熱行為
thermal black	热炭黑	熱碳黑
thermal boundary layer	温度边界层	溫度邊界層
thermal capacity	热容量	熱容量
thermal characterization	热特性	熱示性[法]
thermal conduction	热传导	熱傳導
thermal conductivity coefficient	导热系数	熱傳導係數
thermal conductivity detector	热导检测器	熱傳導[係數]偵檢器
thermal constitutive equation	热本构方程	熱本質方程式
thermal contact resistance	接触热阻	接觸熱阻
thermal convection	热对流	熱對流
thermal cracking	热裂化	熱裂解
thermal cycle	热循环	熱循環
thermal deactivation	热失活	熱失活
thermal decomposition	热分解	熱分解
thermal degradation	热降解	熱降解
thermal depolymerization	热解聚	熱解聚合[反應]
thermal devolatilization	热脱挥发物作用	熱脫揮發物作用
thermal diffusion	热扩散	熱擴散
thermal diffusion coefficient	热扩散系数	熱擴散係數
thermal diffusivity	热扩散系数	熱擴散係數
thermal dissipation	热耗散	熱耗散
thermal dissociation	热解离[反应]	熱解離[反應]
thermal efficiency	热效率	熱效率
thermal endurance	耐热性	耐熱性
thermal energy	热能	熱能
thermal energy equation	热能方程式	熱能方程式
thermal equilibrium	热平衡	熱平衡
thermal expansion	热膨胀	熱膨脹

英　文　名	大　陆　名	台　湾　名
thermal expansion coefficient	热膨胀系数	熱[膨]脹係數
thermal fission	热裂变	熱[中子]分裂
thermal insulation	隔热	隔熱
thermal method	热方法	熱方法
thermal pollution	热污染	熱污染
thermal polymerization	热聚合	熱聚合[反應]
thermal precipitation	热沉降	熱沈澱[作用]
thermal radiation	热辐射	熱輻射
thermal reforming	热重整	熱重組[反應]
thermal resistance	热阻	熱阻
thermal shock	热冲击	熱衝擊
thermal shock test	热冲击试验	熱衝擊試驗
thermal source	热源	熱源
thermal stability	热稳定性	熱安定性
thermal strain	热应变	熱應變
thermal stress	热应力	熱應力
thermal synthesis	热合成	熱合成
thermal technique	热方法	熱方法
thermal unit	热单位	熱單位
thermal value	热值	熱值
thermal value test	热值试验	熱值試驗
thermal wave	热波	熱波
thermal well	热套管	熱套管
thermistor	热敏电阻器	熱[敏電]阻器
thermite	铝热剂	鋁熱劑
thermoacoustimetry	热传声法	熱聲學
thermoammeter	热偶安培计	熱偶安培計
thermoanalysis	热分析	熱分析
thermoanalytical method	热分析法	熱分析法
thermoanalytical microscopy	热分析显微镜	熱分析顯微術
thermoanalyzer	热分析仪	熱分析儀
thermoatomic process	热原子法	熱原子法
thermobalance	热天平	熱天平
thermobattery	热电池	熱電池
thermochemistry	热化学	熱化學
thermo-conductivity cell	热传导系数测定池	熱傳導係數測定槽
thermocouple	热电偶	熱[電]偶
thermocouple pyrometer	热电偶高温计	熱[電]偶高溫計

英　文　名	大　陆　名	台　湾　名
thermocouple reference junction	热电偶参考接点	熱[電]偶參考接點
thermocouple type anemometer	热电偶式流速计	熱[電]偶式流速計
thermocurrent	热电流	熱電流
thermodiffusion	热扩散	熱擴散
thermodynamic analysis	热力学分析	熱力學分析
thermodynamic analysis of process	过程热力学分析	程序熱力學分析
thermodynamic characteristic function	热力学特性函数	熱力學特徵函數
thermodynamic consistency	热力学一致性	熱力學一致性
thermodynamic consistency test	热力学一致性检验	熱力學一致性試驗[法]
thermodynamic constitutive equation	热力学本构方程	熱力學本質方程式
thermodynamic diagram	热力学图	熱力學圖
thermodynamic efficiency	热力学效率	熱力學效率
thermodynamic equilibrium	热力学平衡	熱力學平衡
thermodynamic flux	热力学通量	熱力學通量
thermodynamic function	热力学函数	熱力學函數
thermodynamic probability	热力学概率	熱力學機率
thermodynamic property	热力学性质	熱力學性質
thermodynamic property table	热力学性质表	熱力學性質表
thermodynamic relation	热力学关系	熱力學關係
thermodynamics	热力学	熱力學
thermodynamic system	热力学系统	熱力學系統
thermodynamic temperature	热力学温度	熱力學溫度
thermodynamic temperature scale	热力学温标	熱力學溫標
thermoelectric effect	热电效应	熱電效應
thermoelectricity	热电	熱電
thermoelectric junction	热电偶接点	熱電[偶]接點
thermoelectric potentiometer	热电式电位计	熱電[式]電位計
thermoelectric pyrometer	热电高温计	熱電高溫計
thermoelectric thermometer	热电温度计	熱電溫度計
thermoelectrometer	热电计	熱電計
thermoelectrometry	热电法	熱電法
thermofixation	热固定	熱固定
thermofor catalytic cracking process	热催化裂化工艺	熱觸媒裂解法
thermofor catalytic reforming process	热催化重整工艺	熱觸媒重組法
thermofor continuous percolation process	热连续渗滤工艺	熱連續滲濾法
thermofor pyrolytic cracking process	高温热裂解工艺	高熱裂解法
thermoforming	热成型	熱壓成型

英 文 名	大 陆 名	台 湾 名
thermogalvanometer	热电流计	熱電流計
thermogravimetric analysis	热重分析[法]	熱重分析[法]
thermogravimetric analyzer	热重量分析仪	熱重量分析儀
thermogravimetry	热重分析法	熱重分析術
thermokinematics	热运动学	熱運動學
thermoluminescence	热[致]发光	熱[致]發光
thermomagnetic analysis	热磁分析	熱磁分析[法]
thermomagnetism	热磁性	熱磁性
thermomechanical analysis	热机械分析	熱機械分析[法]
thermometer	温度计	溫度計
thermometric titration	热滴定[法]	溫度滴定[法]
thermometric titrimetry	温度滴定法	溫度滴定[分析]法
thermometry	测温法	測溫法
thermonuclear reaction	热核反应	熱核反應
thermoosmosis	热渗透	熱滲透[作用]
thermooxidative degradation	热氧化降解	熱氧化降解
thermopane	双层隔热玻璃板	雙層隔熱玻璃板
thermoparticulate analysis	热颗粒分析	熱顆粒分析[法]
thermophilic bacteria	嗜热细菌	嗜熱菌
thermophoresis	热电泳	熱泳法
thermophotometry	热光分析[法]	熱光分析[法]
thermophysical property	热物理性质	熱物理性質
thermopile	温差电堆	熱電堆
thermoplasticity	热塑性	熱塑性
thermoplastic plastics	热塑性塑料	熱塑性塑膠
thermoplastic polymer	热塑性聚合物	熱塑性聚合物
thermoplastic resin	热塑性树脂	熱塑性樹脂
thermopolymerization	热聚合	熱聚合[反應]
thermoreflectometry	热反射法	熱反應[分析]法
thermo regulating valve	温度调节阀	熱調閥
thermoregulator	温度调节器	調溫器
thermoset	热固性塑料	熱固物
thermosetting	热固化	熱固性
thermosetting resin	热固性树脂	熱固性樹脂
thermo siphon	热虹吸	熱虹吸
thermosiphon reboiler	热虹吸式再沸器	熱虹吸再沸器
thermosonimetry	热发声法	熱聲[分析]法
thermostability	热稳性	熱安定性

英　文　名	大　陆　名	台　湾　名
thermostat	恒温器	恆溫器
thermostatics	热静力学	熱靜力學
thermostatic trap	恒温阱	恆溫阱
thermotropic liquid crystal	热致液晶	熱致液晶
thermotropism	向热性	向熱性
thermowell	温度计套管	溫度計套管
thickener	增稠器	增濃器
thickening	增稠	增濃,增黏
thickening agent	增稠剂	增稠劑
thickening capacity	增稠能力	增稠能力
thickening process	增稠过程	增濃程序,增黏程序
thickening sludge	污泥增稠	增黏污泥
thickness	稠度	厚度
thick slurry process	浓浆法	濃漿法
Thiele modulus	蒂勒模数	蒂勒模數
thin-boiling starch	稀糊化淀粉	稀糊化澱粉
thin-film evaporator	薄膜蒸发器	薄膜蒸發器
thin layer chromatography	薄层色谱法	薄層層析法
thinner	减黏剂,稀释剂	減黏劑,稀釋劑
thiohydrogenolysis	硫氢解反应	硫氫解[反應]
thiohydrolysis	硫代水解	硫化氫解[反應]
thioindigo	硫靛蓝	硫靛
thiokol	聚硫橡胶	多硫橡膠
thiokol polymer	聚硫聚合物	多硫聚合物
thiol	硫醇	硫醇
thiolmodified rubber	硫醇改性橡胶	硫醇改質橡膠
thiostrepton	硫链丝菌肽	硫鏈絲菌肽
thiourea resin	硫脲树脂	硫脲樹脂
third law of thermodynamics	热力学第三定律	熱力學第三定律
thixotrope	触变胶	觸變減黏膠,搖變減黏膠
thixotropic fluid	触变性流体	觸變減黏流體,搖變減黏流體
thixotropic property	触变性质	觸變減黏性質,搖變減黏性質
thixotropy	触变性	觸變減黏性,搖變減黏性
thorite	硅酸钍矿	矽酸釷礦

英　文　名	大　陆　名	台　湾　名
three-mode control	三式控制	三式控制
three-mode controller	三式控制器	三式控制器
three-phase fluidization	三相流态化	三相流體化
three-phase fluidized bed	三相流化床	三相流[體]化床
three-phase reactor	三相反应器	三相反應器
threshold	阈[值]	閾值,低限
threshold concentration	浓度极限	閾濃度,低限濃度
threshold energy	阈能	閾能,低限能
threshold frequency	阈频[率]	閾頻,低限頻率
threshold limit value	阈限值	閾值,低限值
throat	喉管	喉道
throat velocity	喉道流速	喉道流速
throttle valve	节流阀	節流閥
throttling	节流	節流
throttling band	节流带	節流帶
throttling calorimeter	节流量热计	節流卡計
throttling device	节流装置	節流裝置
throttling process	节流过程	節流程序
throttling valve	节流阀	節流閥
throughput	产量	生產量
throwing	拉坯	拉坯
throwing power	均镀能力	均鍍力
tie line	结线	連結線
time and motion study	工时学	工時學
time constant	时间常量	時間常數
time control	时间控制	時間控制,時控
time delay	时延	時[間遲]延
time dependence	时间关连	時間關連[性]
time derivative	时间导数	時間導數
time domain	时域	時域
time invariant system	非时变系统	非時變系統
time invarying system	不变时间系统	非時變系統
time lag	时滞	時[間遲]滯
time line	时间线	時間線
time minimum control	时间最短控制	時間最短控制
timer	计时器	計時器,定時器
time ratio method	时间比率法	時間比率法
time scaling	时间标度	時間標度

英　文　名	大　陆　名	台　湾　名
time schedule	时程表	時程表
time series	时间序列	時間序列
time series model	时间序列模型	時間數列模型
time sharing	分时	分時系統
time switch	定时开关	定時開關
time value of money	资金的时间价值	貨幣時間價值
time variant system	时变系统	時變系統
time varying system	时变系统	時變系統
tinfoil	锡箔	錫箔
tin mordant	锡媒染剂	錫媒染劑
tinning	镀锡	鍍錫
tin plate	镀锡板	馬口鐵
tin plating	镀锡	鍍錫
tin soap	锡皂	錫皂
tip speed	浆尖速度	梢速
tire	轮胎	輪胎
tire cord	轮胎帘线	輪胎簾布
tire yarn	轮胎砂	輪胎紗
tissue	组织	組織
titania	二氧化钛	氧化鈦
titania porcelain	钛瓷	鈦瓷
titania whiteware	钛白陶	鈦白陶[器]
titanium alloy	钛合金	鈦合金
titanium enamel	钛搪瓷	鈦搪瓷
titanium sponge	海绵钛	海綿鈦
titanium white	钛白	鈦白
titration	滴定	滴定[法]
titrimeter	滴定计	滴定計
titrimetric method	滴定[分析]法	滴定[分析]法
tolerance interval	容许区间	容許區間
top phase	顶相	頂[層]相
topping	拔顶[蒸馏]	直鎦
top plate	顶板	頂板
torque	转矩	扭矩
torque rheometer	转矩流变仪	扭矩流變儀
torque tube	转矩管	扭矩管
torr	托(压力单位)	托
torsion	扭转	扭轉,扭力

英 文 名	大 陆 名	台 湾 名
torsional braid analysis	扭辫分析	扭辮分析
torsional creep	扭转蠕变	扭轉潛變
torsional flow	扭转流动	扭轉流動
torsional stress	扭转应力	扭轉應力
torsion balance	扭力天平	扭力天平
torsion viscometer	扭力黏度计	扭力黏度計
tortuosity	曲折因子	扭曲度
tortuosity factor	弯曲因子	扭曲因數
total capital investment	总资本投资额	總資本投資額
total derivative	全导数	全導數
total emissivity	总发射系数	總發射係數
total energy	总能	總能
total hardness	总硬度	總硬度
total head	总压头	總高差
total organic carbon	总有机碳	總有機碳
total pressure	总压力	總壓力
total pressure method	总压法	總壓法
total recycle process	全部循环过程	全部循環程序
total reflux	全回流	全回流
total solid	总固体	總固體[量]
toughened glass	钢化玻璃	韌化玻璃
toughened polystyrene	韧性聚苯乙烯	韌化聚苯乙烯
toughness	①韧性 ②韧度	①韌性 ②韌度
tower	塔	塔[器]
tower hold-up	塔贮留量	塔滯留量
tower packing	塔填充物	塔填充物
toxic chemical	毒性化学品	毒性化學品
toxic gas	毒气	毒氣
toxicity	毒性	毒性
toxic materials	有毒材料	有毒材料,毒物
toxic smoke	毒烟	毒煙
toxin	毒素	毒素
T-piece	T形件,三通	T形件,三通
trace element	痕量元素	痕量元素,微量元素
tracer	示踪剂	示蹤劑
tracer chemistry	示踪化学	示蹤[劑]化學
tracer curve	示踪曲线	示蹤劑曲線
tracer information	示踪信息	示蹤劑資訊

英　文　名	大　陆　名	台　湾　名
tracer isotope	示踪同位素	示蹤同位素
tracer response curve	示踪响应曲线	示蹤劑應答曲線
tracer technique	示踪技术	示蹤劑法
tracing paper	描图纸	描圖紙
tracing steam	随管加热蒸汽	隨管加熱蒸汽
tragacanth gum	黄芪胶	黄芪膠
trailing vortex	尾随涡	拖尾漩渦
trans addition	反式加成	反式加成[反應]
transalkylation	烷基转移	轉烷化[作用]
transcrystallization	横晶	横穿結晶
transducer	传感器	[信號]轉換器
transesterification	酯交换	轉酯化[作用]
transfer	传递	傳送,轉移
transfer coefficient	传递系数	傳送係數
transfer function	传递函数	轉換函數
transfer lag	传递滞后	傳送滯延
transfer molding	传递成型	轉注成型
transfer unit	传递单元	傳送單元
transform	变换	變換[式]
transformer	变压器	變壓器
transient behavior	瞬态行为	瞬態行為,暫態行為
transient diffusion	瞬态扩散	瞬態擴散,暫態擴散
transient phenomenon	瞬态现象	瞬態現象,暫態現象
transient response	瞬态响应	瞬態應答,暫態應答
transient response analysis	瞬态响应分析	瞬態應答分析,暫態應答分析
transistor	晶体管	電晶體
transition element	过渡元素	過渡元素
transition flow	过渡流	過渡流
transition length	过渡长度	過渡長度
transition metal	过渡金属	過渡金屬
transition point	转变点	轉變點
transition range	转变范围	轉變範圍
transition region	过渡区	過渡區
transition state	过渡态	過渡狀態
transition-state theory	过渡态理论	過渡狀態理論
transition temperature	转变温度	轉變溫度,轉移溫度
translational energy	平移位能	平移能,移動能

英 文 名	大 陆 名	台 湾 名
translational partition function	平动配分函数	移動分配函數
translation of function	函数平移	函數平移
translucence	半透明度	半透明性
transmethylation	甲基转移作用	轉甲基化[作用]
transmission lag	传动滞后	傳送滯延
transmission line	传输线	傳輸線
transmissivity	透射率	透射係數
transmittance	透光率	透光率,透射係數
transmitter	变送器	傳送器,[信號]發射器
transmitter gain	变送器增益	傳送器增益
transparency	透明度	透明度,透明性
transparent glaze	透明釉	透明釉
transparent soap	透明皂	透明皂
transpiration	蒸腾	蒸散[作用]
transpiration cooling	发散冷却	蒸散冷卻
transportation	运输	運輸,輸送
transportation lag	运输滞后	輸送滯延
transport diffusivity	传递扩散系数	輸送擴散係數
transport disengaging height	分离高度	[流體化床]分離高度
transport phenomenon	传递现象	輸送現象
transport property	运输性质	輸送性質
transverse fin	横向翅片	横向鰭片
trap	①阱 ②疏水器	①阱 ②閘
trapezoidal weir	梯形堰	梯形堰
traveling screen	移动筛	移動篩
tray	塔板	塔板
tray column	板式塔	板[式]塔
tray drier	厢式干燥器	盤[式]乾燥器
tray efficiency	板效率	板效率
tray spacing	塔板间距	塔板間距
treatment device	处理设备	處理設備
treatment plant	处理工厂	處理工場
trench	海沟	溝
triangular coordinates	三角坐标	三角坐標
triangular diagram	三角图	三角圖
triangular notch	三角缺口	三角凹槽
triaxial stress	三轴应力	三軸應力
tricarboxylic acid cycle	三羧酸循环	三羧酸循環

英　文　名	大　陆　名	台　湾　名
trickle bed	滴流床	滴流床
trickle-bed reactor	滴流床反应器	滴流床反應器
trickling	滴流	滴流
trickling filter	滴滤池	滴濾池
trickling filtration	滴滤	滴濾
tridiagonal matrix	三对角矩阵	三對角矩陣
trimer	三聚体	三聚物
trimolecular reaction	三分子反应	三分子反應
triple point	三相点	三相點
triple salt	三合盐	三合鹽
tripper	①倾卸装置 ②跳开装置	跳動装置
trip wire	拉发线	引線
trommel	圆筒筛	礦石篩,轉筒篩
trona	天然碱	碳酸鈉石
trough	料槽	槽
true boiling point	真沸点	真沸點
true boiling point curve	实沸点蒸馏曲线	實沸點曲線
true density	真密度	真密度
true mean	实平均	實平均
true rate of return	实报酬率	實報酬率
tube bank	管排	管排
tube bundle	管束	管束
tube drier	管束干燥器	管式乾燥器
tube furnace	管式炉	管形爐
tube gage	管式气压计	管式壓力計
tube mill	管磨机	管磨機
tube pass	管程	管程
tube sheet	管板	管板
tube side pass	管程	管程
tube still	管式蒸馏釜	管餾器
tubing machine	制管机	製管機
tubular condenser	管式冷凝器	管式冷凝器
tubular evaporator	管式蒸发器	管式蒸發器
tubular heater	管式加热器	管式加熱器
tubular module	管式组件	管式組件
tubular reactor	管式反应器	管式反應器
tubular still	管式蒸馏釜	管餾器
tumbler mixer	转鼓混合机	轉鼓混合機

英　文　名	大　陆　名	台　湾　名
tumbling mill	滚磨机	滚磨機
tung oil	桐油	桐油
tuning	调谐	調諧
tuning parameter	调谐参数	調諧參數
tunnel drier	隧道干燥器	隧道乾燥器
tunnel kiln	隧道窑	隧道窯
turbidimeter	①浊度计 ②比浊法	濁度計
turbidity	浊度	濁度
turbidometer	浊度计	濁度計
turbine	涡轮	渦輪機
turbine agitator	涡轮搅拌器	渦輪攪拌器
turbine centrifugal pump	涡轮离心泵	渦輪離心泵
turbine flowmeter	涡轮流量计	渦輪[式]流量計
turbine impeller	涡轮叶轮	渦輪[式]葉輪
turbine pump	涡轮泵	渦輪泵
turbo-blower	涡轮鼓风机	渦輪鼓風機
turbo-compressor	涡轮压缩机	渦輪壓縮機
turbo-dryer	涡轮干燥机	渦輪乾燥機
turbo-grid tray	穿流栅板	穿流栅板
turbo-machine	涡轮机	渦輪機
turbo-pump	涡轮泵	渦輪泵
turbulence promoter	湍流促进器	紊流促進元件
turbulence scale	湍流标度	紊流標度
turbulent boundary layer	湍流边界层	紊流邊界層
turbulent core	湍流核心	紊流核心
turbulent diffusion	湍流扩散	紊流擴散
turbulent diffusivity	湍流扩散系数	紊流擴散係數
turbulent-energy spectrum	湍流能谱	紊流能譜
turbulent flow	①湍流 ②紊流	紊流
turbulent flow reactor	湍流反应器	紊流反應器
turbulent fluidized bed	湍动流化床	紊動流[體]化床
turbulent motion	湍流运动	紊流運動
turbulent relaxation phenomenon	湍流松弛现象	紊流弛豫現象
turbulent shear stress	湍流剪应力	紊流剪應力
turbulent stress	湍流应力	紊流應力
turndown ratio	[燃烧器]调节比	[燃燒器]調節比
turnover number	转化数	[觸]媒轉化數
turnover ratio	周转率	周轉率

英　文　名	大　陆　名	台　湾　名
turpentine oil	松节油	松節油
tuyere distributor	风帽分布板	風帽分佈器
twin-screw extruder	双螺杆挤出机	雙螺桿擠出機
two-cell stabilization pond	双室稳定化池	雙室穩定化池
two-cycle engine	二行程引擎	二行程引擎
two-dimensional chromatography	二向色谱法	二維層析[法]
two-dimensional model	二维模型	二維模式
two-film theory	双膜理论	雙膜理論
two-fluid theory	两流体理论	雙流體說
two-liquid theory	两液体理论	兩液體理論
two-mode control	双式控制	雙式控制
two-phase flow	两相流	兩相流
two-phase flow pattern	两相流动型式	兩相流動型式
two-phase heat transfer	两相传热	兩相熱傳
two-phase model	两相模型	兩相模式
two-phase region	两相区	兩相區[域]
two-phase system	两相系统	兩相系統
two-position actuator	双位驱动器	雙位致動器
two-position controller	双位控制器	雙位控制器
two-resistance theory	双阻力理论	雙阻力理論
two-stage cascade system	两段串级系统	二階串級系統
two-stage distillation	两段蒸馏	二階蒸餾
two-stage pilot valve	两级导向阀	二階導引閥
two tier approach	双层联立模块法	二階法
Tyler standard sieve	泰勒标准筛	泰勒標準篩
tylosin	泰乐菌素	泰黴素
tyrocidine	短杆菌酪肽	短桿菌酪肽
tyrothricin	短杆菌素	短桿菌素

U

英　文　名	大　陆　名	台　湾　名
ultimate disposal	最终处理	最終處理
ultimate period	最终周期	最終週期
ultimate periodic response	最终周期应答	最終週期應答
ultimate periodic solution	最终周期解	最終週期解
ultimate sensitivity	极限灵敏度	極限靈敏度
ultimate stress	极限应力	極限應力

英　文　名	大　陆　名	台　湾　名
ultra accelerator	超加速剂	超催速劑
ultracentrifugation	超速离心	超離心[作用]
ultracentrifuge	超[速]离心机	超離心機
ultrafilter	超滤机	超過濾器
ultrafiltration	超滤	超過濾
ultragrinder	超细研磨机	超細研磨機
ultramarine	群青	群青
ultramarine blue	群青蓝	群青藍
ultramarine brown	群青棕	群青棕
ultramarine green	群青绿	群青綠
ultramarine violet	群青紫	群青紫
ultramarine yellow	群青黄	群青黃
ultra microchemistry	超微量化学	超微量化學
ultra microscope	超显微镜	超顯微鏡
ultra microscopy	超显微法	超顯微法
ultrapurification	超净化	超淨化
ultrasonic agglomeration	超声附聚	超音波團聚
ultrasonic cleaning	超声净化	超音波清潔
ultrasonic flow	超声波流动	超音波流
ultrasonic meter	超声波计	超音波計
ultrasonic precipitation	超声波沉淀	超音波沈澱
ultrasonics	超声波学	超音波學
ultrasonic vibrator	超声波振动器	超音波振動器
ultrasonic viscometer	超声波黏度计	超音波黏度計
ultrathin film	超薄膜	超薄膜
ultraviolet absorber	紫外线吸收剂	紫外線吸收劑
ultraviolet absorption	紫外线吸收	紫外線吸收
ultraviolet colorimeter	紫外线比色计	紫外線比色計
ultraviolet curing	紫外线熟化	紫外線熟化
ultraviolet fluorescence	紫外线荧光	紫外線螢光
ultraviolet generator	紫外线发生器	紫外線產生器
ultraviolet lamp	紫外灯	紫外線燈
ultraviolet light	紫外光	紫外光
ultraviolet ray	紫外线	紫外線
ultraviolet spectrophotometer	紫外[线]分光光度计	紫外線分光光度計
umber	棕土(颜料)	富錳棕土
unbound water	非结合水分	非結合水分
uncompetitive inhibition	反竞争性抑制	反競爭抑制[作用]

英　文　名	大　陆　名	台　湾　名
unconstrained optimization	无约束优化	無約束最適化
uncoupling	解偶联	解偶
undamped natural frequency	无阻尼自然频率	無阻尼自然頻率
undamped response	无阻尼响应	無阻尼應答
undamped system	无阻尼系统	無阻尼系統
undercoat	底漆	底漆
under cure	欠硫	低硫化
under cured rubber	欠硫化橡胶	低硫化橡膠
underdamped response	欠阻尼响应	欠阻尼應答
underdamped system	欠阻尼系统	欠阻尼系統
underdamping	欠阻尼	欠阻尼
underdrainage system	地下排水系统	排水系統
under-driven buhrstone mill	下传动式石磨机	底動石磨機
underflow	底流	底流
underflow locus	底流轨迹	底流軌跡
underflow rate	底流速率	底流速率
underflow velocity	底流速度	底流速度
underloading	负荷不足	負荷不足
undershoot	低于额定值	低於額定值,不足量
uniaxial stress	单轴应力	單軸應力
unicellular growth	单细胞生长	單細胞生長
unidirectional reaction	单向反应	單向反應
uniflux tray	S 形塔板	單向流塔板
uniform coking	均匀结焦	均匀結焦
uniform conversion	均匀转化	均匀轉化
uniform conversion model	均匀转化模型	均匀轉化模型
uniform crystal	均匀晶体	均匀晶體
uniform distribution	均匀分布	均匀分佈
uniform flow	均匀流	均匀流動
uniformity	均匀性	均匀性
uniformity coefficient	均匀系数	均匀性係數
uniform loading	均匀载荷	均匀負荷
uniform poisoning	均匀中毒	均匀中毒
uniform stream	均匀流动	均匀流
unilateral slit	单向狭缝	單向狹縫
unimolecular reaction	单分子反应	單分子反應
union	活[管]接头	活[管]接頭
union of asymptotic stability	渐近稳定性联合	漸近穩定聯合域

英　文　名	大　陆　名	台　湾　名
uniqueness of steady state	稳态唯一性	穩態唯一性
unique steady state	单一稳态	單一穩態
unit computation	单元计算	單元計算
unit cost	单位成本	單位成本
unit cost estimation	单位成本估计	單位成本估計
unit feedback	单位回馈	單位回饋
unit operation	单元操作	單元操作
unit process	单元过程	單元程序
universal condenser	通用冷凝器	通用冷凝器
universal gas constant	普适气体常量	通用氣體常數
universal indicator	通用指示剂	通用指示劑
universal joint	万向接头	萬向接頭
universal quasi-chemical activity coefficient method(UNIQUAC method)	活度系数通用准化学关联法	UNIQUAC 法
universal quasi-chemical functional group activity coefficient method(UNIFAC method)	活度系数通用准化功能团贡献法	UNIFAC 法
universal velocity distribution	泛速度分布	泛速度分佈
unleaded gasoline	无铅汽油	無鉛汽油
unloading	卸载	卸載
unmixed flow reactor	无混合流动反应器	無混合流動反應器
unox process	纯氧活性污泥法	純氧活性污泥法
unpacked-tube reactor	空管反应器	空管反應器
unreacted core model	未反应核模型	未反應核模型
unsaponified matter	不皂化物	不皂化物
unsaturated fatty acid	不饱和脂肪酸	不飽和脂肪酸
unsaturated hydrocarbon	不饱和烃	不飽和烴
unsaturated polyester	不饱和聚酯	不飽和聚酯
unsaturation	不饱和	不飽和
unstable operating point	不稳定操作点	不穩定操作點
unstable response	不稳定响应	不穩定應答
unstable state	非稳态	不穩定狀態
unstable steady state	不稳定状态	不安定穩態
unstable system	不稳定系统	不穩定系統
unsteady flow	非定常流	非穩態流動
unsteady state	非定态	非穩[定]態
unsteady state heat transfer	非定态传热	非穩態熱傳
unsteady state operation	非定态操作	非穩態操作

英 文 名	大 陆 名	台 湾 名
unsteady state process	非定态过程	非穩態程序
unstructured model	非结构模型	非結構化模式
unsupported catalyst	无载体催化剂	無擔載觸媒
upflow column	上流塔	上向流塔,升流塔
upgrading	提升	[品質]提升,升級
upper bound	上限	上限
upper control limit	上控制限	管制上限
upper limit	上极限	上限
upstream	上游	上游
upstream pressure	上游压力	上游壓力
uptake	摄取	攝取
uptake rate	摄取速率	攝取速率
uranite	云母铀矿	瀝青鈾礦
urea	尿素	尿素,脲
urea-formaldehyde resin	脲醛树脂	脲甲醛樹脂
urease	脲酶	脲酶,尿素酶
urethane	尿烷	胺甲酸乙酯
utility	公用设施	公用設施
utility check list	公共设施检查报表	公共設施檢查報表
utility flowsheet	公共设施流程图	公用設施流程圖
utility service	公用服务事业	公用設施
U-tube	U 型管	U 形管
U-tube heat exchanger	U 型管换热器	U 形管熱交換器
U-type crystallizer	U 型结晶器	U 形結晶器
UV spectrophotometer	紫外光谱仪	紫外線光譜儀

V

英 文 名	大 陆 名	台 湾 名
vaccine	疫苗	疫苗
vacuometer	真空计	真空計
vacuum	真空	真空
vacuum column	真空蒸馏塔	真空蒸餾塔
vacuum concentration	真空浓缩	真空濃縮
vacuum crystallization	真空结晶	真空結晶
vacuum crystallizer	真空结晶器	真空結晶器
vacuum dehydrator	真空脱水器	真空脱水器
vacuum desiccator	真空干燥器	真空乾燥器

英 文 名	大 陆 名	台 湾 名
vacuum discharge	真空放电	真空放電
vacuum distillation	真空蒸馏	真空蒸餾
vacuum drum filter	真空鼓式过滤器	真空濾桶
vacuum drying	真空干燥	真空乾燥
vacuum drying apparatus	真空干燥器	真空乾燥装置
vacuum drying oven	真空干燥箱	真空乾燥烘箱
vacuum evaporation	真空蒸发	真空蒸發
vacuum evaporator	真空蒸发器	真空蒸發器
vacuum filter	真空过滤机	真空過濾機
vacuum flash vaporization	真空闪蒸	真空驟汽化
vacuum flotation	真空浮选	真空浮選
vacuum forming	真空成型	真空成型
vacuum fractionator	真空分馏器	真空分餾器
vacuum gauge	真空计	真空計
vacuum grease	真空润滑油	真空潤滑脂
vacuum head	真空高差	真空高差
vacuum manometer	真空压力计	真空[壓力]計
vacuum molding	真空成型	真空成型
vacuum oil	真空油	真空油
vacuum oven	真空烘箱	真空烘箱
vacuum pan	真空锅	真空罐
vacuum press	真空压制机	真空壓製機
vacuum pressing	真空压制	真空壓製
vacuum pressure	真空压	真空壓力
vacuum process	真空过程	真空程序
vacuum pump	真空泵	真空泵
vacuum relief	真空解除	破真空
vacuum retort	真空甑	真空甑
vacuum seal	真空封接	真空密封
vacuum shelf drier	真空盘架干燥器	真空箱乾燥器
vacuum still	真空蒸馏釜	真空蒸餾器
vacuum tar	真空焦油	真空焦油
vacuum tray drier	真空盘架干燥器	真空盤式乾燥器
vacuum tube	真空管	真空管
vacuum valve	真空阀门	真空閥
valve actuator	阀驱动器	製動閥
valve body	阀体	閥[主]體
valve capacity	阀容量	閥容量

英　文　名	大　陆　名	台　湾　名
valve characteristics	阀特性	閥特性
valve control	阀控制	閥控制
valve equivalent pipe length	阀当量管长	閥相當管長
valve gain	阀增益	閥增益
valve hysteresis	阀滞后	閥遲滯
valve identification code	阀鉴定编码	閥識別碼
valve loss	阀损失	閥損失
valve plug	阀塞	閥塞
valve positioner	阀门定位器	閥定位器
valve rangeability	阀范围性	閥範圍性
valve resistance	阀阻力	閥阻力
valve response	阀响应	閥回應
valve sequencing	阀序列	閥定序
valve stem	阀杆	閥桿
valve tray	浮阀塔板	閥板
vanadium contact process	钒接触法	釩接觸法
vancomycin	万古霉素	萬古黴素
van der Waals adsorption	范德瓦耳斯吸附	凡得瓦吸附[作用]
van der Waals constant	范德瓦耳斯常数	凡得瓦常數
van der Waals equation	范德瓦耳斯方程	凡得瓦方程式
van der Waals equation of state	范德瓦耳斯状态方程	凡得瓦狀態方程式
van der Waals force	范德瓦耳斯力	凡得瓦力
vane flow	叶片式流动	葉輪流動
vane type blower	叶片式鼓风机	葉輪式鼓風機
vanilla	香草	香草
vanillin	香草醛	香草精
van Laar equation	范拉尔方程	范拉爾方程
vanner	淘矿机	淘洗器
van't Hoff equation	范托夫方程式	凡特何夫方程式
van't Hoff factor	范托夫因子	凡特何夫因子
van't Hoff law	范托夫定律	凡特何夫定律
vapor	蒸气	蒸氣,汽
vapor bypass	蒸气旁路	蒸氣旁路
vapor compression	蒸气压缩	蒸氣壓縮
vapor deposition	气相沉积	氣相沈積[法]
vapor diffusion pump	蒸气扩散泵	蒸氣擴散泵
vaporization	汽化	汽化
vaporization curve	汽化曲线	汽化曲線

英 文 名	大 陆 名	台 湾 名
vaporizer	汽化器	汽化器
vapor-liquid equilibrium	汽液平衡	汽液平衡
vapor-liquid equilibrium ratio	汽液平衡比	汽液平衡比
vapor-liquid separator	汽液分离器	汽液分離器
vapor phase association	汽相缔合	汽相締合
vapor pressure	蒸气压	蒸氣壓
variability	变异性	變異性
variable	变量	變數
variable cost	变动成本	變動成本
variable orifice meter	可变孔板流量计	可變孔流量計
variable-speed pump	变速泵	變速泵
variable transformer	调压变压器	可調變壓器
variant	变体	變異體
variational calculus	变分微积分	變分學
varnish	清漆	清漆
vaseline	凡士林	凡士林
vat dye	还原染料	甕染料
vat leaching	桶式浸取	槽瀝取
vegetable oil	植物油	植物油
vehicle	①载体 ②漆料	①載具 ②媒液, 展色劑
velocity	速度	速度
velocity boundary layer	速度边界层	速度邊界層
velocity boundary layer thickness	速度边界层厚度	速度邊界層厚度
velocity-distance lag	速度距离滞后	速度距離滯延
velocity distribution	速度分布	速度分佈
velocity gradient	速度梯度	速度梯度
velocity head	速度压头	速度高差
velocity meter	速度计	速度計
velocity potential	速度势	速度勢
velocity profile	速度[分布]剖面[图]	速度剖面[圖], 速度分布
vena contracta	流颈, 缩脉	收縮口
vena contracta tap	流颈接头	縮口分接頭
vent	①排空 ②排气孔	通氣孔, 排氣孔
vent and drain seal	排气及排泄口密封	排氣卸液口密封
ventilation	通风	通風
ventilation requirement	通风需求	通風需求
ventilator	通风机	通風器

英　文　名	大　陆　名	台　湾　名
vent line	排气管线	排放管線
vent pipe	排气管	通風管
vent ratio	通风比例	通風比率
venture analysis	风险分析	風險分析
venture profit	风险利润	風險利潤
Venturi	文丘里	文氏管
Venturi flowmeter	文丘里流量计	文氏流量計
Venturi meter	文丘里流量计	文氏計
Venturi nozzle	文丘里喷嘴	文氏噴嘴
Venturi scrubber	文丘里洗涤器	文氏管洗滌器
Venturi tube	文丘里管	文氏管
Vernier caliper	游标卡尺	游標卡尺
Vernier scale	游标尺	游標尺
vertical digester	立式蒸煮器	直立式消化器
vertical retort	竖甑	豎甑
vertical sieve tray	垂直筛板	垂直篩板
vertical tube evaporator	竖管式蒸发器	豎管蒸發器
very large scale integrated circuit（VLSI）	超大规模集成电路	超大型積體電路
vibrated fluidized bed	振动流化床	振動流[體]化床
vibrating screen	振动筛	振動篩
vibrational energy	振动能	振動能
vibrational energy level	振动能级	振動能階
vibration partition function	振动配分函数	振動分配函數
vibrator	振动器	振動器
vibratory feeder	振动给料器	振動進料機
vibrometer	测振仪	測振儀
view factor	视因数	視因數
vinyl chloride monomer	氯乙烯单体	氯乙烯單體
vinyl compound	乙烯基化合物	乙烯基化合物
vinylon	维纶	維尼綸
vinyl pipe	聚氯乙烯管	聚氯乙烯管
vinyl polymer	烯类聚合物	乙烯系聚合物
vinyl resin	乙烯系树脂	乙烯系樹脂
viomycin	紫霉素	紫黴素
virial coefficient	位力系数	位力系數
virial equation	位力方程	位力方程
virus	病毒	病毒
visbreaking	减黏裂化	減黏裂煉

英　文　名	大　陆　名	台　湾　名
viscoelastic fluid	黏弹性流体	黏彈性流體
viscoelasticity	黏弹性	黏彈性
viscometer	黏度计	黏度計
viscometric flow	测黏流动	測黏[度]流動
viscometry	黏度测定法	黏度測定法
viscoplastic fluid	黏塑性流体	黏塑性流體
viscoplasticity	黏塑性	黏塑性
viscose rayon	黏胶人造丝	嫘縈
viscose staple	黏胶人造短纤维	嫘縈棉
viscosimeter	黏度计	黏度計
viscosimetry	黏度测定法	黏度測定法
viscosity	黏度	黏度
viscosity average molecular weight	黏度平均分子量	黏度平均分子量
viscosity blending chart	黏度掺和图	黏度摻配圖
viscosity coefficient	黏度系数	黏度係數
viscosity index	黏度指数	黏度指數
viscosity number	黏度数	黏度值
viscosity test	黏度试验	黏度試驗
viscous damper	黏性减震器	黏性阻尼器
viscous dissipation	黏性耗散	黏性耗散
viscous flow	黏性流动	黏性流[動]
viscous force	黏性力	黏滯力
viscous motion	黏性运动	黏性運動
viscous resistance	黏性阻力	黏性阻力
viscous stress	黏性应力	黏性應力
viscous sublayer	黏性底层	黏性次層
vitamin	维生素	維生素,維他命
vitreous china	玻化瓷器	玻化瓷器
vitreous phase	玻化相	玻化相
vitrification	玻璃固化	玻[璃]化
vitrification period	玻璃固化期	玻[璃]化期
vitrified brick	玻璃固化砖	玻[璃]化磚
V notch	V 型缺口	V 形凹槽
void	空隙	空隙
voidage	空隙率	空隙[率]
void fraction	空隙分数	空隙分率
void volume	空隙体积	空隙體積
volatile suspended solid	挥发性悬浮固体	揮發性懸浮固體

英　文　名	大　陆　名	台　湾　名
volatility	挥发度	揮發度
volatilization	挥发	揮發[作用]
volclay	膨润土	膨土,漿土
voltage	电压	電壓
voltmeter	伏特计	伏特計,電壓計
volume	体积	體積,容積
volume average velocity	体积平均速度	體積平均速度
volume expansivity	体膨胀率	體[積]膨脹係數
volume fraction	体积分率	體積分率
volume index	体积指数	體積指數
volume mean diameter	体积平均直径	體積平均直徑
volume modulus	体积模数	體積[彈性]模數
volume shrinkage	体积收缩	體積收縮[率]
volume surface diameter	体积表面直径	體積表面[相當]直徑
volumetric average boiling point	体积平均沸点	體積平均沸點
volumetric efficiency	体积效率	體積效率
volumetric flow rate	体积流率	體積流量
volumetric hourly space velocity	[小时]体积空[间]速[度]	體積空間時速
volumetric oxygen transfer coefficient	容积传氧系数,体积传氧系数	體積氧質傳係數
volumetric titration	容量滴定法	容量滴定[法]
volumetry	容量分析法	容量分析法
volume work	体积功	體積功
volute	蜗壳	渦捲
volute pump	涡壳泵	渦捲泵
von Karman analogy	冯卡门类比	馮卡門類比
von Karman boundary layer theory	冯卡门边界层理论	馮卡門邊界層理論
von Karman integral method	冯卡门积分法	馮卡門積分法
von Karman number	冯卡门数	馮卡門數
von Karman vortex street	冯卡门涡街	馮卡門渦列
von Weimarn equation	冯韦曼方程式	馮韋曼方程式
vortex	涡旋	漩渦
vortex agitator	涡旋搅拌器	漩渦攪拌器
vortex breaker	涡流消除器	漩渦碎機
vortex cavity	旋涡空穴	漩渦空洞
vortex line	涡线	漩渦線
vortex motion	涡旋运动	漩渦運動

英 文 名	大 陆 名	台 湾 名
vortex potential	涡旋势	漩涡势
vortex shedding	涡旋脱落	漩涡剥離
vortex sheet	涡片	漩涡片
vortex trail	涡旋尾迹	漩涡尾跡
vorticity	涡度	漩涡度
votator heat exchanger	套管冷却结晶器	刮刀式熱交換器
vulcanite	硬橡胶	硬橡膠
vulcanizate	硫化橡胶	硫化橡膠
vulcanization	硫化	硫化,交聯
vulcanization coefficient	硫化系数	硫化係數
vulcanized fiber paper	硫化纤维纸	硬化紙
vulcanized oil	硫化油	硫化油
vulcanizer	硫化器	硫化器
vulcanizing agent	硫化剂	硫化劑
vulcanizing chamber	硫化室	硫化室
vulcanizing pan	硫化锅	硫化鍋
vycor glass	高硅氧玻璃	耐熱玻璃

W

英 文 名	大 陆 名	台 湾 名
wafer	晶片	晶片,晶圆
wage	工资	工资
wake	尾流,尾涡	尾流
wall effect	壁效应	壁效應
wall thickness	壁厚	壁厚
wall turbulence	壁湍流	壁紊流
warpe	翘曲	翹曲[變形]
washable ointment base	可洗软膏基	可洗軟膏基
washing rate	洗涤速率	洗滌速率
washing soda	洗涤碱	洗滌鹼,碳酸鈉
washing tower	洗涤塔	洗滌塔
washout	冲洗	沖盡
waste	废弃物	廢棄物
waste acid	废酸	廢酸
waste disposal	废物处理	廢棄物處理
waste effluent	废液	廢液
waste fuel	废燃料	廢燃料

英　文　名	大　陆　名	台　湾　名
waste gas	废气	廢氣
waste heat	废热	廢熱
waste heat boiler	废热锅炉	廢熱鍋爐
waste heat drier	废热干燥器	廢熱乾燥器
waste heat recovery	废热回收	廢熱回收
waste liquid	废液	廢液
waste liquor	废液	廢液
waste management	废物管理	廢棄物管理
waste treatment	废物处理	廢棄物處理
waste water	废水	廢水
waste water treatment	废水处理	廢水處理
water absorption	吸水率	吸水率
water base paint	水性漆	水性漆
water bath	水浴	水浴
water conditioning	水调理	水調理
water content	含水量	含水量
water cooled crystallizer	水冷结晶器	水冷結晶器
water cooler	水冷却器	水冷卻器
water cooling	水冷却	水冷卻
water curtain	水帘	水簾
water gas	水煤气	水煤氣
water gas shift reaction	水煤气变换反应	水煤氣[轉化]反應
water glass	水玻璃	水玻璃
water hammer	水锤	水鎚
water hardness	水质硬度	水質硬度
water head	水头	水位高差
water heater	热水器	水加熱器
water in oil emulsion	油包水乳状液	油包水乳液
water jacket	水套	水[夾]套
water jet	喷水器	噴水器,水刀
water of crystallization	结晶水	結晶水
water paint	水性漆	水性漆
water permeability	透水性	透水性
water polisher	水磨光机	水純化器
water pollutant	水污染物	水污染物
water pollution	水污染	水污染
water pollution abatement	水污染降低	水污染減量
water pollution control	水污染控制	水污染控制

英 文 名	大 陆 名	台 湾 名
water pollution engineering	水污染工程	水污染工程
waterproof	不透水的	防水
waterproof cement	防水水泥	防水水泥
waterproof grease	防水润滑脂	防水潤滑脂
water proofing	防水处理	防水加工
water purification	水净化	水淨化
water quality	水质	水質
water quality analysis	水质分析	水[質]分析
water quality criterion	水质判据	水質準則
water repellence agent	抗水剂	撥水劑
water repellent	抗水[作用]	撥水劑
water seal	水封	水[密]封
water separator	水分离器	水分離器
water softener	软水剂	水軟化劑
water softening	水软化	水軟化
water-soluble ointment base	水溶性软膏基	水溶性軟膏基
water tolerance	耐水度	水容限
water treatment	水处理	水處理
water valve	水阀	水閥
wavelength	波长	波長
wave mechanics	波动力学	波動力學
wave motion	波状运动	波動
wave number	波数	波數
wave propagation	波传播	波傳播
wavy flow	波状流	波狀流
wax	蜡	蠟
wax distillate	蜡馏出物	蠟餾出物
wax fractionation	含蜡分馏	蠟分餾
wax oil	含蜡油	蠟油
wax paper	蜡纸	蠟紙
wax residuum	含蜡渣油	蒸餘蠟
waxy crude	含蜡原油	含蠟原油
waxy oil	含蜡油	含蠟油
weak acid	弱酸	弱酸
weak base	弱碱	弱鹼
weak electrolyte	弱电解质	弱電解質
weak reversibility	弱可逆性	弱可逆性
wearing ring	耐磨圈	耐磨圈

英　文　名	大　陆　名	台　湾　名
weatherability	耐候性	耐候性
weathering	风化	風化[作用]
weathering test	耐候性试验	耐候試驗
weaving	编织	梭織
weaving machine	编织机	梭織機
Weber	韦[伯]（磁通量单位）	韋伯
Weber number	韦伯数	韋伯數
wedge	楔	楔形,楔形體
weed killer	除草剂	除草劑
weeping	漏液	滴流
weeping hole	漏液孔	滴流孔
Wegstein method	韦格斯坦法	韋格斯坦法
weighing feeder	喂料装置	稱量進料機
weight average boiling point	重均沸点	重量平均沸點
weight average molecular weight	重均分子量	重量平均分子量
weight distribution	权重分布	重量分佈
weighted mean	加权平均[值]	加權平均[值]
weighted residual	加权残差	加權殘差
weighted temperature difference	加权温差	加權溫差
weight factor	权[重]因子	權重[因數]
weight fraction	重量分率	重量分率
weight function	权函数	權函數
weighting factor	加权因子	加權因數
weighting function	加权函数	加權函數
weight loss	重量损失	失重
weight mean diameter	重量平均直径	重量平均直徑
weightometer	自动皮带秤	自動稱重計
weir	堰	堰
weir height	堰高	堰高
Weisz modulus	韦斯模数	維茲模數
welding	焊接	焊接
well type nozzle	井式喷嘴	井型噴嘴
wet and dry bulb hygrometer	干湿球湿度	乾濕球濕度計
wet-bulb depression	湿球下降	濕球下降
wet bulb temperature	干湿球温度表	濕球溫度
wet-bulb thermometer	湿球温度表	濕球溫度計
wet combustion	湿式燃烧	濕式燃燒
wet gas	湿气	含油氣

英　文　名	大　陆　名	台　湾　名
wet gas meter	湿气计	濕氣計
wet grinding	湿磨	濕磨
wet oxidation	湿式氧化	濕式氧化法
wet pan mill	湿辗机	濕輾機
wet process	湿法过程	濕法
wet screening	湿筛选	濕篩選
wet seal holder	湿式密闭容器	濕式密封器
wet separation	湿法分离	濕法分離
wet spinning	湿纺	濕紡[絲]
wet steam	湿蒸汽	濕蒸汽
wet surface	湿表面	濕面
wettability effect	润湿效应	潤濕效應
wettable agent	可湿性助剂	潤濕劑
wetted area	湿面积	潤濕面積
wetted perimeter	润湿周边	潤濕周邊
wetted surface area	润湿表面积	潤濕表面積
wetted-wall column	湿壁塔	濕壁塔
wetted-wall tower	湿壁塔	濕壁塔
wetting agent	湿润剂	潤濕劑
wet well	湿井	濕井
whale oil	鲸油	鯨油
whale tallow	鲸脂	鯨脂
whipping	打泡	攪打
white cement	白水泥	白水泥
white charcoal	白炭	白碳,白煙
white gold	白金	白色[飾]金
white liquor	白液	白液,燒鹼液
white phosphorus	白磷	白磷,黃磷
wild stream	未控流	未控流
wilkinite	膨润土	膠膨潤土
willemite	硅锌矿	矽鋅礦
Wilson equation	威尔逊方程	威爾森方程
wind tunnel	风洞	風洞
windup	卷绕	繞緊
winterizing	冬化	冬化,低溫脫脂
wire glass	夹丝玻璃	夾網玻璃
W/O emulsion ointment base	水/油乳剂软膏基	水/油乳劑軟膏基
Wohl expansion	沃尔展开式	沃爾展開式

英 文 名	大 陆 名	台 湾 名
wollastonite	硅灰石	矽灰石
wood alcohol	木精	木精
wood creosote	木杂酚油	木雜酚油
wood-derived chemical	木材化学品	木材衍生化學品
wood distillation	木材蒸馏	木材蒸餾
wood extractive	木材提取物	木材萃取物
wood gas	木煤气	木煤氣
wood preservative	木材防腐剂	木材防腐劑
wood pyrolysis	木材热解	木材熱解
wood rosin	木松香	松香
wood saccharification	木材糖化	木材糖化
wood spirit	木精	木精
wood sugar	木糖	木糖
wood tar	木焦油	木溚
wool alcohol	羊毛脂醇	羊毛醇
wool fat	羊毛脂	羊毛脂
wool fiber	毛纤维	毛纖維
wool grease	羊毛脂	[粗]羊毛脂
wool oil	羊毛油	羊毛油
wool wax	羊毛蜡	羊毛蠟
work function	功函数	功函數
work index	功指数	功指數
working capital	流动资金	營運資金
working fluid	工作流体	工作流體
working stress	工作应力	工作應力
work of deformation	变形功	變形功
work requirement	功需要量	功需要量
worksheet	工作单	表單
wort	麦芽汁	麥芽漿
wrought iron	熟铁	熟鐵
Wulff-Bock crystallizer	摆动连续结晶槽	擺動連續結晶槽,沃柏结晶槽

X

英 文 名	大 陆 名	台 湾 名
xanthation	黄原酸化［反应］	黃酸化［作用］
xerogel	干凝胶	乾凝膠
X-ray diffraction	X 射线衍射	X 射線繞射
X-ray fluorescence	X 射线荧光	X 射線螢光
X-ray spectrum	X 射线谱	X 射線譜
xylene	二甲苯	二甲苯
xylose	木糖	木糖
xylulose	木酮糖	木酮糖

Y

英 文 名	大 陆 名	台 湾 名
yarn	纱	紗
yarn dyeing machine	纱染机	紗染機
yeast	酵母	酵母
yellow cake	黄饼	黃餅
yellow lead	黄丹	黃丹
yield	收率	產率
yield coefficient	收率系数	產率係數
yield limit	屈伏极限	降伏極限
yield per path	单程收率	單程產率
yield point	屈服点	降伏點
yield strength	屈服强度	降伏強度
yield stress	屈服应力	降伏應力
Y-X diagram	Y–X 图	汽–液莫耳分率圖,Y–X 圖

Z

英 文 名	大 陆 名	台 湾 名
zeolite	沸石	沸石
zeolite catalyst	沸石催化剂	沸石觸媒
zero-flux surface	零通量表面	零通量表面

英　文　名	大　陆　名	台　湾　名
zero frequency gain	零频率增益	零頻率增益
zero memory system	零记忆系统	零記憶系統
zero-order reaction	零级反应	零階反應
zero-pressure state	零压状态	零壓狀態
zinc alloy	锌合金	鋅合金
zinc blende	闪锌矿	閃鋅礦
zinc bronze	锌青铜	鋅青銅
zinc chrome	锌铬黄	鋅鉻黃
zinc-crown glass	锌冕玻璃	鋅冕玻璃
zinc flower	锌华	鋅華
zinc green	锌绿	鋅綠
zincite	红锌矿	紅鋅礦
zinc plating	镀锌	鍍鋅
zinc spar	菱锌矿	菱鋅礦
zinc sponge	锌海绵	鋅海綿
zinc storage battery	锌蓄电池	鋅蓄電池
zinc white	锌白	鋅白
zinc yellow	锌铬黄	鋅黃
zircon	锆石	鋯英石
zirconia	二氧化锆	氧化鋯
zirconia glass	氧化锆玻璃	鋯玻璃
zirconium oxide refractory	氧化锆耐火材料	氧化鋯耐火物
zircon porcelain	锆石瓷	鋯瓷
zircon refractory	锆英石耐火材料	鋯英石耐火物
zone electrophoresis	区带电泳	帶域電泳
zone freezing	区域冷冻	帶域冷凍
zone melting	区域熔炼	帶域熔煉
zone melting method	区域熔炼法	區域熔融法
zone refining method	区域精制	帶域精煉［法］
zone settling	区域沉降	帶域沈降
Z transformation	Z 转换	Z 變換
zymase	酿酶	解醣酶